Robust Adaptive Beamforming
Jian Li and Petre Stoica, Editors

Expert System Applications for Telecommunications
Jay Liebowitz

Polynomial Signal Processing
V. John Mathews and Giovanni L. Sicuranza

Digital Signal Estimation
Robert J. Mammone, Editor

Digital Communication Receivers: Synchronization, Channel Estimation, and Signal Processing
Heinrich Meyr, Marc Moeneclaey, and Stefan A. Fechtel

Synchronization in Digital Communications, Volume I
Heinrich Meyr and Gerd Ascheid

Business Earth Stations for Telecommunications
Walter L. Morgan and Denis Rouffet

Wireless Information Networks, 2nd Edition
Kaveh Pahlavan and Allen H. Levesque

Satellite Communications: The First Quarter Century of Service
David W. E. Rees

Fundamentals of Telecommunication Networks
Tarek N. Saadawi, Mostafa Ammar, with Ahmed El Hakeem

Analogue and Digital Microwave Links: Theory and Design
Carlos Salema

Microwave Radio Links: From Theory to Design
Carlos Salema

Meteor Burst Communications: Theory and Practice
Donald L. Schilling, Editor

Digital Communication over Fading Channels, 2nd Edition
Marvin K. Simon and Mohamed-Slim Alouini

Digital Signal Processing: A Computer Science Perspective
Jonathan (Y) Stein

Vector Space Projections: A Numerical Approach to Signal and Image Processing, Neural Nets, and Optics
Henry Stark and Yongyi Yang

Signaling in Telecommunication Networks
John G. van Bosse

Telecommunication Circuit Design, 2nd Edition
Patrick D. van der Puije

Worldwide Telecommunications Guide for the Business Manager
Walter H. Vignault

Robust Adaptive Beamforming

Robust Adaptive Beamforming

Edited by

Jian Li and Petre Stoica

WILEY-INTERSCIENCE

A JOHN WILEY & SONS, INC., PUBLICATION

Published by John Wiley & Sons, Inc., Hoboken, New Jersey.
Published simultaneously in Canada.

Library of Congress Cataloging-in-Publication Data:

Robust adaptive beamforming/edited by Jian Li and Petre Stoica.
p. cm.
Includes bibliographical references and index.
ISBN-13 978-0-471-67850-2 (cloth)
ISBN-10 0-471-67850-3 (cloth)
1. Adaptive antennas. 2. Antenna radiation patterns. I. Li, Jian. II. Stoica, Petre.

TK7871.67.A33R63 2005
621.382′4--dc22 2004065908

Printed in the United States of America

10 9 8 7 6 5 4 3 2 1

CONTENTS

CONTRIBUTORS

STEPHEN P. BOYD, Information Systems Laboratory, Stanford University, Stanford, CA 94305

ELIO D. DI CLAUDIO, INFOCOM Department, University of Roma "La Sapienza," Via Eudossiana 18, I-00184 Roma, Italy

YONINA C. ELDAR, Department of Electrical Engineering, Technion—Israel Institute of Technology, Haifa 32000, Israel

ALEX B. GERSHMAN, Darmstadt University of Technology, Institute of Telecommunications, Merckstrasse 25, 64283 Darmstadt, Germany

MIGUEL A. LAGUNAS, Centre Tecnològic de Telecomunicacions de Catalunya, NEXUS 1 Building, Gran Capita 2–4, 08034 Barcelona, Spain

AMIR LESHEM, School of Engineering, Bar-Ilan University, 52900 Ramat-Gan, Israel

JIAN LI, Department of Electrical and Computer Engineering, Engineering Bldg., Center Drive, University of Florida, Gainesville, FL 32611

ROBERT G. LORENZ, Beceem Communications, Santa Clara, CA 95054

ZHI-QUAN LUO, Department of Electrical and Computer Engineering, University of Minnesota, Minneapolis, MN 55455

XAVIER MESTRE, Centre Tecnològic de Telecomunicacions de Catalunya, NEXUS 1 Building, Gran Capita 2–4, 08034 Barcelona, Spain

ARYE NEHORAI, Department of Electrical Engineering and Computer Science, University of Illinois at Chicago, Chicago, IL 60607

RAFFAELE PARISI, INFOCOM Department, University of Roma "La Sapienza," Via Eudossiana 18, I-00184 Roma, Italy

SHAHRAM SHAHBAZPANAHI, McMaster University, Hamilton, Ontario L8S 4L8, Canada

PETRE STOICA, Division of Systems and Control, Department of Information Technology, Uppsala University, SE-75105 Uppsala, Sweden

ALLE-JAN VAN DER VEEN, Department of Electrical Engineering, Delft University of Technology, 2628 Delft, The Netherlands

ZHISONG WANG, Department of Electrical and Computer Engineering, Engineering Bldg., Center Drive, University of Florida, Gainesville, FL 32611

PREFACE

Beamforming is a ubitiquitous task in array signal processing with applications, among others, in radar, sonar, acoustics, astronomy, seismology, communications, and medical-imaging. The standard data-independent beamformers include the delay-and-sum approach as well as methods based on various weight vectors for sidelobe control. The data-dependent or adaptive beamformers select the weight vector as a function of the data to optimize the performance subject to various constraints. The adaptive beamformers can have better resolution and much better interference rejection capability than the data-independent beamformers. However, the former are much more sensitive to errors, such as the array steering vector errors caused by imprecise sensor calibrations, than the latter. As a result, much effort has been devoted over the past three decades to devise robust adaptive beamformers.

The primary goal of this edited book is to present the latest research developments on robust adaptive beamforming. Most of the early methods of making the adaptive beamformers more robust to array steering vector errors are rather ad hoc in that the choice of their parameters is not directly related to the uncertainty of the steering vector. Only recently have some methods with a clear theoretical background been proposed, which, unlike the early methods, make explicit use of an uncertainty set of the array steering vector. The application areas of robust adaptive beamforming are also continuously expanding. Examples of new areas include smart antennas in wireless communications, hand-held ultrasound imaging systems, and directional hearing aids. The publication of this book will hopefully provide timely information to the researchers in all the aforementioned areas.

The book is organized as follows. The first three chapters (Chapter 1 by Robert G. Lorenz and Stephen P. Boyd; Chapter 2 by Alex B. Gershman, Zhi-Quan Luo, and Shahram Shahbazpanahi; and Chapter 3 by Jian Li, Petre Stoica, and Zhisong Wang) discuss how to address directly the array steering vector uncertainty within a clear theoretical framework. Specifically, the robust adaptive beamformers in these chapters couple the standard Capon beamformers with a spherical or ellipsoidal uncertainty set of the array steering vector. The fourth chapter (by Xavier Mestre and Miguel A. Lagunas) concentrates on alleviating the finite sample size effect. Two-dimensional asymptotics are considered based on the assumptions that both the number of sensors and the number of observations are large and that they have the same order of magnitude. The fifth chapter (by Yonina C. Eldar and Arye Nehorai) considers the signal waveform estimation. The mean-squared error rather than the signal-to-interference-plus-noise ratio is used as a performance measure. Two cases are treated, including the case of known steering vectors and the case of random steering vectors with known second-order statistics. The sixth chapter (by Alle-Jan van der Veen and Amir Leshem) focuses on constant modulus algorithms. Two constant modulus algorithms are put into a common framework with further discussions on their iterative and adaptive implementations and their direction finding applications. Finally, the seventh chapter (by Elio D. Di Claudio and Raffaele Parisi) is devoted to robust wideband beamforming. Based on a constrained stochastic maximum likelihood error functional, a steered adaptive beamformer is presented to adapt the weight vector within a generalized sidelobe canceller formulation.

We are grateful to the authors who have contributed to the chapters of this book for their excellent work. We would also like to acknowledge the contributions of several other people and organizations to the completion of this book. Most of our work in the area of robust adaptive beamforming is an outgrowth of our research programs in array signal processing. We would like to thank those who have supported our research in this area: the National Science Foundation, the Swedish Science Council (VR), and the Swedish Foundation for International Cooperation in Research and Higher Education (STINT). We also wish to thank George Telecki (Associate Publisher) and Rachel Witmer (Editorial Assistant) at Wiley for their effort on the publication of this book.

JIAN LI AND PETRE STOICA

1

ROBUST MINIMUM VARIANCE BEAMFORMING

Robert G. Lorenz
Beceem Communications, Santa Clara, CA 95054

Stephen P. Boyd
Information Systems Laboratory, Stanford University, Stanford, CA 94305

1.1 INTRODUCTION

Consider the n dimensional sensor array depicted in Figure 1.1. Let $a(\theta) \in \mathbf{C}^n$ denote the response of the array to a plane wave of unit amplitude arriving from direction θ; we shall refer to $a(\cdot)$ as the *array manifold*. We assume that a narrow-band source $s(t)$ is impinging upon the array from angle θ and that the source is in the far-field of the array. The vector array output $y(t) \in \mathbf{C}^n$ is then

$$y(t) = a(\theta)s(t) + v(t), \tag{1.1}$$

where $a(\theta)$ includes effects such as coupling between elements and subsequent amplification; $v(t)$ is a vector of additive noises representing the effect of undesired signals, such as thermal noise or interference. We denote the sampled array output by $y(k)$. Similarly, the combined beamformer output is given by

$$y_c(k) = w^*y(k) = w^*a(\theta)s(k) + w^*v(k)$$

where $w \in \mathbf{C}^n$ is a vector of weights, that is, design variables, and $(\cdot)^*$ denotes the conjugate transpose.

Robust Adaptive Beamforming, Edited by Jian Li and Petre Stoica

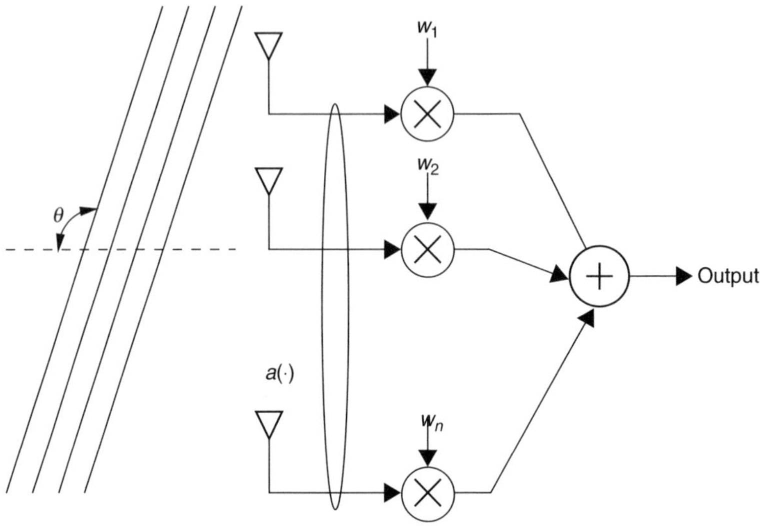

Figure 1.1 Beamformer block diagram.

The goal is to make $w^*a(\theta) \approx 1$ and $w^*v(t)$ small, in which case, $y_c(t)$ recovers $s(t)$, that is, $y_c(t) \approx s(t)$. The gain of the weighted array response in direction θ is $|w^*a(\theta)|$; the expected effect of the noise and interferences at the combined output is given by $w^*R_v w$, where $R_v = \mathbf{E}\, vv^*$ and $\mathbf{E}$ denotes the expected value. If we presume $a(\theta)$ and R_v are known, we may choose w as the optimal solution of

$$\begin{array}{ll} \text{minimize} & w^*R_v w \\ \text{subject to} & w^*a(\theta_d) = 1. \end{array} \tag{1.2}$$

Minimum variance beamforming is a variation on (1.2) in which we replace R_v with an estimate of the received signal covariance derived from recently received samples of the array output, for example,

$$R_y = \frac{1}{N} \sum_{i=k-N+1}^{k} y(i)y(i)^* \in \mathbf{C}^{n\times n}. \tag{1.3}$$

The minimum variance beamformer (MVB) is chosen as the optimal solution of

$$\begin{array}{ll} \text{minimize} & w^*R_y w \\ \text{subject to} & w^*a(\theta) = 1. \end{array} \tag{1.4}$$

This is commonly referred to as Capon's method [1]. Equation (1.4) has an analytical solution given by

$$w_{\text{mv}} = \frac{R_y^{-1}a(\theta)}{a(\theta)^* R_y^{-1} a(\theta)}. \tag{1.5}$$

Equation (1.4) also differs from (1.2) in that the power expression we are minimizing includes the effect of the desired signal plus noise. The constraint $w^*a(\theta) = 1$ in (1.4) prevents the gain in the direction of the signal from being reduced.

A measure of the effectiveness of a beamformer is given by the signal-to-interference-plus-noise ratio, commonly abbreviated as SINR, given by

$$\text{SINR} = \frac{\sigma_{\text{d}}^2 |w^* a(\theta)|^2}{w^* R_v w}, \tag{1.6}$$

where σ_{d}^2 is the power of the signal of interest. The assumed value of the array manifold $a(\theta)$ may differ from the actual value for a host of reasons including imprecise knowledge of the signal's angle of arrival θ. Unfortunately, the SINR of Capon's method can degrade catastrophically for modest differences between the assumed and actual values of the array manifold. We now review several techniques for minimizing the sensitivity of MVB to modeling errors in the array manifold.

1.1.1 Previous Work

One popular method to address uncertainty in the array response or angle of arrival is to impose a set of unity-gain constraints for a small spread of angles around the nominal look direction. These are known in the literature as point mainbeam constraints or neighboring location constraints [2]. The beamforming problem with point mainbeam constraints can be expressed as

$$\begin{array}{ll} \text{minimize} & w^* R_y w \\ \text{subject to} & C^* w = f, \end{array} \tag{1.7}$$

where C is a $n \times L$ matrix of array responses in the L constrained directions and f is an $L \times 1$ vector specifying the desired response in each constrained direction. To achieve wider responses, additional constraint points are added. We may similarly constrain the derivative of the weighted array output to be zero at the desired look angle. This constraint can be expressed in the same framework as (1.7); in this case, we let C be the derivative of the array manifold with respect to look angle and $f = 0$. These are called *derivative mainbeam constraints*; this derivative may be approximated using regularization methods. Point and derivative mainbeam constraints may also be used in conjunction with one another. The minimizer of (1.7) has an analytical solution given by

$$w_{\text{opt}} = R_y^{-1} C (C^* R_y^{-1} C)^{-1} f. \tag{1.8}$$

Each constraint removes one of the remaining degrees of freedom available to reject undesired signals; this is particularly significant for an array with a small number of elements. We may overcome this limitation by using a low-rank approximation to the constraints [3]. The best rank k approximation to C, in a least squares sense, is given by $U\Sigma V^*$, where Σ is a diagonal matrix consisting of the largest k singular values, U is a $n \times k$ matrix whose columns are the corresponding left singular vectors of C, and V is a $L \times k$ matrix whose columns are the corresponding right singular vectors of C. The reduced rank constraint equations can be written as $V\Sigma^T U^* w = f$, or equivalently

$$U^* w = \Sigma^\dagger V^* f, \tag{1.9}$$

where $\dagger$ denotes the Moore–Penrose pseudoinverse. Using (1.8), we compute the beamformer using the reduced rank constraints as

$$w_{\text{epc}} = R_y^{-1} U (U^* R_y^{-1} U)^{-1} \Sigma^\dagger V^* f.$$

This technique, used in source localization, is referred to as minimum variance beamforming with environmental perturbation constraints (MV-EPC), see Krolik [2] and the references contained therein.

Unfortunately, it is not clear how best to pick the additional constraints, or, in the case of the MV-EPC, the rank of the constraints. The effect of additional constraints on the design specifications appears difficult to predict.

Regularization methods [4] have also been used in beamforming. One technique, referred to in the literature as diagonal loading, chooses the beamformer to minimize the sum of the weighted array output power plus a penalty term, proportional to the square of the norm of the weight vector. The gain in the assumed angle of arrival (AOA) of the desired signal is constrained to be unity. The beamformer is chosen as the optimal solution of

$$\begin{array}{ll} \text{minimize} & w^* R_y w + \mu w^* w \\ \text{subject to} & w^* a(\theta) = 1. \end{array} \tag{1.10}$$

The parameter $\mu > 0$ penalizes large values of w and has the general effect of *detuning* the beamformer response. The regularized least squares problem (1.10) has an analytical solution given by

$$w_{\text{reg}} = \frac{(R_y + \mu I)^{-1} a(\theta)}{a(\theta)^* (R_y + \mu I)^{-1} a(\theta)}. \tag{1.11}$$

Gershman [5] and Johnson and Dudgeon [6] provide a survey of these methods; see also the references contained therein. Similar ideas have been used in adaptive algorithms, see Haykin [7].

Beamformers using eigenvalue thresholding methods to achieve robustness have also been used; see Harmanci et al. [8]. The beamformer is computed according to Capon's method, using a covariance matrix which has been modified to ensure no eigenvalue is less than a factor μ times the largest, where $0 \le \mu \le 1$. Specifically, let $V \Lambda V^*$ denote the eigenvalue/eigenvector decomposition of R_y, where Λ is a diagonal matrix, the ith entry (eigenvalue) of which is given by λ_i, that is,

$$\Lambda = \begin{bmatrix} \lambda_1 & & \\ & \ddots & \\ & & \lambda_n \end{bmatrix}.$$

Without loss of generality, assume $\lambda_1 \ge \lambda_2 \ldots \ge \lambda_n$. We form the diagonal matrix Λ_{thr}, the ith entry of which is given by $\max\{\mu\lambda_1, \lambda_i\}$; viz,

$$\Lambda_{\text{thr}} = \begin{bmatrix} \lambda_1 & & & \\ & \max\{\mu\lambda_1, \lambda_2\} & & \\ & & \ddots & \\ & & & \max\{\mu\lambda_1, \lambda_n\} \end{bmatrix}.$$

The modified covariance matrix is computed according to $R_{\text{thr}} = V \Lambda_{\text{thr}} V^*$. The beamformer using eigenvalue thresholding is given by

$$w_{\text{thr}} = \frac{R_{\text{thr}}^{-1} a(\theta)}{a(\theta)^* R_{\text{thr}}^{-1} a(\theta)}. \tag{1.12}$$

The parameter μ corresponds to the reciprocal of the condition number of the covariance matrix. A variation on this approach is to use a fixed value for the minimum eigenvalue threshold. One interpretation of this approach is to incorporate a priori knowledge of the presence of additive white noise when the sample covariance is unable to observe said white noise floor due to short observation time [8]. The performance of this beamformer appears similar to that of the regularized beamformer using diagonal loading; both usually work well for an appropriate choice of the regularization parameter μ.

We see two limitations with regularization techniques for beamformers. First, it is not clear how to efficiently pick μ. Second, this technique does not take into account any knowledge we may have about variation in the array manifold, for example, that the variation may not be isotropic.

In Section 1.1.3, we describe a beamforming method that explicitly uses information about the variation in the array response $a(\cdot)$, which we model explicitly as an uncertainty ellipsoid in $\mathbf{R}^{2n}$. Prior to this, we introduce some notation for describing ellipsoids.

1.1.2 Ellipsoid Descriptions

An n-dimensional ellipsoid can be defined as the image of an n-dimensional Euclidean ball under an affine mapping from $\mathbf{R}^n$ to $\mathbf{R}^n$; that is,

$$\mathcal{E} = \{Au + c \mid \|u\| \leq 1\}, \tag{1.13}$$

where $A \in \mathbf{R}^{n \times n}$ and $c \in \mathbf{R}^n$. The set $\mathcal{E}$ describes an ellipsoid whose center is c and whose *principal semiaxes* are the unit-norm left singular vectors of A scaled by the corresponding singular values. We say that an ellipsoid is *flat* if this mapping is not injective, that is, one-to-one. Flat ellipsoids can be described by (1.13) in the proper affine subspaces of $\mathbf{R}^n$. In this case, $A \in \mathbf{R}^{n \times l}$ and $u \in \mathbf{R}^l$. An interpretation of a flat uncertainty ellipsoid is that some linear combinations of the array manifold are known exactly [9].

Unless otherwise specified, an ellipsoid in $\mathbf{R}^n$ will be parameterized in terms of its center $c \in \mathbf{R}^n$ and a symmetric non-negative definite configuration matrix $Q \in \mathbf{R}^{n \times n}$ as

$$\mathcal{E}(c, Q) = \{Q^{1/2}u + c \mid \|u\| \leq 1\} \tag{1.14}$$

where $Q^{1/2}$ is any matrix square root satisfying $Q^{1/2}(Q^{1/2})^T = Q$. When Q is full rank, the nondegenerate ellipsoid $\mathcal{E}(c, Q)$ may also be expressed as

$$\mathcal{E}(c, Q) = \{x \mid (x - c)^T Q^{-1}(x - c) \leq 1\} \tag{1.15}$$

or by the equivalent quadratic function

$$\mathcal{E}(c, Q) = \{x \mid T(x) \leq 0\}, \tag{1.16}$$

where $T(x) = x^T Q^{-1} x - 2c^T Q^{-1} x + x_c^T Q^{-1} x_c - 1$. The first representation (1.14) is more natural when $\mathcal{E}$ is degenerate or poorly conditioned. Using the second description (1.15), one may easily determine whether a point lies within the ellipsoid. The third representation (1.16) will be used in Section 1.6.1 to compute the minimum-volume ellipsoid covering the union of ellipsoids.

We will express the values of the array manifold $a \in \mathbf{C}^n$ as the direct sum of its real and imaginary components in $\mathbf{R}^{2n}$; that is,

$$z_i = [\mathbf{Re}(a_1) \cdots \mathbf{Re}(a_n)\ \mathbf{Im}(a_1) \cdots \mathbf{Im}(a_n)]^T. \tag{1.17}$$

While it is possible to cover the field of values with a complex ellipsoid in $\mathbf{C}^n$, doing so implies a symmetry between the real and imaginary components which generally results in a larger ellipsoid than if the direct sum of the real and imaginary components are covered in $\mathbf{R}^{2n}$.

1.1.3 Robust Minimum Variance Beamforming

A generalization of (1.4) that captures our desire to minimize the weighted power output of the array in the presence of uncertainties in $a(\theta)$ is then:

$$\begin{array}{ll} \text{minimize} & w^* R_y w \\ \text{subject to} & \mathbf{Re}\ w^* a \geq 1\ \forall a \in \mathcal{E}, \end{array} \tag{1.18}$$

where **Re** denotes the real part. Here, $\mathcal{E}$ is an ellipsoid that covers the possible range of values of $a(\theta)$ due to imprecise knowledge of the array manifold $a(\cdot)$, uncertainty in the angle of arrival θ, or other factors. We shall refer to the optimal solution of (1.18) as the robust minimum variance beamformer (RMVB).

We use the constraint $\mathbf{Re}\ w^* a \geq 1$ for all $a \in \mathcal{E}$ in (1.18) for two reasons. First, while normally considered a semi-infinite constraint, we show in Section 1.3 that it can be expressed as a second-order cone constraint. As a result, the robust minimum variance beamforming problem (1.18) can be solved reliably and efficiently. Second, the real part of the response is an efficient lower bound for the magnitude of the response, as the objective $w^* R_y w$ is unchanged if the weight vector w is multiplied by an arbitrary shift $e^{j\phi}$. This is particularly true when the uncertainty in the array response is relatively small. It is unnecessary to constrain the imaginary part of the response to be nominally zero.

Our approach differs from the previously mentioned beamforming techniques in that the weight selection uses the a priori uncertainties in the array manifold in a precise way; the RMVB is guaranteed to satisfy the minimum gain constraint for all values in the uncertainty ellipsoid.

Recently, several papers have addressed uncertainty in a similar framework. Wu and Zhang [10] observe that the array manifold may be described as a polyhedron and that the robust beamforming problem can be cast as a quadratic program. While the polyhedron approach is less conservative, the size of the description and hence the complexity of solving the problem grows with the number of vertices. Vorobyov et al. [11, 12] and Gershman [13] describe the use of second-order cone programming for robust beamforming in the case where the uncertainty is in the array response is isotropic, that is, a Euclidean ball. Our method, while derived differently, yields the same beamformer as proposed by Li et al. [14–16].

In this chapter, we consider the case in which the uncertainty is anisotropic [17–19]. We also show how the beamformer weights can be computed efficiently.

1.1.4 Outline of the Chapter

The rest of this chapter is organized as follows. In Section 1.2, we motivate the need for robustness with a simple array which includes the effect of coupling between antenna elements. In Section 1.3 we discuss the RMVB. A numerically efficient technique based on Lagrange multiplier methods is described; we will see that the RMVB can be computed with the same order of complexity as its nonrobust counterpart. A numerical example is given in Section 1.4. In Section 1.5 we describe ellipsoidal modeling methods which make use of simulated or measured values of the array manifold. In Section 1.6 we discuss more sophisticated techniques, based on ellipsoidal calculus, for propagating uncertainty ellipsoids. In particular, we describe a numerically efficient method for approximating the numerical range of the Hadamard (element-wise) product of two ellipsoids. This form of uncertainty arises when the array outputs are subject to multiplicative uncertainties. A numerical beamforming example considering multiplicative uncertainties is given in Section 1.7. Our conclusions are given in Section 1.8.

1.2 A PRACTICAL EXAMPLE

Our goals for this section are twofold:

- To make the case that antenna elements may behave *very* differently in free space than as part of closely spaced arrays, and
- To motivate the need for robustness in beamforming.

Consider the four-element linear array of half-wave dipole antennas depicted in Figure 1.2. Let the frequency of operation be 900 MHz and the diameter of the

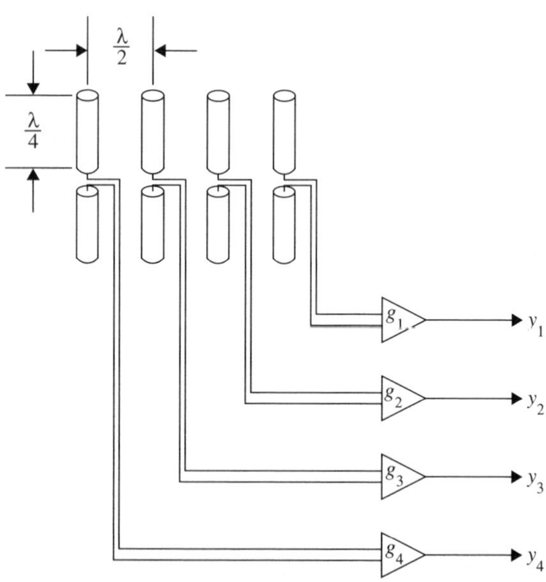

Figure 1.2 The four-element array. For this array, we simulate the array response which includes the effect of coupling between elements. In this example, the gains $g_1, \ldots, g_4$ are all assumed nominal. Later we consider the effect of multiplicative uncertainties.

elements be 1.67 mm. Assume each dipole is terminated into a 100 ohm load. The length of the dipole elements was chosen such that an isolated dipole in free space matched this termination impedance.

The array was simulated using the Numerical Electromagnetics Code, version 4 (NEC-4) [20]. Each of the radiating elements was modeled with six wire segments. The nominal magnitude and phase responses are given in Figures 1.3 and 1.4, respectively. Note that the amplitude is *not* constant for all angles of arrival or the same for all elements. This will generally be the case with closely spaced antenna elements due to the high level of interelement coupling.

In Figure 1.5, we see that the vector norm of the array response is *not* a constant function of AOA, despite the fact that the individual elements, in isolation, have an isotropic response.

Next, let us compare the performance of the RMVB with Capon's method using this array, with nominal termination impedances. Assume the desired signal impinges on the array from an angle $\theta_{\text{sig}} = 127^\circ$ and has a signal-to-noise ratio (SNR) of 20 decibels (dB). We assume that an interfering signal arrives at an

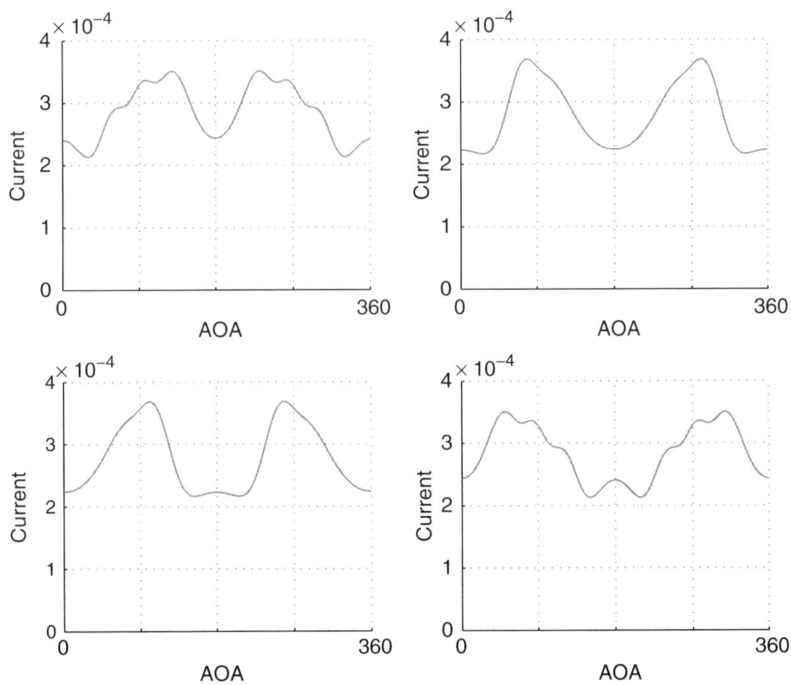

Figure 1.3 Magnitude of response of four-element array consisting of half-wave dipoles with uniform spacing of $\lambda/2$. The currents have units of amperes for a field strength of 1 volt/meter. The angle of arrival (AOA) is in degrees. Note the symmetry of the response. The outer elements correspond to the top left and bottom right plots; the inner elements, top right and lower left.

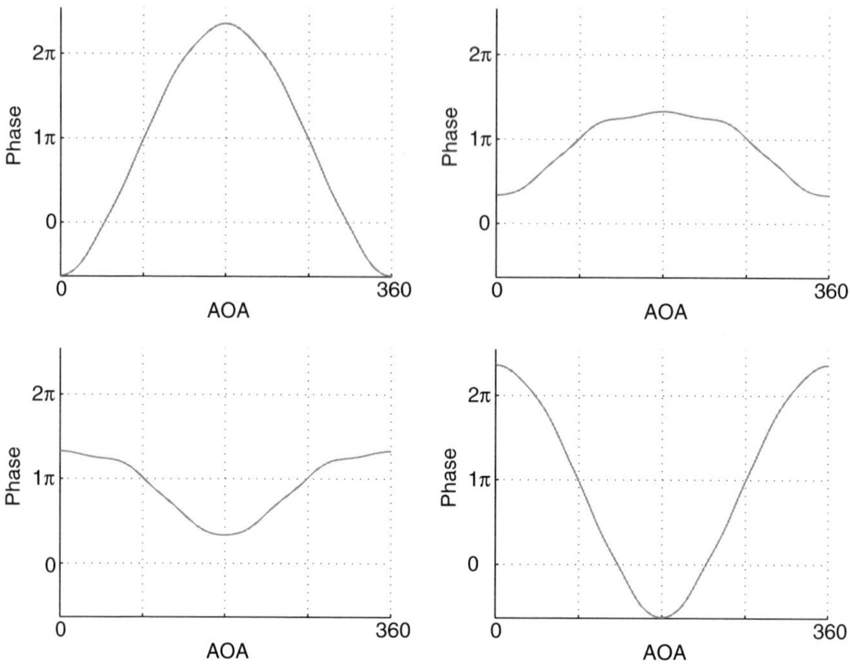

Figure 1.4 Phase response, in radians, of the four-element half-wave dipole array. The angle of arrival is in degrees. Again, note the symmetry in the response.

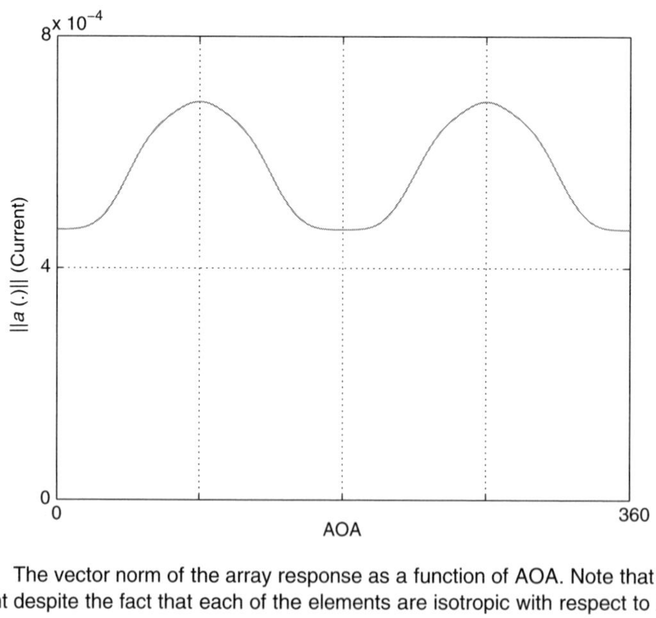

Figure 1.5 The vector norm of the array response as a function of AOA. Note that the norm is not constant despite the fact that each of the elements are isotropic with respect to AOA.

angle of $\theta_{\text{int}} = 150^\circ$ with amplitude twice that of the desired signal. For Capon's method, we assume an AOA of $\theta_{\text{nom}} = 120^\circ$. For the RMVB, we compute a minimum-volume ellipsoid covering the numerical range of the array manifold for all angles of arrival between 112° and 128°. The details of this calculation will be described in Section 1.5. Let $w_{\text{mv}} \in \mathbf{C}^4$ denote the beamformer vector produced by Capon's method and $w_{\text{rmvb}} \in \mathbf{C}^4$ the robust minimum-variance beamformer, that is, the optimal solution of (1.18).

A plot of the response of the minimum-variance beamformer (MVB) and the robust minimum-variance beamformer (RMVB) as a function of angle of arrival is shown in Figure 1.6. By design, the response of the MVB has unity gain in the direction of the assumed AOA, that is, $w_{\text{mv}}^* a(\theta_{\text{nom}}) = 1$, where $a : \mathbf{R} \to \mathbf{C}^4$ denotes the array manifold. The MVB produces a deep null in the direction of the interference: $w_{\text{mv}}^* a(\theta_{\text{int}}) = -0.0061 + 0i$. Unfortunately, the MVB also strongly attenuates the desired signal, with $w_{\text{mv}}^* a(\theta_{\text{sig}}) = -0.0677 + 0i$. The resulting post-beamforming signal-to-interference-plus-noise ratio (SINR) is -10.5 dB, appreciably worse than the SINR obtained using a single antenna *without* beamforming.

While the robust beamformer does not cast as deep a null in the direction of the interfering signal, that is, $w_{\text{rmvb}}^* a(\theta_{\text{int}}) = -0.0210 + 0i$, it maintains greater than unity gain for all angles of arrival in our design specification. The SINR obtained using the RMVB is 12.4 dB.

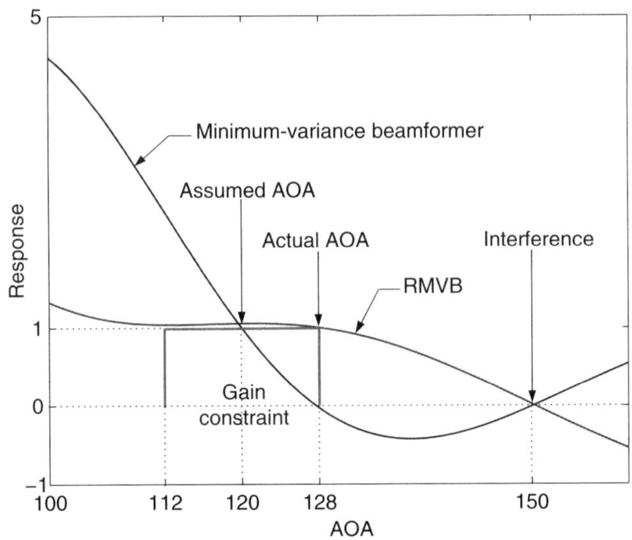

Figure 1.6 The response of the minimum-variance beamformer (Capon's method) and the robust minimum-variance beamformer (RMVB). The a priori uncertainty in the angle of arrival (AOA) was $\pm 8^\circ$. We see that the RMVB maintains at least unity gain for all angles in this range, whereas Capon's method fails for an AOA of approximately 127°.

When the actual AOA of the desired signal equals the assumed 120°, the SINR of the MVB is an impressive 26.5 dB, compared to 10.64 dB for the RMVB. It is tempting then to consider methods to reduce the uncertainty and potentially realize this substantial improvement in SINR. Such efforts are unlikely to be fruitful. For example, a 1° error in the assumed AOA reduces the SINR of Capon's method by more than 20 dB to 4.0 dB. Also, the mathematical values of the array model differ from the actual array response for a number of reasons, of which error in the assumed AOA is but one. In the presence of array calibration errors, variations due to termination impedances, and multiplicative gain uncertainties, nonrobust techniques simply do not work reliably.

In our example, we considered only uncertainty in the angle of arrival; verifying the performance for the nonrobust method involved evaluating points in a one-dimensional interval. Had we considered the additional effect of multiplicative gain variations, for example, the numerical cost of verifying the performance of the beamformer for a dense grid of possible array values could dwarf the computational complexity of the robust method. The approach of the RMVB is different; it makes specific use of the uncertainty in the array response. We compute either a worst-case optimal vector for the ellipsoidal uncertainty region or a proof that the design specification is infeasible. No subsequent verification of the performance is required.

1.3 ROBUST WEIGHT SELECTION

Recall from Section 1.1 that the RMVB was the optimal solution to

$$\begin{array}{ll} \text{minimize} & w^* R_y w \\ \text{subject to} & \mathbf{Re}\; w^* a \geq 1 \;\; \forall a \in \mathcal{E}. \end{array} \tag{1.19}$$

For purposes of computation, we will express the weight vector w and the values of the array manifold a as the direct sum of the corresponding real and imaginary components

$$x = \begin{bmatrix} \mathbf{Re}\; w \\ \mathbf{Im}\; w \end{bmatrix} \quad \text{and} \quad z = \begin{bmatrix} \mathbf{Re}\; a \\ \mathbf{Im}\; a \end{bmatrix}. \tag{1.20}$$

The real and imaginary components of the product w^*a can be expressed as

$$\mathbf{Re}\; w^* a = x^T z \tag{1.21}$$

and

$$\mathbf{Im}\; w^* a = x^T U z, \tag{1.22}$$

where U is the orthogonal matrix

$$U = \begin{bmatrix} 0 & I_n \\ -I_n & 0 \end{bmatrix},$$

and I_n is an $n \times n$ identity matrix. The quadratic form $w^* R_y w$ may be expressed in terms of x as $x^T R x$, where

$$R = \begin{bmatrix} \mathbf{Re}\, R_y & -\mathbf{Im}\, R_y \\ \mathbf{Im}\, R_y & \mathbf{Re}\, R_y \end{bmatrix}.$$

Assume R is positive definite; with sufficient sample support, it is with probability - one.

Let $\mathcal{E} = \{Au + c \mid \|u\| \leq 1\}$ be an ellipsoid covering the possible values of x, that is, the real and imaginary components of a. The ellipsoid $\mathcal{E}$ is centered at c; the matrix A determines its size and shape. The constraint $\mathbf{Re}\ w^* a \geq 1$ for all $a \in \mathcal{E}$ in (1.18) can be expressed

$$x^T z \geq 1 \quad \forall z \in \mathcal{E}, \tag{1.23}$$

which is equivalent to

$$u^T A^T x \leq c^T x - 1 \text{ for all } u \text{ s.t.,} \quad \|u\| \leq 1. \tag{1.24}$$

Now, (1.24) holds for all $\|u\| \leq 1$ if and only if it holds for the value of u that maximizes $u^T A^T x$, namely $u = A^T x / \|A^T x\|$. By the Cauchy-Schwartz inequality, we see that (1.23) is equivalent to the constraint

$$\|A^T x\| \leq c^T x - 1, \tag{1.25}$$

which is called a *second-order cone constraint* [21]. We can then express the robust minimum-variance beamforming problem (1.18) as

$$\begin{array}{ll} \text{minimize} & x^T R x \\ \text{subject to} & \|A^T x\| \leq c^T x - 1, \end{array} \tag{1.26}$$

which is a second-order cone program. See references [21–23]. The subject of robust convex optimization is covered in references [9, 24–28].

By assumption, R is positive definite and the constraint $\|A^T x\| \leq c^T x - 1$ in (1.26) precludes the trivial minimizer of $x^T R x$. Hence, this constraint will be tight for any optimal solution and we may express (1.26) in terms of real-valued quantities as

$$\begin{array}{ll} \text{minimize} & x^T R x \\ \text{subject to} & c^T x = 1 + \|A^T x\|. \end{array} \tag{1.27}$$

Compared to the MVB, the RMVB adds a margin that scales with the size of the uncertainty. In the case of no uncertainty where $\mathcal{E}$ is a singleton whose center is $c = [\mathbf{Re}\, a(\theta_d)^T \;\; \mathbf{Im}\, a(\theta_d)^T]^T$, (1.27) reduces to Capon's method and admits an analytical solution given by the MVB (1.5). Unlike the use of additional point or derivative main-beam constraints or a regularization term, the RMVB is guaranteed to satisfy the minimum gain constraint for all values in the uncertainty ellipsoid. In the case of isotropic array uncertainty, the optimal solution of (1.18) yields the same weight vector (to a scale factor) as the regularized beamformer for the proper the proper choice of μ.

1.3.1 Lagrange Multiplier Methods

We may compute the RMVB efficiently using Lagrange multiplier methods. See, for example, references [29–30], [31, §12.1.1], and [32]. The RMVB is the optimal solution of

$$\begin{array}{ll} \text{minimize} & x^T R x \\ \text{subject to} & \|A^T x\|^2 = (c^T x - 1)^2 \end{array} \tag{1.28}$$

if we impose the additional constraint that $c^T x \geq 1$. We define the *Lagrangian* L: $\mathbf{R}^n \times \mathbf{R} \to \mathbf{R}$ associated with (1.28) as

$$\begin{aligned} L(x, \lambda) &= x^T R x + \lambda\big(\|A^T x\|^2 - (c^T x - 1)^2\big) \\ &= x^T (R + \lambda Q) x + 2\lambda c^T x - \lambda, \end{aligned} \tag{1.29}$$

where $Q = AA^T - cc^T$. To calculate the stationary points, we differentiate $L\,(x, y)$ with respect to x and λ; setting these partial derivatives equal to zero yields the Lagrange equations:

$$(R + \lambda Q)x = -\lambda c \tag{1.30}$$

and

$$x^T Q x + 2c^T x - 1 = 0. \tag{1.31}$$

To solve for the Lagrange multiplier λ, we note that equation (1.30) has an analytical solution given by

$$x = -\lambda (R + \lambda Q)^{-1} c;$$

applying this to (1.31) yields

$$f(\lambda) = \lambda^2 c^T (R + \lambda Q)^{-1} Q (R + \lambda Q)^{-1} c - 2\lambda c^T (R + \lambda Q)^{-1} c - 1. \tag{1.32}$$

The optimal value of the Lagrange multiplier λ^* is then a zero of (1.32). We proceed by computing the eigenvalue/eigenvector decomposition

$$V \Gamma V^T = R^{-1/2} Q (R^{-1/2})^T$$

to diagonalize (1.32), that is,

$$f(\lambda) = \lambda^2 \bar{c}^T (I + \lambda\Gamma)^{-1} \Gamma (I + \lambda\Gamma)^{-1} \bar{c} - 2\lambda \bar{c}^T (I + \lambda\Gamma)^{-1} \bar{c} - 1, \tag{1.33}$$

where $\bar{c} = V^T R^{-1/2} c$. Equation (1.33) reduces to the following scalar *secular* equation:

$$f(\lambda) = \lambda^2 \sum_{i=1}^{n} \frac{\bar{c}_i^2 \gamma_i}{(1 + \lambda\gamma_i)^2} - 2\lambda \sum_{i=1}^{n} \frac{\bar{c}_i^2}{(1 + \lambda\gamma_i)} - 1, \tag{1.34}$$

where $\gamma \in \mathbf{R}^n$ are the diagonal elements of Γ. The values of γ are known as the *generalized eigenvalues* of Q and R and are the roots of the equation $\det(Q - \lambda R) = 0$. Having computed the value of λ^* satisfying $f(\lambda^*) = 0$, the RMVB is computed according to

$$x^* = -\lambda^* (R + \lambda^* Q)^{-1} c. \tag{1.35}$$

Similar techniques have been used in the design of filters for radar applications; see Stutt and Spafford [33] and Abramovich and Sverdlik [34].

In principle, we could solve for all the roots of (1.34) and choose the one that results in the smallest objective value $x^T R x$ *and* satisfies the constraint $c^T x > 1$, assumed in (1.28). In the next section, however, we show that this constraint is only met for values of the Lagrange multiplier λ greater than a minimum value, $\lambda_{\min}$. We will see that there is a single value of $\lambda > \lambda_{\min}$ that satisfies the Lagrange equations.

1.3.2 A Lower Bound on the Lagrange Multiplier

We begin by establishing the conditions under which (9) has a solution. Assume $R = R^T \succ 0$, that is, R is symmetric and positive definite.

Lemma 1. For $A \in \mathbf{R}^{n \times n}$ full rank, there exists an $x \in \mathbf{R}^n$ for which $\|A^T x\| = c^T x - 1$ if and only if $c^T (AA^T)^{-1} c > 1$.

Proof. To prove the if direction, define

$$x(\lambda) = (cc^T - AA^T - \lambda^{-1} R)^{-1} c. \tag{1.36}$$

By the matrix inversion lemma, we have

$$\begin{aligned} c^T x(\lambda) - 1 &= c^T (cc^T - AA^T - \lambda^{-1} R)^{-1} c - 1 \\ &= \frac{1}{c^T (AA^T + \lambda^{-1} R)^{-1} c - 1}. \end{aligned} \tag{1.37}$$

For $\lambda > 0$, $c^T (AA^T + \lambda^{-1} R)^{-1} c$ is a monotonically increasing function of λ; therefore, for $c^T (AA^T)^{-1} c > 1$, there exists a $\lambda_{\min} \in \mathbf{R}^+$ for which

$$c^T (AA^T + \lambda_{\min}^{-1} R)^{-1} c = 1. \tag{1.38}$$

This implies that the matrix $(R + \lambda_{\min} Q)$ is singular. Since

$$\begin{aligned} \lim_{\lambda \to \infty} c^T x(\lambda) - 1 &= -c^T (AA^T - cc^T)^{-1} c - 1 \\ &= \frac{1}{c^T (AA^T)^{-1} c - 1} > 0, \end{aligned}$$

$c^T x(\lambda) - 1 > 0$ for all $\lambda > \lambda_{\min}$.

As in (1.32) and (1.34), let $f(\lambda) = \|A^T x\|^2 - (c^T x - 1)^2$. Examining (1.32), we see

$$\begin{aligned} \lim_{\lambda \to \infty} f(\lambda) &= -c^T (AA^T - cc^T)^{-1} c - 1 \\ &= \frac{1}{c^T (AA^T)^{-1} c - 1} > 0. \end{aligned}$$

Evaluating (1.32) or (1.34), we see $\lim_{\lambda \to \lambda_{\min}^+} f(\lambda) = -\infty$. For all $\lambda > \lambda_{\min}$, $c^T x > 1$ and $f(\lambda)$ is continuous. Hence $f(\lambda)$ assumes the value of 0, establishing the existence of a $\lambda > \lambda_{\min}$ for which $c^T x(\lambda) - 1 = \|A^T x(\lambda)\|$.

To show the only if direction, assume x satisfies $\|A^T x\| \leq c^T x - 1$. This condition is equivalent to

$$z^T x \geq 1 \forall z \in \mathcal{E} = \{Au + c \mid \|u\| \leq 1\}. \tag{1.39}$$

For (1.39) to hold, the origin cannot be contained in ellipsoid $\mathcal{E}$, which implies $c^T (AA^T)^{-1} c > 1$. □

REMARK. The constraints $(c^T x - 1)^2 = \|A^T x\|^2$ and $c^T x - 1 > 0$ in (1.28), taken together, are equivalent to the constraint $c^T x - 1 = \|A^T x\|$ in (1.27). For $R = R^T \succ 0$, A full rank and $c^T (AA^T)^{-1} c > 1$, (1.27) has a unique minimizer x^*. For $\lambda > \lambda_{\min}$, $(\lambda^{-1} R + Q)$ is full rank, and the Lagrange equation (1.30)

$$(\lambda^{-1} R + Q) x^* = -c$$

holds for only a single value of λ. This implies there is a unique value of $\lambda > \lambda_{\min}$, for which the secular equation (1.34) equals zero.

Lemma 2. For $x = -\lambda(R + \lambda Q)^{-1}c \in \mathbf{R}^n$ with $A \in \mathbf{R}^{n\times n}$ full rank, $c^T(AA^T)^{-1}c > 1$, and $\lambda > 0, c^T x > 1$ if and only if the matrix $\big(R + \lambda\big(AA^T - cc^T\big)\big)$ has a negative eigenvalue.

Proof. Consider the matrix

$$M = \begin{bmatrix} \lambda^{-1}R + AA^T & c \\ c^T & 1 \end{bmatrix}.$$

We define the inertia of M as the triple $In\{M\} = \{n_+, n_-, n_0\}$, where n_+ is the number of positive eigenvalues, n_- is the number of negative eigenvalues, and n_0 is the number of zero eigenvalues of M. See Kailath et al. [35, pp. 729–730].

Since both block diagonal elements of M are invertible,

$$\begin{aligned} In\{M\} &= In\{\lambda^{-1}R + AA^T\} + In\{\Delta_1\} \\ &= In\{1\} + In\{\Delta_2\}, \end{aligned} \tag{1.40}$$

where $\Delta_1 = 1 - c^T(\lambda^{-1}R + AA^T)^{-1}c$, the Schur complement of the (1,1) block in M, and $\Delta_2 = \lambda^{-1}R + AA^T - cc^T$, the Schur complement of the (2,2) block in M. We conclude $c^T(\lambda^{-1}R + AA^T)^{-1}c > 1$ if and only if the matrix $(\lambda^{-1}R + AA^T - cc^T)$ has a negative eigenvalue. By the matrix inversion lemma,

$$\frac{1}{c^T(\lambda^{-1}R + AA^T)^{-1}c - 1} = -c^T(\lambda^{-1}R + AA^T - cc^T)^{-1}c - 1. \tag{1.41}$$

Inverting a scalar preserves its sign, therefore,

$$c^T x - 1 = -c^T(\lambda^{-1}R + AA^T - cc^T)^{-1}c - 1 > 0 \tag{1.42}$$

if and only if $\lambda^{-1}R + AA^T - cc^T$ has a negative eigenvalue. □

Remark. Applying Sylvester's law of inertia to equations (1.32) and (1.34), we see that

$$\lambda_{\min} = -\frac{1}{\gamma_j}, \tag{1.43}$$

where γ_j is the single negative generalized eigenvalue. Using this fact and (1.34), we can readily verify $\lim_{\lambda \to \lambda_{\min}^+} f(\lambda) = -\infty$, as stated in Lemma 1.

Two immediate consequences follow from Lemma 2. First, we may exclude from consideration any value of λ less than $\lambda_{\min}$. Second, for all $\lambda > \lambda_{\min}$, the matrix $R + \lambda Q$ has a single negative eigenvalue. We now use these facts to obtain a tighter lower bound on the value of the optimal Lagrange multiplier. We begin by rewriting (1.34) as

$$\sum_{i=1}^{n} \frac{\bar{c}_i^2(-2 - \lambda\gamma_i)}{(1 + \lambda\gamma_i)^2} = \frac{1}{\lambda}. \tag{1.44}$$

Recall exactly one of the generalized eigenvalues γ in the secular equation (1.44) is negative. We rewrite (1.44) as

$$\lambda^{-1} = \frac{\bar{c}_j^2(-2 - \lambda\gamma_j)}{(1 + \lambda\gamma_j)^2} - \sum_{i \neq j} \frac{\bar{c}_i^2(2 + \lambda\gamma_i)}{(1 + \lambda\gamma_i)^2} \tag{1.45}$$

where j denotes the index associated with this negative eigenvalue.

A lower bound on λ can be found by ignoring the terms involving the non-negative eigenvalues in (1.45) and solving

$$\lambda^{-1} = \frac{\bar{c}_i^2(-2 - \lambda\gamma_j)}{(1 + \lambda\gamma_j)^2}.$$

This yields a quadratic equation in λ

$$\lambda^2(\bar{c}_j^2\gamma_j + \gamma_j^2) + 2\lambda(\gamma_j + \bar{c}_j^2) + 1 = 0, \tag{1.46}$$

the roots of which are given by

$$\lambda = \frac{-1 \pm |\bar{c}_j|(\gamma_j + \bar{c}_j^2)^{-1/2}}{\gamma_j}. \tag{1.47}$$

By Lemma 2, the constraint $c^T x^* \geq 1$ implies $R + \lambda^* Q$ has a negative eigenvalue, since

$$\begin{aligned} c^T x^* &= c^T(-\lambda^*(R + \lambda Q)^{-1})c \geq 1 \\ &= -\lambda^* \bar{c}^T(I + \lambda^*\Gamma)^{-1}\bar{c} \end{aligned}$$

Hence, $\lambda^* > -1/\gamma_j$ where γ_j is the single negative eigenvalue. We conclude $\lambda^* > \hat{\lambda}$, where

$$\hat{\lambda} = \frac{-1 - |\bar{c}_j|(\gamma_j + \bar{c}_j^2)^{-1/2}}{\gamma_j}. \tag{1.48}$$

In Figure 1.7 we see a plot of the secular equation and the improved lower bound $\hat{\lambda}$ found in (1.48).

1.3.3 Some Technical Details

In this section, we show that the parenthetical quantity in (1.48) is always nonnegative for any feasible beamforming problem. We also prove that the lower bound on the Lagrange multiplier in (1.48) is indeed that.

Recall that for any feasible beamforming problem, $Q = AA^T - cc^T$ has a negative eigenvalue. Note that $\bar{c}_j = v_j^T R^{-1/2} c$, where v_j is the eigenvector associated with the negative eigenvalue γ_j. Hence, $v_j \in \mathbf{R}^n$ can be expressed as the optimal solution of

$$\begin{array}{ll} \text{minimize} & v^T R^{-1/2}(AA^T - cc^T)(R^{-1/2})^T v \\ \text{subject to} & \|v\| = 1 \end{array} \tag{1.49}$$

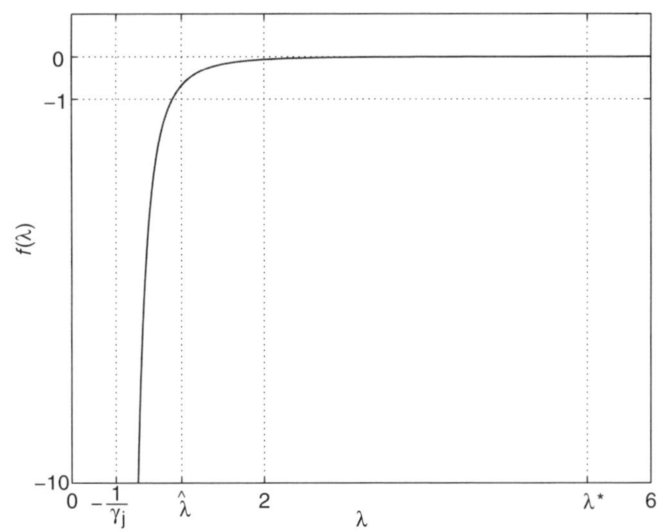

Figure 1.7 Plot of the secular equation from the Section 1.2 example. Here γ_j is the (single) negative eigenvalue of $R^{-1/2}(AA^T - cc^T)(R^{-1/2})^T$, $\hat{\lambda}$ is the computed lower bound on the Lagrange multiplier, and λ^* the solution to the secular equation.

and $\gamma_j = v_j^T R^{-1/2}(AA^T - cc^T)(R^{-1/2})^T v_j$, the corresponding objective value. Since

$$\bar{c}_j^2 = v_j^T R^{-1/2} c \left(v_j^T R^{-1/2} c \right)^T = v_j^T R^{-1/2} cc^T \left(R^{-1/2} \right)^T v_j, \tag{1.50}$$

we conclude $(\gamma_j + \bar{c}_j^2) = v_j^T R^{-1/2} AA^T (R^{-1/2})^T v_j > 0$.

To show that there exists no root between $\lambda_{\min}$ and $\hat{\lambda}$, we rewrite the secular equation (1.34)

$$f(\lambda) = g(\lambda) + h(\lambda), \tag{1.51}$$

where

$$\begin{aligned} g(\lambda) &= \lambda^2 \frac{\bar{c}_j^2 \gamma_j}{(1+\lambda\gamma_j)^2} - 2\lambda \frac{\bar{c}_j^2}{(1+\lambda\gamma_j)} - 1 \\ &= \frac{\lambda^2(\bar{c}_j^2\gamma_j + \gamma_j^2) + 2\lambda(\gamma_j + \bar{c}_j^2) + 1}{-(1+\lambda\gamma_j)^2} \end{aligned} \tag{1.52}$$

and

$$\begin{aligned} h(\lambda) &= \lambda^2 \sum_{i \neq j} \frac{\bar{c}_j^2 \gamma_j}{(1+\lambda\gamma_j)^2} - 2\lambda \sum_{i \neq j} \frac{\bar{c}_j^2}{(1+\lambda\gamma_j)} \\ &= -\lambda \sum_{i \neq j} \frac{(\lambda\gamma_i + 2)(\gamma_i + \bar{c}_i^2)}{(1+\lambda\gamma_i)^2} \end{aligned} \tag{1.53}$$

Comparing (1.46) and (1.52), we see the roots of $g(\lambda)$ are given by (1.47). Since $g'(\lambda) < 0$ for all $\lambda < -1/\gamma_j$ and $\lim_{\lambda \to 0} g(\lambda) = -1$, there exists no solution to the secular equation for $\gamma \in [0, -1/\gamma_j)$. Hence the unique root of $g(\lambda)$ is given by (1.48).

Since all of the eigenvalues γ_i, $i \neq j$ in (1.52) are non-negative, $h(\lambda)$ is continuous, bounded and differentiable for all $\lambda > 0$. The derivative of the h with respect to λ is given by

$$h'(\lambda) = -2 \sum_{i \neq j} \frac{\bar{c}_i^2}{(1+\lambda\gamma_i)^3} \tag{1.54}$$

By inspection, $h'(\lambda) < 0$ for all $\lambda > 0$.

We now show that $\hat{\lambda}$ is a strict lower bound for the root of the secular equation (1.34).

Define t: $\mathbf{R} \times \mathbf{R} \to \mathbf{R}$, according to:

$$t(\lambda, \theta) = g(\lambda) + \theta h(\lambda), \tag{1.55}$$

where $\theta \in [0, 1]$. For $\theta = 0, t(\lambda, \theta) = g(\lambda)$; hence $t(\hat{\lambda}, 0)$, where $\hat{\lambda}$ is as in (1.48). As $g(\lambda)$ and $h(\lambda)$ are locally smooth and bounded, the total differential of t is given by

$$dt = \left(\frac{\partial g}{\partial \lambda} + \theta \frac{\partial h}{\partial \lambda}\right) d\lambda + h(\lambda) d\theta.$$

The first order condition for the root t is given by:

$$\left(\frac{\partial g}{\partial \lambda} + \theta \frac{\partial h}{\partial \lambda}\right) d\lambda = -h(\lambda) d\theta.$$

Since $f(\lambda)$ is an increasing function of λ for all $\lambda \in [-1/\gamma_j, \lambda^*]$ and $h'(\lambda) < 0$ for all $\lambda > 0, (\partial g/\partial\lambda + \theta(\partial h/\partial\lambda)) > 0$ for all $\theta \in [0, 1]$ and $\lambda \in [-1/\gamma_j, \lambda^*]$. Recall $h(\lambda) < 0$ for all $\lambda > 0$. Hence, as θ is increased, the value of λ satisfying $t(\theta, \lambda)$ increases. The value of λ satisfying $t(1, \lambda)$ is the solution to the secular equation, establishing that the (1.48) is a lower bound.

1.3.4 Solution of the Secular Equation

The secular equation (1.34) can be efficiently solved using the Newton–Raphson method. This method enjoys quadratic convergence if started sufficiently close to the root λ^*; see Dahlquist and Björck [36, §6] for details. The derivative of this secular equation with respect to λ is given by

$$f'(\lambda) = -2 \sum_{i=1}^{n} \frac{\bar{c}_i^2}{(1 + \lambda\gamma_i)^3}. \tag{1.56}$$

The secular equation (1.34) is not necessarily a monotonically increasing function of λ. A plot showing the convergence of the secular equation, from the Section 1.2 example, is shown in Figure 1.8.

1.3.5 Summary and Computational Complexity of the RMVB Computation

We summarize the algorithm below. In parentheses are approximate costs of each of the numbered steps; the actual costs will depend on the implementation and problem size [37]. As in reference [31], we will consider a flop to be any single floating-point operation.

RMVB Computation

Given R, strictly feasible A and c.

1. *Calculate* $Q \leftarrow AA^T - cc^T$. $(2n^2)$
2. *Change coordinates.* $(2n^3)$
   (a) Compute Cholesky factorization $LL^T = R$.
   (b) Compute $L^{-1/2}$.
   (c) $\tilde{Q} \leftarrow L^{-1/2} Q (L^{-1/2})^T$.

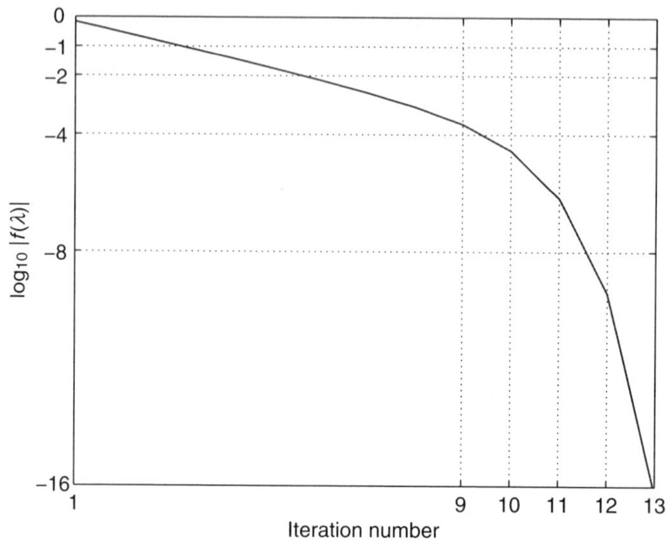

Figure 1.8 Iterations of the secular equation. For λ sufficiently close to λ^*, in this case, after nine iterations, the iterates converge quadratically, doubling the number of bits of precision at every iteration.

3. *Eigenvalue/eigenvector computation.* $(10n^3)$
   (a) Compute $V\Gamma V^T = \tilde{Q}$.
4. *Change coordinates.* $(4n^2)$
   (a) $\bar{c} \leftarrow V^T R^{-1/2} c$.
5. *Secular equation solution.* $(80n)$
   (a) Compute initial feasible point $\hat{\lambda}$.
   (b) Find $\lambda^* < \hat{\lambda}$ for which $f(\lambda) = 0$.
6. *Compute* $x^* \leftarrow (R + \lambda^* Q)^{-1} c$. (n^3)

The computational complexity of these steps is discussed below:

1. Forming the matrix product AA^T is expensive and should be avoided. If the parameters of the uncertainty ellipsoid are stored, the shape parameter may be stored as AA^T. In the event that an aggregate ellipsoid is computed using the methods of Section 1.6, the quantity AA^T is produced. In either case, only the subtraction of the quantity cc^T need be performed, requiring $2n^2$ flops.
2. Computing the Cholesky factor L in step 2 requires $n^3/3$ flops. The resulting matrix is triangular, hence computing its inverse requires $n^3/2$ flops. Forming the matrix $\tilde{Q}$ in step 2(c) requires n^3 flops.

3. Computing the eigenvalue/eigenvector decomposition is the most expensive part of the algorithm. In practice, it takes approximately $10n^3$ flops.
5. Solution of the secular equation requires minimal effort. The solution of the secular equation converges quadratically. In practice, the starting point $\hat{\lambda}$ is close to λ^*; hence, the secular equation generally converges in 7 to 10 iterations, independent of problem size.
6. Accounting for the symmetry in R and Q, computing x^* requires n^3 flops.

In comparison, the regularized beamformer requires n^3 flops. Hence the RMVB requires approximately 12 times the computational cost of the regularized beamformer. Note that this factor is independent of problem size.

In Section 1.6, we extend the methods of this section to the case of multiplicative uncertainties by computing an outer approximation to the element-wise or Hadamard product of ellipsoids. Using this approximation, no subsequent verification of the performance is required.

1.4 A NUMERICAL EXAMPLE

Consider a 10-element uniform linear array, centered at the origin, in which the spacing between the elements is half of a wavelength. Assume the response of each element is isotropic and has unit norm. If the coupling between elements is ignored, the response of the array a: $\mathbf{R} \to \mathbf{C}^{10}$ is given by:

$$a(\theta) = [\, e_j^{-9\phi/2} \quad e_j^{-7\phi/2} \quad \cdots \quad e_j^{7\phi/2} \quad e_j^{9\phi/2} \,]^T,$$

where $\phi = \pi \cos(\theta)$, $j = \sqrt{-1}$, and θ is the angle of arrival. As seen in Section 1.2, the responses of closely spaced antenna elements may differ substantially from this model.

In this example, three signals impinge upon the array: a desired signal $s_{\text{d}}(t)$ and two uncorrelated interfering signals $s_{\text{int1}}(t)$ and s_{int2}. The signal-to-noise ratio (SNR) of the desired signal at each element is 20 dB. The angles of arrival of the interfering signals, θ_{int1} and θ_{int2}, are 30° and 75°; the SNRs of these interfering signals, 40 dB and 20 dB, respectively. We model the received signals as:

$$y(t) = a_{\text{d}} s_{\text{d}}(t) + a(\theta_{\text{int1}}) s_{\text{int1}}(t) + a(\theta_{\text{int2}}) s_{\text{int2}}(t) + v(t), \tag{1.57}$$

where a_{d} denotes the array response of the desired signal, $a(\theta_{\text{int1}})$ and $a(\theta_{\text{int2}})$, the array responses for the interfering signals, $s_{\text{d}}(t)$ denotes the complex amplitude of the desired signal, $s_{\text{int1}}(t)$ and $s_{\text{int2}}(t)$, the interfering signals, and $v(t)$ is a complex vector of additive white noises.

Let the noise covariance $\mathbf{E}\, vv^* = \sigma_n^2 I$, where I is an $n \times n$ identity matrix and n is the number of antennas, namely, 10. Similarly define the powers of the desired signal and interfering signals to be $\mathbf{E}\, s_d s_d^* = \sigma_d^2$, $\mathbf{E}\, s_{\text{int1}} s_{\text{int1}}^* = \sigma_{\text{int1}}^2$, and $\mathbf{E}\, s_{\text{int1}} s_{\text{int2}}^* = \sigma_{\text{int2}}^2$, where

$$\frac{\sigma_d^2}{\sigma_n^2} = 10^2, \qquad \frac{\sigma_{\text{int1}}^2}{\sigma_n^2} = 10^4, \qquad \frac{\sigma_{\text{int2}}^2}{\sigma_n^2} = 10^2.$$

If we assume the signals $s_d(t), s_{\text{int1}}(t), s_{\text{int2}}(t)$, and $v(t)$ are all uncorrelated, the estimated covariance, which uses the actual array response, is given by

$$\mathbf{E}\, R = \mathbf{E}\, yy^* = \sigma_d^2 a_d a_d^* + \sigma_{\text{int1}}^2 a(\theta_{\text{int1}}) a(\theta_{\text{int1}})^* + \sigma_{\text{int2}}^2 a(\theta_{\text{int2}}) a(\theta_{\text{int2}})^* + \sigma_n^2 I. \quad (1.58)$$

In practice, the covariance of the received signals plus interference is often neither known nor stationary and hence must be estimated from recently received signals. As a result, the performance of beamformers is often degraded by errors in the covariance due to either small sample size or movement in the signal sources.

We will compare the performance of the robust beamformer with beamformers using two regularization techniques: diagonal loading and eigenvalue thresholding. In this example, we assume a priori, that the nominal AOA, θ_{nom}, is 45°. The actual array response is contained in an ellipsoid $\mathcal{E}(c, P)$, whose center and configuration matrix are computed from N equally-spaced samples of the array response at angles between 40° and 50° according to

$$c = \frac{1}{N}\sum_{i=1}^{N} a(\theta_i) \qquad P = \frac{1}{\alpha N}\sum_{i=1}^{N} (a(\theta_i) - c)(a(\theta_i) - c)^*, \quad (1.59)$$

where

$$\theta_i = \theta_{\text{nom}} + \left(-\frac{1}{2} + \frac{i-1}{N-1}\right)\Delta_\theta, \quad \text{for } i \in [1, N], \quad (1.60)$$

and

$$\alpha = \sup\, (a(\theta_i) - c)^* P^{-1} (a(\theta_i) - c) \quad i \in [1, N]$$

Here, $\Delta_\theta = 10°$, and $N = 64$.

In Figure 1.9, we see the reception pattern of the array employing the MVB, the regularized beamformer (1.10), and the RMVB, all computed using the nominal AOA and the corresponding covariance matrix R. The regularization term used in the regularized beamformer was chosen to be 1/100 of the largest eigenvalue of the received covariance matrix. By design, both the MVB and the regularized beamformer have unity gain at the nominal AOA. The response of the regularized beamformer is seen to be a detuned version of the MVB. The RMVB maintains greater-than-unity gain for all AOAs covered by the uncertainty ellipsoid $\mathcal{E}(c, P)$.

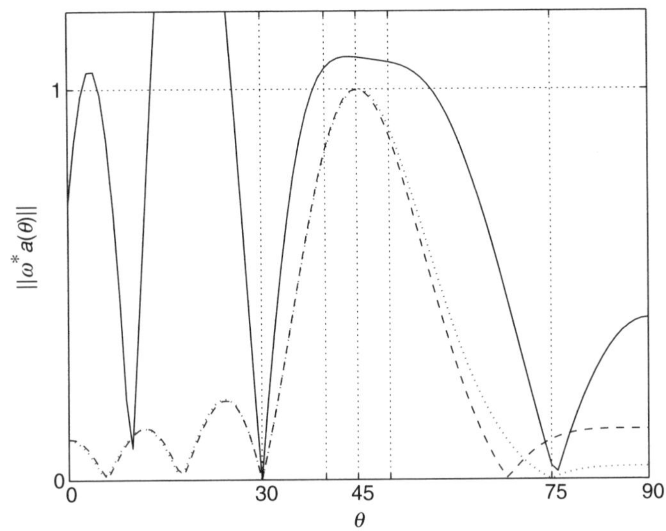

Figure 1.9 The response of the MVB (Capon's method, dashed trace), the regularized beamformer employing diagonal loading (dotted trace), and the RMVB (solid trace) as a function of angle of arrival θ. Note that the RMVB preserves greater than unity gain for all angles of arrival in the design specification of $\theta \in [40^\circ, 50^\circ]$.

In Figure 1.10 we see the effect of changes in the regularization parameter μ on the worst-case SINRs for the regularized beamformers using diagonal loading and eigenvalue thresholding, and the effect of scaling the uncertainty ellipsoid on the RMVB. Using the definition of SINR (1.6), we define the worst case SINR is as the minimum objective value of the following optimization problem:

$$\begin{array}{ll} \text{minimize} & \dfrac{\sigma_{\text{d}}^2 \|w^* a\|^2}{\mathbf{E}\; w^* R_v w} \\ \text{subject to} & a \in \mathcal{E}(c, P), \end{array}$$

where the expected covariance of the interfering signals and noises is given by

$$\mathbf{E}\; R_v = \sigma_{\text{int1}}^2 a(\theta_{\text{int1}}) a(\theta_{\text{int1}})^* + \sigma_{\text{int1}}^2 a(\theta_{\text{int2}}) a(\theta_{\text{int2}})^* + \sigma_{\text{n}}^2 I.$$

The weight vector w and covariance matrix of the noise and interfering signals R_v used in its computation reflect the chosen value of the array manifold.

For diagonal loading, the parameter μ is the scale factor multiplying the identity matrix added to the covariance matrix, divided by the largest eigenvalue of the covariance matrix R. For small values of μ, that is, 10^{-6}, the performance of the regularized beamformer approaches that of Capon's method; the worst-case SINR for Capon's method is -29.11 dB. As $\mu \to \infty$, $w_{\text{reg}} \to a(\theta_{\text{nom}})$.

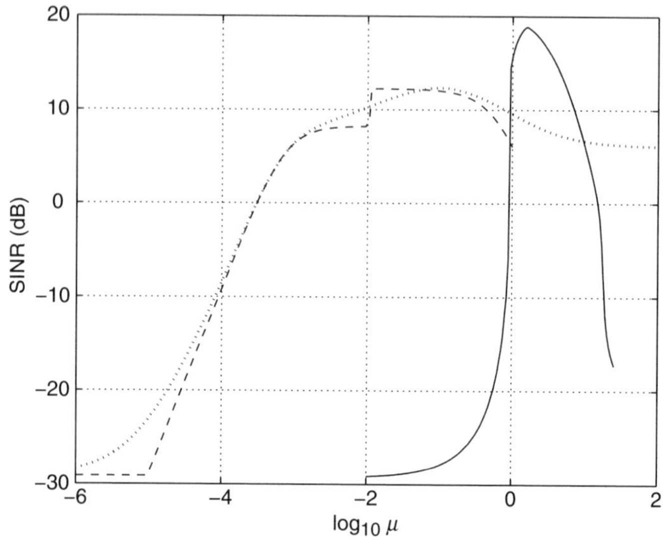

Figure 1.10 The worst-case performance of the regularized beamformers based on diagonal loading (dotted) and eigenvalue thresholding (dashed) as a function of the regularization parameter μ. The effect of scaling of the uncertainty ellipsoid used in the design of the RMVB (solid) is seen; for $\mu = 1$ the uncertainty used in designing the robust beamformer equals the actual uncertainty in the array manifold.

The beamformer based on eigenvalue thresholding performs similarly to the beamformer based on diagonal loading. In this case, μ is defined to be the ratio of the threshold to the largest eigenvalue of R; as such, the response of this beamformer is only computed for $\mu \leq 1$.

For the robust beamformer, we use μ to define the ratio of the size of the ellipsoid used in the beamformer computation $\mathcal{E}_{\text{design}}$ divided by size of the actual array uncertainty $\mathcal{E}_{\text{actual}}$. Specifically, if $\mathcal{E}_{\text{actual}} = \{Au + c \mid \|u\| \leq 1\}, \mathcal{E}_{\text{design}} = \{\mu Av + c \mid \|v\| \leq 1\}$. When the design uncertainty equals the actual, the worst-case SINR of the robust beamformer is seen to be 15.63 dB. If the uncertainty ellipsoid used in the RMVB design significantly overestimates or underestimates the actual uncertainty, the worst-case SINR is decreased.

For comparison, the worst-case SINR of the MVB with (three) unity mainbeam constraints at 40°, 45°, and 50° is 1.85 dB. The MV-EPC beamformer was computed using the same 64 samples of the array manifold as the computation of the uncertainty ellipsoid (1.59); the design value for the response in each of these directions was unity. The worst-case SINRs of the rank-1 through rank-4 MV-EPC beamformers were found to be -28.96 dB, -3.92 dB, 1.89 dB, and 1.56 dB, respectively. The worst-case response for the rank-5 and rank-6 MV-EPC beamformers is *zero*; that is, it can fail completely.

1.4.1 Power Estimation

If the signals and noises are all uncorrelated, the sample covariance, as computed in (1.3), equals its expected value, and the uncertainty ellipsoid contains the actual array response, the RMVB is guaranteed to have greater than unity magnitude response for all values of the array manifold in the uncertainty ellipsoid $\mathcal{E}$. In this case, an upper bound on the power of the desired signal, $\sigma_{\rm d}^2$, is simply the weighted power out of the array, namely

$$\hat{\sigma}_{\rm d}^2 = w^* R_y w. \tag{1.61}$$

In Figure 1.11, we see the square of the norm of the weighted array output as a function of the hypothesized angle of arrival $\theta_{\rm nom}$ for the RMVB using uncertainty ellipsoids computed according to (1.59) and (1.60) with $\Delta\theta = 10^\circ$, 4°, and 0°. If the units of the array output correspond to volts or amperes, the square of the magnitude of the weighted array output has units of power. This plot is referred to in the literature as a spatial ambiguity function [15, 16]; its resolution is seen to decrease with increasing uncertainty ellipsoid size. The RMVB computed for $\Delta\theta = 0^\circ$ corresponds to the Capon beamformer. The spatial ambiguity function using the Capon beamformer provides an accurate power estimate only when the assumed array manifold equals the actual.

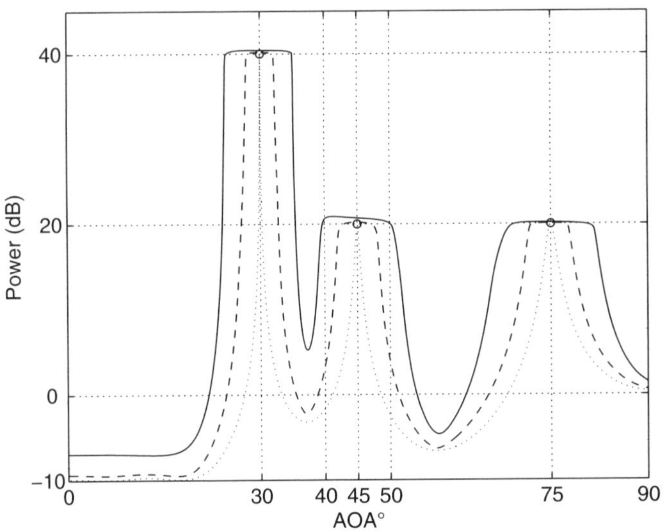

Figure 1.11 The ambiguity function for RMVB beamformer using an uncertainty ellipsoid computed from a beamwidth of 10° (solid), 2° (dashed) and the Capon beamformer (dotted). The true powers of the signal of interest and interfering signals are denoted with circles. In this example, the additive noise power at each element has unit variance; hence, the ambiguity function corresponds to SNR.

We summarize the effect of differences between assumed and actual uncertainty regions on the performance of the RMVB:

- If the assumed uncertainty ellipsoid equals the actual uncertainty, the gain constraint is met and no other choice of gain vector yields better worst-case performance over all values of the array manifold in the uncertainty ellipsoid.
- If the assumed uncertainty ellipsoid is smaller than the actual uncertainty, the minimum gain constraint will generally not be met for all possible values if the array manifold. If the uncertainty ellipsoid used in computing the RMVB is much smaller than the actual uncertainty, the performance may degrade substantially. The power estimate, computed using the RMVB as in (1.61) is not guaranteed to be an upper bound, even when an accurate covariance is used in the computation.
- If assumed uncertainty is greater than the actual uncertainty, the performance is generally degraded, but the minimum gain in desired look direction is maintained. Given accurate covariance, the appropriately scaled weighted power out of the array yields an upper bound on the power of the received signal.

The performance of the RMVB is not optimal with respect to SINR; it is optimal in the following sense. For a fixed covariance matrix R and an array response contained in an ellipsoid $\mathcal{E}$, no other vector achieves a lower weighted power out of the array while maintaining the real part of the response greater than unity for all values of the array contained in $\mathcal{E}$.

In the next section, we describe two methods for computing ellipsoids covering a collection of points.

1.5 ELLIPSOIDAL MODELING

The uncertainty in the response of an antenna array to a plane wave arises principally from three sources:

- Uncertainty in the angle of arrival (AOA),
- Uncertainty in the array manifold given perfect knowledge of the AOA (also called calibration errors), and
- Variations in the gains of the signal-processing paths.

In this section, we describe methods to compute an ellipsoid that approximates or covers the range of possible values of the array manifold, given these uncertainties.

1.5.1 Ellipsoid Computation Using Mean and Covariance of Data

If the array manifold is measured in a controlled manner, the ellipsoid describing it may be generated from the mean and covariance of the measurements from repeated trials. In the case where the array manifold is not measured but rather predicted from numerical simulations, the uncertainty may take into account variation in the array response due to manufacturing tolerance, termination impedance, and similar effects. If the underlying distribution is multivariate normal, the k standard deviation ($k\sigma$) ellipsoid would be expected to contain a fraction of points equal to $1 - \chi^2(k^2, n)$, where n is the dimension of the random variable and $\chi^2(k^2, n)$ denotes the cumulative distribution function of a chi-squared random variable with n degrees of freedom evaluated at k^2.

In Figure 1.12, we see a two-dimensional ellipsoid generated according to $\mathcal{E} = \{Au \mid \|u\| \leq 1\}$, where

$$A = \begin{bmatrix} 1 & 2 \\ -1 & 3 \end{bmatrix}.$$

The one-, two-, and three-standard deviation ellipsoids are shown along with the minimum-volume ellipsoid containing these points.

We may generate an ellipsoid that covers a collection of points by using the mean as the center and an inflated covariance. While this method is very efficient numerically, it is possible to generate 'smaller' ellipsoids using the methods of the next section.

1.5.2 Minimum-Volume Ellipsoid (MVE)

Let $\mathcal{S} = \{s_1, \ldots, s_m\} \in \mathbf{R}^{2n}$ be a set of possible values of the array manifold $a(\cdot)$. Assume $\mathcal{S}$ is bounded. In the case of a full rank ellipsoid, the problem of finding

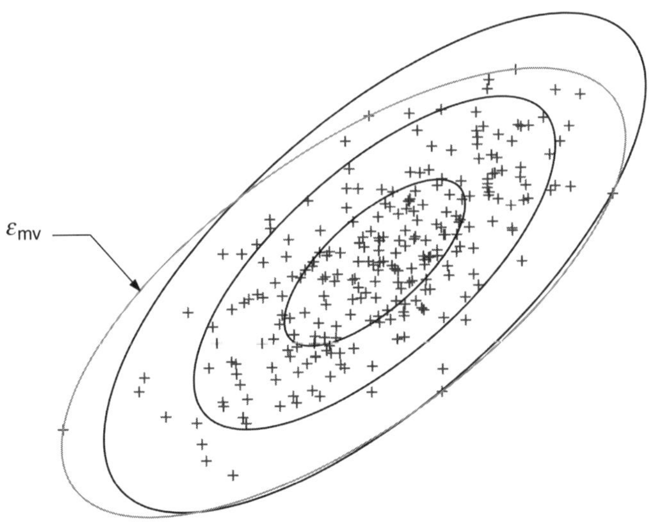

Figure 1.12 A minimum-volume ellipsoid $\mathcal{E}_{\text{mv}}$ covering points drawn from a bivariate normal distribution. The one-, two-, and three-standard deviation ellipsoids calculated from the first and second moments of the data are also shown.

the minimum-volume ellipsoid containing the convex hull of $\mathcal{S}$ can be expressed as the following semidefinite program:

$$\begin{array}{ll} \text{minimize} & \log\det F^{-1} \\ \text{subject to} & F = F^T \succ 0 \\ & \|Fs_i - g\| \le 1, \quad i = 1, \ldots, m. \end{array} \tag{1.62}$$

See Vandenberghe and Boyd [38] and Wu and Boyd [39]. The minimum-volume ellipsoid $\mathcal{E}$ containing $\mathcal{S}$ is called the *Löwner–John ellipsoid.* Equation (1.62) is a convex problem in variables F and g. For A full rank,

$$\{x \mid \|Fx - g\| \le 1\} \equiv \{Au + c \mid \|u\| \le 1\} \tag{1.63}$$

with $A = F^{-1}$ and $c = F^{-1}g$. The choice of A is not unique; in fact, any matrix of the form $F^{-1}U$ will satisfy (1.63), where U is any orthogonal matrix.

Commonly, $\mathcal{S}$ is well approximated by an affine set of dimension $l < 2n$ and (1.62) will be poorly conditioned numerically. We proceed by first applying a rank-preserving affine transformation f: $\mathbf{R}^{2n} \to \mathbf{R}^l$ to the elements of $\mathcal{S}$, with $f(s) = U_1^T(s - s_1)$. The matrix U_1 consists of the l left singular vectors, corresponding to the significant singular values, of the $2 \times (m-1)$ matrix

$$[(s_2 - s_1)(s_3 - s_1) \cdots (s_m - s_1)].$$

We may then solve (1.62) for the minimum-volume, nondegenerate ellipsoid in $\mathbf{R}^l$ that covers the image of $\mathcal{S}$ under f. The resulting ellipsoid can be described in $\mathbf{R}^{2n}$ as $\mathcal{E} = \{Au + c \mid \|u\| \le 1\}$, with

$$A = U_1 F^{-1}$$

and

$$c = U_1 F^{-1} g + s_1.$$

For an l-dimensional ellipsoid description, a minimum of $l + 2$ points are required; that is, $m \ge l + 2$.

Compared to an ellipsoid based on the first- and second-order statistics, a minimum-volume ellipsoid is robust in the sense that it is guaranteed to cover all the data points used in the description; the MVE is *not* robust to data outliers. The computation of the covering ellipsoid is relatively complex; see Vandenberghe et al. [40]. In applications where a real-time response is required, the covering ellipsoid calculations may be profitably performed in advance and stored in a table.

In the next section, our philosophy is different. Instead of computing ellipsoid descriptions to describe collections of points, we consider operations on ellipsoids. While it is possible to develop tighter ellipsoidal approximations using the methods just described, the computational burden of these methods often precludes their use.

1.6 UNCERTAINTY ELLIPSOID CALCULUS

1.6.1 Union of Ellipsoids

Suppose the actual AOA could assume one of p values and associated with each of these AOAs was an uncertainty ellipsoid. The possible values of the array manifold would be covered by the union of these ellipsoids. The resulting problem is then to find the 'smallest' ellipsoid $\mathcal{E}_0$ that covers the union of ellipsoids, $\mathcal{E}_1(c_1, Q_1), \ldots,$ $\mathcal{E}_p(c_p, Q_p)$. As in (1.16), we will describe these ellipsoids in terms of the associated quadratic functions

$$T_i(x) = x^T F_i x + 2x^T g_i + h_i,$$

where

$$F_i(x) = Q^{-1}, \quad g_i = -Q^{-1}c, \quad \text{and} \quad h_i = c^T Q^{-1} c - 1.$$

By the $\mathcal{S}$-procedure [41, pp. 23–24], $\mathcal{E}_i \subseteq \mathcal{E}_0$ for $i = 1, \ldots, p$ if and only if there exists non-negative scalars $\tau_1, \ldots, \tau_p$ such that

$$T_0(x) - \tau_i T_i(x) \leq 0, \quad i = 1, \ldots, p,$$

or equivalently, such that

$$\begin{bmatrix} F_0 & g_0 & 0 \\ g_0^T & -1 & g_0^T \\ 0 & g_0 & -F_0 \end{bmatrix} - \tau_i \begin{bmatrix} F_i & g_i & 0 \\ g_i^T & h_i & 0 \\ 0 & 0 & 0 \end{bmatrix} \leq 0,$$

for $i = 1, \ldots, p$. We can find the MVE containing the union of ellipsoids $\mathcal{E}_1, \ldots, \mathcal{E}_p$ by solving the matrix completion problem:

$$\begin{array}{ll} \text{minimize} & \log \det F_0^{-1} \\ \text{subject to} & F_0 > 0, \\ & \tau_1 \geq 0, \ldots, \tau_p \geq 0, \end{array}$$

$$\begin{bmatrix} F_0 & g_0 & 0 \\ g_0^T & -1 & g_0^T \\ 0 & g_0 & -F_0 \end{bmatrix} - \tau_i \begin{bmatrix} F_i & g_i & 0 \\ g_i^T & h_i & 0 \\ 0 & 0 & 0 \end{bmatrix} \leq 0,$$

for $i = 1, \ldots, p$, with variables F_0, g_0, and $\tau_1, \ldots, \tau_p$ [41, pp. 43–44]. The MVE covering the union of ellipsoids is then given by $\mathcal{E}(-F_0^{-1}g_0, F_0^{-2})$. An example of the minimum-volume ellipsoid covering the union of two ellipsoids in $\mathbf{R}^2$ is shown in Figure 1.13.

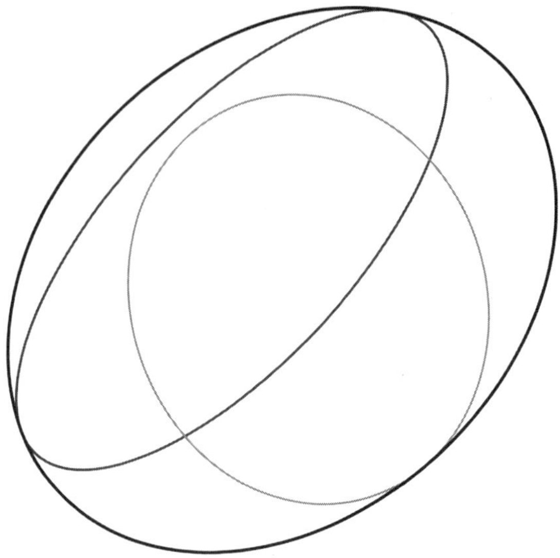

Figure 1.13 A minimum-volume ellipsoid covering the union of two ellipsoids.

1.6.2 The Sum of Two Ellipsoids

Recall that we can parameterize an ellipsoid in $\mathbf{R}^n$ in terms of its center $c \in \mathbf{R}^n$ and a symmetric non-negative definite configuration matrix $Q \in \mathbf{R}^{n\times n}$ as

$$\mathcal{E}(c, Q) = \{Q^{1/2}u + c \mid \|u\| \leq 1\},$$

where $Q^{1/2}$ is any matrix square root satisfying $Q^{1/2}(Q^{1/2})^T = Q$. Let $x \in \mathcal{E}_1 = \mathcal{E}(c_1, Q_1)$ and $y \in \mathcal{E}_2 = \mathcal{E}(c_2, Q_2)$. The range of values of the geometrical (or Minkowski) sum $z = x + y$ is contained in the ellipsoid

$$\mathcal{E} = \mathcal{E}(c_1 + c_2, Q(p)) \tag{1.64}$$

for all $p > 0$, where

$$Q(p) = (1 + p^{-1})Q_1 + (1 + p)Q_2; \tag{1.65}$$

see Kurzhanski and Vályi [42]. The value of p is commonly chosen to minimize either the determinant of $Q(p)$ or the trace of $Q(p)$. An example of the geometrical sum of two ellipses for various values of p is shown in Figure 1.14.

1.6.2.1 Minimum Volume. If $Q_1 \succ 0$ and $Q_2 \succ 0$, there exists a unique ellipsoid of minimal volume that contains the sum $\mathcal{E}_1 + \mathcal{E}_2$. It is described by

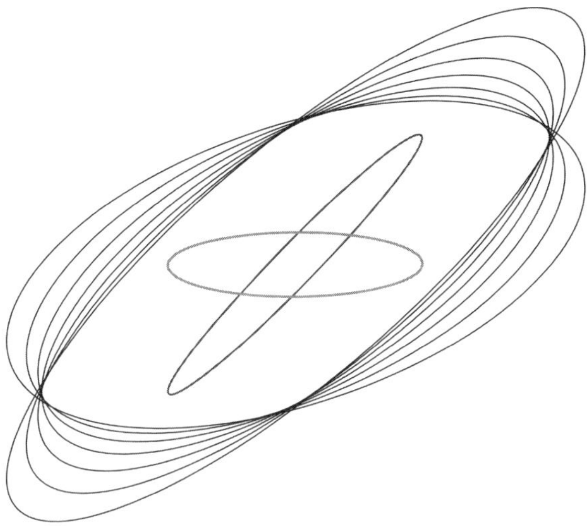

Figure 1.14 Outer approximations of the sum of two ellipses (center) for different configuration matrices $Q(p) = (1 + 1/p)Q_1 + (1 + p)Q_2$.

$\mathcal{E}(c_1 + c_2, Q(p^*))$, where $p^* \in (0, \infty)$ is the unique solution of the equation

$$f(p) = \sum_{i=1}^{n} \frac{1}{\lambda_i + p} - \frac{n}{p(p+1)} = 0. \tag{1.66}$$

Here, $0 < \lambda_i < \infty$ are the roots of the equation $\det(Q_1 - \lambda Q_2) = 0$, that is, the generalized eigenvalues of Q_1 and Q_2 [42, pp. 133–135]. The generalized eigenvalues can be determined by computing the eigenvalues of the matrix $Q_2^{-1/2} Q_1 (Q_2^{-1/2})^T$. Using the methods of Section 1.3, the solution of (1.66) may be found efficiently using Newton's method. In the event that neither Q_1 nor Q_2 is positive definite, but their sum is, a line search in p may be used to find the minimum-volume ellipsoid.

1.6.2.2 Minimum Trace.

There exists an ellipsoid of minimum trace, that is, sum of squares of the semiaxes, that contains the sum $\mathcal{E}_1(c_1, Q_q) + \mathcal{E}_2(c_2, Q_2)$; it is described by $\mathcal{E}(c_1 + c_2, Q(p^*))$, where $Q(p)$ is as in (1.65),

$$p^* = \sqrt{\frac{\mathbf{Tr}\, Q_1}{\mathbf{Tr}\, Q_2}}, \tag{1.67}$$

and **Tr** denotes trace. This fact, noted by Kurzhanski and Vályia [42, §2.5], may be verified by direct calculation.

Minimizing the trace of Q in equation (1.65) affords two computational advantages over minimizing the determinant. First, computing the optimal value of p can be done with $\mathcal{O}(n)$ operations; minimizing the determinant requires $\mathcal{O}(n^3)$. Second, the minimum-trace calculation is well-posed with degenerate ellipsoids.

1.6.3 An Outer Approximation to the Hadamard Product of Two Ellipsoids

In practice, the output of the antenna array is often subject to uncertainties that are multiplicative in nature. These may be due to gains and phases of the electronics paths that are not precisely known. The gains may be known to have some formal uncertainty; in other applications, these quantities are estimated in terms of a mean vector and covariance matrix. In both cases, this uncertainty is well-described by an ellipsoid; this is depicted schematically in Figure 1.15.

Assume that the range of possible values of the array manifold is described by an ellipsoid $\mathcal{E}_1 = \{Au + b \mid \|u\| \leq 1\}$. Similarly assume the multiplicative uncertainties lie within a second ellipsoid $\mathcal{E}_2 = \{Cv + d \mid \|v\| \leq 1\}$. The set of possible values of the array manifold in the presence of multiplicative uncertainties is described by the numerical range of the Hadamard, that is, element-wise product of $\mathcal{E}_1$ and $\mathcal{E}_2$. We will develop outer approximations to the Hadamard product of two ellipsoids. In Section 1.6.5, we consider the case where both ellipsoids describe real numbers; the case of complex values is considered in Section 1.6.6. Prior to this, we will review some basic facts about Hadamard products.

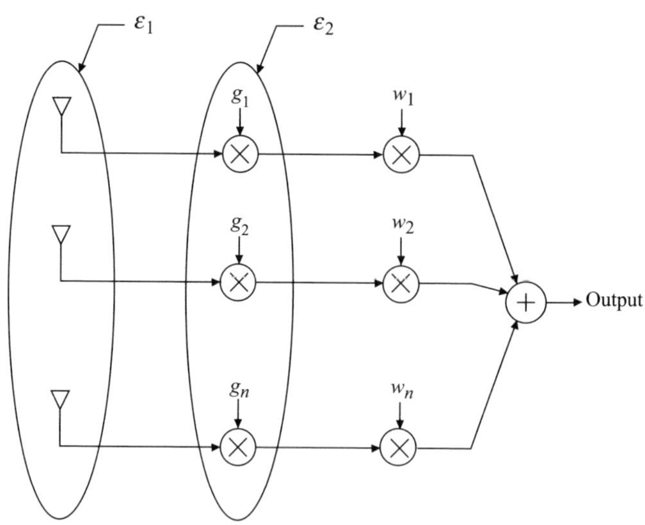

Figure 1.15 The possible values of array manifold are contained in ellipsoid $\mathcal{E}_1$; the values of gains are described by ellipsoid $\mathcal{E}_2$. The design variable w needs to consider the multiplicative effect of these uncertainties.

1.6.4 Preliminaries

Lemma 3. For any $x, y \in \mathbf{R}^n$,

$$(x \circ y)(x \circ y)^T = (xx^T) \circ (yy^T).$$

Proof. Direct calculation shows that the i,j entry of the product is $x_i y_i x_j y_j$, which can be regrouped as $x_i x_j y_i y_j$. □

Lemma 4. Let $x \in \mathcal{E}_x = \{Au \mid \|u\| \leq 1\}$ and $y \in \mathcal{E}_y = \{Cv \mid \|v\| \leq 1\}$. The field of values of the Hadamard product $x \circ y$ is contained in the ellipsoid

$$\mathcal{E}_{xy} = \left\{ \left(AA^T \circ CC^T\right)^{1/2} w \mid \|w\| \leq 1 \right\}.$$

Proof. By Lemma 3 we have

$$(x \circ y)(x \circ y)^T = (xx^T) \circ (yy^T).$$

In particular,

$$(Au \circ Cv)(Au \circ Cv)^T = (Auu^T A^T) \circ (Cvv^T C^T).$$

Expanding $AA^T \circ CC^T$ as:

$$\begin{aligned} AA^T \circ CC^T &= A\left(uu^T\right)A^T \circ C\left(vv^T\right)C^T + A\left(uu^T\right)A^T \circ C\left(I_n - vv^T\right)C^T \\ &\quad + A\left(I_n - uu^T\right)A^T \circ C\left(vv^T\right)C^T + A\left(I_n - uu^T\right)A^T \circ C\left(I_n - vv^T\right)C^T, \end{aligned} \tag{1.68}$$

we see the Hadamard product of two positive semidefinite matrices is also positive semidefinite [43, pp. 298–301]. Therefore,

$$\left(AA^T \circ CC^T\right) \succeq (Au \circ Cv)(Au \circ Cv)^T \; \forall \|u\| \leq 1, \|v\| \leq 1.$$ □

Lemma 5. Let $\mathcal{E}_1 = \{Au \mid \|u\| \leq 1\}$ and let d be any vector in $\mathbf{R}^n$. The Hadamard product of $\mathcal{E}_1 \circ d$ is contained in the ellipsoid

$$\mathcal{E} = \left\{ \left(AA^T \circ dd^T\right)^{1/2} w \mid \|w\| \leq 1 \right\}.$$

Proof. This is simply a special case of Lemma 3. □

1.6.5 Outer Approximation

Let $\mathcal{E}_1 = \{Au + b \mid \|u\| \leq 1\}$ and $\mathcal{E}_2 = \{Cv + d \mid \|v\| \leq 1\}$ be ellipsoids in $\mathbf{R}^n$. Let x and y be n dimensional vectors taken from ellipsoids $\mathcal{E}_1$ and $\mathcal{E}_2$, respectively. Expanding the Hadamard product $x \circ y$, we have:

$$x \circ y = b \circ d + Au \circ Cv \ + Au \circ d + b \circ Cv. \tag{1.69}$$

By Lemmas 4 and 5, the field of values of the Hadamard product

$$x \circ y \in \{(Au + b) \circ (Cv + d) \mid \|u\| \leq 1, \|v\| \leq 1\}$$

is contained in the geometrical sum of three ellipsoids

$$\mathcal{S} = \mathcal{E}\big(b \circ d, AA^T \circ CC^T\big) + \mathcal{E}\big(0, AA^T \circ dd^T\big) + \mathcal{E}\big(0, bb^T \circ CC^T\big). \tag{1.70}$$

Ignoring the correlations between terms in the above expansion, we find that $\mathcal{S} \subseteq \mathcal{E}(b \circ d, Q)$, where

$$\begin{aligned} Q = {} & (1 + 1/p_1)(1 + 1/p_2)AA^T \circ CC^T + (1 + p_1)(1 + 1/p_2)AA^T \circ dd^T \\ & + (1 + p_1)(1 + p_2)CC^T \circ bb^T \end{aligned} \tag{1.71}$$

for all $p_1 > 0$ and $p_2 > 0$. The values of p_1 and p_2 may be chosen to minimize the trace or the determinant of Q. The trace metric requires far less computational effort and is numerically more reliable; if either b or d has a very small entry, the corresponding term in expansion (1.71) will be poorly conditioned.

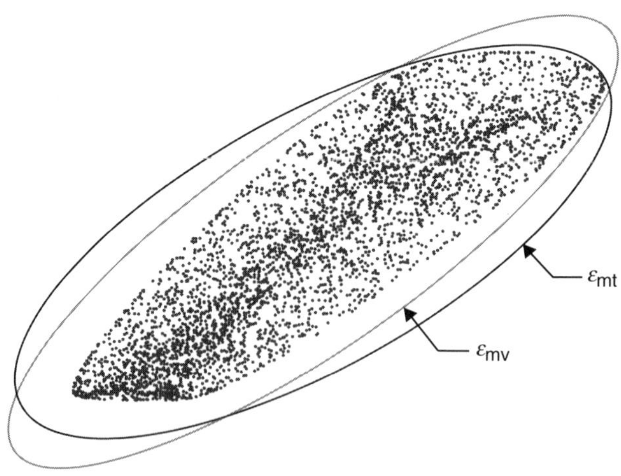

Figure 1.16 Samples of the Hadamard product of two ellipsoids. The outer approximations based on the minimum-volume and minimum-trace metrics are labeled $\mathcal{E}_{\text{mv}}$ and $\mathcal{E}_{\text{mt}}$.

As a numerical example, we consider the Hadamard product of two ellipsoids in $\mathbf{R}^2$. The ellipsoid $\mathcal{E}_1$ is described by

$$A = \begin{bmatrix} -0.6452 & -1.5221 \\ 0.2628 & 2.2284 \end{bmatrix}, \quad b = \begin{bmatrix} -5.0115 \\ 1.8832 \end{bmatrix};$$

the parameters of $\mathcal{E}_2$ are

$$C = \begin{bmatrix} -1.0710 & 0.7919 \\ 0.8744 & 0.7776 \end{bmatrix}, \quad d = \begin{bmatrix} -9.5254 \\ 9.7264 \end{bmatrix}.$$

Samples of the Hadamard product of $\mathcal{E}_1 \circ \mathcal{E}_2$ are shown in Figure 1.16 along with the outer approximations based on the minimum-volume and minimum-trace metrics; more Hadamard products of ellipsoids and outer approximations are shown in Figures 1.17 and 1.18.

1.6.6 The Complex Case

We now extend the results of Section 1.6.5 to the case of complex values. For numerical efficiency, we compute the approximating ellipsoid using the minimum-trace metric. As before, we represent complex numbers by the direct sum of their real and imaginary components. Let $x \in \mathbf{R}^{2n}$ and $y \in \mathbf{R}^{2n}$ be the direct-sum

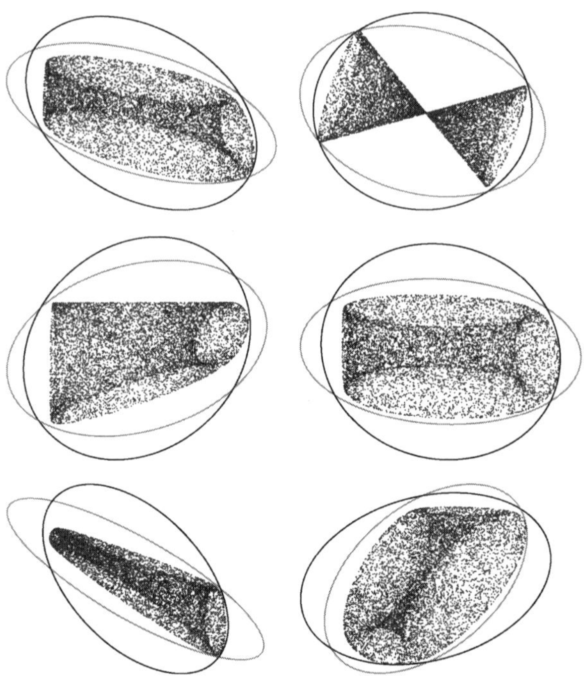

Figure 1.17 The Hadamard product of ellipsoids.

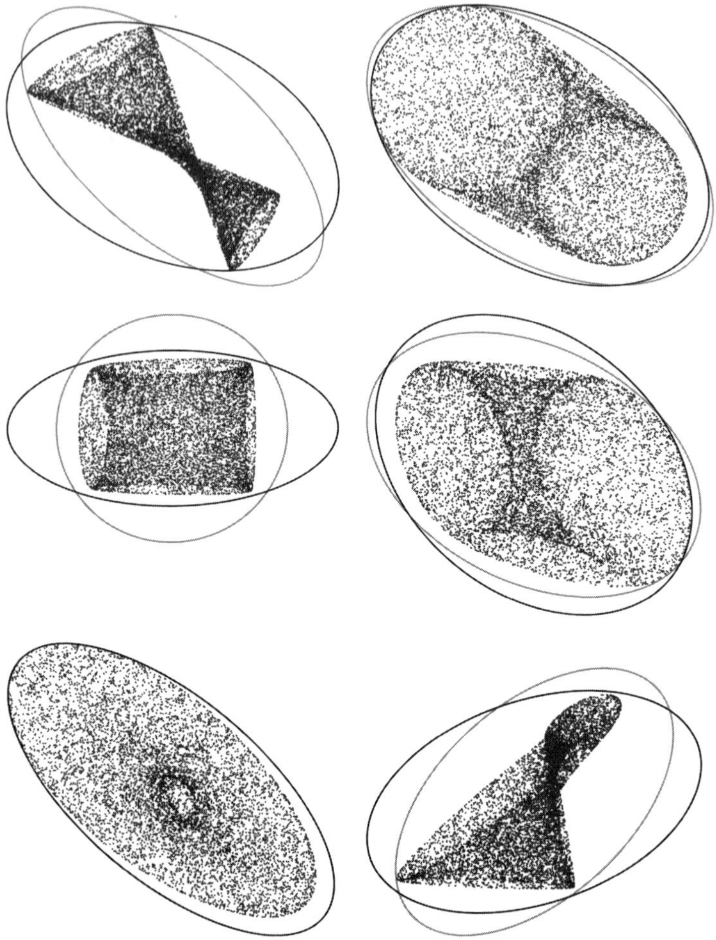

Figure 1.18 More Hadamard products of ellipsoids.

representations of $\alpha \in \mathbf{C}^n$ and $\beta \in \mathbf{C}^n$, respectively; that is,

$$x = \begin{bmatrix} \operatorname{Re} \alpha \\ \operatorname{Im} \alpha \end{bmatrix}, \quad y = \begin{bmatrix} \operatorname{Re} \beta \\ \operatorname{Im} \beta \end{bmatrix}.$$

We can represent the real and imaginary components of $\gamma = \alpha \circ \beta$ as

$$\begin{aligned} z &= \begin{bmatrix} \operatorname{Re} \gamma \\ \operatorname{Im} \gamma \end{bmatrix} \\ &= \begin{bmatrix} \operatorname{Re} \alpha \circ \operatorname{Re} \beta - \operatorname{Im} \alpha \circ \operatorname{Im} \beta \\ \operatorname{Im} \alpha \circ \operatorname{Re} \beta + \operatorname{Re} \alpha \circ \operatorname{Im} \beta \end{bmatrix} \qquad (1.72) \\ &= F_1 x \circ F_2 y + F_3 x \circ F_4 y, \end{aligned}$$

where

$$F_1 = \begin{bmatrix} I_n & 0 \\ 0 & I_n \end{bmatrix}, \quad F_2 = \begin{bmatrix} I_n & 0 \\ I_n & 0 \end{bmatrix},$$

and

$$F_3 = \begin{bmatrix} 0 & -I_n \\ I_n & 0 \end{bmatrix}, \quad F_4 = \begin{bmatrix} 0 & I_n \\ 0 & I_n \end{bmatrix}.$$

The multiplications associated with matrices $F_1, \ldots, F_4$ are achieved with a reordering of the calculations. Applying (1.72) to $x \in \mathcal{E}_1 = \{Au + b | \|u\| \leq 1\}$ and $y \in \mathcal{E}_2 = \{Cv + d \mid \|v\| \leq 1\}$ yields:

$$\begin{aligned} z = F_1 b \circ F_2 d + F_3 b \circ F_4 d + F_1 Au \circ F_2 Cv + F_1 Au \circ F_2 d + F_1 b \circ F_2 Cv \\ + F_3 Au \circ F_4 Cv + F_3 Au \circ F_4 d + F_3 b \circ F_4 Cv. \end{aligned} \tag{1.73}$$

The direct-sum representation of the field of values of the complex Hadamard product $\alpha \circ \beta$ is contained in the geometrical sum of ellipsoids

$$\begin{aligned} \mathcal{S} = \mathcal{E}(F_1 b \circ F_2 d, F_1 AA^T F_1^T \circ F_2 CC^T F_2^T) + \mathcal{E}(F_3 b \circ F_4 d, F_1 AA^T F_1^T \circ F_2 dd^T F_2^T) \\ + \mathcal{E}(0, F_1 bb^T F_1^T \circ F_2 CC^T F_2^T) + \mathcal{E}(0, F_3 AA^T F_3^T \circ F_4 CC^T F_4^T) \\ + \mathcal{E}(0, F_3 AA^T F_3^T \circ F_4 dd^T F_4^T) + \mathcal{E}(0, F_3 bb^T F_3^T \circ F_4 CC^T F_4^T). \end{aligned} \tag{1.74}$$

We compute $\mathcal{E}(c, Q) \supseteq \mathcal{S}$, where the center of the covering ellipsoid is given by the sum of the first two terms of (1.73). The configuration matrix Q is calculated by repeatedly applying (1.64) and (1.65) to the remaining terms of (1.73), where p is chosen according to (1.67).

1.6.7 An Improved Approximation

We now make use of two facts that generally lead to tighter approximations. First, the ellipsoidal outer approximation ignores any correlation between the terms in expansion (1.73); hence, it is productive to reduce the *number* of these terms.

Consider a Givens rotation matrix of the form:

$$T = \begin{bmatrix} \cos\theta_1 & & & \sin\theta_1 & & \\ & \ddots & & & \ddots & \\ & & \cos\theta_n & & & \sin\theta_n \\ -\sin\theta_1 & & & \cos\theta_1 & & \\ & \ddots & & & \ddots & \\ & & -\sin\theta_n & & & \cos\theta_n \end{bmatrix}. \tag{1.75}$$

The effect of premultiplying a direct sum-representation of a complex vector by T is to shift the phase of each component by the corresponding angle θ_i. It follows that for all T_x and T_y of the form (1.75) we have

$$T_x^{-1} T_y^{-1} (F_1 T_x x \circ F_2 T_y y + F_3 T_x x \circ F_4 T_y y) = F_1 x \circ F_2 y + F_3 x \circ F_4 y, \tag{1.76}$$

which does not hold for unitary matrices in general.

We now compute rotation matrices T_b and T_d such that the entries associated with the imaginary components of products $T_b b$ and $T_d d$ are zero. In computing T_b, we choose the values of θ in (1.75) according to $\theta_i = \angle[b(i) + \sqrt{-1} \times b(n+i)]$. T_y is similarly computed using the values of d; that is, $\theta_i = \angle[d(i) + \sqrt{-1} \times d(n+i)]$. We change coordinates according to

$$\begin{aligned} A &\leftarrow T_b A \\ b &\leftarrow T_b b \\ C &\leftarrow T_d C \\ d &\leftarrow T_d d. \end{aligned}$$

The rotated components associated with the ellipsoid centers have the form

$$T_b b = \begin{bmatrix} \tilde{b}_1 \\ \vdots \\ \tilde{b}_n \\ 0 \\ \vdots \\ 0 \end{bmatrix}, \quad T_d d = \begin{bmatrix} \tilde{d}_1 \\ \vdots \\ \tilde{d}_n \\ 0 \\ \vdots \\ 0 \end{bmatrix}, \tag{1.77}$$

zeroing the term $F_3 T_b A A^T T_b^T F_3^T \circ (F_4 T_d d d^T T_d^T F_4^T)$ in (1.73). The desired outer approximation is computed as the geometrical sum of outer approximations to the remaining five terms. That is,

$$\begin{aligned} \mathcal{E}(c, Q) \supseteq\ & \mathcal{E}(F_1 b \circ F_2 d, F_1 A A^T F_1^T \circ F_2 C C^T F_2^T) + \mathcal{E}(F_3 b \circ F_4 d, F_1 A A^T F_1^T \circ F_2 d d^T F_2^T) \\ & + \mathcal{E}(0, F_1 b b^T F_1^T \circ F_2 C C^T F_2^T) + \mathcal{E}(0, F_3 A A^T F_3^T \circ F_4 C C^T F_4^T) \\ & + \mathcal{E}(0, F_3 b b^T F_3^T \circ F_4 C C^T F_4^T). \end{aligned} \tag{1.78}$$

Second, while the Hadamard product is commutative, the outer approximation based on covering the individual terms in the expansion (1.73) *is* sensitive to ordering; simply interchanging the dyads $\{A, b\}$ and $\{C, d\}$ results in different qualities of approximations. The ellipsoidal approximation associated with this interchanged ordering is given by:

$$\begin{aligned}\mathcal{E}(c, Q) \supseteq\, & \mathcal{E}(F_1 d \circ F_2 b, F_1 CC^T F_1^T \circ F_2 AA^T F_2^T) + \mathcal{E}(F_3 d \circ F_4 b, F_1 CC^T F_1^T \circ F_2 bb^T F_2^T) \\ & + \mathcal{E}(0, F_1 dd^T F_1^T \circ F_2 AA^T F_2^T) + \mathcal{E}(0, F_3 CC^T F_3^T \circ F_4 AA^T F_4^T) \\ & + \mathcal{E}(0, F_3 dd^T F_3^T \circ F_4 AA^T F_4^T).\end{aligned} \tag{1.79}$$

Since our goal is to find the smallest ellipsoid covering the numerical range of z, we compute the trace associated with both orderings and choose the smaller of the two. This determination can be made without computing the minimum-trace ellipsoids explicitly. Let $\mathcal{E}_0$ be the minimum-trace ellipsoid covering $\mathcal{E}_1 + \cdots + \mathcal{E}_p$. The trace of $\mathcal{E}_0$ is given by:

$$\mathbf{Tr}\, \mathcal{E}_0 = \left(\sqrt{\mathbf{Tr}\, \mathcal{E}_1} + \sqrt{\mathbf{Tr}\, \mathcal{E}_2} + \cdots + \sqrt{\mathbf{Tr}\, \mathcal{E}_p}\right)^2,$$

which may be verified by direct calculation. Hence, determining which of (1.78) and (1.79) yields the smaller trace can be performed in $\mathcal{O}(n)$ calculations. After making this determination, we perform the remainder of the calculations to compute the desired configuration matrix Q. We then transform Q back to the original coordinates according to:

$$Q \leftarrow (T_b^{-1} T_d^{-1}) Q (T_b^{-1} T_d^{-1})^T.$$

1.7 BEAMFORMING EXAMPLE WITH MULTIPLICATIVE UNCERTAINTIES

Consider a six-element uniform linear array, centered at the origin, in which the spacing between the elements is half of a wavelength whose response is given by:

$$a(\theta) = [\, e_j^{-5\phi/2} \quad e_j^{-3\phi/2} \quad \cdots \quad e_j^{3\phi/2} \quad e_j^{5\phi/2}\,]^T,$$

where $\phi = \pi \cos(\theta)$ and θ is the angle of arrival and $j = \sqrt{-1}$.

As in the previous example, three signals impinge upon the array: a desired signal $s_{\mathrm{d}}(t)$ and two uncorrelated interfering signals $s_{\mathrm{int1}}(t)$ and s_{int2}. The signal-to-noise ratio (SNR) of the desired signal at each element is 20 dB. The angles of arrival of the interfering signals, θ_{int1} and θ_{int2}, are 30° and 75°; the SNRs of these interfering signals, 40 dB and 20 dB, respectively. The received signals are modeled as in (1.57). The signals pass through an amplification stage as depicted in Fig. 1.15. The gain vector $g \in \mathbf{C}^6$ is chosen from the ellipsoid which we represent, in terms of the direct sum of the real and imaginary components in $\mathbf{R}^{12}$ according to

$\mathcal{E}_g = \mathcal{E}(Q_g, c_g)$, where

$$Q_g = \begin{bmatrix} Q_d & \\ & Q_d \end{bmatrix}, \quad c_g = [1 \ \ \ldots \ \ 1 \ \ 0 \ \ \ldots \ \ 0]^T,$$

and Q_d is a diagonal matrix, the ith diagonal element of which equals 10^{-i}. Given the symmetry in the uncertainty region of the present example, the set of possible values of $g \in \mathbf{C}^6$ also satisfy $(g^* - \mathbf{1})Q_d^{-1}(g^* - \mathbf{1})$, where $\mathbf{1}$ is a vector of ones.

As in Section 1.4, the actual array response is contained in an ellipsoid $\mathcal{E}_a(c, P)$, whose center and configuration matrix are computed from 64 equally-spaced samples of the array response at angles between 40° and 50° according to (1.59), (1.60).

The aggregate uncertainty in the Hadamard product of the array manifold and the gain vector is then given by the (complex) Hadamard product of the above uncertainty ellipsoids. We compute an ellipsoidal outer approximation to this aggregate uncertainty ellipsoid, using the methods of Sections 1.6.6 and 1.6.7, namely, $\mathcal{E}_a(c, P) \subset \mathcal{E}_g \circ \mathcal{E}_a$.

We will use an analytically computed, expected covariance which again uses the actual array response and which assumes that the signals $s_{\text{d}}(t), s_{\text{int1}}(t), s_{\text{int2}}(t)$, and $v(t)$ are all uncorrelated and that the additive noise is applied at the output of the amplification stage. The covariance is modeled as:

$$\begin{aligned} \mathbf{E}\ R = \mathbf{E}\ yy^* &= \sigma_{\text{d}}^2(g \circ a_{\text{d}})(g \circ a_{\text{d}})^* + \sigma_{\text{int1}}^2(g \circ a(\theta_{\text{int1}}))(g \circ a(\theta_{\text{int1}}))^* \\ &\quad + \sigma_{\text{int2}}^2(g \circ a(\theta_{\text{int2}}))(g \circ a(\theta_{\text{int2}}))^* + \sigma_{\text{n}}^2 I. \end{aligned} \tag{1.80}$$

The worst-case SINR is the minimum objective value of the following optimization problem:

$$\begin{array}{ll} \text{minimize} & \dfrac{\sigma_{\text{d}}^2 \|w^*(g \circ a)\|^2}{\mathbf{E}\ w^* R_v w} \\ \text{subject to} & a \in \mathcal{E}(c, P); \end{array}$$

where the expected covariance of the interfering signals and noises is given by

$$\mathbf{E}\ R_v = \sigma_{\text{int1}}^2(g \circ a(\theta_{\text{int1}}))(g \circ a(\theta_{\text{int1}}))^* + \sigma_{\text{int1}}^2(g \circ a(\theta_{\text{int2}}))(g \circ a(\theta_{\text{int2}}))^* + \sigma_{\text{n}}^2 I.$$

The weight vector w and covariance matrix of the noise and interfering signals R_v used in its computation reflect the chosen values of the gain vector and array manifold.

We will consider four cases:

1. The assumed and actual gains are nominal (unity).
2. The gain, assumed and actual, can assume any value in $\mathcal{E}_g$.
3. The gain is assumed to vary within $\mathcal{E}_a$; the actual gain is nominal.
4. The gain is assumed nominal, but can assume any value in $\mathcal{E}_g$.

The beamformers and worst-case SINRs for these cases were computed to be:

$$\text{Case 1:} \quad w_1 = \begin{bmatrix} -0.1760 + 0.1735i \\ -1.1196 + 0.5592i \\ -0.4218 + 0.4803i \\ -0.4245 - 0.4884i \\ -1.1173 - 0.5598i \\ -0.1720 - 0.1767i \end{bmatrix}, \quad \text{SINR} = 14.26\,\text{dB}.$$

$$\text{Case 2:} \quad w_2 = \begin{bmatrix} 0.0350 + 0.0671i \\ -0.6409 - 0.0109i \\ 0.2388 - 0.3422i \\ 1.1464 - 1.1488i \\ 0.2749 - 2.1731i \\ 0.0201 - 1.2138i \end{bmatrix}, \quad \text{SINR} = 11.22\,\text{dB}.$$

$$\text{Case 3:} \quad w_3 = \begin{bmatrix} 0.0418 + 0.0740i \\ -0.6248 + 0.0241i \\ 0.2579 - 0.3097i \\ 1.1192 - 1.1111i \\ 0.2445 - 2.0927i \\ 0.0317 - 1.1681i \end{bmatrix}, \quad \text{SINR} = 11.30\,\text{dB}.$$

$$\text{Case 4:} \quad w_4 = \begin{bmatrix} 0.9141 + 2.6076i \\ 2.4116 + 1.6939i \\ -0.1105 - 0.1361i \\ -0.6070 + 1.2601i \\ -0.4283 - 0.8408i \\ -1.1158 - 1.0300i \end{bmatrix}, \quad \text{SINR} = -2.81\,\text{dB}.$$

In the first case, the gains nominal and actual are unity; the worst-case SINR is seen to be 14.26 dB. In the second case, the gain is allowed to vary; not surprisingly, the worst-case SINR decreases to 11.22 dB. In the third case, the beamformer is computed assuming possible variation in the gains when in fact, there is none. The worst-case SINR in this case is 11.30 dB, quite close to that of the second case. The interpretation is that robustness comes at the expense of nominal performance. In the last case, the uncertainty ellipsoid used in the beamformer computation underestimated the aggregate uncertainty; this optimism is seen to be punished.

The uncertainty in the gain for the first antenna element is large, for the last, small, and for the middle elements, somewhere in between. When this possible gain variation is factored into the aggregate uncertainty ellipsoid, the RMVB based on this aggregate ellipsoid discounts the information in the less reliable measurements by assigning to them small (in absolute value) weights. This is seen in the first and (to a lesser extent) the second entries of beamformer vectors w_2 and w_3.

1.8 SUMMARY

The main ideas of our approach are as follows:

- The possible values of the manifold are approximated or covered by an ellipsoid that describes the uncertainty.
- The robust minimum variance beamformer is chosen to minimize the weighted power out of the array subject to the constraint that the gain is greater than unity for all array manifold values in the ellipsoid.
- The RMVB can be computed very efficiently using Lagrange multiplier techniques.
- Ellipsoidal calculus techniques may be used to efficiently propagate the uncertainty ellipsoid in the presence of multiplicative uncertainties.

APPENDIX: NOTATION AND GLOSSARY

$\mathbf{R}$	The set of real numbers.
$\mathbf{R}^m$	The set of real m-vectors.
$\mathbf{R}^{m\times n}$	The set of real $m \times n$ matrices.
$\mathbf{C}$	The set of complex numbers.
$\mathbf{C}^m$	The set of complex m-vectors.
$\mathbf{C}^{m\times n}$	The set of complex $m \times n$ matrices.
$\mathbf{Tr}\, X$	The trace of X.
$\mathbf{E}X$	The expected value of X.
$\det X$	The determinant of X.
$\|x\|$	The Euclidean (l_2) norm of x.
I	The identity matrix (of appropriate dimensions).
$x \circ y$	The Hadamard or element-wise product of x and y.
$X \succ 0 (X \succeq 0)$	X is positive (semi-)definite, that is $X = X^T$ and $z^T X z > 0$ ($z^T X z \geq 0$) for all nonzero z.
$X \succ Y (X \succeq Y)$	$X - Y$ is positive (semi-)definite.
AOA	Angle of arrival
dB	Decibel
MVE	Minimum-volume ellipsoid

MVB	Minimum-variance beamformer
NEC	Numerical electromagnetics code
RMVB	Robust minimum variance beamformer
SINR	Signal-to-interference-plus-noise ratio
SNR	Signal-to-noise ratio

REFERENCES

1. J. Capon. High-resolution frequency-wavenumber spectrum analysis. *Proc. IEEE*, 57(8), 1408–1418 (1969).
2. J. L. Krolik. The performance of matched-field beamformers with Mediterranean vertical array data. *IEEE Transactions on Signal Processing*, 44(10), 2605–2611 (1996).
3. J. L. Krolik. Matched-field minimum variance beamforming. *J. Acoust. Soc. Am.*, 92(3), 1406–1419 (1992).
4. A. N. Tikhonov and Y. V. Arsenin. *Solution of Ill-Posed Problems*. V. H. Winston and Sons, 1977. Translated from Russian.
5. A. B. Gershman. Robust adaptive beamforming in sensor arrays. *AEU-International Journal of Electronics and Communication*, 53(6), 305–314 (1999).
6. D. Johnson and D. Dudgeon. *Array Signal Processing: Concepts and Techniques*. Signal Processing Series, Prentice Hall, Englewood Cliffs, 1993.
7. S. Haykin. *Adaptive Filter Theory*. Prentice Hall Information and System Sciences Series, Prentice Hall, Englewood Cliffs, 1996.
8. K. Harmanci, J. Tabrikian, and J. L. Krolik. Relationships between adaptive minimum variance beamforming and optimal source localization. *IEEE Transactions on Signal Processing*, 48(1), 1–13 (2000).
9. A. Ben-Tal and A. Nemirovski. *Lectures on Modern Convex Optimization: Analysis, Algorithms, and Engineering Applications*. MPS/SIAM Series on Optimization, SIAM, Philadelphia, 2001.
10. S. Q. Wu and J. Y. Zhang. A new robust beamforming method with antennae calibration errors. In *1999 IEEE Wireless Communications and Networking Conference, New Orleans, LA, USA 21–24 Sept.*, Vol. 2, pp. 869–872, 1999.
11. S. A. Vorobyov, A. B. Gershman, and Z.-Q. Luo. Robust adaptive beamforming using worst-case performance optimization via second-order cone programming. In *Proc. IEEE International Conf. on Acoustics, Speech and Signal Processing*, Vol. III, 2002.
12. S. A. Vorobyov, A. B. Gershman, and Z.-Q. Luo. Robust adaptive beamforming using worst-case performance optimization. *IEEE Transactions on Signal Processing*, 51(2), 313–324 (2003).
13. A. B. Gershman, Z.-Q. Luo, S. Shahbazpanahi, and S. Vorobyov. Robust adaptive beamforming based on worst-case performance optimization. In *The Thirty-Seventh Asilomar Conference on Signals, Systems, and Computers*, Pacific Grove, CA, pp. 1353–1357, 2003.
14. P. Stoica, Z. Wang, and J. Li. Robust Capon beamforming. *IEEE Signal Processing Letters*, 10(6), 172–175 (2003).

15. J. Li, P. Stoica, and Z. Wang. On robust Capon beamforming and diagonal loading. *IEEE Transactions on Signal Processing*, 51(7), 1702–1715 (2003).
16. J. Li, P. Stoica, and Z. Wang. Doubly constrained robust Capon beamformer. *IEEE Transactions on Signal Processing*, 52, 2407–2423 (2004).
17. R. G. Lorenz and S. P. Boyd. Robust minimum variance beamforming. *IEEE Transactions on Signal Processing*, 53(5), 1684–1696 (2005).
18. R. G. Lorenz and S. P. Boyd. Robust beamforming in GPS arrays. In *Proc. Institute of Navigation, National Technical Meeting*, Jan. 2002.
19. R. Lorenz and S. Boyd. Robust minimum variance beamforming. In *The Thirty-Seventh Asilomar Conference on Signals, Systems, and Computers*, Vol. 2, pp. 1345–1352, 2003.
20. G. J. Burke. Numerical electromagnetics code—NEC-4 method of moments. Technical Report UCRL-MA-109338, Lawrence Livermore National Laboratory, Jan. 1992.
21. M. S. Lobo, L. Vandenberghe, S. P. Boyd, and H. Lebret. Applications of second-order cone programming. *Linear Algebra and Applications*, 284(1–3), 193–228 (1998).
22. A. Ben-Tal and A. Nemirovski. Robust solutions of uncertain linear programs. *Operations Research Letters*, 25(1), 1–13 (1999).
23. H. Lebret and S. P. Boyd. Antenna array pattern synthesis via convex optimization. *IEEE Trans. Antennas Propag.*, 45(3), 526–532 (1997).
24. S. P. Boyd and L. Vandenberghe. *Convex Optimization*. Cambridge University Press, Cambridge, UK, 2004.
25. A. L. Soyster. Convex programming with set-inclusive constraints and applications to inexact linear programming. *Operations Research*, 21(5), 1154–1157 (1973).
26. L. El Ghaoui and H. Lebret. Robust solutions to least-squares problems with uncertain data. *SIAM J. Matrix Anal. Appl.*, 18(4), 1035–1064 (1997).
27. A. Ben-Tal and A. Nemirovski. Robust convex optimization. *Mathematics of Operations Research*, 23(4), 769–805 (1998).
28. A. Ben-Tal, L. El Ghaoui, and A. Nemirovski. Robustness. In *Handbook on Semidefinite Programming*, Chapter 6, pp. 138–162. Kluwer, Boston, 2000.
29. W. Gander. Least squares with a quadratic constraint. *Numerische Mathematik*, 36(3), 291–307 (1981).
30. G. H. Golub and U. von Matt. Quadratically constrained least squares and quadratic problems. *Numerische Mathematik*, 59(1), 561–580 (1991).
31. G. H. Golub and C. Van Loan. *Matrix Computations*. Johns Hopkins University Press, Baltimore, 2nd edition, 1989.
32. D. P. Bertsekas. *Constrained Optimization and Lagrange Multiplier Methods*. Athena Scientific, Belmont, MA, 1996.
33. C. A. Stutt and L. J. Spafford. A "best" mismatched filter response for radar clutter discrimination. *IEEE Transactions on Information Theory*, IT-14(2), 280–287 (1968).
34. Y. I. Abromovich and M. B. Sverdlik. Synthesis of a filter which maximizes the signal-to-noise ratio under additional quadratic constraints. *Radio Eng. and Electron. Phys.*, 15(11), 1977–1984 (1970).
35. T. Kailath, A. H. Sayed, and B. Hassibi. *Linear Estimation*. Information and System Sciences Series, Prentice Hall, Upper Saddle River, NJ, 2000.
36. G. Dahlquist and Å. Björck. *Numerical Methods*. Series in Automatic Computation, Prentice Hall, Englewood Cliffs, 1974.

37. J. W. Demmel. *Applied Numerical Linear Algebra.* SIAM, Philadelphia, 1997.

38. L. Vandenberghe and S. P. Boyd. Semidefinite programming. *Siam Review* (1995).

39. S.-P. Wu and S. P. Boyd. SDPSOL: A parser/solver for semidefinite programs with matrix structure. In L. El Ghaoui and S.-I. Niculescu, Eds. *Advances in Linear Matrix Inequality Methods in Control*, Chapter 4, pp. 79–91. SIAM, Philadelphia, 2000.

40. L. Vandenberghe, S. P. Boyd, and S.-P. Wu. Determinant maximization with linear matrix inequality constraints. *SIAM J. Matrix Anal. Appl.*, 19(2), 499–533 (1998).

41. S. P. Boyd, L. El Ghaoui, E. Feron, and V. Balakrishnan. Linear Matrix Inequalities in System and Control Theory, Vol. 15, *Studies in Applied Mathematics*, SIAM, Philadelphia, June 1994.

42. A. Kurzhanski and I. Vályi. Ellipsoidal calculus for estimation and control. In *Systems & Control: Foundations & Applications*. Birkhauser, Boston, 1997.

43. R. Horn and C. Johnson. *Topics in Matrix Analysis.* Cambridge University Press, Cambridge, 1991.

2

ROBUST ADAPTIVE BEAMFORMING BASED ON WORST-CASE PERFORMANCE OPTIMIZATION

Alex B. Gershman
Darmstadt University of Technology, Darmstadt, Germany

Zhi-Quan Luo
Department of Electrical and Computer Engineering, University of Minnesota, Minneapolis, MN 55455

Shahram Shahbazpanahi
McMaster University, Hamilton, Ontario, Canada

2.1 INTRODUCTION

Adaptive beamforming is a versatile approach to detect and estimate the signal-of-interest at the output of a sensor array by means of data-adaptive spatial filtering and interference rejection. It has a long and rich history of interdisciplinary theoretical research [1–8] and practical applications to numerous areas such as sonar [9–14], radar and remote sensing [15–18], wireless communications [19–23], global positioning [24–26], radio astronomy [27, 28], microphone array speech processing [29–31], seismology [32, 33], biomedicine [34, 35], and other fields. In a few recent years, there has been a renewed interest to this area in application to wireless communications where smart (adaptive) antennas have emerged as one of the key technologies for the third and higher generations of mobile radio systems [23].

The traditional approach to the design of adaptive array algorithms assumes that there is no desired signal component in the beamformer training cell data [2, 4, 8].

Robust Adaptive Beamforming, Edited by Jian Li and Petre Stoica

Although this assumption may be relevant in several specific cases (for example, in certain radar and active sonar problems), in most applications the interference and noise observations are 'contaminated' by the signal component [36–38]. Such applications include, for example, passive sonar, wireless communications, microphone array processing, and radioastronomy. If signal-free beamformer training snapshots are available, adaptive array algorithms are known to be quite robust against errors in the steering vector of the desired signal and limited training sample size [2–8, 39, 40]. However, the situation is completely different in the case when the desired signal is present in the training data snapshots. It is well known that in the latter case, traditional adaptive beamforming methods suffer from the signal cancellation phenomenon, that is, they degrade severely in their performance and convergence rate. Such a degradation can take place even when the signal steering vector is precisely known at the beamformer but the sample size is limited [36, 38, 41].

In practical scenarios, the performance degradation of traditional adaptive beamforming techniques may become even more pronounced because most of these techniques are based on the assumption of an accurate knowledge of the array response to the desired signal. Moreover, these methods often use quite restrictive assumptions on the environment and interferences, for example, they assume that the received array data are stationary and/or that the interferers can be described using a low-rank model. As a result, such techniques can become severely degraded in scenarios when the exploited assumptions on the environment, antenna array and/or sources are wrong or inaccurate [36, 38].

One of the most typical reasons of performance degradation of adaptive beamformers is a mismatch between the presumed and the actual array responses to the desired signal. Such a mismatch can be caused by look direction/pointing errors [42–45], an imperfect array calibration (distorted antenna shape) [46], unknown wavefront distortions and signal fading [11, 47–49], near-far wavefront mismodeling [50], local scattering [51], as well as other effects [36, 52]. Traditional adaptive array algorithms are known to be extremely sensitive even to slight mismatches of such type because in the presence of them, an adaptive beamformer tends to mix up the signal and interference components, that is, it interprets the desired signal component in array observations as an additional interfering source and, consequently, suppresses the desired signal instead of maintaining distortionless response to it [36, 41]. This phenomenon is sometimes called *self-nulling* in the adaptive beamforming literature [38, 53].

Another cause of performance degradation of adaptive beamformers is a nonstationarity of the environment, antenna array, and/or sources. Such nonstationarity effects can be induced by rapid variations of the propagation channel, interferer and antenna motion and/or vibration, and are quite typical for radar, sonar, and wireless communications [54–58]. They may cause a substantial performance degradation of adaptive beamformers because they limit the training sample size and may lead to *interference undernulling*. When such nonstationarity effects are combined with the effect of the presence of the desired signal in the training cell, the aforementioned degradation can become much stronger than in the case of signal-free beamformer training data [56].

One typical example of negative effects of nonstationarity is the case when the interfering sources move rapidly. In such case, the array weights may not be able to adapt fast enough to compensate for this motion. That is, the interferers tend to be always located outside the narrow areas of the adapted beampattern nulls and to leak to the output of adaptive beamformer through the beampattern sidelobes [56]. The same situation may occur when moving or vibrating antenna arrays are employed, for example, towed arrays in sonar [14] or airborne antenna arrays [54].

In many practical sonar and wireless communications scenarios, the signal and interference wavefronts may suffer from a multiplicative noise and angular spreading. In sonar, this type of noise is caused by a long-distance propagation through a randomly inhomogeneous medium [10, 47, 48]. In wireless communications, the array signal response may suffer from fading and local scattering [49, 51].

In the presence of multiplicative noise, higher-rank signal source models have to be used instead of the point (rank-one) model because in this case, each source results into multiple rank-one components in the array covariance matrix [38]. It can be shown that, in such scenarios, the array response should be characterized by the signal covariance matrix rather than the signal steering vector [36, 38, 59]. As a result, the robustness of adaptive beamformers against mismatches between the presumed and actual signal covariance matrices (rather than the mismatches between the corresponding steering vectors) must be considered.

In this chapter, we provide an overview of traditional ad hoc robust adaptive beamforming techniques and give a detailed introduction to a recently emerged rigorous approach to robust minimum variance beamforming based on worst-case performance optimization [59–62]. This approach represents the current state of the art of robust adaptive beamforming. It is shown that it provides efficient solutions to the aforementioned robustness problems including the array response mismatch and data nonstationarity problems.

The remainder of this chapter is organized as follows. In the next section, some background on adaptive arrays is given and the traditional (robust and nonrobust) adaptive beamforming techniques are discussed. Then, in Section 2.3, the worst-case performance optimization-based adaptive beamformers are considered. In Section 2.4, simulation results are presented that demonstrate an improved robustness of these worst-case optimization-based beamformers as compared to the earlier robust and nonrobust techniques. Conclusions are given in Section 2.5.

2.2 BACKGROUND AND TRADITIONAL APPROACHES

The generic scheme of a narrowband beamformer is shown in Figure 2.1. The beamformer output signal can be written as

$$y(k) = \mathbf{w}^H \mathbf{x}(k)$$

where k is the time index, $\mathbf{x}(k) = [x_1(k), \ldots, x_M(k)]^T$ is the $M \times 1$ complex vector of array observations, $\mathbf{w} = [w_1, \ldots, w_M]^T$ is the $M \times 1$ complex vector of beamformer

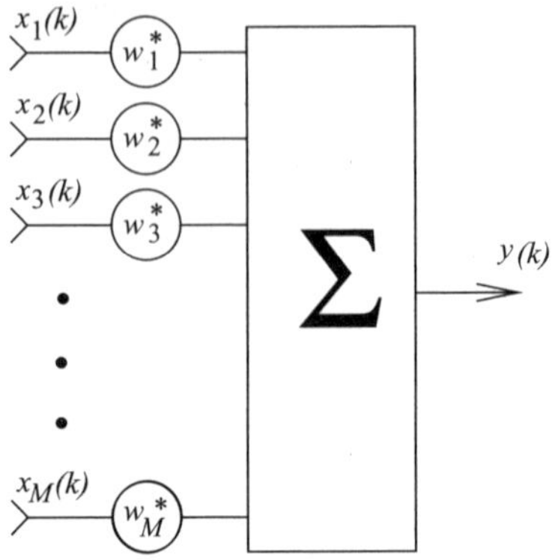

Figure 2.1 The generic scheme of a narrowband beamformer.

weights, M is the number of array sensors, and $(\cdot)^T$ and $(\cdot)^H$ denote the transpose and Hermitian transpose, respectively. The training snapshot (array observation vector) is given by

$$\mathbf{x}(t) = \beta\mathbf{s}_s(t) + \mathbf{i}(t) + \mathbf{n}(t) \tag{2.1}$$

where $\mathbf{s}_s(t)$, $\mathbf{i}(t)$, and $\mathbf{n}(t)$ are the statistically independent components of the desired signal, interference, and sensor noise, respectively, and the binary parameter β is equal to zero if the training cell snapshots are signal-free and is equal to one otherwise. In what follows, mostly the case $\beta = 1$ will be considered. If the desired signal is a point source and has a time-invariant wavefront, we obtain that

$$\mathbf{s}_s(t) = s(t)\mathbf{a}_s$$

where $s(t)$ is the complex signal waveform and $\mathbf{a}_s$ is its $M \times 1$ steering vector. Then, taking into account that $\beta = 1$, (2.1) can be written as

$$\mathbf{x}(t) = s(t)\mathbf{a}_s + \mathbf{i}(t) + \mathbf{n}(t)$$

The optimal weight vector can be obtained by means of maximizing the signal-to-interference-plus-noise ratio (SINR) [4, 8]

$$\text{SINR} = \frac{\mathbf{w}^H\mathbf{R}_s\mathbf{w}}{\mathbf{w}^H\mathbf{R}_{i+n}\mathbf{w}} \tag{2.2}$$

where

$$\mathbf{R}_s \triangleq \text{E}\{\mathbf{s}_s(t)\mathbf{s}_s^H(t)\}$$
$$\mathbf{R}_{i+n} \triangleq \text{E}\{(\mathbf{i}(t) + \mathbf{n}(t))\,(\mathbf{i}(t) + \mathbf{n}(t))^H\}$$

are the $M \times M$ signal and interference-plus-noise covariance matrices, respectively, and $E\{\cdot\}$ denotes the statistical expectation. Note that the matrix $\mathbf{R}_s$ can have an arbitrary rank, that is, $1 \leq \text{rank}\{\mathbf{R}_s\} \leq M$.

In many practical situations, $\text{rank}\{\mathbf{R}_s\} > 1$. Typical examples of such situations are scenarios with incoherently scattered sources or signals with randomly fluctuating wavefronts which frequently occur in sonar and wireless communications.

In the incoherently scattered source case, $\mathbf{R}_s$ has the following form [63, 64]:

$$\mathbf{R}_s = \sigma_s^2 \int_{-\pi/2}^{\pi/2} \rho(\theta)\, \mathbf{a}(\theta)\mathbf{a}^H(\theta)\, d\theta \tag{2.3}$$

where $\rho(\theta)$ is the normalized angular power density $[\int_{-\pi/2}^{\pi/2} \rho(\theta)\, d\theta = 1]$, σ_s^2 is the signal power, and $\mathbf{a}(\theta)$ is the array steering vector.

In the case of randomly fluctuating wavefronts, the signal covariance matrix takes another form [47, 48, 65]

$$\mathbf{R}_s = \sigma_s^2\, \mathbf{B} \odot \{\mathbf{a}_s\mathbf{a}_s^H\} \tag{2.4}$$

where $\mathbf{B}$ is the $M \times M$ coherence loss matrix and $\odot$ is the Schur-Hadamard (elementwise) matrix product.

There are two commonly used models for the coherence loss matrix [47, 48, 63, 65]:

$$[\mathbf{B}]_{m,n} = \exp\{-(m-n)^2\zeta\} \tag{2.5}$$

$$[\mathbf{B}]_{m,n} = \exp\{-|m-n|\zeta\} \tag{2.6}$$

where ζ is the coherence loss parameter.

Obviously, the rank of $\mathbf{R}_s$ in (2.3) and (2.4) can be higher than one. It is important to stress that in practice, both $\rho(\theta)$ and $\mathbf{B}$ may be uncertain [11,51]. Therefore, in the both cases of spatially spread and imperfectly coherent sources, we may expect a substantial mismatch between the presumed and actual signal covariance matrices [59].

In the special case of a point signal source, we have

$$\mathbf{R}_s = \sigma_s^2 \mathbf{a}_s\mathbf{a}_s^H$$

In this case, $\text{rank}\{\mathbf{R}_s\} = 1$ and (2.2) can be simplified to

$$\text{SINR} = \frac{\sigma_s^2\, |\mathbf{w}^H\mathbf{a}_s|^2}{\mathbf{w}^H\mathbf{R}_{i+n}\mathbf{w}} \tag{2.7}$$

To find the optimal solution for the weight vector, we should maximize the SINR in (2.2) or, alternatively, in (2.7). These optimization problems are equivalent to maintaining distortionless response to the desired signal while minimizing the output interference-plus-noise power, that is,

$$\min_{\mathbf{w}} \mathbf{w}^H \mathbf{R}_{\mathrm{i+n}} \mathbf{w} \quad \text{subject to } \mathbf{w}^H \mathbf{R}_{\mathrm{s}} \mathbf{w} = 1 \tag{2.8}$$

$$\min_{\mathbf{w}} \mathbf{w}^H \mathbf{R}_{\mathrm{i+n}} \mathbf{w} \quad \text{subject to } \mathbf{w}^H \mathbf{a}_{\mathrm{s}} = 1 \tag{2.9}$$

in the general-rank and rank-one signal cases, respectively. This approach is usually referred to as the minimum variance distortionless response (MVDR) beamforming [4, 8].

The solution to (2.8) can be found by means of minimization of the function

$$H(\mathbf{w}, \lambda) = \mathbf{w}^H \mathbf{R}_{\mathrm{i+n}} \mathbf{w} + \lambda(1 - \mathbf{w}^H \mathbf{R}_{\mathrm{s}} \mathbf{w}) \tag{2.10}$$

where λ is a Lagrange multiplier. Taking the gradient of (2.10) and equating it to zero, we obtain that the solution to (2.8) is given by the following generalized eigenvalue problem [36, 59]:

$$\mathbf{R}_{\mathrm{i+n}} \mathbf{w} = \lambda \mathbf{R}_{\mathrm{s}} \mathbf{w} \tag{2.11}$$

where the Lagrange multiplier λ can be interpreted as a corresponding generalized eigenvalue. It is easy to prove that all generalized eigenvalues in (2.11) are nonnegative real numbers. Indeed, using (2.11) we have that $\mathbf{w}^H \mathbf{R}_{\mathrm{i+n}} \mathbf{w} = \lambda \mathbf{w}^H \mathbf{R}_{\mathrm{s}} \mathbf{w}$. Using the fact that the matrices $\mathbf{R}_{\mathrm{i+n}}$ and $\mathbf{R}_{\mathrm{s}}$ are positive semidefinite, we prove that λ is always real and non-negative.

The solution to the problem (2.8) is the generalized eigenvector that corresponds to the minimal generalized eigenvalue of the matrix pencil $\{\mathbf{R}_{\mathrm{i+n}}, \mathbf{R}_{\mathrm{s}}\}$. Multiplying (2.11) by $\mathbf{R}_{\mathrm{i+n}}^{-1}$, we can write this equation as

$$\mathbf{R}_{\mathrm{i+n}}^{-1} \mathbf{R}_{\mathrm{s}} \mathbf{w} = \frac{1}{\lambda} \mathbf{w} \tag{2.12}$$

which can be identified as the characteristic equation for the matrix $\mathbf{R}_{\mathrm{i+n}}^{-1} \mathbf{R}_{\mathrm{s}}$. From the fact of non-negativeness of λ, it follows that the minimal generalized eigenvalue λ_{min} in (2.11) corresponds to the maximal eigenvalue $1/\lambda_{\mathrm{min}}$ in (2.12). Using the latter fact, the optimal weight vector can be explicitly written as

$$\mathbf{w}_{\mathrm{opt}} = \mathcal{P}\{\mathbf{R}_{\mathrm{i+n}}^{-1} \mathbf{R}_{\mathrm{s}}\} \tag{2.13}$$

where $\mathcal{P}\{\cdot\}$ is the operator which returns the *principal eigenvector* of a matrix, that is, the eigenvector that corresponds to its maximal eigenvalue. According to (2.8) and the fact that any eigenvector can be normalized arbitrarily, the resulting weight has to be normalized to satisfy the constraint $\mathbf{w}_{\mathrm{opt}}^H \mathbf{R}_{\mathrm{s}} \mathbf{w}_{\mathrm{opt}} = 1$ in (2.8). However, it is clear that multiplying the weight vector by any nonzero constant, we do not affect the output SINR (2.2). Hence, such normalization is immaterial [38].

The optimal solution (2.13) will not change if the interference-plus-noise covariance matrix $\mathbf{R}_{\rm i+n}$ would be replaced by the training data covariance matrix

$$\mathbf{R} = \mathrm{E}\{\mathbf{x}(t)\mathbf{x}^H(t)\} = \mathbf{R}_{\rm i+n} + \mathbf{R}_{\rm s} \tag{2.14}$$

Therefore, we have

$$\mathbf{w}_{\rm opt} = \mathcal{P}\{\mathbf{R}_{\rm i+n}^{-1}\mathbf{R}_{\rm s}\} = \mathcal{P}\{\mathbf{R}^{-1}\mathbf{R}_{\rm s}\} \tag{2.15}$$

Note that (2.15) directly follows from (2.8) and (2.14).

In the rank-one signal source case, $\mathbf{R}_{\rm s} = \sigma_{\rm s}^2\mathbf{a}_{\rm s}\mathbf{a}_{\rm s}^H$ and we have that equation (2.13) can be rewritten as

$$\begin{aligned}\mathbf{w}_{\rm opt} &= \mathcal{P}\{\mathbf{R}_{\rm i+n}^{-1}\mathbf{a}_{\rm s}\mathbf{a}_{\rm s}^H\} \\ &= \alpha\mathbf{R}_{\rm i+n}^{-1}\mathbf{a}_{\rm s}\end{aligned} \tag{2.16}$$

where the constant α can be obtained from the MVDR constraint $\mathbf{w}_{\rm opt}^H\mathbf{a}_{\rm s} = 1$ in (2.9) and is equal to [4]

$$\alpha = \frac{1}{\mathbf{a}_{\rm s}^H\mathbf{R}_{\rm i+n}^{-1}\mathbf{a}_{\rm s}}$$

However, as has been noted before, this constant does not affect the output SINR and, therefore, is omitted in the sequel. Equation (2.16) is the classic Wiener solution for the weight vector of the optimal beamformer in the rank-one signal case [2, 4].

In practical applications, the true matrices $\mathbf{R}_{\rm i+n}$ and $\mathbf{R}$ are unavailable but can be estimated from the received data or obtained from a priori information about the sources. Usually, the sample covariance matrix [2, 4]

$$\hat{\mathbf{R}} = \frac{1}{N}\sum_{n=1}^{N}\mathbf{x}(n)\mathbf{x}^H(n) \tag{2.17}$$

is used in the optimization problems (2.8) and (2.9) instead of $\mathbf{R}_{\rm i+n}$, where N is the training sample size. The solutions to these modified problems are usually referred to as the sample matrix inverse (SMI) beamformers [2]

$$\mathbf{w}_{\rm SMI} = \mathcal{P}\{\hat{\mathbf{R}}^{-1}\mathbf{R}_{\rm s}\} \tag{2.18}$$

$$\mathbf{w}_{\rm SMI} = \hat{\mathbf{R}}^{-1}\mathbf{a}_{\rm s} \tag{2.19}$$

for the general-rank and rank-one cases, respectively.

The use of the sample covariance matrix $\hat{\mathbf{R}}$ instead of the exact array covariance matrix $\mathbf{R}$ in (2.19) is known to lead to a substantial performance degradation in the case when the signal component is present in the beamformer training data. It is well known that in the signal-free training data case the output SINR of the SMI beamformer (2.19) converges to the optimal SINR

$$\mathrm{SINR}_{\mathrm{opt}} = \sigma_{\mathrm{s}}^2 \mathbf{a}_{\mathrm{s}}^H \mathbf{R}_{\mathrm{i+n}}^{-1} \mathbf{a}_{\mathrm{s}} \tag{2.20}$$

so that the mean losses relative to (2.20) are less than 3 dB if the following condition is satisfied [2]:

$$N \geq 2M \tag{2.21}$$

However, this rule is no longer applicable when the desired signal contaminates the beamformer training data. In the latter case, the same performance loss can be achieved only when [41]

$$N \geq \mathrm{SINR}_{\mathrm{opt}}(M-1) \gg M \tag{2.22}$$

where the SNR is assumed to be high. According to (2.22), in the presence of the desired signal in the beamformer training data, the SMI algorithm has much slower convergence and weaker robustness against finite sample effects than in the signal-free training data case.

In practice, the situation is further complicated by the fact that the signal covariance matrix is usually known imprecisely, that is, there is always a certain mismatch between the presumed signal covariance matrix $\mathbf{R}_{\mathrm{s}}$ and its actual value which is hereafter denoted as $\tilde{\mathbf{R}}_{\mathrm{s}}$. The main objective of the remainder of this section is to overview traditional ad hoc robust approaches to adaptive beamforming that aim to improve the beamformer performance in scenarios with arbitrary errors in the array response to the desired signal (i.e., the errors between the matrices $\mathbf{R}_{\mathrm{s}}$ and $\tilde{\mathbf{R}}_{\mathrm{s}}$), small training sample size, and training data nonstationarity.

One of the most popular approaches to robust adaptive beamforming in the presence of such array response errors and small training sample size is the diagonal loading technique which was developed independently in [37, 66–68]. The central idea of this approach is to regularize the problem (2.8) by adding a quadratic penalty term to the objective function [68]. Then, in the finite sample case we obtain the following regularized problem [38]:

$$\min_{\mathbf{w}} \mathbf{w}^H \hat{\mathbf{R}} \mathbf{w} + \gamma \mathbf{w}^H \mathbf{w} \quad \text{subject to } \mathbf{w}^H \mathbf{R}_{\mathrm{s}} \mathbf{w} = 1 \tag{2.23}$$

where γ is the penalty weight (also called the diagonal loading factor). We will refer to the solution to (2.23) as the loaded SMI (LSMI) beamformer whose weight vector has the following form [38, 59]:

$$\mathbf{w}_{\mathrm{LSMI}} = \mathcal{P}\{(\hat{\mathbf{R}} + \gamma \mathbf{I})^{-1}\mathbf{R}_{\mathrm{s}}\} \tag{2.24}$$

where $\mathbf{I}$ is the identity matrix. In the rank-one signal source case (rank$\{\mathbf{R}_{\mathrm{s}}\} = 1$), (2.24) reduces to [37, 66, 68]

$$\mathbf{w}_{\mathrm{LSMI}} = (\hat{\mathbf{R}} + \gamma \mathbf{I})^{-1}\mathbf{a}_{\mathrm{s}} \tag{2.25}$$

From (2.24) and (2.25), it is clear that adding the penalty term $\gamma \mathbf{w}^H \mathbf{w}$ to the objective function in (2.23) amounts to loading the diagonal of the sample covariance matrix $\hat{\mathbf{R}}$ by the value of γ. This means that the diagonal loading operation can be interpreted in terms of injecting an artificial amount of white noise into the main diagonal of this matrix. An important property of diagonal loading is that it warrants invertibility of the diagonally loaded matrix $\hat{\mathbf{R}} + \gamma \mathbf{I}$ irrespectively whether $\hat{\mathbf{R}}$ is singular or not.

Moreover, the diagonal loading approach is known to improve the performance of the SMI beamformer in scenarios with mismatched array response [36, 37, 45, 60]. However, the main shortcoming of traditional diagonal loading-based techniques is that there is no rigorous way of choosing the loading parameter γ. In [37], it was proposed to choose this parameter using the following white noise gain constraint:

$$|\mathbf{w}^H \mathbf{a}_{\mathrm{s}}|^2 = \kappa \, \|\mathbf{w}\|^2 \tag{2.26}$$

where hereafter $\|\mathbf{w}\|$ denotes the two-norm of a vector or a matrix, and the parameter κ determines the required white noise gain. This constraint can be added to the MVDR beamformer as follows [37]:

$$\min_{\mathbf{w}} \mathbf{w}^H \mathbf{R}_{\mathrm{i+n}} \mathbf{w} \quad \text{subject to } \mathbf{w}^H \mathbf{a}_{\mathrm{s}} = 1, \quad |\mathbf{w}^H \mathbf{a}_{\mathrm{s}}|^2 = \kappa \, \mathbf{w}^H \mathbf{w} \tag{2.27}$$

The solution to the problem (2.27) is given by [37]

$$\mathbf{w} = \frac{(\mathbf{R}_{\mathrm{i+n}} + \gamma \mathbf{I})^{-1}\mathbf{a}_{\mathrm{s}}}{\mathbf{a}_{\mathrm{s}}^H (\mathbf{R}_{\mathrm{i+n}} + \gamma \mathbf{I})^{-1}\mathbf{a}_{\mathrm{s}}}$$

which, after replacing $\mathbf{R}_{\mathrm{i+n}}$ by $\hat{\mathbf{R}}$ and ignoring the immaterial constant $(\mathbf{a}_{\mathrm{s}}^H (\mathbf{R}_{\mathrm{i+n}} + \gamma \mathbf{I})^{-1}\mathbf{a}_{\mathrm{s}})^{-1}$, becomes equivalent to the LSMI beamformer (2.25) whose diagonal loading parameter should satisfy the white noise gain constraint (2.26).

Unfortunately, it is not quite clear how to choose the white noise gain parameter κ and, as a rule, this parameter is chosen is a somewhat ad hoc way [37]. Also, there is no simple relationship between the parameters κ and γ. Hence, an iterative procedure is required to obtain γ for any given κ [37].

A much simpler and more common ad hoc way of choosing the parameter γ is based on estimating the noise power (e.g., using the noise-subspace eigenvalues or the minimal eigenvalue of the sample covariance matrix) and choosing γ of the same or higher order of magnitude [8, 36–38, 45, 59, 66]. A typical choice of γ is $10 \div 15$ dB higher than the noise power.

As the optimal choice of the diagonal loading factor is well known to be scenario-dependent [38], such a method of choosing fixed γ is only suboptimal and may cause a substantial performance degradation of adaptive beamformers [59–62, 69].

Another popular robust adaptive beamforming technique in the rank-one signal case (i.e., in the presence of steering vector errors) and in situations with small sample size is the eigenspace-based beamformer [41, 70]. In contrast to the LSMI beamformer, this approach is only applicable to the rank-one signal case. The key idea of this technique is to reduce steering vector errors by projecting the signal steering vector onto the estimated signal-plus-interference subspace obtained via the eigendecomposition of the sample covariance matrix (2.17). This eigendecomposition can be written as

$$\hat{\mathbf{R}} = \hat{\mathbf{E}}\hat{\mathbf{\Lambda}}\hat{\mathbf{E}}^H + \hat{\mathbf{G}}\hat{\mathbf{\Gamma}}\hat{\mathbf{G}}^H$$

where the $M \times (L+1)$ matrix $\hat{\mathbf{E}}$ contains the $L+1$ signal-plus-interference subspace eigenvectors of $\hat{\mathbf{R}}$, and the $(L+1) \times (L+1)$ diagonal matrix $\hat{\mathbf{\Lambda}}$ contains the corresponding eigenvalues of this matrix. Similarly, the $M \times (M-L-1)$ matrix $\hat{\mathbf{G}}$ contains the $M-L-1$ noise-subspace eigenvectors of $\hat{\mathbf{R}}$, and the $(M-L-1) \times (M-L-1)$ diagonal matrix $\hat{\mathbf{\Gamma}}$ contains the corresponding eigenvalues. The rank of the interference subspace, L, is assumed to be known. The weight vector of the eigenspace-based beamformer can be written as

$$\mathbf{w}_{\text{eig}} = \hat{\mathbf{R}}^{-1}\mathbf{P}_{\hat{\mathbf{E}}}\,\mathbf{a}_{\text{s}} \tag{2.28}$$

where

$$\mathbf{P}_{\hat{\mathbf{E}}} = \hat{\mathbf{E}}(\hat{\mathbf{E}}^H\hat{\mathbf{E}})^{-1}\hat{\mathbf{E}}^H = \hat{\mathbf{E}}\hat{\mathbf{E}}^H$$

is the orthogonal projection matrix onto the estimated signal-plus-interference subspace. The weight vector (2.28) can be alternatively written as

$$\mathbf{w}_{\text{eig}} = \hat{\mathbf{E}}\hat{\mathbf{\Lambda}}^{-1}\hat{\mathbf{E}}^H\mathbf{a}_{\text{s}} \tag{2.29}$$

If the rank of signal-plus-interference subspace is low and if the parameter L is exactly known, the eigenspace-based beamformer is known to provide excellent robustness against arbitrary steering vector errors [70]. Unfortunately, this approach may degrade severely if the low-rank interference-plus-signal assumption is violated or if the subspace dimension L is uncertain or known imprecisely. For example, in the presence of incoherently scattered (spatially dispersed) interfering sources, interferers with randomly fluctuating wavefronts, and moving interferers, the low-rank interference assumption may become violated and L can be uncertain. Therefore, the eigenspace-based beamformer may be not a proper method of choice in such cases [38]. Moreover, even if the low-rank model assumption remains relevant, the eigenspace-based beamformer can be only used in scenarios where the signal-to-noise ratio (SNR) is sufficiently high because, otherwise, subspace swap effects become dominant and may cause a severe performance degradation of the eigen-space-based beamformer [60]. All these shortcomings make it very difficult to use this beamformer in practice where the dimension of the signal-plus-interference sub-space may be uncertain and relatively high due to the source scattering and fading effects as well as training data nonstationarity [10, 11, 14, 47, 49, 51, 54–58].

In the past decade, several advanced methods have been developed to mitigate performance degradation of adaptive beamformers in the case of nonstationary training data (e.g., in scenarios with moving interferers or rotating antenna) [54–58]. For example, several authors independently used the idea of artificial broadening the adaptive beampattern nulls to improve the robustness of adaptive beamforming, see [55–58, 71, 72].

One approach to broaden the adaptive beampattern nulls has been proposed in [55] and [56] using the data-dependent derivative constraints (DDCs). The essence of this approach is to replace the sample covariance matrix $\hat{\mathbf{R}}$ in the SMI and LSMI beamformers by the modified covariance matrix

$$\breve{\mathbf{R}} = \hat{\mathbf{R}} + \sum_{k=1}^{K} \zeta_k \mathbf{B}^k \hat{\mathbf{R}} \mathbf{B}^k \tag{2.30}$$

where $\mathbf{B}$ is the known diagonal matrix whose entries are determined by the array geometry, K is the highest order of the data-dependent constraints used, and the coefficients ζ_k determine the tradeoff between the constraints of different order. In practical applications, $K = 1$ is shown in [56] to be sufficient to provide satisfactory robustness against interferer motion. Using $K = 1$, (2.30) can be simplified as

$$\breve{\mathbf{R}} = \hat{\mathbf{R}} + \zeta_1 \mathbf{B} \hat{\mathbf{R}} \mathbf{B}$$

where ζ_1 determines the tradeoff between the null depth and the null width. Under a few mild conditions, the optimal value of ζ_1 becomes independent of the source parameters and can be easily computed from the known array parameters [56].

Another way to broaden the adaptive beampattern nulls is based on point constraints and is referred to as the so-called covariance matrix tapering (MT) method [57, 58, 71–73]. The essence of this approach is to replace the sample covariance matrix $\hat{\mathbf{R}}$ in the SMI or LSMI beamformer by the following tapered covariance matrix:

$$\hat{\mathbf{R}}_{\mathrm{T}} = \hat{\mathbf{R}} \odot \mathbf{T}$$

where $\mathbf{T}$ is the so-called $M \times M$ taper matrix and $\odot$ denotes the Schur–Hadamard matrix product. Using the taper matrix introduced in [71] and [72], we can express the elements of $\mathbf{T}$ as

$$[\mathbf{T}]_{il} = \frac{\sin(i-l)\xi}{(i-l)\xi} \tag{2.31}$$

where the parameter ξ determines the required beampattern null width. Another type of matrix taper is proposed in [57].

An interesting link between the MT and DDC approaches was discovered in [73]. In this work, it has been proven that the matrix (2.30) can be viewed as a tapered covariance matrix with particular choice of $\mathbf{T}$. Hence, the DDC approach can be interpreted and implemented using the MT method. However, a serious shortcoming of the MT approach with respect to the DDC technique is that, in the general case, the former approach does not have computationally efficient on-line implementations [38].

The performance of both these methods has been studied thoroughly by means of computer simulations [56–58] and real sonar data processing [14]. The results of this study have shown that these two approaches provide an additional robustness relative to the SMI and LSMI beamformers in slowly moving interference cases, but their performance can become degraded is situations with rapidly moving interferers. Moreover, both these techniques exploit the assumptions of known array geometry and plane interferer wavefronts. Therefore, they may degrade in the case when the array is imperfectly calibrated (e.g., has a distorted shape or unknown sensor gains and phases) or when the wavefronts of the interferers deviate from the plane wavefront form because of multiplicative noise and signal fading/multipath effects or due to interferers located in the near field.

2.3 ROBUST MINIMUM VARIANCE BEAMFORMING BASED ON WORST-CASE PERFORMANCE OPTIMIZATION

In the previous section, main ad hoc approaches to robust adaptive beamforming have been discussed. In this section, we discuss a more powerful and theoretically rigorous worst-case performance optimization-based approach to robust adaptive beamforming that has been recently developed in [59–62].

2.3.1 Rank-One Signal Case

First of all, let us consider the simplest case of a rank-one desired signal with mismatched steering vector. Let the vector of unknown mismatch between the actual steering vector $\tilde{\mathbf{a}}_s$ and its presumed value $\mathbf{a}_s$ be denoted as

$$\boldsymbol{\delta} = \tilde{\mathbf{a}}_s - \mathbf{a}_s$$

Following the idea of [60], we assume that the unknown mismatch vector $\boldsymbol{\delta}$ is norm-bounded by some known constant ϵ, that is,

$$\|\boldsymbol{\delta}\| \leq \epsilon \tag{2.32}$$

To incorporate robustness into the MVDR beamforming problem, let us maximize the worst-case SINR by solving the following problem:

$$\max_{\mathbf{w}} \min_{\boldsymbol{\delta}} \frac{\sigma_s^2 |\mathbf{w}^H(\mathbf{a}_s + \boldsymbol{\delta})|^2}{\mathbf{w}^H \mathbf{R}_{i+n} \mathbf{w}} \qquad \text{subject to } \|\boldsymbol{\delta}\| \leq \epsilon$$

This problem is equivalent to the following robust MVDR beamforming problem [60]:

$$\min_{\mathbf{w}} \mathbf{w}^H \mathbf{R}_{i+n} \mathbf{w} \quad \text{subject to } |\mathbf{w}^H(\mathbf{a}_s + \boldsymbol{\delta})| \geq 1 \quad \text{for all } \|\boldsymbol{\delta}\| \leq \epsilon \tag{2.33}$$

The main modification in (2.33) with respect to the original problem (2.9) is that instead of requiring fixed distortionless response towards the single presumed steering vector $\mathbf{a}_s$, such distortionless response is now maintained in (2.33) by means of inequality constraints for a continuum of all possible steering vectors that belong to the spherical uncertainty set

$$\mathcal{A} \triangleq \{\mathbf{c} \mid \mathbf{c} = \mathbf{a}_s + \boldsymbol{\delta}, \quad \|\boldsymbol{\delta}\| \leq \epsilon\}$$

The constraints in (2.33) guarantee that the distortionless response will be maintained in the worst case, that is, for the particular vector $\boldsymbol{\delta}$ which corresponds to the smallest value of $|\mathbf{w}^H(\mathbf{a}_s + \boldsymbol{\delta})|$ provided that $\|\boldsymbol{\delta}\| \leq \epsilon$.

In the finite sample case, $\mathbf{R}_{i+n}$ should be replaced by $\hat{\mathbf{R}}$. Doing so and replacing the infinite number of constraints in (2.33) by the aforementioned single worst-case constraint, the problem (2.33) becomes

$$\min_{\mathbf{w}} \mathbf{w}^H \hat{\mathbf{R}} \mathbf{w} \quad \text{subject to } \min_{\|\boldsymbol{\delta}\| \leq \epsilon} |\mathbf{w}^H \mathbf{a}_s + \mathbf{w}^H \boldsymbol{\delta}| \geq 1 \tag{2.34}$$

Note that the inequality constraint in (2.34) is equivalent to the equality constraint

$$\min_{\|\boldsymbol{\delta}\| \leq \epsilon} |\mathbf{w}^H \mathbf{a}_s + \mathbf{w}^H \boldsymbol{\delta}| = 1 \tag{2.35}$$

The equivalence of the equality constraint (2.35) and the inequality constraint in (2.34) can be easily proved by contradiction as follows [60]. If they are not equivalent to each other then the minimum of the objective function in (2.34) is achieved when $\chi \triangleq \min_{\|\boldsymbol{\delta}\| \leq \epsilon} |\mathbf{w}^H \mathbf{a}_s + \mathbf{w}^H \boldsymbol{\delta}| > 1$. However, replacing $\mathbf{w}$ with $\mathbf{w}/\sqrt{\chi}$, we can decrease the objective function $\mathbf{w}^H \hat{\mathbf{R}} \mathbf{w}$ by the factor of $\chi > 1$ while the constraint in (2.34) will be still satisfied. This is an obvious contradiction to the original statement that the objective function is minimized when $\chi > 1$. Therefore, the minimum of the objective function is achieved at $\chi = 1$ and this means that the inequality constraint in (2.34) is equivalent to the equality constraint (2.35).

If the sequel, we will use this constraint in both its inequality and equality equivalent forms. The following lemma [60] can be proved.

Lemma 1. If

$$|\mathbf{w}^H \mathbf{a}_s| \geq \varepsilon \|\mathbf{w}\| \tag{2.36}$$

then

$$\min_{\|\boldsymbol{\delta}\| \leq \epsilon} |\mathbf{w}^H (\mathbf{a}_s + \boldsymbol{\delta})| = |\mathbf{w}^H \mathbf{a}_s| - \epsilon \|\mathbf{w}\|$$

Proof. See Appendix 2.A. □

Note that, according to (2.26), the condition (2.36) is used in Lemma 1 to guarantee a sufficient white noise gain [37].

Assuming that this condition is satisfied and using Lemma 1, we can rewrite problem (2.34) as the following quadratic minimization problem with a *single* nonlinear constraint:

$$\min_{\mathbf{w}} \mathbf{w}^H \hat{\mathbf{R}} \mathbf{w} \quad \text{subject to } |\mathbf{w}^H \mathbf{a}_s| - \epsilon \|\mathbf{w}\| \geq 1 \tag{2.37}$$

The nonlinear constraint in (2.37) is still nonconvex due to the absolute value operation on the left-hand side. To convert this problem to a convex one, we can use the fact that the cost function in (2.37) is *unchanged* when $\mathbf{w}$ undergoes an arbitrary phase rotation [60]. Therefore, if $\mathbf{w}_0$ is an optimal solution to (2.37), we can always rotate, without affecting the objective function value, the phase of $\mathbf{w}_0$ so that $\mathbf{w}^H \mathbf{a}_s$ is real. Thus, without any loss of generality, $\mathbf{w}$ can be chosen such that

$$\text{Re}\,\{\mathbf{w}^H \mathbf{a}_s\} \geq 0 \tag{2.38}$$

$$\text{Im}\,\{\mathbf{w}^H \mathbf{a}_s\} = 0 \tag{2.39}$$

Using this observation, the problem can be written as [60]

$$\min_{\mathbf{w}} \mathbf{w}^H \hat{\mathbf{R}} \mathbf{w} \quad \text{subject to } \mathbf{w}^H \mathbf{a}_s \geq \epsilon \|\mathbf{w}\| + 1 \tag{2.40}$$

where, according to the aforementioned fact that the constraint in (2.40) is satisfied with equality, (2.39) can be ignored because from $\mathbf{w}^H \mathbf{a}_s = \epsilon \|\mathbf{w}\| + 1$ it follows that the value of $\mathbf{w}^H \mathbf{a}_s$ is real-valued and positive.

Comparing the white noise gain constraint (2.26) and the constraint in (2.40), we see that they have a high degree of similarity, although the latter constraint contains an additional constant term in the right-hand side. This observation helps us to understand the relationship between the white noise gain constraint based beamformer (2.27) and the robust beamformer (2.40).

It is also important to stress that the original problem (2.33) appears to be computationally intractable (NP-hard), whereas the robust MVDR beamformer (2.40) of [60] belongs to the class of convex second-order cone (SOC) programming problems [74] which can be easily solved using standard and highly efficient interior point method software [75]. For example, using the primal-dual potential reduction method [74], the complexity of solving (2.40) is $O(M^3)$ per iteration, and the algorithm converges typically in less than 10 iterations (a well-known and widely accepted fact in the optimization community). Therefore, the overall computational complexity of the SOC programming based beamformer is $O(M^3)$ [60]. This complexity is comparable to that of the SMI and LSMI algorithms. An alternative way to solve problem (2.40) with the complexity $O(M^3)$ is to use the Newton-type algorithms developed in [62] and [76].

Let us overview the algorithm of [76]. As the constraint in (2.40) is satisfied with equality, we can rewrite this problem as

$$\min_{\mathbf{w}} \ \mathbf{w}^H \hat{\mathbf{R}} \mathbf{w} \quad \text{subject to } \mathbf{w}^H \mathbf{a}_s - \epsilon \|\mathbf{w}\| = 1$$

Using the Lagrange multiplier method, we can write the Lagrangian function as

$$L(\mathbf{w}, \lambda) = \mathbf{w}^H \hat{\mathbf{R}} \mathbf{w} - \lambda(\mathbf{w}^H \mathbf{a}_s - \epsilon \|\mathbf{w}\| - 1) \tag{2.41}$$

where λ is the Lagrange multiplier. Differentiating (2.41) and equating the result to zero, we obtain the following equation:

$$\hat{\mathbf{R}} \mathbf{w} + \lambda \epsilon \frac{\mathbf{w}}{\|\mathbf{w}\|} = \lambda \mathbf{a}_s \tag{2.42}$$

To solve (2.42), we need to know the Lagrange multiplier λ. However, using the fact that multiplying the weight vector by any arbitrary constant does not change the output SINR, we can transform this equation to [76]

$$\hat{\mathbf{R}} \mathbf{w} + \epsilon \frac{\mathbf{w}}{\|\mathbf{w}\|} = \mathbf{a}_s \tag{2.43}$$

so that (2.43) does not contain the Lagrange multiplier anymore. For the sake of simplicity, the same notation **w** is used in (2.43) for the rescaled weight vector as for the original one in (2.42).

Equation (2.43) can be rewritten as

$$\left(\hat{\mathbf{R}} + \frac{\epsilon}{\|\mathbf{w}\|} \mathbf{I}\right) \mathbf{w} = \mathbf{a}_s \tag{2.44}$$

From (2.44), it can be seen that the robust MVDR beamformer (2.40) belongs to the class of diagonal loading techniques. Note that this beamformer uses *adaptive* diagonal loading because the diagonal loading factor $\epsilon/\|\mathbf{w}\|$ depends on the norm of the weight vector and, therefore, is scenario-dependent. It should be stressed that, in contrast to the fixed diagonal loading approach used in the LSMI beamformer, such an adaptive diagonal loading technique optimally matches the diagonal loading factor to the known amount of uncertainty in the signal steering vector [60, 76].

A noteworthy observation following from (2.44) is that, if $\|\mathbf{w}\|$ is available, then we can use (2.44) to calculate the weight vector of the robust MVDR beamformer. To determine $\|\mathbf{w}\|$, the following simple method can be used [76]. Rewriting (2.44) as

$$\mathbf{w} = \left(\hat{\mathbf{R}} + \frac{\epsilon}{\|\mathbf{w}\|}\mathbf{I}\right)^{-1} \mathbf{a}_s \tag{2.45}$$

and taking the norm squared of the both sides of (2.45), we have

$$\|\mathbf{w}\|^2 = \left\| \left(\hat{\mathbf{R}} + \frac{\epsilon}{\|\mathbf{w}\|}\mathbf{I}\right)^{-1} \mathbf{a}_s \right\|^2 \tag{2.46}$$

Introducing $\tau \triangleq \|\mathbf{w}\| > 0$, we obtain that solving (2.46) is equivalent to finding a positive value of τ such that

$$\tau^2 = \left\| \left(\hat{\mathbf{R}} + \frac{\epsilon}{\tau}\mathbf{I}\right)^{-1} \mathbf{a}_s \right\|^2 \tag{2.47}$$

To simplify (2.47), let us use the eigendecomposition[1] of $\hat{\mathbf{R}}$,

$$\hat{\mathbf{R}} = \mathbf{U}\boldsymbol{\Xi}\mathbf{U}^H \tag{2.48}$$

where $\mathbf{U}$ is the $M \times M$ unitary matrix whose columns are the eigenvectors of $\hat{\mathbf{R}}$ and $\boldsymbol{\Xi}$ is the diagonal matrix of eigenvalues of $\hat{\mathbf{R}}$ given by

$$\boldsymbol{\Xi} = \mathrm{diag}\{\xi_1, \ldots, \xi_M\}$$

Here, diag $\{\cdot\}$ denotes a diagonal matrix and $\{\xi_i\}_{i=1}^M$ are the real positive eigenvalues of $\hat{\mathbf{R}}$. Without loss of generality, we assume that $\xi_1 \geq \xi_2 \geq \cdots \geq \xi_M > 0$.

Using (2.48), we can rewrite (2.47) as

$$\|\mathbf{U}\boldsymbol{\Psi}^{-1}(\tau)\mathbf{U}^H\mathbf{a}_s\|^2 - \tau^2 = 0 \tag{2.49}$$

[1]Note that the eigendecomposition is also used in [69] in a similar way to derive a Newton-type algorithm.

where

$$\boldsymbol{\Psi}(\tau) \triangleq \boldsymbol{\Xi} + \frac{\epsilon}{\tau}\mathbf{I}$$

Introducing the $M \times 1$ vector $\mathbf{g}$ as

$$\mathbf{g} = [g_1, \ldots, g_M]^T \triangleq \mathbf{U}^H \mathbf{a}_s \tag{2.50}$$

and taking into account that $\mathbf{U}$ is a unitary matrix, we can rewrite the left-hand side of (2.49) as

$$\begin{aligned} \|\mathbf{U}\boldsymbol{\Psi}^{-1}(\tau)\mathbf{U}^H\mathbf{a}_s\|^2 - \tau^2 &= \|\boldsymbol{\Psi}^{-1}(\tau)\mathbf{g}\|^2 - \tau^2 \\ &= \sum_{i=1}^{L}\left(\frac{|g_i|}{\xi_i + \dfrac{\epsilon}{\tau}}\right)^2 - \tau^2 \\ &= \left[\sum_{i=1}^{L}\left(\frac{|g_i|}{\epsilon + \tau\xi_i}\right)^2 - 1\right]\tau^2 \end{aligned} \tag{2.51}$$

Using (2.51) and taking into account that $\tau > 0$, we obtain that solving (2.49) is equivalent to finding a positive value for τ such that

$$f(\tau) \triangleq \sum_{i=1}^{M}\left(\frac{|g_i|}{\epsilon + \tau\xi_i}\right)^2 - 1 = 0 \tag{2.52}$$

Note that (2.52) may not always have a real and positive solution. The following lemma [76] states the necessary and sufficient conditions under which (2.52) has a unique positive solution.

Lemma 2. Equation (2.52) has a unique real-valued and positive solution if and only if

$$\|\mathbf{a}_s\| > \epsilon \tag{2.53}$$

Proof. See Appendix 2.B. □

The condition similar to (2.53) has been also used in [69] and yields an intuitively appealing interpretation. As the parameter ϵ characterizes the maximal norm of the mismatch between the presumed and the actual signal steering vectors, equation (2.53) simply states that the approach we are going to develop is applicable only if the maximum norm of such a mismatch does not exceed the norm of the presumed signal steering vector itself. In the sequel, we assume that (2.53) is always satisfied.

Using (2.52), we can upper-bound the function $f(\tau)$ as

$$\begin{aligned} f(\tau) &< \frac{\sum_{i=1}^{M} |g_i|^2}{(\epsilon + \tau\xi_M)^2} - 1 \\ &= \frac{\|\mathbf{g}\|^2}{(\epsilon + \tau\xi_M)^2} - 1 \\ &= \frac{\|\mathbf{a}_\mathrm{s}\|^2}{(\epsilon + \tau\xi_M)^2} - 1 \triangleq f_\mathrm{up}(\tau) \end{aligned} \tag{2.54}$$

Noting that $f(\tau)$ and $f_\mathrm{up}(\tau)$ are both decreasing functions for positive values of τ and that, according to Lemma 2, the root τ of $f(\tau)$ is positive, we obtain from (2.54) that this root is always smaller than the root

$$\tau_\mathrm{up} = \frac{\|\mathbf{a}_\mathrm{s}\| - \epsilon}{\xi_M}$$

of $f_\mathrm{up}(\tau)$. Therefore, the value of τ lies in the interval $(0, \tau_\mathrm{up})$. With this condition, the problem of computing τ becomes standard. For example, the algorithm of [77] can be used for this purpose [76]. The latter algorithm consists of a binary search followed by Newton–Raphson iterations. The binary search technique is used in this algorithm to obtain a proper initialization for the subsequent Newton–Raphson iterations. As shown in [77], this algorithm converges to a ν-neighborhood of τ in $\mathcal{O}(\log\log(\tau_\mathrm{up}/\nu))$ iterations. The algorithm to compute $\|\mathbf{w}\|$ can be summarized as follows [76]:

1. Use binary search to find $\tau_0 \in (0, \tau_\mathrm{up})$ such that $f(\tau_0) > 0$ and $f\left(\frac{13}{12}\tau_0\right) < 0$ (see [77] for details).
2. Set $l = 1$ and select a small positive value of ξ which will be used in the algorithm stopping criterion.
3. Obtain τ_l as

$$\tau_l = \tau_{l-1} - \frac{f(\tau_{l-1})}{f'(\tau_{l-1})}$$

   where $f'(\tau_{l-1})$ is the derivative of $f(\tau)$ at $\tau = \tau_{l-1}$.
4. If $|f(\tau_l)| < \xi$, go to the next step. Otherwise, repeat steps 2 and 3.
5. Compute $\|\mathbf{w}\|$ as $\tau = \tau_l$.

The value of $\|\mathbf{w}\|$ which is computed by means of this procedure can be then substituted to (2.45) to obtain the resulting weight vector which solves the problem (2.40) [76].

The dominant computational complexity of this algorithm is determined by that of the eigendecomposition and inversion of the matrix $\hat{\mathbf{R}}$ and is equal to $O(M^3)$ [76]. It is worth noting that this complexity is equivalent to that of the SMI and LSMI algorithms.

Several further extensions of the robust MVDR beamformer of [60] have been recently developed by different authors. In [62], this beamformer has been extended to the case of ellipsoidal (anisotropic) uncertainty. The authors of [62] considered the following problem:

$$\min_{\mathbf{w}} \mathbf{w}^H \hat{\mathbf{R}} \mathbf{w} \quad \text{subject to} \quad \mathrm{Re}\{\mathbf{w}^H \mathbf{a}_s\} \geq 1, \qquad \text{for all } \mathbf{a}_s \in \mathcal{E} \tag{2.55}$$

where $\mathcal{E}$ is an ellipsoid that covers the possible range of uncertainty of the steering vector $\mathbf{a}_s$. In [62], some opportunities to estimate optimal parameters of $\mathcal{E}$ from the received array data are discussed.

In [69], a covariance fitting-based interpretation of the robust MVDR problems of [60] and [62] has been developed. Although the problem in [69] is formulated in a different form as compared to that of [60] and [62], the authors of [69] have shown that such reformulated problem (which is referred to as a *robust Capon beamformer* in [69]) leads to exactly the same beamforming solutions as those in [60] and [62]. An additional useful feature of the approach of [69] is its ability to estimate the mismatched signal steering vector. An alternative Newton-type algorithm is derived in [69] to compute the weight vectors of the robust MVDR beamformers of [60] and [62]. The problem formulation of [69] is further modified in [78] by adding an ad hoc quadratic constraint.

In [76], the approach of [60] has been extended to robust multiuser detection problems. In [79], an efficient Kalman filter-based on-line implementation of the robust MVDR beamformer of [60] with the complexity of $O(M^2)$ per step has been developed.

In [61], the approach of [60] is extended to a more general case where, apart from the steering vector mismatch, there is a nonstationarity of the training data (which, as mentioned before, may be caused by the nonstationarity of interference and propagation channel, as well as antenna motion or vibration). To explain the results of [61], let us define the *data matrix* as

$$\mathbf{X} = [\mathbf{x}(1), \mathbf{x}(2), \ldots, \mathbf{x}(N)] \tag{2.56}$$

Using (2.56), the sample covariance matrix (2.17) can be expressed as

$$\hat{\mathbf{R}} = \frac{1}{N} \mathbf{X}\mathbf{X}^H$$

The approach of [61] suggests to model the uncertainty which is caused by nonstationarities of the training data by means of adding this uncertainty to the data matrix.

Towards this end, let us introduce the mismatch matrix

$$\mathbf{\Delta} = \tilde{\mathbf{X}} - \mathbf{X}$$

where $\tilde{\mathbf{X}}$ and $\mathbf{X}$ are, respectively, the actual and presumed data matrices in the test cell (at the beamforming sample). The presumed data matrix corresponds to the *measured* training cell data. In real-time adaptive beamforming problems, such training cell data correspond to the measurements that are made prior to the test cell. Thus, because of possible data nonstationarity effects, such past data snapshots may inadequately model the current test cell, where the actual (but unknown) data matrix is $\tilde{\mathbf{X}}$ rather than $\mathbf{X}$. Hence, in the nonstationary case, the actual sample covariance matrix can be expressed as

$$\begin{aligned}\hat{\tilde{\mathbf{R}}} &= \frac{1}{N}\tilde{\mathbf{X}}\tilde{\mathbf{X}}^H \\ &= \frac{1}{N}(\mathbf{X}+\mathbf{\Delta})(\mathbf{X}+\mathbf{\Delta})^H \end{aligned} \tag{2.57}$$

According to (2.57), the matrix $\hat{\tilde{\mathbf{R}}}$ is Hermitian and non-negative definite. However, this matrix is unknown because the mismatch $\mathbf{\Delta}$ is unknown.

The authors of [61] proposed to combine the robustness against interference nonstationarity and steering vector errors using the ideas similar to that originally proposed in [60]. They assume that the norms of both the steering vector mismatch $\boldsymbol{\delta}$ and the data matrix mismatch $\mathbf{\Delta}$ are bounded by some known constants, that is,

$$\|\boldsymbol{\delta}\| \le \epsilon, \qquad \|\mathbf{\Delta}\|_{\mathcal{F}} \le \eta$$

where $\|\cdot\|_{\mathcal{F}}$ denotes the Frobenius norm of a matrix. Then, the weight vector can be found from maximizing the worst-case SINR, that is, by solving the following problem:

$$\max_{\mathbf{w}} \min_{\boldsymbol{\delta},\mathbf{\Delta}} \frac{\sigma_{\mathrm{s}}^2|\mathbf{w}^H(\mathbf{a}_{\mathrm{s}}+\boldsymbol{\delta})|^2}{\mathbf{w}^H\hat{\tilde{\mathbf{R}}}\mathbf{w}} \quad \text{subject to } \|\boldsymbol{\delta}\| \le \epsilon, \quad \|\mathbf{\Delta}\|_{\mathcal{F}} \le \eta \tag{2.58}$$

Using (2.58) and (2.57), the robust formulation of the MVDR beamforming problem takes the following form [61]:

$$\begin{aligned}&\min_{\mathbf{w}} \max_{\|\mathbf{\Delta}\|_{\mathcal{F}}\le\eta} \|(\mathbf{X}+\mathbf{\Delta})^H\mathbf{w}\| \\ &\text{subject to } |\mathbf{w}^H(\mathbf{a}_{\mathrm{s}}+\boldsymbol{\delta})| \ge 1 \quad \text{for all } \|\boldsymbol{\delta}\| \le \epsilon \end{aligned} \tag{2.59}$$

Note that this problem represents a further extension of (2.33) with additional robustness against nonstationary training data. The key idea of (2.59) is to minimize the beamformer output power in the scenario with the worst-case nonstationarity mismatch of the data matrix subject to the constraint which maintains the distortionless response for the worst-case steering vector mismatch. Note that the latter constant is the same as in (2.33), while the objective function is further modified with respect to (2.33).

To simplify the problem (2.59), the authors of [61] replaced the infinite number of constraints by a single worst-case constraint

$$\min_{\|\boldsymbol{\delta}\|\leq\epsilon} |\mathbf{w}^H\mathbf{a}_\mathrm{s} + \mathbf{w}^H\boldsymbol{\delta}| \geq 1 \tag{2.60}$$

in the same way as it was done in (2.34) and made use of Lemma 1 and the following Lemma.

Lemma 3.

$$\max_{\|\boldsymbol{\Delta}\|_\mathcal{F}\leq\eta} \|(\mathbf{X}+\boldsymbol{\Delta})^H\mathbf{w}\| = \|\mathbf{X}^H\mathbf{w}\| + \eta\|\mathbf{w}\|$$

Proof. See Appendix 2.C. □

Using Lemmas 1 and 3 along with (2.60), and taking into account that the cost function in (2.59) remains unchanged when $\mathbf{w}$ undergoes an arbitrary phase rotation [61], the problem (2.59) can be converted to

$$\min_{\mathbf{w}} \|\mathbf{X}^H\mathbf{w}\| + \eta\|\mathbf{w}\| \quad \text{subject to } \mathbf{w}^H\mathbf{a}_\mathrm{s} \geq \epsilon\|\mathbf{w}\| + 1 \tag{2.61}$$

where, similar to (2.40), the constraint is satisfied with equality. This guarantees that (2.38) and (2.39) are satisfied automatically and, hence, there is no need to add them as additional constraints to (2.61).

Problem (2.61) can be viewed as an extended version of (2.40). Note that (2.61) also belongs to the class of SOC programming problems and can be efficiently solved using standard interior point method software [75]. Clearly, the robust beamformer (2.40) is a particular case of (2.61), because if we set $\eta = 0$ in (2.61) then it transforms to (2.40).

To further improve the robustness against moving interferers, the beamformer (2.61) can be combined with the MT method [61]. For that purpose, one should replace the matrix $\mathbf{X}$ in (2.61) by $\sqrt{N}\,\hat{\mathbf{R}}_\mathrm{T}^{1/2}$.

2.3.2 General-Rank Signal Case

Now, let us consider the general-rank signal case and consider the robust MVDR beamformer that has been recently derived in [59]. Following the philosophy of this work, we take into account that in practical situations, both the signal and interference-plus-noise covariance matrices are known with some errors. In other words, there is always a certain mismatch between the *actual* and *presumed* values of these matrices. This yields

$$\tilde{\mathbf{R}}_{\mathrm{s}} = \mathbf{R}_{\mathrm{s}} + \mathbf{\Delta}_1$$
$$\tilde{\mathbf{R}}_{\mathrm{i+n}} = \mathbf{R}_{\mathrm{i+n}} + \mathbf{\Delta}_2$$

where the presumed signal and interference-plus-noise covariance matrices are denoted as $\mathbf{R}_{\mathrm{s}}$ and $\mathbf{R}_{\mathrm{i+n}}$, respectively, while their actual values are denoted as $\tilde{\mathbf{R}}_{\mathrm{s}}$ and $\tilde{\mathbf{R}}_{\mathrm{i+n}}$, respectively. Here, $\mathbf{\Delta}_1$ and $\mathbf{\Delta}_2$ are the unknown matrix mismatches. These mismatches may occur because of a limited number of data snapshots that are used to estimate the signal and interference-plus-noise covariance matrices, environmental nonstationarities (such as rapid motion of the desired signal and interferers), signal location errors, and, moreover, due to the fact that in many applications, signal- and interference-free samples are usually unavailable.

In the presence of the mismatches $\mathbf{\Delta}_1$ and $\mathbf{\Delta}_2$, equation (2.2) for the output SINR of an adaptive beamformer must be rewritten as

$$\mathrm{SINR} = \frac{\mathbf{w}^H \tilde{\mathbf{R}}_{\mathrm{s}} \mathbf{w}}{\mathbf{w}^H \tilde{\mathbf{R}}_{\mathrm{i+n}} \mathbf{w}}$$

Let the unknown mismatch matrices $\mathbf{\Delta}_1$ and $\mathbf{\Delta}_2$ be bounded in their norm by some known constants as [59]

$$\|\mathbf{\Delta}_1\|_{\mathcal{F}} \leq \varepsilon, \qquad \|\mathbf{\Delta}_2\|_{\mathcal{F}} \leq \gamma$$

To provide robustness against such norm-bounded mismatches, the authors of [59] used the idea similar to [60], that is, they obtained the beamformer weight vector via maximizing the worst-case output SINR. This corresponds to the following optimization problem [59]

$$\max_{\mathbf{w}} \min_{\mathbf{\Delta}_1,\, \mathbf{\Delta}_2} \frac{\mathbf{w}^H(\mathbf{R}_{\mathrm{s}} + \mathbf{\Delta}_1)\mathbf{w}}{\mathbf{w}^H(\mathbf{R}_{\mathrm{i+n}} + \mathbf{\Delta}_2)\mathbf{w}} \quad \text{subject to } \|\mathbf{\Delta}_1\|_{\mathcal{F}} \leq \varepsilon,\ \|\mathbf{\Delta}_2\|_{\mathcal{F}} \leq \gamma \tag{2.64}$$

where $\mathbf{\Delta}_1$ and $\mathbf{\Delta}_2$ are Hermitian matrices.

This problem can be rewritten as

$$\max_{\mathbf{w}} \frac{\min_{\|\mathbf{\Delta}_1\|_{\mathcal{F}} \leq \varepsilon} \mathbf{w}^H(\mathbf{R}_{\mathrm{s}} + \mathbf{\Delta}_1)\mathbf{w}}{\max_{\|\mathbf{\Delta}_2\|_{\mathcal{F}} \leq \gamma} \mathbf{w}^H(\mathbf{R}_{\mathrm{i+n}} + \mathbf{\Delta}_2)\mathbf{w}} \tag{2.65}$$

To solve (2.65), the following result can be used [59].

Lemma 4

$$\min_{\|\mathbf{\Delta}_1\|_{\mathcal{F}} \le \varepsilon} \mathbf{w}^H(\mathbf{R}_s + \mathbf{\Delta}_1)\mathbf{w} = \mathbf{w}^H(\mathbf{R}_s - \varepsilon\mathbf{I})\mathbf{w}$$

$$\max_{\|\mathbf{\Delta}_2\|_{\mathcal{F}} \le \gamma} \mathbf{w}^H(\mathbf{R}_{i+n} + \mathbf{\Delta}_2)\mathbf{w} = \mathbf{w}^H(\mathbf{R}_{i+n} + \gamma\mathbf{I})\mathbf{w}$$

where the worst-case mismatch matrices $\mathbf{\Delta}_1$ and $\mathbf{\Delta}_2$ are given by

$$\mathbf{\Delta}_1 = -\varepsilon\frac{\mathbf{w}\mathbf{w}^H}{\|\mathbf{w}\|^2}, \qquad \mathbf{\Delta}_2 = \gamma\frac{\mathbf{w}\mathbf{w}^H}{\|\mathbf{w}\|^2}$$

respectively.

Proof. See Appendix 2.D. □

Using Lemma 4, the problem (2.65) can be converted to

$$\max_{\mathbf{w}} \frac{\mathbf{w}^H(\mathbf{R}_s - \varepsilon\mathbf{I})\mathbf{w}}{\mathbf{w}^H(\mathbf{R}_{i+n} + \gamma\mathbf{I})\mathbf{w}}$$

which, in turn, is equivalent to the following modified MVDR problem:

$$\min_{\mathbf{w}} \mathbf{w}^H(\mathbf{R}_{i+n} + \gamma\mathbf{I})\mathbf{w} \qquad \text{subject to } \mathbf{w}^H(\mathbf{R}_s - \varepsilon\mathbf{I})\mathbf{w} = 1 \tag{2.67}$$

Note that the problems (2.64) and (2.67) are equivalent if ε is smaller than the maximal eigenvalue of $\mathbf{R}_s$. In the opposite case (when ε is larger than the maximal eigenvalue of $\mathbf{R}_s$), the matrix $\mathbf{R}_s - \varepsilon\mathbf{I}$ is negative definite and (2.67) does not have any solution because the constraint in (2.67) cannot be satisfied. Therefore, the parameter ε which is smaller than the maximal eigenvalue of $\mathbf{R}_s$ has to be chosen. A simple interpretation of this condition is that the allowed uncertainty in the signal covariance matrix should be sufficiently small.

Clearly, the structure of the problem (2.67) is similar to that of the problems (2.8) and (2.23). Using this fact, the solution to (2.67) can be expressed in the following form [59]:

$$\mathbf{w}_{\text{rob}} = \mathcal{P}\{(\mathbf{R}_{i+n} + \gamma\mathbf{I})^{-1}(\mathbf{R}_s - \varepsilon\mathbf{I})\} \tag{2.68}$$

In practical situations, the matrix $\mathbf{R}_{i+n}$ is not available and the sample covariance matrix $\hat{\mathbf{R}}$ should be used in lieu of $\mathbf{R}_{i+n}$ in (2.67). The solution to such a modified problem yields the following sample version of the robust beamformer (2.68):

$$\mathbf{w}_{\text{rob}} = \mathcal{P}\{(\hat{\mathbf{R}} + \gamma\mathbf{I})^{-1}(\mathbf{R}_s - \varepsilon\mathbf{I})\} \tag{2.69}$$

In the rank-one signal case, assuming without loss of generality that $\sigma_s^2 = 1$ (i.e., absorbing the constant $1/\sigma_s^2$ in ε), we obtain that the robust MVDR beamformer (2.69) can be rewritten as

$$\mathbf{w}_{\text{rob}} = \mathcal{P}\{(\hat{\mathbf{R}} + \gamma\mathbf{I})^{-1}(\mathbf{a}_s\mathbf{a}_s^H - \varepsilon\mathbf{I})\} \tag{2.70}$$

From (2.69) it follows that the worst-case performance optimization approach of [59] leads to a new diagonal loading-based beamformer which naturally combines both the *negative* and *positive* types of diagonal loading, where the negative loading is applied to the presumed covariance matrix of the desired signal $\mathbf{R}_s$, while the positive loading is applied to the sample covariance matrix $\hat{\mathbf{R}}$.

Setting $\varepsilon = 0$, we obtain that in this case (2.69) converts to the conventional LSMI beamformer (2.24). Hence, this beamformer can be interpreted as a solution to the worst-case performance optimization problem involving errors in the sample covariance matrix. This explains a commonly known fact that diagonal loading can be efficiently applied to a substantially broader class of problems than the small sample size problem (which, however, was originally one of the main arguments why to use diagonal loading).

Interestingly, the robust beamformer (2.69) offers a simpler and somewhat more motivated way of choosing the parameters ε and γ as compared to the way of choosing γ in the diagonal loading method based on the white noise gain constraint. Indeed, the choice of ε and γ in (2.69) is dictated by the *physical* parameters of the environment (upper bounds on the covariance matrix mismatches). It appears that in many practical situations it is relatively easy to obtain the parameters γ and ε based on some preliminary knowledge of the type of environment considered [38].

An important difference between the general-rank robust MVDR beamformer (2.69) and rank-one robust MVDR beamformers (2.40) and (2.61) is that (2.69) is not able to take into account the constraint that the actual signal covariance matrix $\tilde{\mathbf{R}}_s$ must be non-negative definite, while the techniques (2.61) and (2.69) take into account this constraint. To clarify this point, note that the matrix $\tilde{\mathbf{R}}_s$ in (2.69) is not necessarily positive semidefinite. From the form of (2.70) it also becomes clear that in the rank-one signal case, this matrix always has negative eigenvalues if $\varepsilon > 0$. As a result, the aforementioned non-negative definiteness constraint is not satisfied in the problem (2.67). Ignoring this constraint may, in fact, lead to an overly conservative approach (when more robustness than necessary is provided) [38], although from the simulation results of [59] it follows that this does not affect seriously the performance of (2.69).

An interesting interpretation of the robust beamformer (2.69) in terms of positive-only diagonal loading has been obtained in [59]. According to (2.69), the weight vector $\mathbf{w}_{\text{rob}}$ satisfies the following characteristic equation

$$(\hat{\mathbf{R}} + \gamma\mathbf{I})^{-1}(\mathbf{R}_s - \varepsilon\mathbf{I})\mathbf{w}_{\text{rob}} = \mu\mathbf{w}_{\text{rob}} \tag{2.71}$$

where μ is the maximal eigenvalue of the matrix $(\hat{\mathbf{R}} + \gamma\mathbf{I})^{-1}(\mathbf{R}_s - \varepsilon\mathbf{I})$ and $\mathbf{w}_{\text{rob}}$ plays the role of the principal eigenvector of this matrix. Equation (2.71) can be rewritten as

$$(\mu\hat{\mathbf{R}} + (\mu\gamma + \varepsilon)\mathbf{I})\mathbf{w}_{\text{rob}} = \mathbf{R}_s\mathbf{w}_{\text{rob}}$$

The latter equation is equivalent to

$$\left[\hat{\mathbf{R}} + \left(\gamma + \frac{\varepsilon}{\mu}\right)\mathbf{I}\right]^{-1}\mathbf{R}_s\mathbf{w}_{\text{rob}} = \mu\mathbf{w}_{\text{rob}} \tag{2.73}$$

which implies that the robust beamformer (2.69) can be reinterpreted in terms of traditional (positive-only) diagonal loading with the *adaptive* loading factor $\gamma + \varepsilon/\mu$. However, it should be stressed that (2.73) is not a characteristic equation for the matrix $(\hat{\mathbf{R}} + (\gamma + \varepsilon/\mu)\mathbf{I})^{-1}\mathbf{R}_s$ because μ is involved in both left- and right-hand sides of (2.73). This fact poses major obstacles to find the weight vector $\mathbf{w}_{\text{rob}}$ directly from equation (2.73) and clarifies that (2.69) yields an easy way to solve equation (2.73) indirectly and in a closed form. However, equation (2.73) shows that the robust beamformer (2.69) that uses both the negative and positive types of diagonal loading is equivalent to the traditional diagonal loading method (with positive diagonal loading only) whose loading factor is selected adaptively, to optimally match to the given amount of uncertainty in the signal and data covariance matrices.

An efficient on-line implementation of the robust MVDR beamformer (2.69) has been developed in [59] where the following lemma has been proved.

Lemma 5. For arbitrary $M \times M$ Hermitian matrix $\mathbf{Y}$ and arbitrary $M \times M$ full-rank Hermitian matrix $\mathbf{Z}$ the following relationship holds

$$\mathcal{P}\{\mathbf{YZ}\} = \mathbf{Z}^{-1/2}\mathcal{P}\{\mathbf{Z}^{1/2}\mathbf{Y}\mathbf{Z}^{1/2}\} \tag{2.74}$$

Proof. See Appendix 2.E. □

Applying this lemma to the beamformer (2.69), we rewrite it as

$$\begin{aligned}\mathbf{w}_{\text{rob}} &= (\mathbf{R}_s - \varepsilon\mathbf{I})^{-1/2}\mathcal{P}\{(\mathbf{R}_s - \varepsilon\mathbf{I})^{1/2}(\hat{\mathbf{R}} + \gamma\mathbf{I})^{-1}(\mathbf{R}_s - \varepsilon\mathbf{I})^{1/2}\} \\ &= (\mathbf{R}_s - \varepsilon\mathbf{I})^{-1/2}\,\mathcal{P}\{\mathbf{G}^{-1}\}\end{aligned} \tag{2.75}$$

where the matrix $\mathbf{G}$ is defined as

$$\mathbf{G} \triangleq (\mathbf{R}_s - \varepsilon\mathbf{I})^{-1/2}(\hat{\mathbf{R}} + \gamma\mathbf{I})(\mathbf{R}_s - \varepsilon\mathbf{I})^{-1/2} \tag{2.76}$$

It is noteworthy that even if the matrix $\mathbf{R}_s$ is singular or ill-conditioned, the matrix $\mathbf{R}_s - \varepsilon\mathbf{I}$ can be made full-rank (well-conditioned) by a proper choice of the parameter ε. Furthermore, for any nonzero ε, $\text{rank}\{\mathbf{R}_s - \varepsilon\mathbf{I}\} = M$ *almost surely*.

To develop an on-line implementation of the beamformer (2.69), let us consider the case of *rectangular* sliding window of the length N where the update of the matrix $\hat{\mathbf{R}}_{\rm dl} = \hat{\mathbf{R}} + \gamma\mathbf{I}$ in the nth step can be computed as [80]

$$\hat{\mathbf{R}}_{\rm dl}(n) = \hat{\mathbf{R}}_{\rm dl}(n-1) + \frac{1}{N}\mathbf{x}(n)\mathbf{x}^H(n) - \frac{1}{N}\mathbf{x}(n-N)\mathbf{x}^H(n-N) \tag{2.77}$$

Note that (2.77) represents the so-called rank-two update [80]. The diagonal load should be added to the initialization step of (2.77), that is, $\gamma\mathbf{I}$ should be chosen to initialize the matrix $\hat{\mathbf{R}}_{\rm dl}$. Using (2.77), we can rewrite the corresponding update of the matrix (2.76) as

$$\mathbf{G}(n) = \mathbf{G}(n-1) + \tilde{\mathbf{x}}(n)\tilde{\mathbf{x}}^H(n) - \tilde{\mathbf{x}}(n-N)\tilde{\mathbf{x}}^H(n-N) \tag{2.78}$$

where the *transformed* training snapshots are defined as

$$\tilde{\mathbf{x}}(i) = \frac{1}{\sqrt{N}}(\mathbf{R}_{\rm s} - \varepsilon\mathbf{I})^{-1/2}\mathbf{x}(i)$$

and, according to (2.76), $\gamma(\mathbf{R}_{\rm s} - \varepsilon\mathbf{I})^{-1}$ should be chosen to initialize the matrix $\mathbf{G}$.

According to equations (2.75) and (2.78), on-line algorithms for updating the weight vector $\mathbf{w}_{\rm rob}$ should be based on combining the matrix inversion lemma and some subspace tracking algorithm to track the principal eigenvector of the matrix $\mathbf{G}^{-1}$. Any of subspace tracking algorithms available in the literature can be used for this purpose [80, 81]. As the complexities of the existing subspace tracking techniques lie between $O(M)$ and $O(M^2)$ per step, the total complexity of this on-line implementation of the robust MVDR beamformer (2.69) is $O(M^2)$ per step [59]. This conclusion can be made because, regardless of the complexity of the subspace tracking algorithm used, $O(M^2)$ operations per step are required to update the weight vector (2.75).

Further extensions of the worst-case approach of [59] to the robust blind multi-user detection problem can be found in [82].

2.4 NUMERICAL EXAMPLES

In all numerical examples, we assume a uniform linear array (ULA) of $M = 20$ omnidirectional sensors spaced half-wavelength apart. All the results are averaged over 100 simulation runs. Throughout all examples, we assume that there is one desired and one interfering source. The desired signal is assumed to be always present in the training data cell and the interference-to-noise ratio (INR) is equal to 20 dB. We compare the performances of the benchmark SMI beamformer, conventional SMI beamformer, LSMI beamformer with fixed diagonal loading, and our robust MVDR beamformers (2.40) and (2.69) with adaptive diagonal loading (these techniques are referred to as the *rank-one* and *general-rank* robust beamformers, respectively). Note that the *benchmark* SMI beamformer corresponds to the ideal case when the matrix $\mathbf{R}_{\rm s}$ in (2.18) is known exactly. This algorithm does not correspond to any real situation and is included in our simulations for the sake of comparison only. All other beamformers tested use a mismatched covariance matrix (or steering vector) of the desired signal. Following [59], the diagonal loading parameter $\gamma = 30$ is chosen for the LSMI algorithm (2.24) and our robust algorithm (2.69) in all examples.

Additionally, the optimal SINR curve is displayed in each figure.

In our first example, we consider a point source scenario where $\text{rank}\{\tilde{\mathbf{R}}_s\} = \text{rank}\{\mathbf{R}_s\} = 1$. Both the desired signal and interferer are assumed to be plane waves impinging on the array from the directions 20° and -20°, respectively, while the presumed signal direction is equal to 22°. That is, there is the 2° signal look direction mismatch in this scenario.

Figure 2.2 displays the output SINRs of the beamformers tested versus N for $\text{SNR} = 0\,\text{dB}$. The SINRs of the same beamformers are shown in Figure 2.3 versus SNR for $N = 100$. The parameters $\epsilon = 4$ and $\varepsilon = 16$ are chosen for the robust beamformers (2.40) and (2.69), respectively.[2]

In the second example, again a point source scenario is considered where the steering vector of the desired signal and interferer are plane wavefronts impinging on the array from 30° and -30°, respectively, and are additionally distorted in phase. For both wavefronts and in each run, these phase distortions have been independently and randomly drawn from a Gaussian random generator with zero mean and the variance of 0.2. Note that the distortions change from run to run but remain fixed from snapshot to snapshot. The presumed signal steering vector does not take into account any distortions, that is, it corresponds to a plane wave with the DOA of 30°. This example models the case of coherent scattering, imperfectly calibrated array, or wavefront perturbation in an inhomogeneous medium [60]. In wireless communications, such scenario may be used to model the case of spatial signature estimation errors caused by a limited amount of pilot symbols.

Figure 2.4 displays the output SINRs of the beamformers tested versus N for the fixed $\text{SNR} = 0\,\text{dB}$ in the second example. The performance of the same methods versus the SNR for the fixed training data size $N = 100$ is shown in Figure 2.5.

In the third example, a scenario with non-point full-rank sources is considered. In this example, we assume locally incoherently scattered desired signal and interferer with Gaussian and uniform angular power densities characterized by the central angles of 30° and -30°, respectively. Each of these sources is assumed to have the same angular spread equal to 4°. The presumed signal covariance matrix, however, ignores local scattering effects and corresponds to the case of a point (rank-one) plane wavefront source with the DOA of 32°. The parameters $\epsilon = 3$ and $\varepsilon = 9$ are chosen in this example.

Figure 2.6 shows the performances of the methods tested versus N for the fixed $\text{SNR} = 0\,\text{dB}$. The performance of the same methods versus the SNR for the fixed training data size $N = 100$ is displayed in Figure 2.7.

[2]Note that the choice of $\epsilon = 4$ is consistent to the choice of $\varepsilon = 16$ because ϵ is related to the Euclidean norm of the signal steering vector mismatch, whereas ε is related to the Frobenius norm of the signal covariance matrix mismatch.

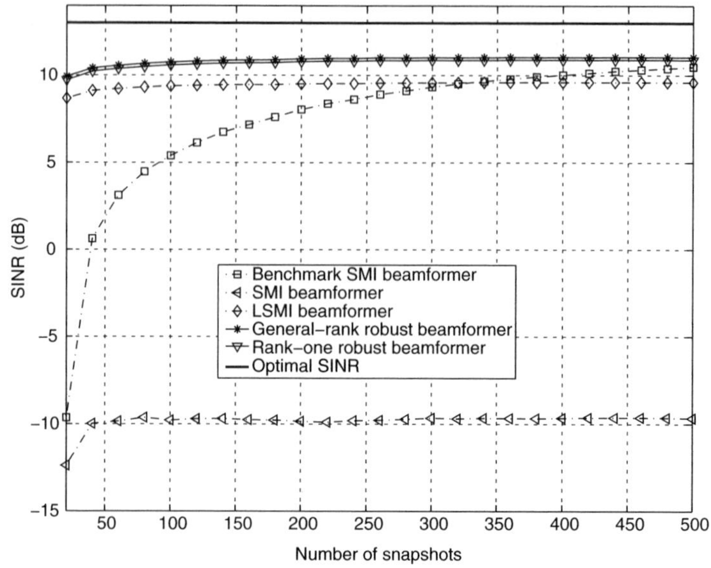

Figure 2.2 Output SINRs versus N; first example.

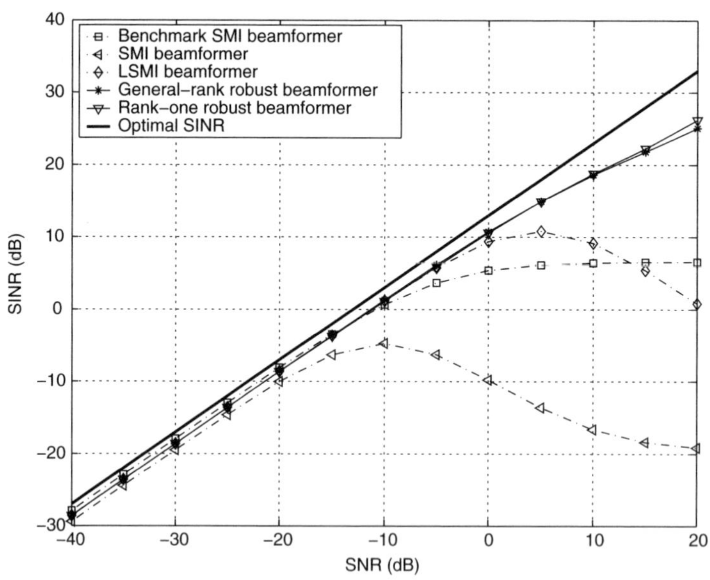

Figure 2.3 Output SINRs versus SNR; first example.

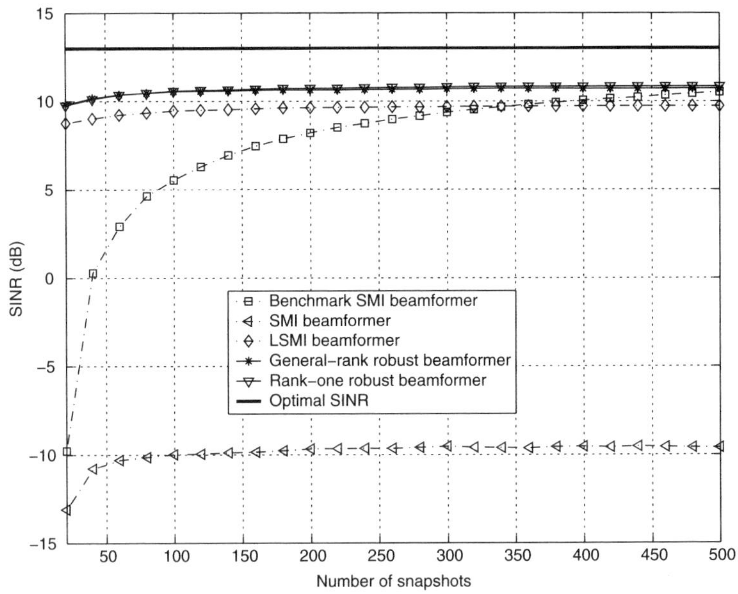

Figure 2.4 Output SINRs versus N; second example.

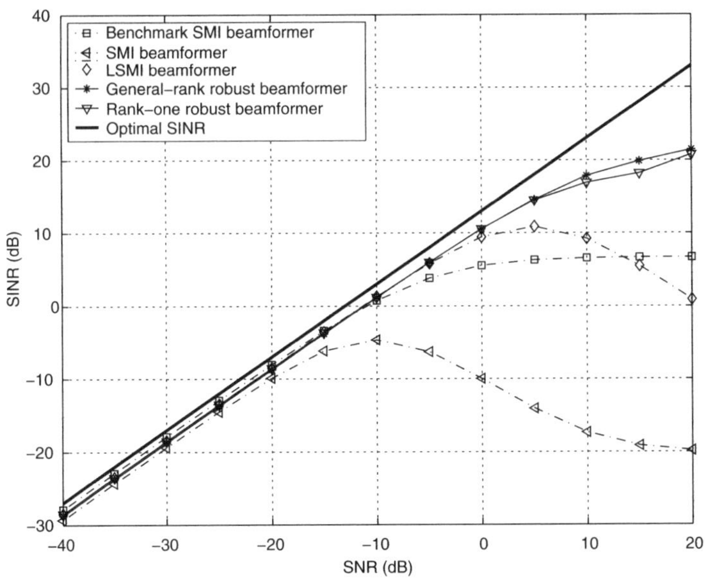

Figure 2.5 Output SINRs versus SNR; second example.

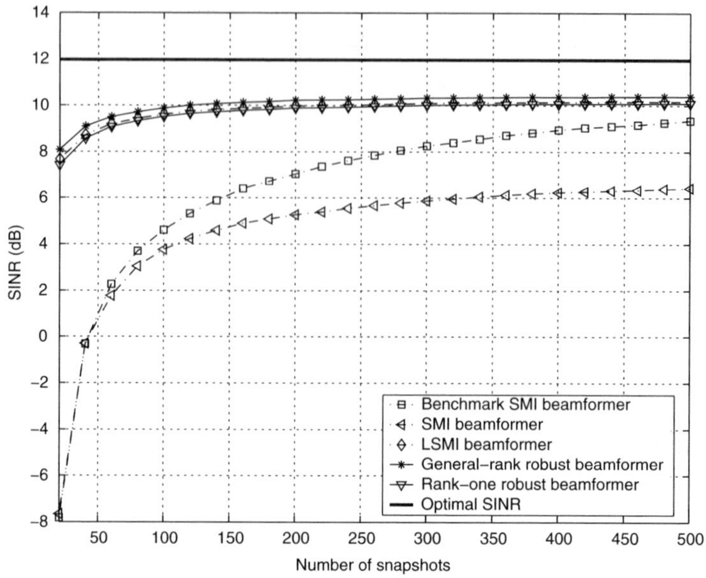

Figure 2.6 Output SINRs versus N; third example.

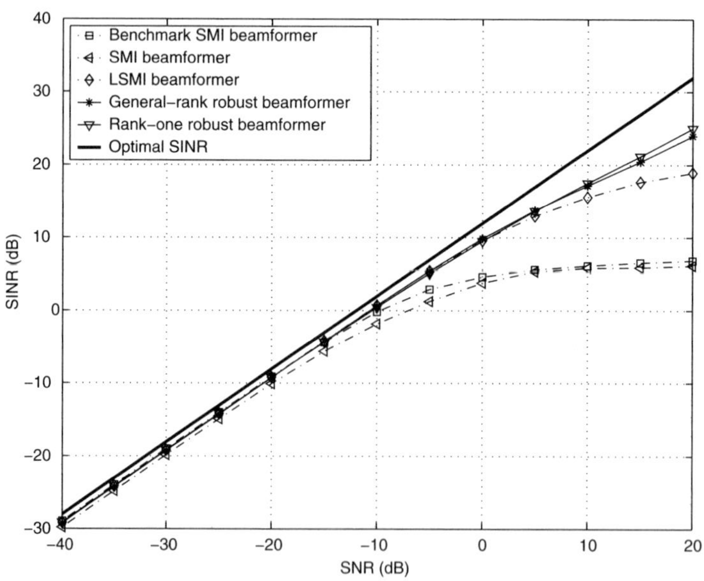

Figure 2.7 Output SINRs versus SNR; third example.

Similar to the third example, in our last example we assume a scenario with nonpoint full-rank sources. We model incoherently scattered desired signal and interferer with the Gaussian and uniform angular power densities and the central angles of 20° and -20°, respectively. Each of these sources is assumed to have the same angular spread equal to 4°. In contrast to the previous example, the presumed covariance matrix is also full rank and corresponds to a Gaussian incoherently distributed source with the central angle of 22° and angular spread of 6°. That is, there is a signal mismatch both in the central angle and angular spread. In this example, $\varepsilon = 9$ is taken (note that the rank-one robust MVDR beamformer (2.40) is not applicable to this example and its performance is not shown).

Figure 2.8 depicts the performance of the methods tested versus N for the fixed SNR $= 0$ dB. The performance of these methods versus the SNR for the fixed training data size $N = 100$ is shown in Figure 2.9.

2.4.1 Discussion

Figures 2.2–2.9 clearly demonstrate that in all our simulation examples, the robust MVDR beamformers (2.40) and (2.69) consistently outperform the other beamformers tested and achieve the SINR that is close to the optimal one for all tested values of SNR and N. This conclusion holds true for both the rank-one and full-rank signal scenarios considered in our examples and shows that the performance losses remain small compared to the ideal (nonmismatched) case.

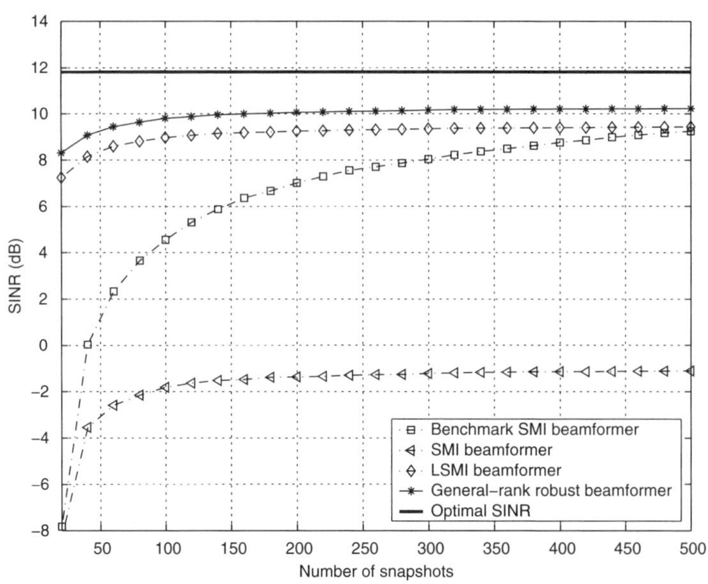

Figure 2.8 Output SINRs versus N; fourth example.

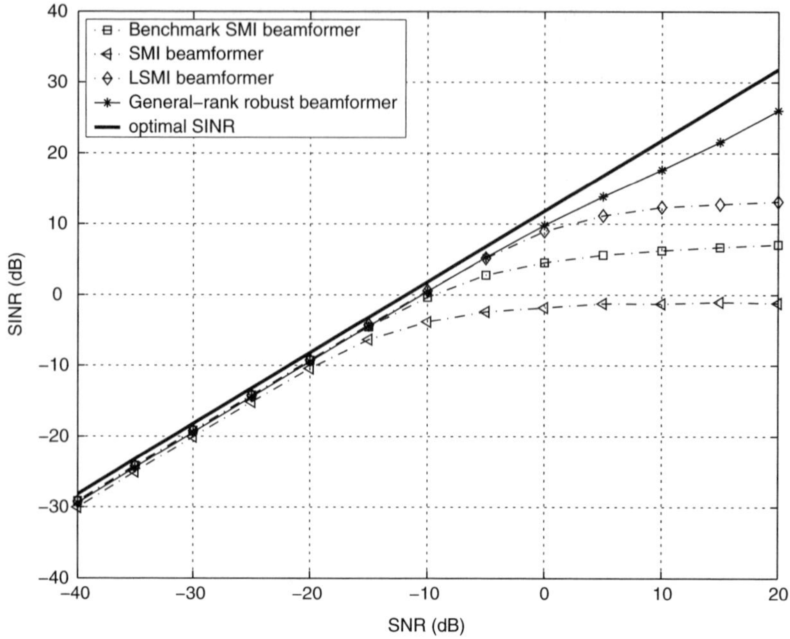

Figure 2.9 Output SINRs versus SNR; fourth example.

In all examples where both the beamformers (2.40) and (2.69) are tested, their performance can be observed to be nearly identical. However, in all examples these robust MVDR techniques outperform the SMI and LSMI beamformers. These performance improvements are especially pronounced at high SNRs.

Interestingly, the robust MVDR beamformers (2.40) and (2.69) not only substantially outperform the SMI and LSMI beamformers, but also perform better than the benchmark SMI beamformer. This can be explained by the fact that, although the benchmark SMI beamformer perfectly knows the signal covariance matrix $\mathbf{R}_s$, it exploits the sample estimate $\hat{\mathbf{R}}$ of the interference-plus-noise covariance matrix and, because of this, it suffers from severe signal self-nulling.

2.5 CONCLUSIONS

This chapter has provided an overview of the main advances in the area of robust adaptive beamforming. After reviewing the required types of robustness and known ad hoc solutions, a recently emerged rigorous approach to robust adaptive beamforming based on the worst-case performance optimization has been addressed in detail. This approach greatly improves the robustness of traditional minimum variance beamformers in the presence of various types of unknown mismatches and nonidealities. Both the rank-one and general-rank signal cases have been investigated in detail. Several state-of-the-art robust MVDR beamformers that are able to achieve different robustness tradeoffs have been introduced and studied in these cases. These algorithms include both closed-form solutions and convex optimization-based techniques which can be efficiently implemented using modern convex optimization algorithms and software and whose order of computational complexity is similar to that of the traditional SMI and LSMI adaptive beamformers.

APPENDIX 2.A: Proof of Lemma 1

Using the triangle and Cauchy-Schwarz inequalities along with the inequality (2.32) yields

$$|\mathbf{w}^H\mathbf{a}_\mathrm{s} + \mathbf{w}^H\boldsymbol{\delta}| \geq |\mathbf{w}^H\mathbf{a}_\mathrm{s}| - |\mathbf{w}^H\boldsymbol{\delta}| \geq |\mathbf{w}^H\mathbf{a}_\mathrm{s}| - \epsilon\|\mathbf{w}\| \tag{A.1}$$

Also, it can be readily verified that

$$|\mathbf{w}^H\mathbf{a}_\mathrm{s} + \mathbf{w}^H\boldsymbol{\delta}| = |\mathbf{w}^H\mathbf{a}_\mathrm{s}| - \epsilon\|\mathbf{w}\| \tag{A.2}$$

if $|\mathbf{w}^H\mathbf{a}_\mathrm{s}| > \epsilon\|\mathbf{w}\|$ and if

$$\boldsymbol{\delta} = -\frac{\mathbf{w}}{\|\mathbf{w}\|}\,\epsilon\, e^{j\phi}$$

where

$$\phi = \mathrm{angle}\left\{\mathbf{w}^H\mathbf{a}_\mathrm{s}\right\}$$

Combining (A.1) and (A.2), we prove the lemma.

APPENDIX 2.B: Proof of Lemma 2

We first show that if $\epsilon < \|\mathbf{a}_\mathrm{s}\|$ then the solution of $f(\tau) = 0$ is a positive value. To show this, we note that

$$\begin{aligned} f(0) &= \frac{\sum_{i=1}^{M} |g_i|^2}{\epsilon^2} - 1 \\ &= \frac{\|\mathbf{g}\|^2}{\epsilon^2} - 1 \\ &= \frac{\|\mathbf{a}_\mathrm{s}\|^2}{\epsilon^2} - 1 \end{aligned} \tag{B.1}$$

where in the last row of (B.1) we have used the equation $\|\mathbf{g}\| = \|\mathbf{a}_\mathrm{s}\|$ which follows from (2.50) and the fact that the matrix $\mathbf{U}$ is unitary. If $\epsilon < \|\mathbf{a}_\mathrm{s}\|$, then from (B.1) it is clear that $f(0) > 0$. On the other hand, according to (2.52), $f(+\infty) = -1$ and, since $f(\tau)$ is continuous for positive values of τ, it has a root in the interval $(0, +\infty)$. This completes the proof of the sufficiency part of Lemma 2.

The necessity of the condition $\epsilon < \|\mathbf{a}_s\|$ for $f(\tau) = 0$ to have a positive solution can be proved by contradiction. Assume that the equation $f(\tau) = 0$ has a positive solution while $\epsilon \geq \|\mathbf{a}_s\|$. Since τ and $\{\xi_i\}_{i=1}^M$ are all positive, using the definition of $f(\tau)$ in (2.52), we conclude that for any positive τ

$$\begin{aligned} f(\tau) &< \frac{\sum_{i=1}^M |g_i|^2}{\epsilon^2} - 1 \\ &= \frac{\|\mathbf{g}\|^2}{\epsilon^2} - 1 \\ &= \frac{\|\mathbf{a}_s\|^2}{\epsilon^2} - 1 \end{aligned} \tag{B.2}$$

If $\epsilon \geq \|\mathbf{a}_s\|$, it follows from (B.2) that $f(\tau) < 0$ for all positive values of τ. This is an obvious contradiction to the assumption that $f(\tau)$ is zero for some positive τ. The necessity part of Lemma 2 is proven.

The proof of uniqueness is as follows. Assume that τ_1 and τ_2 are two positive values of τ such that $f(\tau_1) = f(\tau_2)$. Then, using (2.52), we can write

$$\sum_{i=1}^M \left(\frac{|g_i|}{\epsilon + \tau_1 \xi_i}\right)^2 - \sum_{i=1}^M \left(\frac{|g_i|}{\epsilon + \tau_2 \xi_i}\right)^2 = 0$$

which means that

$$(\tau_2 - \tau_1) \sum_{i=1}^M \frac{|g_i|^2 \xi_i [2\epsilon + \xi_i(\tau_2 + \tau_1)]}{[\epsilon + \tau_1 \xi_i]^2 [\epsilon + \tau_2 \xi_i]^2} = 0$$

where, because of the positiveness of τ_1, τ_2 and ξ_i $(i = 1, \ldots, M)$,

$$\sum_{i=1}^M \frac{|g_i|^2 \xi_i [2\epsilon + \xi_i(\tau_2 + \tau_1)]}{[\epsilon + \tau_1 \xi_i]^2 [\epsilon + \tau_2 \xi_i]^2} > 0$$

This means that $\tau_1 = \tau_2$ and, therefore, the solution to $f(\tau) = 0$ is unique. With this statement, the proof of Lemma 2 is complete.

APPENDIX 2.C: Proof of Lemma 3

Let us introduce

$$f(\mathbf{w}) \triangleq \max_{\|\boldsymbol{\Delta}\|_{\mathcal{F}} \leq \eta} \|(\mathbf{X} + \boldsymbol{\Delta})^H \mathbf{w}\|$$

First of all, we will show that

$$f(\mathbf{w}) \leq \|\mathbf{X}^H\mathbf{w}\| + \eta\|\mathbf{w}\| \tag{C.1}$$

For any matrix $\mathbf{\Delta}$, we have that $\|\mathbf{\Delta}\| \leq \|\mathbf{\Delta}\|_{\mathcal{F}}$ (recall here that $\|\cdot\|$ denotes the matrix 2-norm). Therefore, for any $\mathbf{\Delta}$, we obtain

$$\begin{aligned}
\|\mathbf{X}^H\mathbf{w} + \mathbf{\Delta}^H\mathbf{w}\| &\leq \|\mathbf{X}^H\mathbf{w}\| + \|\mathbf{\Delta}^H\mathbf{w}\| \\
&\leq \|\mathbf{X}^H\mathbf{w}\| + \|\mathbf{\Delta}\|\,\|\mathbf{w}\| \\
&\leq \|\mathbf{X}^H\mathbf{w}\| + \|\mathbf{\Delta}\|_{\mathcal{F}}\|\mathbf{w}\| \\
&\leq \|\mathbf{X}^H\mathbf{w}\| + \eta\|\mathbf{w}\|
\end{aligned}$$

and (C.1) is proved.

Next, we show that

$$f(\mathbf{w}) \geq \|\mathbf{X}^H\mathbf{w}\| + \eta\|\mathbf{w}\| \tag{C.2}$$

Introducing

$$\mathbf{\Delta}_* \triangleq \frac{\eta\mathbf{w}\mathbf{w}^H\mathbf{X}}{\|\mathbf{w}\|\,\|\mathbf{X}^H\mathbf{w}\|}$$

and using the property

$$\|\mathbf{\Delta}_*\|_{\mathcal{F}}^2 = \text{trace}\{\mathbf{\Delta}_*^H\mathbf{\Delta}_*\}$$

it is easy to verify that $\|\mathbf{\Delta}_*\|_{\mathcal{F}} = \eta$. Therefore,

$$\begin{aligned}
f(\mathbf{w}) &= \max_{\|\mathbf{\Delta}\|_{\mathcal{F}}\leq\eta} \|(\mathbf{X}+\mathbf{\Delta})^H\mathbf{w}\| \\
&\geq \|(\mathbf{X}+\mathbf{\Delta}_*)^H\mathbf{w}\| \\
&= \left\|\mathbf{X}^H\mathbf{w} + \frac{\eta\mathbf{X}^H\mathbf{w}\mathbf{w}^H}{\|\mathbf{w}\|\,\|\mathbf{X}^H\mathbf{w}\|}\mathbf{w}\right\| \\
&= \left\|\mathbf{X}^H\mathbf{w} + \frac{\eta\|\mathbf{w}\|}{\|\mathbf{X}^H\mathbf{w}\|}\mathbf{X}^H\mathbf{w}\right\| \\
&= \|\mathbf{X}^H\mathbf{w}\| + \eta\|\mathbf{w}\|
\end{aligned} \tag{C.3}$$

With (C.3), equation (C.2) is proved. Comparing (C.1) and (C.2), we finally prove Lemma 3.

APPENDIX 2.D: Proof of Lemma 4

Let us solve the following constrained optimization problems

$$\min_{\mathbf{\Delta}_1} \mathbf{w}^H (\mathbf{R}_s + \mathbf{\Delta}_1)\mathbf{w} \qquad \text{subject to } \|\mathbf{\Delta}_1\|_{\mathcal{F}} \leq \varepsilon \tag{D.1}$$

$$\max_{\mathbf{\Delta}_2} \mathbf{w}^H (\mathbf{R}_{\text{i+n}} + \mathbf{\Delta}_2)\mathbf{w} \qquad \text{subject to } \|\mathbf{\Delta}_2\|_{\mathcal{F}} \leq \gamma \tag{D.2}$$

We observe that the objective functions in (D.1) and (D.2) are linear because they are minimized (or maximized) with respect to $\mathbf{\Delta}_1$ (or $\mathbf{\Delta}_2$) rather than $\mathbf{w}$. From the linearity of these objective functions, it follows that the inequality constraints in (D.1) and (D.2) are satisfied with equality. Therefore, the solutions to (D.1) and (D.2) can be obtained using Lagrange multipliers method, by means of minimizing/maximizing the functions

$$L(\mathbf{\Delta}_1, \lambda) = \mathbf{w}^H (\mathbf{R}_s + \mathbf{\Delta}_1)\mathbf{w} + \lambda(\|\mathbf{\Delta}_1\|_{\mathcal{F}} - \varepsilon)$$

$$L(\mathbf{\Delta}_2, \tilde{\lambda}) = \mathbf{w}^H (\mathbf{R}_{\text{i+n}} + \mathbf{\Delta}_2)\mathbf{w} + \tilde{\lambda}(\|\mathbf{\Delta}_2\|_{\mathcal{F}} - \gamma)$$

respectively, where λ and $\tilde{\lambda}$ are the corresponding Lagrange multipliers. Equating the gradients $\partial L(\mathbf{\Delta}_1, \lambda)/\partial \mathbf{\Delta}_1$ and $\partial L(\mathbf{\Delta}_2, \tilde{\lambda})/\partial \mathbf{\Delta}_2$ to zero yields

$$\mathbf{\Delta}_1 = -\frac{1}{2\lambda}\mathbf{w}\mathbf{w}^H, \qquad \mathbf{\Delta}_2 = -\frac{1}{2\tilde{\lambda}}\mathbf{w}\mathbf{w}^H \tag{D.3}$$

Using $\|\mathbf{\Delta}_1\|_{\mathcal{F}} = \varepsilon$ and $\|\mathbf{\Delta}_1\|_{\mathcal{F}} = \gamma$ along with (D.3), we obtain

$$\mathbf{\Delta}_1 = -\varepsilon \frac{\mathbf{w}\mathbf{w}^H}{\|\mathbf{w}\|^2}, \qquad \mathbf{\Delta}_2 = \gamma \frac{\mathbf{w}\mathbf{w}^H}{\|\mathbf{w}\|^2} \tag{D.4}$$

where the signs in (D.4) are determined by the fact that (D.1) and (D.2) are the minimization and maximization problems, respectively. Using (D.4) yields

$$\begin{aligned}\min_{\|\mathbf{\Delta}_1\| \leq \varepsilon} \mathbf{w}^H (\mathbf{R}_s + \mathbf{\Delta}_1)\mathbf{w} &= \mathbf{w}^H \Big(\mathbf{R}_s - \varepsilon \frac{\mathbf{w}\mathbf{w}^H}{\|\mathbf{w}\|^2}\Big)\mathbf{w} \\ &= \mathbf{w}^H (\mathbf{R}_s - \varepsilon \mathbf{I})\mathbf{w}\end{aligned}$$

$$\begin{aligned}\min_{\|\mathbf{\Delta}_2\| \leq \gamma} \mathbf{w}^H (\mathbf{R}_{\text{i+n}} + \mathbf{\Delta}_2)\mathbf{w} &= \mathbf{w}^H \Big(\mathbf{R}_{\text{i+n}} + \gamma \frac{\mathbf{w}\mathbf{w}^H}{\|\mathbf{w}\|^2}\Big)\mathbf{w} \\ &= \mathbf{w}^H (\mathbf{R}_{\text{i+n}} + \gamma \mathbf{I})\mathbf{w}\end{aligned}$$

respectively, and the proof of Lemma 4 is complete.

APPENDIX 2.E: Proof of Lemma 5

Let us write the characteristic equation for the matrix $\mathbf{YZ}$ as

$$\mathbf{YZu}_i = \mu_i \mathbf{u}_i \tag{E.1}$$

where $\{\mu_i\}_{i=1}^M$ and $\{\mathbf{u}_i\}_{i=1}^M$ are the eigenvalues and corresponding eigenvectors of the matrix $\mathbf{YZ}$. Multiplying this equation by $\mathbf{Z}^{1/2}$ yields

$$\mathbf{Z}^{1/2}\mathbf{Y}\underbrace{\mathbf{Z}^{1/2}\mathbf{Z}^{1/2}}_{\mathbf{Z}}\mathbf{u}_i = \mu_i \mathbf{Z}^{1/2}\mathbf{u}_i \tag{E.2}$$

which is also the characteristic equation for the matrix $\mathbf{Z}^{1/2}\mathbf{YZ}^{1/2}$, that is

$$\mathbf{Z}^{1/2}\mathbf{YZ}^{1/2}\mathbf{v}_i = \mu_i \mathbf{v}_i \tag{E.3}$$

where the eigenvectors of the matrices $\mathbf{YZ}$ and $\mathbf{Z}^{1/2}\mathbf{YZ}^{1/2}$ are related as

$$\mathbf{v}_i = \mathbf{Z}^{1/2}\mathbf{u}_i \tag{E.4}$$

for all $i = 1, 2, \ldots, M$. Applying this result to the principal eigenvectors of the matrix $\mathbf{YZ}$ and $\mathbf{Z}^{1/2}\mathbf{YZ}^{1/2}$, we obtain (2.74) and Lemma 5 is proved.

REFERENCES

1. B. Widrow, P. E. Mantey, J. L. Grffiths, and B. B. Goode, "Adaptive antenna systems," *Proc. IEEE*, Vol. 55, pp. 2143–2159, Dec. 1967.
2. I. S. Reed, J. D. Mallett, and L. E. Brennan, "Rapid convergence rate in adaptive arrays," *IEEE Trans. Aerospace and Electron. Syst.*, Vol. 10, pp. 853–863, Nov. 1974.
3. Special issue on adaptive antennas, *IEEE Trans. Antennas and Propagation*, Vol. 24, May 1976.
4. R. A. Monzingo and T. W. Miller, *Introduction to Adaptive Arrays*, Wiley, NY, 1980.
5. B. Widrow and S. Stearns, *Adaptive Signal Processing*, Prentice Hall, Englewood Cliffs, 1985.
6. R. T. Compton, Jr., *Adaptive Antennas: Concepts and Performance*, Prentice Hall, 1988.
7. J. E. Hudson, *Adaptive Array Principles*, Peter Peregrinus Ltd., Stevenage, UK, 1981.
8. H. L. Van Trees, *Optimum Array Processing*, Wiley, NY, 2002.
9. H. Cox, "Resolving power and sensitivity to mismatch of optimum array processors," *J. Acoust. Soc. Amer.*, Vol. 54, pp. 771–758, 1973.
10. D. R. Morgan and T. M. Smith, "Coherence effects on the detection performance of quadratic array processors with application to large-array matched-field beamforming," *J. Acoust. Soc. Amer.*, Vol. 87, pp. 737–747, Feb. 1988.

11. A. B. Gershman, V. I. Turchin, and V. A. Zverev, "Experimental results of localization of moving underwater signal by adaptive beamforming," *IEEE Trans. Signal Processing*, Vol. 43, pp. 2249–2257, Oct. 1995.
12. J. L. Krolik, "The performance of matched-field beamformers with Mediterranean vertical array data," *IEEE Trans. Signal Processing*, Vol. 44, pp. 2605–2611, Oct. 1996.
13. R. J. Vaccaro et al., "The past, present, and the future of underwater acoustic signal processing," *IEEE Signal Processing Magazine*, Vol. 15, pp. 21–51, July 1998.
14. A. B. Gershman, E. Nemeth, and J. F. Böhme, "Experimental performance of adaptive beamforming in a sonar environment with a towed array and moving interfering sources," *IEEE Trans. Signal Processing*, Vol. 48, pp. 246–250, Jan. 2000.
15. L. E. Brennan, J. D. Mallett, and I. S. Reed, "Adaptive arrays in airborne MTI radar," *IEEE Trans. Antennas and Propagation*, Vol. 24, Sept. 1976.
16. S. Haykin, J. Litva, and T. Shepherd, Eds., *Radar Array Processing*, Springer-Verlag, 1992.
17. R. Klemm, *Space-Time Adaptive Processing: Principles and Applications*, IEEE Press, 1998.
18. H. D. Grifflths and P. Mancini, "Ambiguity suppression in SARs using adaptive array techniques," *Proc. Geoscience and Remote Sensing Symposium*, Vol. 2, pp. 1015–1018, June 1991.
19. R. T. Compton, Jr., et al., "Adaptive arrays for communication systems: An overview of the research at the Ohio State University," *IEEE Trans. Antennas and Propagation*, Vol. 24, pp. 599–607, Sept. 1976.
20. J. Winters, "Spread spectrum in a four-phase communication system employing adaptive antennas," *IEEE Trans. Communications*, Vol. 30, pp. 929–936, May 1982.
21. M. Ganz and R. Compton, Jr., "Protection of a narrow-band BPSK communication system with an adaptive array," *IEEE Trans. Communications*, Vol. 35, pp. 1005–1011, Oct. 1987.
22. L. C. Godara, "Application of antenna arrays to mobile communications. II. Beam-forming and direction-of-arrival considerations," *Proc. IEEE*, Vol. 85, pp. 1195–1245, Aug. 1997.
23. T. S. Rapapport, Ed., *Smart Antennas: Adaptive Arrays, Algorithms, and Wireless Position Location*, IEEE Press, 1998.
24. R. L. Fante and J. J. Vacarro, "Cancellation of jammers and jammer multipath in a GPS receiver," *IEEE Aerospace and Electronic Systems Magazine*, Vol. 13, pp. 25–28, Nov. 1998.
25. J. M. Blas et al., "GPS adaptive array for use in satellite mobile communications," *Proc. 5th Int. Conf. Satellite Systems for Mobile Comm. and Navig.*, pp. 28–31, London, UK, May 1996.
26. M. D. Zoltowski and A. S. Gecan, "Advanced adaptive null steering concepts for GPS," *Proc. MILCOM'95*, San Diego, CA, Vol. 3, pp. 1214–1218, Nov. 1995.
27. P. Mousavi et al., "Feed-reflector design for large adaptive reflector antenna (LAR)," *IEEE Trans. Antennas and Propagation*, Vol. 49, pp. 1142–1154, Aug. 2001.
28. S. W. Ellingson and G. A. Hampson, "A subspace-tracking approach to interference nulling for phased array-based radio telescopes," *IEEE Trans. Antennas and Propagation*, Vol. 50, pp. 25–30, Jan. 2002.
29. Y. Kameda and J. Ohga, "Adaptive microphone-array system for noise reduction," *IEEE Trans. Acoust., Speech, Signal Processing*, Vol. 34, pp. 1391–1400, Dec. 1986.
30. D. P. Welker et al., "Microphone-array hearing aids with binaural output. II: A two-microphone adaptive system," *IEEE Trans. Speech and Audio Processing*, Vol. 5, pp. 543–551, Nov. 1997.

31. Y. R. Zheng, R. A. Goubran, M. El-Tanany, "Experimental evaluation of a nested microphone array with adaptive noise cancellers," *IEEE Trans. Instrumentation and Measurement*, Vol. 53, pp. 777–786, June 2004.
32. J. Capon, R. J. Greenfield, and R. J. Kolker, "Multidimensional maximum-likelihood processing for a large aperture seismic array," *Proc. IEEE*, Vol. 55, pp. 192–211, Feb. 1967.
33. J. Capon, R. J. Greenfield, and R. T. Lacoss, "Long-period signal processing results for the large aperture seismic array," *Geophysics*, Vol. 34, pp. 305–329.
34. K. Sekihara, "Performance of an MEG adaptive-beamformer source reconstruction technique in the presence of additive low-rank interference," *IEEE Trans. Biomedical Engineering*, Vol. 51, pp. 90–99, Jan. 2004.
35. M. Kamran, A. Atalar, and H. Koymen, "VLSI circuits for adaptive digital beamforming in ultrasound imaging," *IEEE Trans. Medical Imaging*, Vol. 12, pp. 711–720, Dec. 1993.
36. A. B. Gershman, "Robust adaptive beamforming in sensor arrays," *AEU—Int. J. Electronics and Communications*, Vol. 53, pp. 305–314, Dec. 1999.
37. H. Cox, R. M. Zeskind, and M. H. Owen, "Robust adaptive beamforming," *IEEE Trans. Acoust., Speech, Signal Processing*, Vol. 35, pp. 1365–1376, Oct. 1987.
38. A. B. Gershman, "Robustness issues in adaptive beamforming and high-resolution direction finding," in *High-Resolution and Robust Signal Processing*, Y. Hua, A. B. Gershman, and Q. Cheng, Eds., Marcel Dekker, 2003.
39. L. J. Griffiths and C. W. Jim, "An alternative approach to linearly constrained adaptive beamforming," *IEEE Trans. Antennas and Propagation*, Vol. 30, pp. 27–34, Jan. 1982.
40. E. K. Hung and R. M. Turner, "A fast beamforming algorithm for large arrays," *IEEE Trans. Aerospace and Electron. Syst.*, Vol. 19, pp. 598–607, July 1983.
41. D. D. Feldman and L. J. Griffiths, "A projection approach to robust adaptive beamforming," *IEEE Trans. Signal Processing*, Vol. 42, pp. 867–876, Apr. 1994.
42. L. C. Godara, "The effect of phase-shift errors on the performance of an antenna-array beamformer," *IEEE J. Ocean. Eng.*, Vol. 10, pp. 278–284, July 1985.
43. L. C. Godara, "Error analysis of the optimal antenna array processors," *IEEE Trans. Aerospace and Electron. Syst.*, Vol. 22, pp. 395–409, July 1986.
44. J. W. Kim and C. K. Un, "An adaptive array robust to beam pointing error," *IEEE Trans. Signal Processing*, Vol. 40, pp. 1582–1584, June 1992.
45. K. L. Bell, Y. Ephraim, and H. L. Van Trees, "A Bayesian approach to robust adaptive beamforming," *IEEE Trans. Signal Processing*, Vol. 48, pp. 386–398, Feb. 2000.
46. N. K. Jablon, "Adaptive beamforming with the generalized sidelobe canceller in the presence of array imperfections," *IEEE Trans. Antennas and Propagation*, Vol. 34, pp. 996–1012, Aug. 1986.
47. A. B. Gershman, C. F. Mecklenbräuker, and J. F. Böhme, "Matrix fitting approach to direction of arrival estimation with imperfect spatial coherence of wavefronts," *IEEE Trans. Signal Processing*, Vol. 45, pp. 1894–1899, July 1997.
48. J. Ringelstein, A. B. Gershman, and J. F. Böhme, "Direction finding in random inhomogeneous media in the presence of multiplicative noise," *IEEE Signal Processing Letters*, Vol. 7, pp. 269–272, Oct. 2000.
49. A. Weiss and B. Friedlander, "Fading effects on antenna arrays in cellular communications," *IEEE Trans. Signal Processing*, Vol. 45, pp. 1109–1117, Sept. 1997.

50. Y. J. Hong, C.-C. Yeh, and D. R. Ucci, "The effect of a finite-distance signal source on a far-field steering Applebaum array – two dimensional array case," *IEEE Trans. Antennas and Propagation*, Vol. 36, pp. 468–475, Apr. 1988.

51. K. I. Pedersen, P. E. Mogensen, and B. H. Fleury, "A stochastic model of the temporal and azimuthal dispersion seen at the base station in outdoor propagation environments," *IEEE Trans. Vehicular Technology*, Vol. 49, pp. 437–447, March 2000.

52. U. Nickel, "On the influence of channel errors on array signal processing methods," *AEU – Int. J. Electronics and Communications*, Vol. 47, No. 4, pp. 209–219, 1993.

53. S. M. Kogon, "Robust adaptive beamforming for passive sonar using eigenvector/beam association and excision," in *Proc. 2nd IEEE Workshop on Sensor Array and Multi-channel Signal Processing*, Rosslyn, VA, August 2002.

54. S. D. Hayward, "Effects of motion on adaptive arrays," *IEE Proc.—Radar, Sonar and Navigation*, Vol. 144, pp. 15–20, Feb. 1997.

55. A. B. Gershman, G. V. Serebryakov, and J. F. Böhme, "Constrained Hung-Turner adaptive beamforming algorithm with additional robustness to wideband and moving jammers," *IEEE Trans. Antennas and Propagation*, Vol. 44, No. 3, pp. 361–367, March 1996.

56. A. B. Gershman, U. Nickel, and J. F. Böhme, "Adaptive beamforming algorithms with robustness against jammer motion," *IEEE Trans. Signal Processing*, Vol. 45, pp. 1878–1885, July 1997.

57. J. Riba, J. Goldberg, and G. Vazquez, "Robust beamforming for interference rejection in mobile communications," *IEEE Trans. Signal Processing*, Vol. 45, pp. 271–275, Jan. 1997.

58. J. R. Guerci, "Theory and application of covariance matrix tapers to robust adaptive beamforming," *IEEE Trans. Signal Processing*, Vol. 47, pp. 977–985, Apr. 2000.

59. S. Shahbazpanahi, A. B. Gershman, Z.-Q. Luo, and K. M. Wong, "Robust adaptive beamforming for general-rank signal models," *IEEE Trans. Signal Processing*, Vol. 51, pp. 2257–2269, Sept. 2003.

60. S. Vorobyov, A. B. Gershman, and Z.-Q. Luo, "Robust adaptive beamforming using worst-case performance optimization: A solution to the signal mismatch problem," *IEEE Trans. Signal Processing*, Vol. 51, pp. 313–324, Feb. 2003.

61. S. Vorobyov, A. B. Gershman, and Z.-Q. Luo, and N. Ma, "Adaptive beamforming with joint robustness against mismatched signal steering vector and interference nonstationarity," *IEEE Signal Processing Letters*, Vol. 11, pp. 108–111, Feb. 2004.

62. R. Lorenz and S. P. Boyd, "Robust minimum variance beamforming," *IEEE Trans. Signal Processing*, Vol. 53, pp. 1684–1696, Jan. 2005 (also see *Proc. 37th Asilomar Conf. on Signals, Systems, and Comp.*, Nov. 2003, Pacific Grove, CA).

63. O. Besson and P. Stoica, "Decoupled estimation of DOA and angular spread for a spatially distributed source," *IEEE Trans. Signal Processing*, Vol. 48, pp. 1872–1882, July 2000.

64. S. Shahbazpanahi, S. Valaee, and A. B. Gershman, "A covariance fitting approach to parametric localization of multiple incoherently distributed sources," *IEEE Trans. Signal Processing*, Vol. 52, pp. 592–600, March 2004.

65. O. Besson, F. Vincent, P. Stoica, and A. B. Gershman, "Maximum likelihood estimation for array processing in multiplicative noise environments," *IEEE Trans. Signal Processing*, Vol. 48, pp. 2506–2518, Sept. 2000.

66. B. D. Carlson, "Covariance matrix estimation errors and diagonal loading in adaptive arrays," *IEEE Trans. Aerospace and Electron. Syst.*, Vol. 24, pp. 397–401, July 1988.
67. W. F. Gabriel, "Spectral analysis and adaptive array superresolution techniques," *Proc. IEEE*, Vol. 68, pp. 654–666, June 1980.
68. Y. I. Abramovich, "Controlled method for adaptive optimization of filters using the criterion of maximum SNR," *Radio Engineering and Electronic Physics*, Vol. 26, pp. 87–95, March 1981.
69. J. Li, P. Stoica, and Z. Wang, "On robust Capon beamforming and diagonal loading," *IEEE Trans. Signal Processing*, Vol. 51, pp. 1702–1715, July 2003 (also see *IEEE Signal Processing Letters*, Vol. 10, pp. 172–175, June 2003).
70. L. Chang and C. C. Yeh, "Performance of DMI and eigenspace-based beamformers," *IEEE Trans. Antennas and Propagation*, Vol. 40, pp. 1336–1347, Nov. 1992.
71. R. J. Mailloux, "Covariance matrix augmentation to produce adaptive array pattern troughs," *IEE Electronics Letters*, Vol. 31, No. 10, pp. 771–772, May 1995.
72. M. A. Zatman, "Production of adaptive array troughs by dispersion synthesis," *IEE Electronics Letters*, Vol. 31, No. 25, pp. 2141–2142, Dec. 1995.
73. M. A. Zatman, Comment on "Theory and application of covariance matrix tapers for robust adaptive beamforming," *IEEE Trans. Signal Processing*, Vol. 48, pp. 1796–1800, June 2000.
74. Yu. Nesterov and A. Nemirovsky, *Interior Point Polynomial Algorithms in Convex Programming*, Society for Industrial and Applied Mathematics, Philadelphia, 1994.
75. J. F. Sturm, "Using SeDuMi 1.02, a MATLAB toolbox for optimization over symmetric cones," *Optim. Meth. Software*, Vol. 11–12, pp. 625–653, Aug. 1999.
76. K. Zarifi, S. Shahbazpanahi, A. B. Gershman, and Z.-Q. Luo, "Robust blind multiuser detection based on the worst-case performance optimization of the MMSE receiver," *IEEE Trans. Signal Processing*, Vol. 53, pp. 295–305, Jan. 2005 (also see *Proc. ICASSP'04*, May 2004, Montreal, Canada).
77. Y. Ye, "Combining binary search and Newton's method to compute real roots for a class of real functions," *Journal of Complexity*, Vol. 10, pp. 271–280, Sept. 1994.
78. J. Li, P. Stoica, and Z. Wang, "Doubly constrained robust Capon beamformer," *IEEE Trans. Signal Processing*, Vol. 52, pp. 2407–2423, Sept. 2004.
79. A. El-Keyi, T. Kirubarajan, and A. B. Gershman, "Robust adaptive beamforming based on the Kalman filter," *IEEE Trans. Signal Processing*, to appear August 2005.
80. K.-B. Yu, "Recursive updating of eigenvalue decomposition of a covariance matrix," *IEEE Trans. Signal Processing*, Vol. 39, pp. 1136–1145, May 1991.
81. B. Yang, "Projection approximation subspace tracking," *IEEE Trans. Signal Processing*, Vol. 44, pp. 95–107, 1995.
82. S. Shahbazpanahi and A. B. Gershman, "Robust blind multiuser detection for synchronous CDMA systems using worst-case performance optimization," *IEEE Trans. Wireless Communications*, Vol. 3, pp. 2232–2245, Nov. 2004 (also see *Proc. ICASSP'03*, May 2003, Hong Kong, China).

3

ROBUST CAPON BEAMFORMING

Jian Li and Zhisong Wang
Department of Electrical and Computer Engineering,
University of Florida, Gainesville, FL 32611

Petre Stoica
Department of Information Technology,
Uppsala University, Uppsala, Sweden

3.1 INTRODUCTION

Beamforming is a ubiquitous task in array signal processing with applications, among others, in radar, sonar, acoustics, astronomy, seismology, communications, and medical imaging. The standard data-independent beamformers include the delay-and-sum approach as well as methods based on various data-independent weight vectors for sidelobe control [1, 2]. The data-dependent Capon beamformer adaptively selects the weight vector to minimize the array output power subject to the linear constraint that the signal-of-interest (SOI) does not suffer from any distortion [3, 4]. The Capon beamformer has better resolution and much better interference rejection capability than the data-independent beamformer, provided that the array steering vector corresponding to the SOI is accurately known.

However, the knowledge of the SOI steering vector can be imprecise, which is often the case in practice due to differences between the assumed signal arrival angle and the true arrival angle or between the assumed array response and the true array response (array calibration errors). Whenever this happens, the Capon beamformer may suppress the SOI as an interference, which results in significantly underestimated SOI power and drastically reduced array output signal-to-interference-plus-noise ratio (SINR). Then the performance of the Capon beamformer may become worse than that of the standard beamformers [5, 6].

Robust Adaptive Beamforming, Edited by Jian Li and Petre Stoica

The same happens when the number of snapshots is relatively small (i.e., about the same as or smaller than the number of sensors). In fact, there is a close relationship between the cases of steering vector errors and small-sample errors (see, e.g., [7]) in the sense that the difference between the sample covariance matrix $\hat{\mathbf{R}}$ (estimated from a finite number of snapshots) and the corresponding theoretical (ensemble) covariance matrix $\mathbf{R}$ can be viewed as due to steering vector errors.

Many approaches have been proposed during the past three decades to improve the robustness of the Capon beamformer and the literature on robust adaptive beamforming is extensive (see, e.g., [2, 8–17] and the many references therein). Among these robust approaches, diagonal loading (including its extended versions) has been a popular and widely used approach to improve the robustness of the Capon beamformer (see, e.g., [18–28] and the references therein for more early suggested methods). One representative of the diagonal loading based approaches is the norm constrained Capon beamformer (NCCB), which uses a norm constraint on the weight vector to improve the robustness against array steering vector errors and control the white noise gain [18–22]. However, for NCCB and most other diagonal loading methods, it is not clear how to choose the diagonal loading level based on information about the uncertainty of the array steering vector.

Only recently have some methods with a clear theoretical background been proposed (see, e.g., [14–17, 29–31] and the first two chapters of this book) which, unlike the early methods, make explicit use of an uncertainty set of the array steering vector. In [29], a polyhedron is used to describe the uncertainty set, whereas spherical and ellipsoidal (including flat ellipsoidal) uncertainty sets are considered in [14–17, 30]. The approaches presented in [14, 15] coupled the spatial filtering formulation of the standard Capon beamformer (SCB) in [3] with a spherical or ellipsoidal uncertainty set of the array steering vector whereas we coupled the covariance fitting formulation of SCB in [32] with an ellipsoidal or spherical uncertainty set to obtain a robust Capon beamformer (RCB) [16, 30] and a doubly constrained robust Capon beamformer (DCRCB) in [17].

Interestingly, the methods in [14–16, 30] turn out to be equivalent and to belong to the extended class of diagonal loading approaches, but the corresponding amount of diagonal loading can be calculated precisely based on the ellipsoidal uncertainty set of the array steering vector. However, our RCB in [16] is simpler and computationally more efficient than its equivalent counterparts and its computational complexity is comparable to that of SCB. Moreover, our RCB gives a simple way of eliminating the scaling ambiguity when estimating the power of the desired signal, while the approaches in [14, 15] did not consider the scaling ambiguity problem.

DCRCB is associated with RCB in that they both result from solving the same problem, which involves a natural extension of the covariance fitting formulation of SCB, to the case of uncertain steering vectors by enforcing a double constraint on the steering vector, namely, a constant norm constraint and a spherical uncertainty set constraint. DCRCB provides an exact solution to the aforementioned constrained optimization problem, which is not convex, while RCB yields an approximate solution by first solving a convex optimization problem without the norm constraint and then imposing the norm constraint by possibly violating the uncertainty set constraint. In terms of the computational load, both RCB and

DCRCB can be efficiently computed at a comparable cost with that of SCB. In terms of performance, numerical examples have demonstrated that, for a reasonably tight spherical uncertainty set of the array steering vector, DCRCB is the preferred choice for applications requiring high SINR, while RCB is the favored one for applications demanding accurate signal power estimation [17].

The main purpose of the chapter is to provide a comprehensive review of our recently proposed robust adaptive beamformers including RCB and DCRCB, with additional discussions on several other related beamformers such as SCB and NCCB. We will also present the applications of these beamformers to various fields. In particular, we introduce constant-beamwidth and constant-powerwidth RCB which are suitable for acoustic imaging; we develop a rank-deficient robust Capon filter-bank spectral estimator for spectral estimation and radar imaging; we also apply the rank-deficient RCB to forward-looking ground penetrating radar (FLGPR) imaging systems for landmine detection. For acoustic imaging, we show that by choosing a frequency-dependent uncertainty set for the steering vector or by combining RCB with a shading scheme, we can achieve consistent sound pressure level (SPL) estimation across the frequency bins. For spectral estimation, we show that by allowing the sample covariance matrix to be rank-deficient, RCB can provide much higher resolution than most existing approaches, which is useful in many applications including radar target detection and feature extraction. For FLGPR imaging, the rank-deficient RCB can be applied to the practical scenarios where the number of snapshots is smaller than the number of sensors in the array and, at the same time, can provide better resolution and much better interference and clutter rejection capability than the standard delay-and-sum (DAS) based imaging method. More applications and analyses on RCB can be found in [33–36].

The chapter is organized as follows. In Section 3.2, we formulate the problem of interest. In Section 3.3, we present two equivalent formulations of the standard Capon beamformer, namely the spatial filtering SCB and the covariance fitting SCB. Section 3.4 is devoted to the RCB algorithms for two cases, that is, nondegenerate ellipsoidal constraints and flat ellipsoidal constraints on the steering vector. In Section 3.5, we provide a complete and thorough analysis of NCCB, which sheds more light on the choice of the norm constraint than what was commonly known. In Section 3.6, we present the DCRCB algorithm and explain how to choose the smallest spherical uncertainty set for the SOI steering vector. We also provide a diagonal loading interpretation of NCCB, RCB and DCRCB. Constant-powerwidth RCB (CPRCB) and constant-beamwidth RCB (CBRCB) for consistent acoustic imaging are treated in Section 3.7. In Section 3.8 we develop the rank-deficient robust Capon filter-bank spectral estimator. In Section 3.9 we use the rank-deficient RCB in two FLGPR imaging systems for landmine detection. Finally, Section 3.10 summarizes the chapter.

3.2 PROBLEM FORMULATION

Consider an array comprising M sensors and let $\mathbf{R}$ denote the theoretical covariance matrix of the array output vector. We assume that $\mathbf{R} > 0$ (positive definite) has the following form:

$$\mathbf{R} = \sigma_0^2 \mathbf{a}_0 \mathbf{a}_0^* + \sum_{k=1}^{K} \sigma_k^2 \mathbf{a}_k \mathbf{a}_k^* + \mathbf{Q} \tag{3.1}$$

where $(\sigma_0^2, \{\sigma_k^2\}_{k=1}^K)$ are the powers of the $(K+1)$ uncorrelated signals impinging on the array, $(\mathbf{a}_0, \{\mathbf{a}_k\}_{k=1}^K)$ are the so-called steering vectors that are functions of the location parameters of the sources emitting the signals [e.g., their directions of arrival (DOAs)], $(\cdot)^*$ denotes the conjugate transpose, and $\mathbf{Q}$ is the noise covariance matrix (the 'noise' comprises nondirectional signals, and hence $\mathbf{Q}$ usually has full rank as opposed to the other terms in (3.1) whose rank is equal to one). In what follows we assume that the first term in (3.1) corresponds to the SOI and the remaining rank-one terms to K interferences. To avoid ambiguities, we assume that

$$\|\mathbf{a}_0\|^2 = M \tag{3.2}$$

where $\|\cdot\|$ denotes the Euclidean norm. We note that the above expression for $\mathbf{R}$ holds for both narrowband and wideband signals; in the former case $\mathbf{R}$ is the covariance matrix at the center frequency, in the latter $\mathbf{R}$ is the covariance matrix at the center of a given frequency bin. Let

$$\mathbf{R} = \mathbf{U}\mathbf{\Gamma}\mathbf{U}^* \tag{3.3}$$

where the columns of $\mathbf{U}$ contain the eigenvectors of $\mathbf{R}$ and the diagonal elements of the diagonal matrix $\mathbf{\Gamma}$, $\gamma_1 \geq \gamma_2 \geq \cdots \geq \gamma_M$, are the corresponding eigenvalues. In practical applications, $\mathbf{R}$ is replaced by the sample covariance matrix $\hat{\mathbf{R}}$, where

$$\hat{\mathbf{R}} = \frac{1}{N} \sum_{n=1}^{N} \mathbf{y}_n \mathbf{y}_n^* \tag{3.4}$$

with N denoting the number of snapshots and $\mathbf{y}_n$ representing the nth snapshot with the form:

$$\mathbf{y}_n = \mathbf{a}_0 s_0(n) + \mathbf{e}_n \tag{3.5}$$

with $s_0(n)$ denoting the waveform of the SOI and $\mathbf{e}_n$ being the interference-plus-noise vector for the nth snapshot.

The *robust adaptive beamforming problem* we will deal with in this chapter can now be briefly stated as follows: extend the Capon beamformer so as to be able to accurately determine the power of SOI even when only an imprecise knowledge of its steering vector, $\mathbf{a}_0$, is available. More specifically, we assume that the only knowledge we have about $\mathbf{a}_0$ is that it belongs to an uncertainty ellipsoid. The nondegenerate ellipsoidal uncertainty set has the following form:

$$[\mathbf{a}_0 - \bar{\mathbf{a}}]^* \mathbf{C}^{-1} [\mathbf{a}_0 - \bar{\mathbf{a}}] \leq 1 \tag{3.6}$$

where $\bar{\mathbf{a}}$ (the assumed steering vector of SOI) and $\mathbf{C}$ (a positive definite matrix) are given. In particular, if $\mathbf{C}$ is a scaled identity matrix, that is, $\mathbf{C} = \epsilon \mathbf{I}$, we have the following uncertainty sphere:

$$\|\mathbf{a}_0 - \bar{\mathbf{a}}\|^2 \leq \epsilon \tag{3.7}$$

where ϵ is a user parameter whose choice will be discussed later on. The case of a flat ellipsoidal uncertainty set is considered in Section 3.4.2. We also assume that the steering vector $\bar{\mathbf{a}}$ satisfies the same norm constraint as $\mathbf{a}_0$ of (3.2):

$$\|\bar{\mathbf{a}}\|^2 = M. \tag{3.8}$$

The assumption that $\|\mathbf{a}_0\|^2 = M$ (there is no restriction in $\|\bar{\mathbf{a}}\|^2 = M$ since $\bar{\mathbf{a}}$ is chosen by the user) is reasonable for many scenarios including the cases of the look direction error and phase perturbations. It is violated when the array response vector also has gain perturbations. However, if the gain perturbations are small, the norm constraint still holds approximately.

In this chapter, we focus on the problem of estimating the SOI power σ_0^2 from $\mathbf{R}$ (or more practically $\hat{\mathbf{R}}$) when the knowledge of $\mathbf{a}_0$ is imprecise. However, the beamforming approaches we present herein can also be used for other applications including signal waveform estimation [14, 15, 37] (see also the first two chapters of this book).

3.3 STANDARD CAPON BEAMFORMING

We present in this section two formulations of the standard Capon beamformer, namely the spatial filtering SCB and the covariance fitting SCB, and demonstrate their equivalence.

3.3.1 Spatial Filtering SCB

The common formulation of the beamforming problem that leads to the spatial filtering form of SCB is as follows (see, e.g., [1, 3, 4]).

1. Determine the $M \times 1$ weight vector $\mathbf{w}_0$ that is the solution to the following linearly constrained quadratic problem:

$$\min_{\mathbf{w}} \mathbf{w}^*\mathbf{R}\mathbf{w} \quad \text{subject to } \mathbf{w}^*\mathbf{a}_0 = 1. \tag{3.9}$$

2. Use $\mathbf{w}_0^*\mathbf{R}\mathbf{w}_0$ as an estimate of σ_0^2.

The solution to (3.9) is easily derived:

$$\mathbf{w}_0 = \frac{\mathbf{R}^{-1}\mathbf{a}_0}{\mathbf{a}_0^*\mathbf{R}^{-1}\mathbf{a}_0}. \tag{3.10}$$

Using (3.10) in Step (2) above yields the following estimate of σ_0^2:

$$\tilde{\sigma}_0^2 = \frac{1}{\mathbf{a}_0^* \mathbf{R}^{-1} \mathbf{a}_0}. \tag{3.11}$$

Note that (3.9) can be interpreted as an adaptive *spatial filtering problem*: given $\mathbf{R}$ and $\mathbf{a}_0$ we wish to determine the weight vector $\mathbf{w}_0$ as a spatial filter that can pass the SOI without distortion and at the same time minimize the undesirable interference and noise contributions in $\mathbf{R}$.

3.3.2 Covariance Fitting SCB

The Capon beamforming problem can also be reformulated into a covariance fitting form. To describe the details of our approach, we first prove that $\tilde{\sigma}_0^2$ in (3.11) is the solution to the following problem (see also [30, 32]):

$$\max_{\sigma^2} \sigma^2 \quad \text{subject to } \mathbf{R} - \sigma^2 \mathbf{a}_0 \mathbf{a}_0^* \geq 0 \tag{3.12}$$

where the notation $\mathbf{A} \geq 0$ (for any Hermitian matrix $\mathbf{A}$) means that $\mathbf{A}$ is positive semidefinite. The previous claim follows from the following readily verified equivalences (here $\mathbf{R}^{-1/2}$ is the Hermitian square root of $\mathbf{R}^{-1}$):

$$\begin{aligned} &\mathbf{R} - \sigma^2 \mathbf{a}_0 \mathbf{a}_0^* \geq 0 \\ &\quad \Leftrightarrow \mathbf{I} - \sigma^2 \mathbf{R}^{-1/2} \mathbf{a}_0 \mathbf{a}_0^* \mathbf{R}^{-1/2} \geq 0 \\ &\quad \Leftrightarrow 1 - \sigma^2 \mathbf{a}_0^* \mathbf{R}^{-1} \mathbf{a}_0 \geq 0 \\ &\quad \Leftrightarrow \sigma^2 \leq \frac{1}{\mathbf{a}_0^* \mathbf{R}^{-1} \mathbf{a}_0} = \tilde{\sigma}_0^2. \end{aligned} \tag{3.13}$$

Hence $\sigma^2 = \tilde{\sigma}_0^2$ is indeed the largest value of σ^2 for which the constraint in (3.12) is satisfied.

Note that (3.12) can be interpreted as a *covariance fitting problem*: given $\mathbf{R}$ and $\mathbf{a}_0$ we wish to determine the largest possible SOI term, $\sigma^2 \mathbf{a}_0 \mathbf{a}_0^*$, that can be a part of $\mathbf{R}$ under the natural constraint that the residual covariance matrix is positive semidefinite.

3.4 ROBUST CAPON BEAMFORMING WITH SINGLE CONSTRAINT

The robust Capon beamformer is derived by a natural extension of the covariance fitting SCB in Section 3.3.2 to the case of uncertain steering vector. In doing so we directly obtain a robust estimate of σ_0^2, without any intermediate calculation of a vector $\mathbf{w}$ [16, 30].

In this section, we first consider the case of nondegenerate ellipsoidal constraints on the steering vector and then the case of flat ellipsoidal constraints. These two cases are treated separately due to the differences in their detailed computational steps as well as in the possible values of the associated Lagrange multipliers.

3.4.1 Nondegenerate Ellipsoidal Uncertainty Set

When the uncertainty set of the steering vector $\mathbf{a}$ is a nondegenerate ellipsoid as in (3.6), the RCB problem has the following form [16, 30]:

$$\max_{\sigma^2,\,\mathbf{a}} \sigma^2 \quad \text{subject to } \mathbf{R} - \sigma^2 \mathbf{a}\mathbf{a}^* \geq 0 \\ (\mathbf{a} - \bar{\mathbf{a}})^* \mathbf{C}^{-1} (\mathbf{a} - \bar{\mathbf{a}}) \leq 1 \tag{3.14}$$

where $\bar{\mathbf{a}}$ and $\mathbf{C}$ are given.

The RCB problem in (3.14) can be readily reformulated as a semidefinite program (SDP) [30]. Indeed, using a new variable $\chi = 1/\sigma^2$ along with the standard technique of Schur complements (see, e.g., [1, 38]) we can rewrite (3.14) as:

$$\min_{\chi,\,\mathbf{a}} \chi \quad \text{subject to} \\ \begin{bmatrix} \mathbf{R} & \mathbf{a} \\ \mathbf{a}^* & \chi \end{bmatrix} \geq 0 \\ \begin{bmatrix} \mathbf{C} & \mathbf{a} - \bar{\mathbf{a}} \\ (\mathbf{a} - \bar{\mathbf{a}})^* & 1 \end{bmatrix} \geq 0. \tag{3.15}$$

The constraints in (3.15) are so-called linear matrix inequalities, and hence (3.15) is an SDP, which requires $O(\varrho M^6)$ flops if the SeDuMi type of software [39] is used to solve it, where ϱ is the number of iterations, usually on the order of $\sqrt{M}$. However, the approach we present below only requires $O(M^3)$ flops.

For any given $\mathbf{a}$, the solution $\hat{\sigma}_0^2$ to (3.14) is indeed given by the counterpart of (3.11) with $\mathbf{a}_0$ replaced by $\mathbf{a}$, as shown in Section 3.3.2. Hence (3.14) can be reduced to the following problem

$$\min_{\mathbf{a}} \mathbf{a}^* \mathbf{R}^{-1} \mathbf{a} \quad \text{subject to } (\mathbf{a} - \bar{\mathbf{a}})^* \mathbf{C}^{-1} (\mathbf{a} - \bar{\mathbf{a}}) \leq 1. \tag{3.16}$$

To exclude the trivial solution $\mathbf{a} = \mathbf{0}$ to (3.14), we assume that

$$\bar{\mathbf{a}}^* \mathbf{C}^{-1} \bar{\mathbf{a}} > 1. \tag{3.17}$$

Note that we can decompose any matrix $\mathbf{C} > 0$ in the form:

$$\mathbf{C}^{-1} = \frac{1}{\epsilon} \mathbf{D}^* \mathbf{D} \tag{3.18}$$

where for some $\epsilon > 0$,

$$\mathbf{D} = \sqrt{\epsilon}\mathbf{C}^{-1/2}. \tag{3.19}$$

Let

$$\breve{\mathbf{a}} = \mathbf{D}\mathbf{a}, \quad \breve{\bar{\mathbf{a}}} = \mathbf{D}\bar{\mathbf{a}}, \quad \breve{\mathbf{R}} = \mathbf{D}\mathbf{R}\mathbf{D}^*. \tag{3.20}$$

Then (3.16) becomes

$$\min_{\breve{\mathbf{a}}} \breve{\mathbf{a}}^* \breve{\mathbf{R}}^{-1} \breve{\mathbf{a}} \quad \text{subject to } \|\breve{\mathbf{a}} - \breve{\bar{\mathbf{a}}}\|^2 \leq \epsilon. \tag{3.21}$$

Hence without loss of generality, we will consider solving (3.16) for $\mathbf{C} = \epsilon\mathbf{I}$, that is, solving the following quadratic optimization problem under a spherical constraint:

$$\min_{\mathbf{a}} \mathbf{a}^* \mathbf{R}^{-1} \mathbf{a} \quad \text{subject to } \|\mathbf{a} - \bar{\mathbf{a}}\|^2 \leq \epsilon. \tag{3.22}$$

To exclude the trivial solution $\mathbf{a} = \mathbf{0}$ to (3.22), we now need to assume that

$$\|\bar{\mathbf{a}}\|^2 > \epsilon. \tag{3.23}$$

Let $\bar{\mathcal{S}}$ be the set defined by the constraints in (3.22). To determine the solution to (3.22) under (3.23), we use the *Lagrange multiplier methodology* and consider the function:

$$h_1(\mathbf{a}, \nu) = \mathbf{a}^* \mathbf{R}^{-1} \mathbf{a} + \nu\left(\|\mathbf{a} - \bar{\mathbf{a}}\|^2 - \epsilon\right) \tag{3.24}$$

where $\nu \geq 0$ is the real-valued Lagrange multiplier satisfying $\mathbf{R}^{-1} + \nu\mathbf{I} > 0$ so that the above function can be minimized with respect to $\mathbf{a}$. Evidently we have $h_1(\mathbf{a}, \nu) \leq \mathbf{a}^* \mathbf{R}^{-1} \mathbf{a}$ for any $\mathbf{a} \in \bar{\mathcal{S}}$ with equality on the boundary of $\bar{\mathcal{S}}$. Equation (3.24) can be written as

$$\begin{aligned} h_1(\mathbf{a}, \nu) = & \left[\mathbf{a} - \left(\frac{\mathbf{R}^{-1}}{\nu} + \mathbf{I}\right)^{-1} \bar{\mathbf{a}}\right]^* (\mathbf{R}^{-1} + \nu\mathbf{I}) \left[\mathbf{a} - \left(\frac{\mathbf{R}^{-1}}{\nu} + \mathbf{I}\right)^{-1} \bar{\mathbf{a}}\right] \\ & - \nu^2 \bar{\mathbf{a}}^* (\mathbf{R}^{-1} + \nu\mathbf{I})^{-1} \bar{\mathbf{a}} + \nu \bar{\mathbf{a}}^* \bar{\mathbf{a}} - \nu\epsilon. \end{aligned} \tag{3.25}$$

Hence the *unconstrained* minimization of $h_1(\mathbf{a}, \nu)$ w.r.t. $\mathbf{a}$, for fixed ν, is given by

$$\hat{\mathbf{a}}_0 = \left(\frac{\mathbf{R}^{-1}}{\nu} + \mathbf{I}\right)^{-1} \bar{\mathbf{a}} \tag{3.26}$$

$$= \bar{\mathbf{a}} - (\mathbf{I} + \nu\mathbf{R})^{-1} \bar{\mathbf{a}} \tag{3.27}$$

where we have used the matrix inversion lemma [1] to obtain the second equality. Clearly, we have

$$h_2(\nu) \triangleq h_1(\hat{\mathbf{a}}_0, \nu) = -\nu^2 \bar{\mathbf{a}}^*(\mathbf{R}^{-1} + \nu \mathbf{I})^{-1}\bar{\mathbf{a}} + \nu \bar{\mathbf{a}}^* \bar{\mathbf{a}} - \nu\epsilon \tag{3.28}$$

$$\leq \mathbf{a}^* \mathbf{R}^{-1} \mathbf{a} \quad \text{for any } \mathbf{a} \in \bar{\mathcal{S}}. \tag{3.29}$$

Maximization of $h_2(\nu)$ with respect to ν gives

$$\left\| \left(\frac{\mathbf{R}^{-1}}{\nu} + \mathbf{I} \right)^{-1} \bar{\mathbf{a}} - \bar{\mathbf{a}} \right\|^2 = \epsilon \tag{3.30}$$

which indeed satisfies

$$\|\hat{\mathbf{a}}_0 - \bar{\mathbf{a}}\|^2 = \epsilon. \tag{3.31}$$

Hence $\hat{\mathbf{a}}_0$ belongs to the boundary of $\bar{\mathcal{S}}$. Therefore, $\hat{\mathbf{a}}_0$ is the sought solution. Using (3.27) in (3.31), the Lagrange multiplier $\nu \geq 0$ is then obtained as the solution to the constraint equation:

$$h_2(\nu) \triangleq \|(\mathbf{I} + \nu \mathbf{R})^{-1} \bar{\mathbf{a}}\|^2 = \epsilon. \tag{3.32}$$

Making use of $\mathbf{R} = \mathbf{U}\boldsymbol{\Gamma}\mathbf{U}^*$ and $\mathbf{z} = \mathbf{U}^*\bar{\mathbf{a}}$, (3.32) can be written as

$$h_2(\nu) = \sum_{m=1}^{M} \frac{|z_m|^2}{(1 + \nu\gamma_m)^2} = \epsilon. \tag{3.33}$$

Note that $h_2(\nu)$ is a monotonically decreasing function of $\nu \geq 0$. According to (3.23) and (3.32), $h_2(0) > \epsilon$ and hence $\nu \neq 0$. From (3.33), it is clear that $\lim_{\nu \to \infty} h_2(\nu) = 0 < \epsilon$. Hence there is a unique solution $\nu > 0$ to (3.33). By replacing the γ_m in (3.33) with γ_M and γ_1, respectively, we can obtain the following tighter upper and lower bounds on the solution $\nu > 0$ to (3.33):

$$\frac{\|\bar{\mathbf{a}}\| - \sqrt{\epsilon}}{\gamma_1 \sqrt{\epsilon}} \leq \nu \leq \frac{\|\bar{\mathbf{a}}\| - \sqrt{\epsilon}}{\gamma_M \sqrt{\epsilon}}. \tag{3.34}$$

By dropping the 1 in the denominator of (3.33), we can obtain another upper bound on the solution ν to (3.33):

$$\nu < \left(\frac{1}{\epsilon} \sum_{m=1}^{M} \frac{|z_m|^2}{\gamma_m^2} \right)^{1/2}. \tag{3.35}$$

The upper bound in (3.35) is usually tighter than the upper bound in (3.34) but not always. In summary, the solution $\nu > 0$ to (3.33) is unique and it belongs to the following interval:

$$\frac{\|\bar{\mathbf{a}}\| - \sqrt{\epsilon}}{\gamma_1 \sqrt{\epsilon}} \leq \nu \leq \min\left\{ \left(\frac{1}{\epsilon} \sum_{m=1}^{M} \frac{|z_m|^2}{\gamma_m^2} \right)^{\frac{1}{2}}, \frac{\|\bar{\mathbf{a}}\| - \sqrt{\epsilon}}{\gamma_M \sqrt{\epsilon}} \right\}. \tag{3.36}$$

Once the Lagrange multiplier ν is determined, $\hat{\mathbf{a}}_0$ is determined by using (3.27) and $\hat{\sigma}_0^2$ is computed by using (3.11) with $\mathbf{a}_0$ replaced by $\hat{\mathbf{a}}_0$. Hence the major computational demand of our RCB comes from the eigendecomposition of the Hermitian matrix $\mathbf{R}$, which requires $O(M^3)$ flops. Therefore, the computational complexity of our RCB is comparable to that of the SCB.

Next observe that both the power and the steering vector of SOI are treated as unknowns in our robust Capon beamforming formulation [see (3.14)], and hence that there is a *scaling ambiguity* in the SOI covariance term in the sense that $(\sigma^2, \mathbf{a})$ and $(\sigma^2/\alpha, \alpha^{1/2}\mathbf{a})$ (for any $\alpha > 0$) give the same term $\sigma^2\mathbf{a}\mathbf{a}^*$. To eliminate this ambiguity, we use the knowledge that $\|\mathbf{a}_0\|^2 = M$ [see (3.2)] and hence estimate σ_0^2 as [30]

$$\hat{\hat{\sigma}}_0^2 = \hat{\sigma}_0^2 \|\hat{\mathbf{a}}_0\|^2 / M \tag{3.37}$$

where $\hat{\sigma}_0^2$ is obtained via replacing $\mathbf{a}_0$ in (3.11) by $\hat{\mathbf{a}}_0$ in (3.26). The numerical examples in [30] confirm that $\hat{\hat{\sigma}}_0^2$ is a (much) more accurate estimate of σ_0^2 than $\hat{\sigma}_0^2$. To summarize, our proposed RCB approach consists of the following steps.

Step 1. Compute the eigendecomposition of $\mathbf{R}$ (or more practically of $\hat{\mathbf{R}}$).

Step 2. Solve (3.33) for ν, for example, by a Newton's method, using the knowledge that the solution is unique and it belongs to the interval in (3.36).

Step 3. Use the ν obtained in Step 2 to get

$$\hat{\mathbf{a}}_0 = \bar{\mathbf{a}} - \mathbf{U}(\mathbf{I} + \nu\boldsymbol{\Gamma})^{-1}\mathbf{U}^*\bar{\mathbf{a}} \tag{3.38}$$

where the inverse of the diagonal matrix $\mathbf{I} + \nu\boldsymbol{\Gamma}$ is easily computed. [Note that (3.38) is obtained from (3.27).]

Step 4. Compute $\hat{\sigma}_0^2$ by using

$$\hat{\sigma}_0^2 = \frac{1}{\bar{\mathbf{a}}^*\mathbf{U}\boldsymbol{\Gamma}\left(\nu^{-2}\mathbf{I} + 2\nu^{-1}\boldsymbol{\Gamma} + \boldsymbol{\Gamma}^2\right)^{-1}\mathbf{U}^*\bar{\mathbf{a}}} \tag{3.39}$$

where the inverse of $\nu^{-2}\mathbf{I} + 2\nu^{-1}\boldsymbol{\Gamma} + \boldsymbol{\Gamma}^2$ is also easily computed. Note that $\mathbf{a}_0$ in (3.11) is replaced by $\hat{\mathbf{a}}_0$ in (3.26) to obtain (3.39). Then use the $\hat{\sigma}_0^2$ in (3.37) to obtain $\hat{\hat{\sigma}}_0^2$ as the estimate of σ_0^2.

We remark that in all of the steps above, we do not need to have $\gamma_m > 0$ for all $m = 1, 2, \ldots, M$. Hence $\mathbf{R}$ or $\hat{\mathbf{R}}$ can be singular, which means that we can allow $N < M$ to compute $\hat{\mathbf{R}}$.

Our approach is different from the recent approaches in [14, 15] (see also the first two chapters of this book). The latter approaches extended Step 1 of the spatial filtering SCB in Section 3.3.1 to take into account the fact that when there is uncertainty in $\mathbf{a}_0$, the constraint on $\mathbf{w}^*\mathbf{a}_0$ in (3.9) should be replaced with a constraint on $\mathbf{w}^*\mathbf{a}$ for any vector $\mathbf{a}$ in the uncertainty set (the constraints on $\mathbf{w}^*\mathbf{a}$ used in [14] (see also Chapter 2 of this book) and [15] (see also Chapter 1 of this book) are different from one another); then the so-obtained $\mathbf{w}$ is used in $\mathbf{w}^*\mathbf{R}\mathbf{w}$ to derive an estimate of σ_0^2, as in Step (2) of the spatial filtering SCB. Unlike our approach, the approaches of [14] and [15] (see also the first two chapters of this book) do not provide any *direct* estimate $\hat{\mathbf{a}}_0$. Hence they do not provide a simple way [such as (3.37)] to eliminate the scaling ambiguity of the SOI power estimation that is likely a problem for all robust beamforming approaches (this problem was in fact ignored in both [14] and [15]). Yet SOI power estimation is often the main goal in many applications including radar, sonar, acoustics and medical imaging.

Despite the apparent differences in formulation, we prove in Appendices 3.A and 3.C that our RCB gives the same weight vector as the approaches presented in [14, 15] (see also the first two chapters of this book), yet our RCB is computationally more efficient. The approach in [14] (see also Chapter 2 of this book) requires $O(\varrho M^3)$ flops [40], where ϱ is the number of iterations, usually on the order of $\sqrt{M}$, whereas our RCB approach requires $O(M^3)$ flops. Moreover, our RCB can be readily modified for recursive implementation by adding a new snapshot to $\hat{\mathbf{R}}$ and possibly deleting an old one. By using a recursive eigendecomposition updating method (see, for example, [41, 42] and the references therein) with our RCB, we can update the power and waveform estimates in $O(M^2)$ flops. No results are available so far for efficiently updating the second-order cone program (SOCP) approach in [14] (see also Chapter 2 of this book). The approach in [15] (see also Chapter 1 of this book) can be implemented recursively by updating the eigendecomposition similarly to our RCB. However, its total computational burden can be higher than for ours, as explained in the next subsection.

We also show in Appendix B that, although this aspect was ignored in [14, 15] (see also the first two chapters of this book), the approaches presented in [14, 15] can also be modified to eliminate the scaling ambiguity problem that occurs when estimating the SOI power σ_0^2.

3.4.2 Flat Ellipsoidal Uncertainty Set

When the uncertainty set of $\mathbf{a}$ is a flat ellipsoid, as is considered in [15, 37] (see also Chapter 1 of this book) to make the uncertainty set as tight as possible (assuming that the available a priori information allows that), (3.14) becomes [16]

$$\max_{\sigma^2,\mathbf{a}} \sigma^2 \quad \text{subject to } \mathbf{R} - \sigma^2\mathbf{a}\mathbf{a}^* \geq 0 \qquad\qquad \mathbf{a} = \mathbf{B}\mathbf{u} + \bar{\mathbf{a}}, \quad \|\mathbf{u}\| \leq 1 \tag{3.40}$$

where $\mathbf{B}$ is an $M \times L$ matrix ($L < M$) with full column rank and $\mathbf{u}$ is an $L \times 1$ vector.

[When $L = M$, the second constraint in (3.40) becomes (3.6) with $\mathbf{C} = \mathbf{B}\mathbf{B}^*$.] Below we provide a separate treatment of the case of $L < M$ due to the differences from the case of $L = M$ in the possible values of the Lagrange multipliers and the detailed computational steps. The RCB optimization problem in (3.40) can be reduced to [see (3.16)]:

$$\min_{\mathbf{u}} (\mathbf{B}\mathbf{u} + \bar{\mathbf{a}})^* \mathbf{R}^{-1} (\mathbf{B}\mathbf{u} + \bar{\mathbf{a}}) \quad \text{subject to } \|\mathbf{u}\| \leq 1. \tag{3.41}$$

Note that

$$(\mathbf{B}\mathbf{u} + \bar{\mathbf{a}})^* \mathbf{R}^{-1} (\mathbf{B}\mathbf{u} + \bar{\mathbf{a}}) = \mathbf{u}^* \mathbf{B}^* \mathbf{R}^{-1} \mathbf{B}\mathbf{u} + \bar{\mathbf{a}}^* \mathbf{R}^{-1} \mathbf{B}\mathbf{u} + \mathbf{u}^* \mathbf{B}^* \mathbf{R}^{-1} \bar{\mathbf{a}} + \bar{\mathbf{a}}^* \mathbf{R}^{-1} \bar{\mathbf{a}}. \tag{3.42}$$

Let

$$\check{\mathbf{R}} = \mathbf{B}^* \mathbf{R}^{-1} \mathbf{B} > 0 \tag{3.43}$$

and

$$\check{\bar{\mathbf{a}}} = \mathbf{B}^* \mathbf{R}^{-1} \bar{\mathbf{a}}. \tag{3.44}$$

Using (3.42)–(3.44) in (3.41) gives

$$\min_{\mathbf{u}} \mathbf{u}^* \check{\mathbf{R}} \mathbf{u} + \check{\bar{\mathbf{a}}}^* \mathbf{u} + \mathbf{u}^* \check{\bar{\mathbf{a}}} \quad \text{subject to } \|\mathbf{u}\| \leq 1. \tag{3.45}$$

To avoid the trivial solution $\mathbf{a} = \mathbf{0}$ to the RCB problem in (3.40), we impose the following condition (assuming $\tilde{\mathbf{u}}$ below exists, otherwise there is no trivial solution). Let $\tilde{\mathbf{u}}$ be the solution to the equation

$$\mathbf{B}\tilde{\mathbf{u}} + \bar{\mathbf{a}} = \mathbf{0}. \tag{3.46}$$

Hence

$$\tilde{\mathbf{u}} = -\mathbf{B}^{\dagger} \bar{\mathbf{a}} \tag{3.47}$$

where $\mathbf{B}^{\dagger}$ denotes the Moore–Penrose pseudo-inverse of $\mathbf{B}$. Then we require that

$$\bar{\mathbf{a}}^* \mathbf{B}^{\dagger *} \mathbf{B}^{\dagger} \bar{\mathbf{a}} > 1. \tag{3.48}$$

The *Lagrange multiplier methodology* can again be used to solve (3.40) [43]. Let

$$\check{h}_1(\mathbf{u}, \check{\nu}) = \mathbf{u}^* \check{\mathbf{R}} \mathbf{u} + \check{\bar{\mathbf{a}}}^* \mathbf{u} + \mathbf{u}^* \check{\bar{\mathbf{a}}} + \check{\nu}(\mathbf{u}^* \mathbf{u} - 1) \tag{3.49}$$

where $\breve{\nu} \geq 0$ is the Lagrange multiplier [44]. Differentiation of (3.49) with respect to $\mathbf{u}$ gives

$$\breve{\mathbf{R}}\hat{\mathbf{u}} + \bar{\breve{\mathbf{a}}} + \breve{\nu}\hat{\mathbf{u}} = 0 \tag{3.50}$$

which yields

$$\hat{\mathbf{u}} = -(\breve{\mathbf{R}} + \breve{\nu}\mathbf{I})^{-1}\bar{\breve{\mathbf{a}}}. \tag{3.51}$$

If $\|\breve{\mathbf{R}}^{-1}\bar{\breve{\mathbf{a}}}\| \leq 1$, then the unique solution in (3.51) with $\breve{\nu} = 0$, which is $\hat{\mathbf{u}} = -\breve{\mathbf{R}}^{-1}\bar{\breve{\mathbf{a}}}$, solves (3.45). If $\|\breve{\mathbf{R}}^{-1}\bar{\breve{\mathbf{a}}}\| > 1$, then $\breve{\nu} > 0$ is determined by solving

$$\breve{h}_2(\breve{\nu}) \triangleq \left\|(\breve{\mathbf{R}} + \breve{\nu}\mathbf{I})^{-1}\bar{\breve{\mathbf{a}}}\right\|^2 = 1. \tag{3.52}$$

Note that $\breve{h}_2(\breve{\nu})$ is a monotonically decreasing function of $\breve{\nu} > 0$. Let

$$\breve{\mathbf{R}} = \breve{\mathbf{U}}\breve{\mathbf{\Gamma}}\breve{\mathbf{U}}^* \tag{3.53}$$

where the columns of $\breve{\mathbf{U}}$ contain the eigenvectors of $\breve{\mathbf{R}}$ and the diagonal elements of the diagonal matrix $\breve{\mathbf{\Gamma}}$, $\breve{\gamma}_1 \geq \breve{\gamma}_2 \geq \cdots \geq \breve{\gamma}_L$, are the corresponding eigenvalues. Let

$$\breve{\mathbf{z}} = \breve{\mathbf{U}}^*\bar{\breve{\mathbf{a}}} \tag{3.54}$$

and let $\breve{z}_l$ denote the lth element of $\breve{\mathbf{z}}$. Then

$$\breve{h}_2(\breve{\nu}) = \sum_{l=1}^{L} \frac{|\breve{z}_l|^2}{(\breve{\gamma}_l + \breve{\nu})^2} = 1. \tag{3.55}$$

Note that $\lim_{\breve{\nu}\to\infty} \breve{h}_2(\breve{\nu}) = 0$ and $\breve{h}_2(0) = \|\breve{\mathbf{R}}^{-1}\bar{\breve{\mathbf{a}}}\| > 1$. Hence there is a unique solution to (3.55) between 0 and ∞. By replacing the $\breve{\gamma}_l$ in (3.55) with $\breve{\gamma}_L$ and $\breve{\gamma}_1$, respectively, we obtain tighter upper and lower bounds on the solution to (3.55):

$$\|\bar{\breve{\mathbf{a}}}\| - \breve{\gamma}_1 \leq \breve{\nu} \leq \|\bar{\breve{\mathbf{a}}}\| - \breve{\gamma}_L. \tag{3.56}$$

Hence the solution to (3.55) can be efficiently determined by using, for example, the Newton's method, in the above interval. Then the solution $\breve{\nu}$ to (3.55) is used in (3.51) to obtain the $\hat{\mathbf{u}}$ that solves (3.45).

To summarize, our proposed RCB approach consists of the following steps.

Step 1. Compute the inverse of $\mathbf{R}$ (or more practically of $\hat{\mathbf{R}}$) and calculate $\breve{\mathbf{R}}$ and $\bar{\breve{\mathbf{a}}}$ using (3.43) and (3.44), respectively.

Step 2. Compute the eigendecomposition of $\breve{\mathbf{R}}$ [see (3.53)].

Step 3. If $\|\check{\mathbf{R}}^{-1}\bar{\check{\mathbf{a}}}\| \leq 1$, then set $\check{\nu} = 0$. If $\|\check{\mathbf{R}}^{-1}\bar{\check{\mathbf{a}}}\| > 1$, then solve (3.55) for $\check{\nu}$, for example, by a Newton's method, using the knowledge that the solution is unique and it belongs to the interval in (3.56).

Step 4. Use the $\check{\nu}$ obtained in Step 3 to get:

$$\hat{\mathbf{u}} = -\check{\mathbf{U}}\left(\check{\mathbf{\Gamma}} + \check{\nu}\mathbf{I}\right)^{-1}\check{\mathbf{U}}^{*}\bar{\check{\mathbf{a}}} \tag{3.57}$$

[which is obtained from (3.51)]. Then use the $\hat{\mathbf{U}}$ to obtain the optimal solution to (3.40) as:

$$\hat{\mathbf{a}}_0 = \mathbf{B}\hat{\mathbf{u}} + \bar{\mathbf{a}}. \tag{3.58}$$

Step 5. Compute $\hat{\sigma}_0^2$ by using (3.11) with $\mathbf{a}_0$ replaced by $\hat{\mathbf{a}}_0$ and then use the $\hat{\sigma}_0^2$ in (3.37) to obtain the estimate of σ_0^2.

Hence, under the flat ellipsoidal constraint the complexity of our RCB is also $O(M^3)$ flops, which is on the same order as for SCB and is mainly due to computing $\mathbf{R}^{-1}$ and the eigendecomposition of $\check{\mathbf{R}}$. If $L \ll M$, then the complexity is mainly due to computing $\mathbf{R}^{-1}$. Note, however, that to compute $\check{\nu}$, we need $O(L^3)$ flops while the approach in [15] (see also Chapter 1 of this book) requires $O(M^3)$ flops (and $L \leq M$).

3.4.3 Numerical Examples

Next, we provide numerical examples to compare the performances of the SCB and RCB. In all of the examples considered below, we assume a uniform linear array with $M = 10$ sensors and half-wavelength sensor spacing, and a spatially white Gaussian noise whose covariance matrix is given by $\mathbf{Q} = \mathbf{I}$.

Example 3.4.1: Comparison of SCB and RCB for the Case of Finite Number of Snapshots Without Look Direction Errors We consider the effect of the number of snapshots N on the SOI power estimate when the sample covariance matrix $\hat{\mathbf{R}}$ in (3.4) is used in lieu of the theoretical array covariance matrix $\mathbf{R}$ in both the SCB and RCB. (Whenever $\hat{\mathbf{R}}$ is used instead of $\mathbf{R}$, the average power estimates from 100 Monte Carlo simulations are given. However, the beampatterns shown are obtained using $\hat{\mathbf{R}}$ from one Monte Carlo realization only.) The power of SOI is $\sigma_0^2 = 10\,\text{dB}$ and the powers of the two ($K = 2$) interferences assumed to be present are $\sigma_1^2 = \sigma_2^2 = 20\,\text{dB}$. We assume that the steering vector uncertainty is due to the uncertainty in the SOI's direction of arrival θ_0, which we assume to be $\theta_0 + \Delta$. We assume that $\mathbf{a}(\theta_0)$ belongs to the uncertainty set

$$\|\mathbf{a}(\theta_0) - \bar{\mathbf{a}}\|^2 \leq \epsilon; \quad \bar{\mathbf{a}} = \mathbf{a}(\theta_0 + \Delta) \tag{3.59}$$

where ϵ is a user parameter. Let $\epsilon_0 = \|\mathbf{a}(\theta_0) - \bar{\mathbf{a}}\|^2$. To show that the choice of ϵ is not a critical issue for our RCB approach, we will present numerical results for several values of ϵ. We assume that the SOI's direction of arrival is $\theta_0 = 0^\circ$ and the directions of arrival of the interferences are $\theta_1 = 60^\circ$ and $\theta_2 = 80^\circ$.

In Figure 3.1, we show $\tilde{\sigma}_0^2$ and $\hat{\hat{\sigma}}_0^2$ versus the number of snapshots N for the no mismatch case; hence $\Delta = 0$ in (3.59) and consequently $\epsilon_0 = 0$. Note that the power estimates obtained by using $\hat{\mathbf{R}}$ approach those computed via $\mathbf{R}$ as N increases, and that our RCB converges much faster than the SCB. The SCB requires that N is greater than or equal to the number of array sensors $M = 10$. However, our RCB works well even when N is as small as $N = 2$.

Figure 3.2 shows the beampatterns of the SCB and RCB using $\mathbf{R}$ as well as $\hat{\mathbf{R}}$ with $N = 10$, 100, and 8000 for the same case as in Figure 3.1. Note that the weight vectors used to calculate the beampatterns of RCB in this example (as well as in the following examples) are obtained by using the scaled estimate of the array steering vector $\sqrt{M}\hat{\mathbf{a}}_0/\|\hat{\mathbf{a}}_0\|$ in (3.10) instead of $\hat{\mathbf{a}}_0$. The vertical dotted lines in the figure denote the directions of arrival of the SOI and the interferences. The horizontal dotted lines in the figure correspond to 0 dB. Note from Figure 3.2(a) that although the RCB beampatterns do not have nulls at the directions of arrival of the interferences as deep as those of the SCB, the interferences (whose powers are 20 dB) are sufficiently suppressed by the RCB to not disturb the SOI power estimation. Regarding the poor performance of SCB for small N, note that the error between $\hat{\mathbf{R}}$ and $\mathbf{R}$ can be viewed as due to a steering vector error [7].

Example 3.4.2: Comparison of SCB and RCB for the Case of Finite Number of Snapshots in the Presence of Look Direction Errors This example is similar to Example 3.4.1 except that now the mismatch is $\Delta = 2^\circ$ and accordingly $\epsilon_0 = 3.2460$. We note from Figure 3.3 that even a relatively small Δ can cause a significant degradation of the SCB performance. As can be seen from Figure 3.4, the SOI is considered to be an interference by SCB and hence it is suppressed. On the other hand, the SOI is preserved by our RCB and the performance of $\hat{\hat{\sigma}}_0^2$ obtained

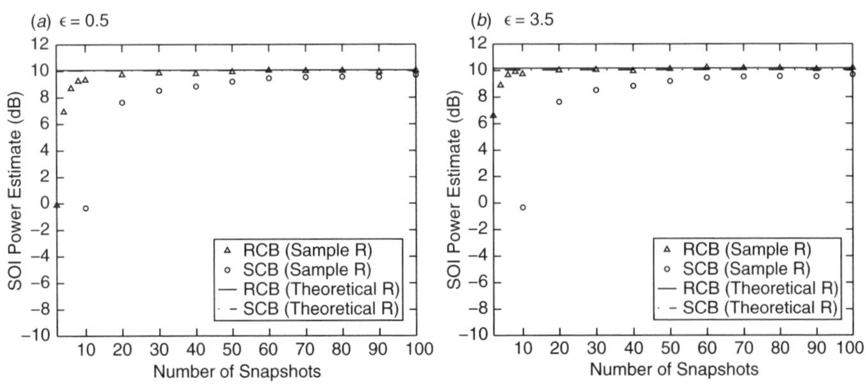

Figure 3.1 $\tilde{\sigma}_0^2$ (SCB using $\hat{\mathbf{R}}$ and $\mathbf{R}$) and $\hat{\hat{\sigma}}_0^2$ (RCB using $\hat{\mathbf{R}}$ and $\mathbf{R}$) versus N for (a) $\epsilon = 0.5$ and (b) $\epsilon = 3.5$. The true SOI power is 10 dB and $\epsilon_0 = 0$ (i.e., no mismatch).

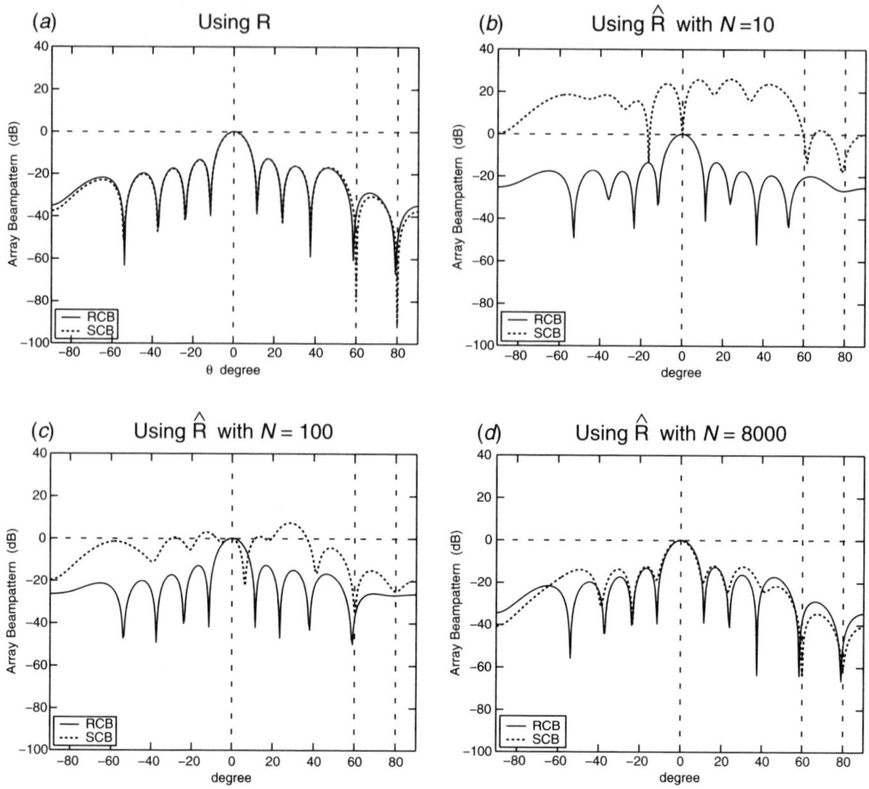

Figure 3.2 Comparison of the beampatterns of SCB and RCB when $\epsilon = 3.5$ for (*a*) using $\mathbf{R}$, (*b*) using $\hat{\mathbf{R}}$ with $N = 10$, (*c*) using $\hat{\mathbf{R}}$ with $N = 100$, and (*d*) using $\hat{\mathbf{R}}$ with $N = 8000$. The true SOI power is 10 dB and $\epsilon_0 = 0$ (i.e., no mismatch).

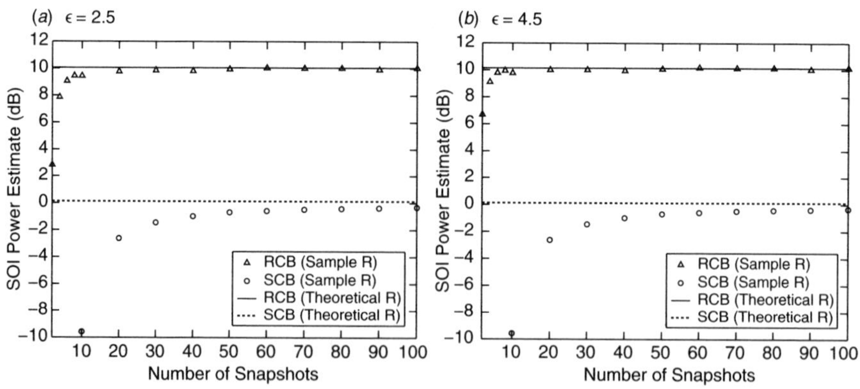

Figure 3.3 $\tilde{\sigma}_0^2$ (SCB using $\hat{\mathbf{R}}$ and $\mathbf{R}$) and $\hat{\hat{\sigma}}_0^2$ (RCB using $\hat{\mathbf{R}}$ and $\mathbf{R}$) versus N for (*a*) $\epsilon = 2.5$ and (*b*) $\epsilon = 4.5$. The true SOI power is 10 dB and $\epsilon_0 = 3.2460$ (corresponding to $\Delta = 2.0^\circ$).

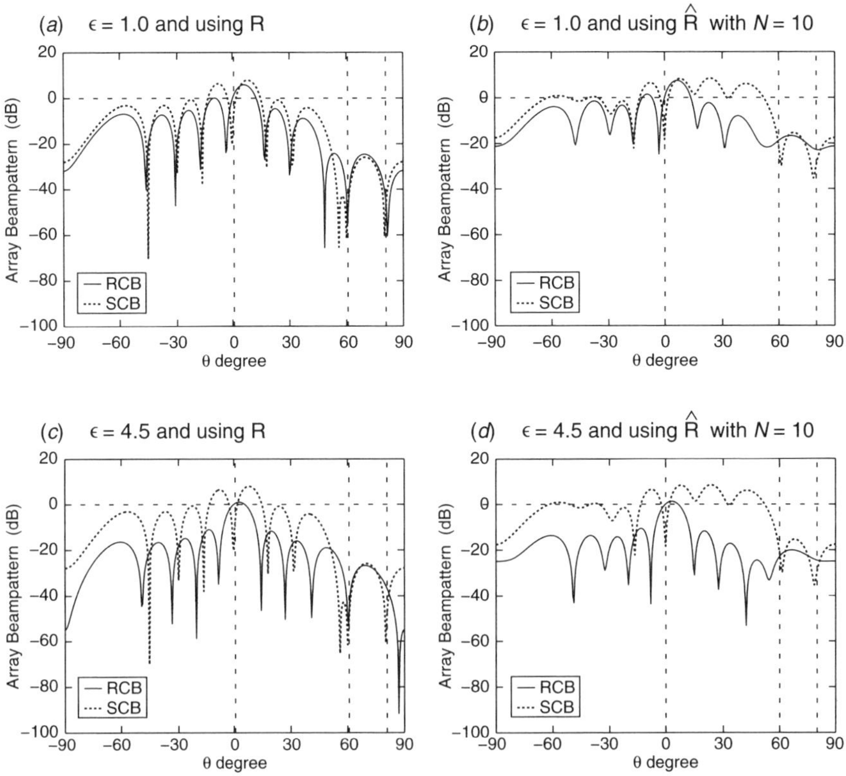

Figure 3.4 Comparison of the beampatterns of SCB and RCB when $\epsilon = 1.0$ for (*a*) using **R** and (*b*) using $\hat{\mathbf{R}}$ with $N = 10$, and when $\epsilon = 4.5$ for (*c*) using **R** and (*d*) using $\hat{\mathbf{R}}$ with $N = 10$. The true SOI power is 10 dB and $\epsilon_0 = 3.2460$ (corresponding to $\Delta = 2.0^\circ$).

via our approach is quite good for a wide range of values of ϵ. Note that the RCB also has a smaller 'noise gain' than the SCB.

Example 3.4.3: Comparison of the RCB Method and a Fixed Diagonal Loading Level Based Approach In Figure 3.5, we compare the performance of our RCB with a fixed diagonal loading level based approach. Specifically, the fixed loading level was chosen equal to 10 times the noise power (assuming the knowledge of the noise power). Consider the same case as Figure 3.4(*d*) except that now we assume that **R** is available and we vary the SNR by changing the SOI or noise power. For Figures 3.5(*a*), 3.5(*c*) and 3.5(*e*), we fix the noise power at 0 dB and vary the SOI power between −10 dB and 20 dB. For Figures 3.5(*b*), 3.5(*d*) and 3.5(*f*), we fix the SOI power at 10 dB and vary the noise power between −10 dB and 20 dB. Figures 3.5(*a*) and 3.5(*b*) show the diagonal loading levels of our RCB as functions of the SNR. Figures 3.5(*c*) and 3.5(*d*) show the SINRs of our RCB and the fixed diagonal loading level approach and Figures 3.5(*e*) and 3.5(*f*)

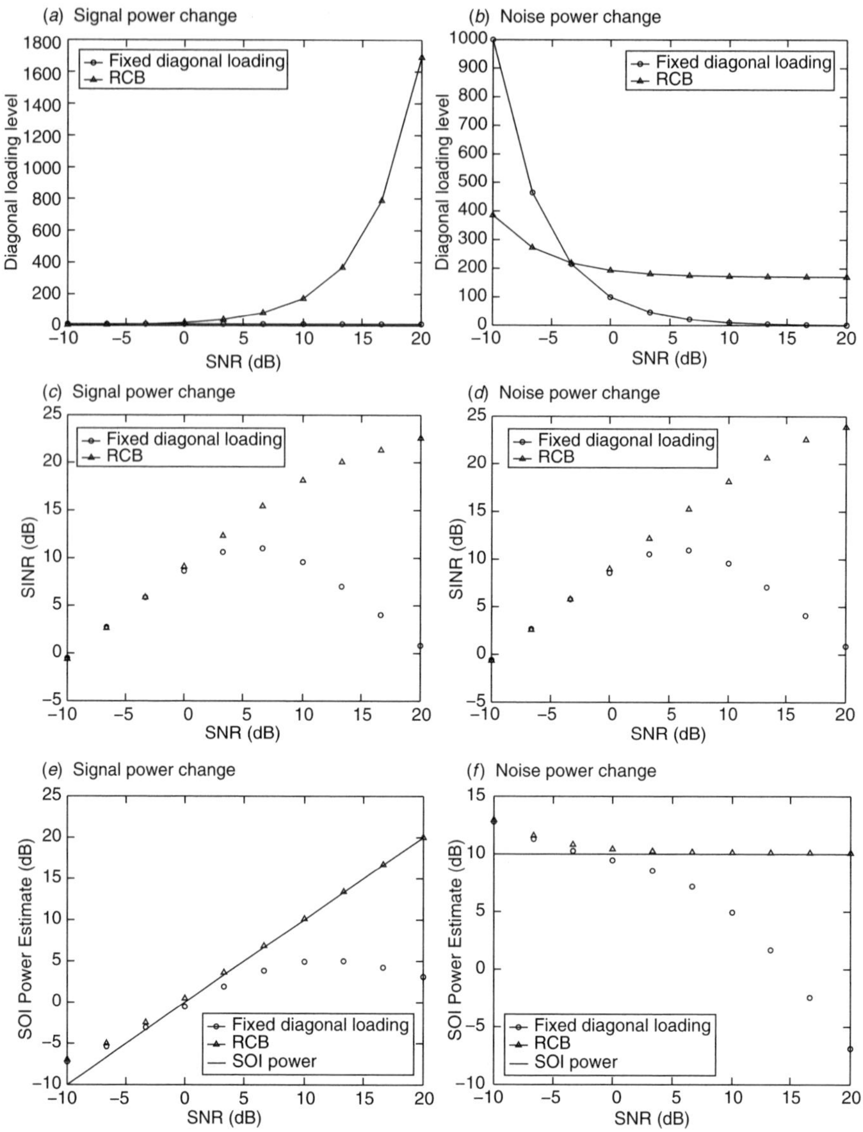

Figure 3.5 Comparison of a fixed diagonal loading level approach and our RCB when $\epsilon = 4.5$ and $\epsilon_0 = 3.2460$ (corresponding to $\Delta = 2.0°$).

show the corresponding SOI power estimates, all as functions of the SNR. Note from Figures 3.5(a) and 3.5(b) that our RCB adjusts the diagonal loading level adaptively as the SNR changes. It is obvious from Figure 3.5 that our RCB significantly outperforms the fixed diagonal loading level approach when the SNR is medium or high.

Example 3.4.4: Comparison of RCB, SCB and the Delay-and-Sum Method in the Presence of Array Calibration Errors We consider an imaging example, where we wish to determine the incident signal power as a function of the steering direction θ. We assume that there are five incident signals with powers 30, 15, 40, 35, and 20 dB from directions -35°, -15°, 0°, 10°, and 40°, respectively. To simulate the array calibration error, each element of the steering vector for each incident signal is perturbed with a zero-mean circularly symmetric complex Gaussian random variable so that the squared Euclidean norm of the difference between the true steering vector and the assumed one is 0.05. The perturbing Gaussian random variables are independent of each other.

Figure 3.6 shows the power estimates of SCB and RCB, obtained using $\mathbf{R}$, as a function of the direction angle, for several values of ϵ. The small circles denote the true (direction of arrival, power)-coordinates of the five incident signals. Figure 3.6 also shows the power estimates obtained with the data-independent beamformer using the assumed array steering vector divided by M as the weight vector. This approach is referred to as the delay-and-sum beamformer. We note that SCB can still give good direction of arrival estimates for the incident signals based on the peak power locations. However, the SCB estimates of the incident signal powers are way off. On the other hand, our RCB provides excellent power estimates of the incident sources and can also be used to determine their directions of arrival based on the peak locations. The delay-and-sum beamformer, however, has much poorer resolution than both SCB and RCB. Moreover, the sidelobes of the former give false peaks.

Example 3.4.5: Comparison of SCB, RCB with Spherical Constraint and RCB with Flat Ellipsoidal Constraint in the Presence of Look Direction Errors We examine now the effects of the spherical and flat ellipsoidal constraints on SOI

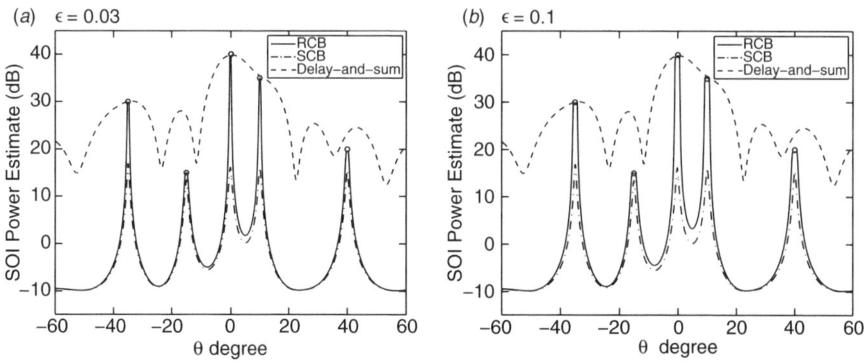

Figure 3.6 Power estimates (using $\mathbf{R}$) versus the steering direction θ when (*a*) $\epsilon = 0.03$ and (*b*) $\epsilon = 0.1$. The true powers of the incident signals from -35°, -15°, 0°, 10°, and 40° are denoted by circles, and $\epsilon_0 = 0.05$.

power estimation. We consider SOI power estimation in the presence of several strong interferences. We will vary the number of interferences from $K = 1$ to $K = 8$. The power of SOI is $\sigma_0^2 = 20\,\text{dB}$ and the interference powers are $\sigma_1^2 = \cdots = \sigma_K^2 = 40\,\text{dB}$. The SOI and interference directions of arrival are $\theta_0 = 10°$, $\theta_1 = -75°$, $\theta_2 = -60°$, $\theta_3 = -45°$, $\theta_4 = -30°$, $\theta_5 = -10°$, $\theta_6 = 25°$, $\theta_7 = 35°$, $\theta_8 = 50°$. We assume that there is a look direction mismatch corresponding to $\Delta = 2°$ and accordingly $\epsilon_0 = 3.1349$.

Figure 3.7 shows the SOI power estimates, as a function of the number of interferences K, obtained by using SCB, RCB (with flat ellipsoidal constraint), and the more conservative RCB (with spherical constraint) all based on the theoretical array covariance matrix $\mathbf{R}$. For RCB with flat ellipsoidal constraint, we let $\mathbf{B}$ contain two columns with the first column being $\mathbf{a}(\theta_0 + \Delta) - \mathbf{a}(\theta_0 + \Delta - \delta)$ and the second column being $\mathbf{a}(\theta_0 + \Delta) - \mathbf{a}(\theta_0 + \Delta + \delta)$. Note that choosing $\delta = \Delta = 2°$ gives the smallest flat ellipsoid that this $\mathbf{B}$ can offer to include $\mathbf{a}(\theta_0)$. However, we do not know the exact look direction mismatch in practice. We choose $\delta = 1.8°$ and $\delta = 2.4°$ in Figures 3.7(*a*) and (*b*), respectively. For RCB with spherical constraint, we choose ϵ to be the larger of $\|\mathbf{a}(\theta_0 + \Delta) - \mathbf{a}(\theta_0 + \Delta - \delta)\|^2$ and $\|\mathbf{a}(\theta_0 + \Delta) - \mathbf{a}(\theta_0 + \Delta + \delta)\|^2$. Note that RCB with flat ellipsoidal constraint and RCB with spherical constraint perform similarly when K is small. However, the former is more accurate than the latter for large K.

Figure 3.8 gives the beampatterns of the SCB and RCBs using $\mathbf{R}$ as well as $\hat{\mathbf{R}}$ with $N = 10$ for various K. For large K, the more conservative RCB with spherical constraint amplifies the SOI while attempting to suppress the interferences, as shown in Figure 3.8. On the other hand, the RCB with flat ellipsoidal constraint maintains an approximate unity gain for the SOI and provides much deeper nulls for the interferences than the RCB with spherical constraint at a cost of worse

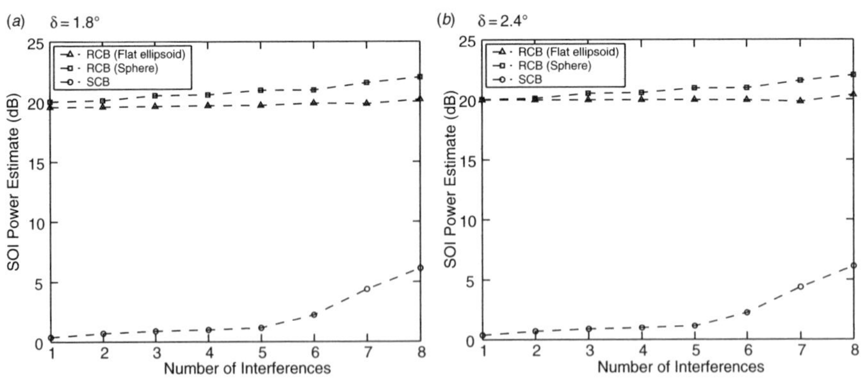

Figure 3.7 $\tilde{\sigma}_0^2$ (SCB), $\hat{\hat{\sigma}}_0^2$ (RCB with flat ellipsoidal constraint with $L = 2$), and $\hat{\hat{\sigma}}_0^2$ (RCB with spherical constraint), based on $\mathbf{R}$, versus the number of interferences K when (*a*) $\delta = 1.8°$ and (*b*) $\delta = 2.4°$. The true SOI power is 20 dB and $\epsilon_0 = 3.1349$ (corresponding to $\Delta = 2°$).

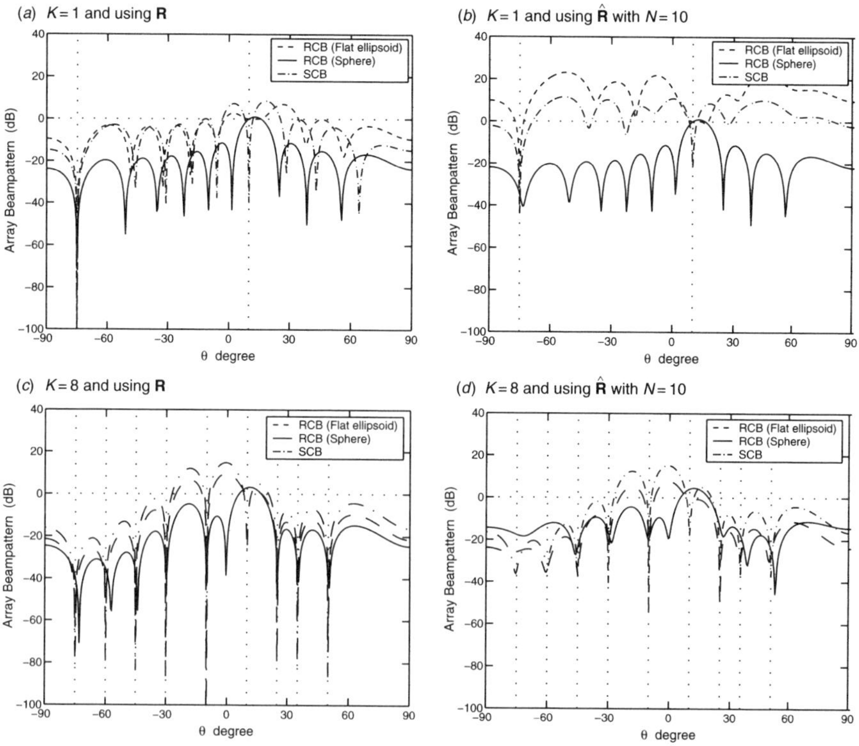

Figure 3.8 Comparison of the beampatterns of SCB, RCB (with flat ellipsoidal constraint) and RCB (with spherical constraint) when $\delta = 2.4^\circ$ for (*a*) $K = 1$ and using $\mathbf{R}$, (*b*) $K = 1$ and using $\hat{\mathbf{R}}$ with $N = 10$, (*c*) $K = 8$ and using $\hat{\mathbf{R}}$, and (*d*) $K = 8$ and using $\hat{\mathbf{R}}$ with $N = 10$. The true SOI power is 20 dB and $\epsilon_0 = 3.1349$ (corresponding to $\Delta = 2^\circ$).

noise gain. As compared to the RCBs, the SCB performs poorly as it attempts to suppress the SOI. Comparing Figures 3.8(*b*) with 3.8(*a*), we note that for small K and N, RCB with spherical constraint has a much better noise gain than RCB with flat ellipsoidal constraint, which has a better noise gain than SCB. From Figure 3.8(*d*), we note that for large K and small N, RCB with flat ellipsoidal constraint places deeper nulls at the interference angles than the more conservative RCB with spherical constraint.

Figure 3.9 shows the SOI power estimates versus the number of snapshots N for $K = 1$ and $K = 8$ when the sample covariance matrix $\hat{\mathbf{R}}$ is used in the beamformers. Note that for small K, RCB with spherical constraint converges faster than RCB with flat ellipsoidal constraint as N increases, while the latter converges faster than SCB. For large K, however, the convergence speeds of RCB with flat ellipsoidal constraint and RCB with spherical constraint are about the same as that of SCB; after convergence, the most accurate power estimate is provided by RCB with flat ellipsoidal constraint.

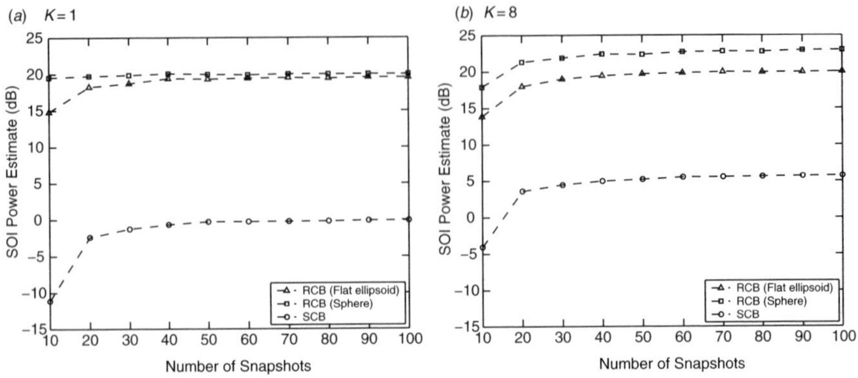

Figure 3.9 Comparison of the SOI power estimates, versus N, obtained using SCB, RCB (with flat ellipsoidal constraint) and RCB (with spherical constraint), all with $\hat{\mathbf{R}}$, when $\delta = 2.4^\circ$ for (*a*) $K = 1$ and (*b*) $K = 8$. The true SOI power is 20 dB and $\epsilon_0 = 3.1349$ (corresponding to $\Delta = 2^\circ$).

3.5 CAPON BEAMFORMING WITH NORM CONSTRAINT

In the presence of array steering vectors, SCB may attempt to suppress the SOI as if it were an interference. Since $\mathbf{a}_0$ is usually close to $\bar{\mathbf{a}}$, the Euclidean norm of the resulting weight vector (which equals the white noise gain at the array output) can become very large in order to satisfy the distortionless constraint $\mathbf{w}^*\bar{\mathbf{a}} = 1$ and at the same time suppress the SOI, that is, $\mathbf{w}^*\mathbf{a}_0 \approx 0$ (note that the previous two conditions on $\mathbf{w}$ imply $\mathbf{w}^*(\bar{\mathbf{a}} - \mathbf{a}_0) \approx 1$, which can only hold if $\|\mathbf{w}\|^2 \gg 1$ whenever $\bar{\mathbf{a}}$ is close to $\mathbf{a}_0$.)

The goal of NCCB is to impose an additional constraint on the Euclidean norm of $\mathbf{w}$ for the purpose of improving the robustness of the Capon beamformer against SOI steering vector errors and control the white noise gain (see, e.g., [17–22] and the references therein). Consequently the beamforming problem is formulated as follows:

$$\min_{\mathbf{w}} \mathbf{w}^*\mathbf{R}\mathbf{w} \quad \text{subject to } \mathbf{w}^*\bar{\mathbf{a}} = 1$$
$$\|\mathbf{w}\|^2 \leq \zeta. \tag{3.60}$$

Note that the quadratic inequality constraint can be interpretted as constraining the white noise gain at the output. The problem with NCCB is that the choice of ζ is not easy to make. In particular, this choice is not directly linked to the uncertainty of the SOI steering vector. The RCB and DCRCB algorithms, on the other hand, do not suffer from this problem.

A solution to (3.60) was found in [18] using the Lagrange multiplier methodology. We present herein (see also [17]) a more thorough analysis of the optimization problem in (3.60), which provides new insights into the choice of ζ and also prepares the grounds for solving the DCRCB optimization problem, which will be discussed in Section 3.6.

Let $\mathcal{S}$ be the set defined by the constraints in (3.60). Also, let

$$g_1(\mathbf{w}, \lambda, \mu) = \mathbf{w}^*\mathbf{R}\mathbf{w} + \lambda(\|\mathbf{w}\|^2 - \zeta) + \mu(-\mathbf{w}^*\bar{\mathbf{a}} - \bar{\mathbf{a}}^*\mathbf{w} + 2) \tag{3.61}$$

where λ and μ are the real-valued Lagrange multipliers with μ being arbitrary and $\lambda \geq 0$ satisfying $\mathbf{R} + \lambda\mathbf{I} > 0$ so that $g_1(\mathbf{w}, \lambda, \mu)$ can be minimized with respect to $\mathbf{w}$. (This optimization problem is somewhat similar to the one in [45].) Then

$$g_1(\mathbf{w}, \lambda, \mu) \leq \mathbf{w}^*\mathbf{R}\mathbf{w} \quad \text{for any } \mathbf{w} \in \mathcal{S} \tag{3.62}$$

with equality on the boundary of $\mathcal{S}$.

Consider the condition

$$\frac{\bar{\mathbf{a}}^*\mathbf{R}^{-2}\bar{\mathbf{a}}}{\left[\bar{\mathbf{a}}^*\mathbf{R}^{-1}\bar{\mathbf{a}}\right]^2} \leq \zeta. \tag{3.63}$$

When the condition in (3.63) is satisfied, the SCB solution in (3.10) with $\mathbf{a}_0$ replaced by $\bar{\mathbf{a}}$, that is,

$$\hat{\mathbf{w}} = \frac{\mathbf{R}^{-1}\bar{\mathbf{a}}}{\bar{\mathbf{a}}^*\mathbf{R}^{-1}\bar{\mathbf{a}}} \tag{3.64}$$

satisfies the norm constraint in (3.60) and hence is also the NCCB solution. For this case, $\lambda = 0$ and the norm constraint in (3.60) is inactive. Otherwise, we have the condition

$$\zeta < \frac{\bar{\mathbf{a}}^*\mathbf{R}^{-2}\bar{\mathbf{a}}}{\left[\bar{\mathbf{a}}^*\mathbf{R}^{-1}\bar{\mathbf{a}}\right]^2} \tag{3.65}$$

which is an upper bound on ζ so that NCCB is different from SCB. To deal with this case, we note that (3.61) can be written as

$$\begin{aligned} g_1(\mathbf{w}, \lambda, \mu) = {} & \left[\mathbf{w} - \mu(\mathbf{R} + \lambda\mathbf{I})^{-1}\bar{\mathbf{a}}\right]^*(\mathbf{R} + \lambda\mathbf{I})\left[\mathbf{w} - \mu(\mathbf{R} + \lambda\mathbf{I})^{-1}\bar{\mathbf{a}}\right] \\ & - \mu^2\bar{\mathbf{a}}^*(\mathbf{R} + \lambda\mathbf{I})^{-1}\bar{\mathbf{a}} - \lambda\zeta + 2\mu. \end{aligned} \tag{3.66}$$

Hence the *unconstrained* minimizer of $g_1(\mathbf{w}, \lambda, \mu)$, for fixed λ and μ, is given by

$$\hat{\mathbf{w}}_{\lambda,\mu} = \mu(\mathbf{R} + \lambda\mathbf{I})^{-1}\bar{\mathbf{a}}. \tag{3.67}$$

Clearly, we have

$$g_2(\lambda, \mu) \triangleq g_1(\hat{\mathbf{w}}_{\lambda,\mu}, \lambda, \mu) = -\mu^2\bar{\mathbf{a}}^*(\mathbf{R} + \lambda\mathbf{I})^{-1}\bar{\mathbf{a}} - \lambda\zeta + 2\mu \tag{3.68}$$

$$\leq \mathbf{w}^*\mathbf{R}\mathbf{w} \quad \text{for any } \mathbf{w} \in \mathcal{S}. \tag{3.69}$$

The maximization of $g_2(\lambda, \mu)$ with respect to μ gives

$$\hat{\mu} = \frac{1}{\bar{\mathbf{a}}^*(\mathbf{R} + \lambda\mathbf{I})^{-1}\bar{\mathbf{a}}} \tag{3.70}$$

and

$$g_3(\lambda) \triangleq g_2(\lambda, \hat{\mu}) = -\lambda\zeta + \frac{1}{\bar{\mathbf{a}}^*(\mathbf{R} + \lambda\mathbf{I})^{-1}\bar{\mathbf{a}}}. \tag{3.71}$$

The maximization of the above function with respect to λ gives

$$\frac{\bar{\mathbf{a}}^*(\mathbf{R} + \lambda\mathbf{I})^{-2}\bar{\mathbf{a}}}{\left[\bar{\mathbf{a}}^*(\mathbf{R} + \lambda\mathbf{I})^{-1}\bar{\mathbf{a}}\right]^2} = \zeta. \tag{3.72}$$

We show in Appendix 3.D (see also [17, 18]) that, under (3.65), we have a unique solution $\lambda > 0$ to (3.72) and also that the left side of (3.72) is a monotonically decreasing function of λ and hence λ can be obtained efficiently via, for example, a Newton's method.

Note that using (3.70) in (3.67) yields

$$\hat{\mathbf{w}} = \frac{(\mathbf{R} + \lambda\mathbf{I})^{-1}\bar{\mathbf{a}}}{\bar{\mathbf{a}}^*(\mathbf{R} + \lambda\mathbf{I})^{-1}\bar{\mathbf{a}}} \tag{3.73}$$

which satisfies

$$\hat{\mathbf{w}}^*\bar{\mathbf{a}} = 1 \tag{3.74}$$

and

$$\|\hat{\mathbf{w}}\|^2 = \zeta. \tag{3.75}$$

Hence $\hat{\mathbf{w}}$ belongs to the boundary of $\mathcal{S}$. Therefore, $\hat{\mathbf{w}}$ is our sought solution. Note that $\hat{\mathbf{w}}$ in (3.73) has the form of a diagonally loaded Capon beamformer. We now provide some insights into the choice of ζ for NCCB. From the distortionless constraint in (3.60), we have

$$1 = |\mathbf{w}^*\bar{\mathbf{a}}|^2 \leq \|\mathbf{w}\|^2\|\bar{\mathbf{a}}\|^2 \leq \zeta M \tag{3.76}$$

and hence we get a lower bound on ζ (see also [2, 17]):

$$\zeta \geq \frac{1}{M}. \tag{3.77}$$

If ζ is less than this lower bound, there is no solution to the NCCB problem. Hence, ζ should be chosen in the interval defined by the inequalities in (3.65) and (3.77).

Next we derive an upper bound on λ. Let

$$\mathbf{z} = \mathbf{U}^* \bar{\mathbf{a}} \tag{3.78}$$

and let z_m denote the mth element of $\mathbf{z}$. Then (3.72) can be written as (see also [17, 18])

$$\frac{\sum_{m=1}^{M} |z_m|^2/(\gamma_m + \lambda)^2}{\left[\sum_{m=1}^{M} |z_m|^2/(\gamma_m + \lambda)\right]^2} = \zeta. \tag{3.79}$$

Hence we have

$$\zeta \leq \frac{\|\bar{\mathbf{a}}\|^2/(\gamma_M + \lambda)^2}{\|\bar{\mathbf{a}}\|^4/(\gamma_1 + \lambda)^2} = \frac{(\gamma_1 + \lambda)^2}{M(\gamma_M + \lambda)^2} \tag{3.80}$$

which gives the following upper bound on λ:

$$\lambda \leq \frac{\gamma_1 - (M\zeta)^{1/2}\gamma_M}{(M\zeta)^{1/2} - 1}. \tag{3.81}$$

We remark that the computations needed by the search for λ via a Newton's method are negligible compared to those required by the eigendecomposition of the Hermitian matrix $\mathbf{R}$ (or $\hat{\mathbf{R}}$). Hence the major computational demand of NCCB comes from the eigendecomposition of $\mathbf{R}$ (or $\hat{\mathbf{R}}$), which requires $O(M^3)$ flops. Therefore, the computational complexity of NCCB is comparable to that of the SCB, which also requires $O(M^3)$ flops.

To summarize, NCCB consists of the following steps.

Step 1. Compute the eigendecomposition of $\mathbf{R}$ (or, in practice, of $\hat{\mathbf{R}}$).

Step 2. If (3.65) is satisfied, solve (3.79) for λ, for example, by a Newton's method, using the knowledge that the solution is unique and it is lower bounded by 0 and upper bounded by (3.81); otherwise, set $\lambda = 0$.

Step 3. Use the λ obtained in Step 2 to get

$$\hat{\mathbf{w}} = \frac{\mathbf{U}(\mathbf{\Gamma} + \lambda\mathbf{I})^{-1}\mathbf{U}^*\bar{\mathbf{a}}}{\bar{\mathbf{a}}^*\mathbf{U}(\mathbf{\Gamma} + \lambda\mathbf{I})^{-1}\mathbf{U}^*\bar{\mathbf{a}}} \tag{3.82}$$

where the inverse of the diagonal matrix $\mathbf{\Gamma} + \lambda\mathbf{I}$ is easily computed and the vector $\mathbf{z} = \mathbf{U}^*\bar{\mathbf{a}}$ is available from Step 2.

Step 4. Compute the SOI power estimate of NCCB

$$\hat{\hat{\sigma}}_0^2 = \frac{\bar{\mathbf{a}}^* \mathbf{U}(\boldsymbol{\Gamma} + \lambda \mathbf{I})^{-2} \boldsymbol{\Gamma} \mathbf{U}^* \bar{\mathbf{a}}}{\left[\bar{\mathbf{a}}^* \mathbf{U}(\boldsymbol{\Gamma} + \lambda \mathbf{I})^{-1} \mathbf{U}^* \bar{\mathbf{a}}\right]^2} \tag{3.83}$$

(which is obtained using $\hat{\hat{\sigma}}_0^2 = \hat{\mathbf{w}}^* \mathbf{R} \hat{\mathbf{w}}$).

Consider the case where $\mathbf{R}$ (or $\hat{\mathbf{R}}$) is singular. Let $\mathbf{U}_n$ denote the submatrix of $\mathbf{U}$ containing the eigenvectors corresponding to the zero eigenvalues of $\mathbf{R}$ (or $\hat{\mathbf{R}}$). Then the upper bound on ζ corresponding to (3.65) becomes

$$\zeta < \frac{1}{\|\mathbf{U}_n^* \bar{\mathbf{a}}\|^2}. \tag{3.84}$$

The above condition on ζ prevents the trivial solution that would give $\mathbf{w}^* \mathbf{R} \mathbf{w} = 0$. To see this, observe that $\mathbf{w} = \mathbf{U}_n \mathbf{U}_n^* \bar{\mathbf{a}} / \|\mathbf{U}_n^* \bar{\mathbf{a}}\|^2$ gives $\mathbf{w}^* \mathbf{R} \mathbf{w} = 0$ and also satisfies $\mathbf{w}^* \bar{\mathbf{a}} = 1$; however, under (3.84), we have $\|\mathbf{w}\|^2 = 1/\|\mathbf{U}_n^* \bar{\mathbf{a}}\|^2 > \zeta$ and hence the previous $\mathbf{w}$ violates the norm constraint in (3.60). When the condition in (3.84) is satisfied, we still have $\lambda > 0$ for NCCB. Moreover, in the steps of NCCB, there is no need that $\gamma_m > 0$ for all $m = 1, 2, \ldots, M$. Hence $\mathbf{R}$ (or $\hat{\mathbf{R}}$) can be singular, under (3.84), and the NCCB is still usable. In particular this means that we can allow $N < M$ to compute $\hat{\mathbf{R}}$.

3.6 ROBUST CAPON BEAMFORMING WITH DOUBLE CONSTRAINTS

3.6.1 The DCRCB Algorithm

To derive DCRCB, we use the covariance fitting SCB in Section 3.3.2, to which we append the spherical uncertainty set in (3.7) and the norm constraint in (3.2). (The extension of DCRCB to an ellipsoidal uncertainty set is possible but it leads to a relatively significant increase of the computational burden, as explained later on.) Proceeding in this way we *directly* obtain a robust estimate of σ_0^2, without any intermediate calculation of a vector $\mathbf{w}$ [17] or any adjustment on $\|\mathbf{a}_0\|$:

$$\begin{aligned} \max_{\sigma^2, \mathbf{a}} \sigma^2 \quad & \text{subject to } \mathbf{R} - \sigma^2 \mathbf{a}\mathbf{a}^* \geq 0 \\ & \|\mathbf{a} - \bar{\mathbf{a}}\|^2 \leq \epsilon \\ & \|\mathbf{a}\|^2 = M \end{aligned} \tag{3.85}$$

where $\bar{\mathbf{a}}$ is given and satisfies (3.8) and ϵ is also given and satisfies $\epsilon > 0$.

Using the fact that, for given $\mathbf{a}$, the solution of (3.85) w.r.t. σ^2 is obtained by $\sigma_0^2 = 1/(\mathbf{a}^*\mathbf{R}^{-1}\mathbf{a})$, the DCRCB problem in (3.85) can be reduced to the following problem

$$\min_{\mathbf{a}} \mathbf{a}^*\mathbf{R}^{-1}\mathbf{a} \quad \text{subject to } \|\mathbf{a} - \bar{\mathbf{a}}\|^2 \leq \epsilon \\ \|\mathbf{a}\|^2 = M. \tag{3.86}$$

Inserting $\|\mathbf{a}\|^2 = \|\bar{\mathbf{a}}\|^2 = M$ in (3.86), we get

$$\min_{\mathbf{a}} \mathbf{a}^*\mathbf{R}^{-1}\mathbf{a} \quad \text{subject to } \mathrm{Re}(\bar{\mathbf{a}}^*\mathbf{a}) \geq M - \frac{\epsilon}{2} \\ \|\mathbf{a}\|^2 = M. \tag{3.87}$$

This optimization problem somewhat resembles the NCCB problem in Section 3.5. Consider first the problem (3.87) without the uncertainty set:

$$\min_{\mathbf{a}} \mathbf{a}^*\mathbf{R}^{-1}\mathbf{a} \quad \text{subject to } \|\mathbf{a}\|^2 = M. \tag{3.88}$$

Let $\mathbf{u}_1$ denote the first eigenvector in $\mathbf{U}$ [see (3.3)]. The solution $\tilde{\mathbf{a}}$ to the above problem is the principal eigenvector $\mathbf{u}_1$ corresponding to the largest eigenvalue of $\mathbf{R}$, scaled so that

$$\|\tilde{\mathbf{a}}\|^2 = M. \tag{3.89}$$

As the eigenvector of a matrix is unique only up to a scalar, we can choose the phase of $\tilde{\mathbf{a}}$ so that $\mathrm{Re}(\bar{\mathbf{a}}^*\tilde{\mathbf{a}})$ is maximum [which is easily done, e.g., $\tilde{\mathbf{a}} = M^{\frac{1}{2}}\mathbf{u}_1 e^{j\phi}$, where $\phi = \arg(\mathbf{u}_1^*\bar{\mathbf{a}})$]. If the so-obtained $\tilde{\mathbf{a}}$ satisfies $\mathrm{Re}(\bar{\mathbf{a}}^*\tilde{\mathbf{a}}) \geq M - \epsilon/2$, then it is our sought solution $\hat{\mathbf{a}}_0$ to (3.87) and the uncertainty set is an inactive constraint.

If not, that is, if

$$\mathrm{Re}(\bar{\mathbf{a}}^*\tilde{\mathbf{a}}) < M - \epsilon/2 \tag{3.90}$$

then $\tilde{\mathbf{a}}$ is not our sought solution. For this case to occur, ϵ must satisfy:

$$\epsilon < 2M - 2\mathrm{Re}(\bar{\mathbf{a}}^*\tilde{\mathbf{a}}) \leq 2M \tag{3.91}$$

where the second inequality above is due to $\mathrm{Re}(\bar{\mathbf{a}}^*\tilde{\mathbf{a}}) \geq 0$. Let $\check{\mathcal{S}}$ be the set defined by the constraints in (3.87). To determine the solution to (3.87) under (3.90), consider the function:

$$f_1(\mathbf{a}, \check{\lambda}, \check{\mu}) = \mathbf{a}^*\mathbf{R}^{-1}\mathbf{a} + \check{\lambda}\big(\|\mathbf{a}\|^2 - M\big) + \check{\mu}(2M - \epsilon - \bar{\mathbf{a}}^*\mathbf{a} - \mathbf{a}^*\bar{\mathbf{a}}) \tag{3.92}$$

where $\check{\lambda}$ and $\check{\mu}$ are the real-valued Lagrange multipliers with $\check{\mu} \geq 0$ and $\check{\lambda}$ satisfying $\mathbf{R}^{-1} + \check{\lambda}\mathbf{I} > 0$ so that the above function can be minimized with respect to $\mathbf{a}$. Evidently we have $f_1(\mathbf{a}, \check{\lambda}, \check{\mu}) \leq \mathbf{a}^*\mathbf{R}^{-1}\mathbf{a}$ for any $\mathbf{a} \in \check{\mathcal{S}}$ with equality on the boundary of $\breve{\mathcal{S}}$. Equation (3.92) can be written as

$$f_1(\mathbf{a}, \breve{\lambda}, \breve{\mu}) = \left[\mathbf{a} - \breve{\mu}(\mathbf{R}^{-1} + \breve{\lambda}\mathbf{I})^{-1}\bar{\mathbf{a}}\right]^*(\mathbf{R}^{-1} + \breve{\lambda}\mathbf{I})\left[\mathbf{a} - \breve{\mu}(\mathbf{R}^{-1} + \breve{\lambda}\mathbf{I})^{-1}\bar{\mathbf{a}}\right]$$
$$- \breve{\mu}^2\bar{\mathbf{a}}^*(\mathbf{R}^{-1} + \breve{\lambda}\mathbf{I})^{-1}\bar{\mathbf{a}} - \breve{\lambda}M + \breve{\mu}(2M - \epsilon). \tag{3.93}$$

Hence the *unconstrained* minimization of $f_1(\mathbf{a}, \breve{\lambda}, \breve{\mu})$ w.r.t. $\mathbf{a}$, for fixed $\breve{\lambda}$ and $\breve{\mu}$, is given by

$$\breve{\mathbf{a}}_{\breve{\lambda}, \breve{\mu}} = \breve{\mu}(\mathbf{R}^{-1} + \breve{\lambda}\mathbf{I})^{-1}\bar{\mathbf{a}} \tag{3.94}$$

Clearly, we have

$$f_2(\breve{\lambda}, \breve{\mu}) \triangleq f_1(\breve{\mathbf{a}}_{\breve{\lambda}, \breve{\mu}}, \breve{\lambda}, \breve{\mu})$$
$$= -\breve{\mu}^2\bar{\mathbf{a}}^*(\mathbf{R}^{-1} + \breve{\lambda}\mathbf{I})^{-1}\bar{\mathbf{a}} - \breve{\lambda}M + \breve{\mu}(2M - \epsilon) \tag{3.95}$$
$$\leq \mathbf{a}^*\mathbf{R}^{-1}\mathbf{a} \quad \text{for any } \mathbf{a} \in \breve{\mathcal{S}}. \tag{3.96}$$

Maximization of $f_2(\breve{\lambda}, \breve{\mu})$ with respect to $\breve{\mu}$ gives

$$\breve{\mu} = \frac{2M - \epsilon}{2\bar{\mathbf{a}}^*(\mathbf{R}^{-1} + \breve{\lambda}\mathbf{I})^{-1}\bar{\mathbf{a}}} \tag{3.97}$$

which indeed satisfies $\breve{\mu} > 0$ [see (3.91)]. Inserting (3.97) into (3.95) we obtain

$$f_3(\breve{\lambda}) \triangleq f_2(\breve{\lambda}, \breve{\mu}) = -\breve{\lambda}M + \frac{\left(M - \frac{\epsilon}{2}\right)^2}{\bar{\mathbf{a}}^*(\mathbf{R}^{-1} + \breve{\lambda}\mathbf{I})^{-1}\bar{\mathbf{a}}}. \tag{3.98}$$

Maximization of the above function with respect to $\breve{\lambda}$ gives

$$\breve{h}(\breve{\lambda}) = \rho \tag{3.99}$$

where

$$\breve{h}(\breve{\lambda}) = \frac{\bar{\mathbf{a}}^*(\mathbf{R}^{-1} + \breve{\lambda}\mathbf{I})^{-2}\bar{\mathbf{a}}}{\left[\bar{\mathbf{a}}^*(\mathbf{R}^{-1} + \breve{\lambda}\mathbf{I})^{-1}\bar{\mathbf{a}}\right]^2} \tag{3.100}$$

and

$$\rho = \frac{M}{\left(M - \frac{\epsilon}{2}\right)^2}. \tag{3.101}$$

Similarly to the proof in Appendix 3.D we can show that, under (3.90), $\check{h}(\check{\lambda})$ is a monotonically decreasing function of $\check{\lambda}$. Moreover, as $\check{\lambda} \to \infty$, $\check{h}(\check{\lambda}) \to 1/M < \rho$ since $\epsilon > 0$. Furthermore, as $\check{\lambda} \to -1/\gamma_1$, $\check{h}(\check{\lambda}) \to 1/|\mathbf{u}_1^*\bar{\mathbf{a}}|^2$. Since $|\mathbf{u}_1^*\bar{\mathbf{a}}|^2 = \mathrm{Re}^2(\tilde{\mathbf{a}}^*\bar{\mathbf{a}})/M < (M - \epsilon/2)^2/M$ [see (3.90)], it follows that $1/|\mathbf{u}_1^*\bar{\mathbf{a}}|^2 > \rho$. Hence there is a unique solution $\check{\lambda} > -1/\gamma_1$ to (3.99) which can be obtained efficiently via, for example, a Newton's method. Using (3.97) in (3.94) yields

$$\hat{\mathbf{a}}_0 = \left(M - \frac{\epsilon}{2}\right) \frac{(\mathbf{R}^{-1} + \check{\lambda}\mathbf{I})^{-1}\bar{\mathbf{a}}}{\bar{\mathbf{a}}^*(\mathbf{R}^{-1} + \check{\lambda}\mathbf{I})^{-1}\bar{\mathbf{a}}} \tag{3.102}$$

which satisfies

$$\mathrm{Re}(\hat{\mathbf{a}}^*\hat{\mathbf{a}}_0) = \hat{\mathbf{a}}^*\bar{\mathbf{a}} = M - \frac{\epsilon}{2} \tag{3.103}$$

and

$$\|\hat{\mathbf{a}}_0\|^2 = M. \tag{3.104}$$

Hence $\hat{\mathbf{a}}_0$ belongs to the boundary of $\check{\mathcal{S}}$. Therefore, $\hat{\mathbf{a}}_0$ is the sought solution. The SOI power estimate is then calculated as

$$\hat{\sigma}_0^2 = \frac{1}{\hat{\mathbf{a}}_0^*\mathbf{R}^{-1}\hat{\mathbf{a}}_0}. \tag{3.105}$$

To derive an upper bound on $\check{\lambda}$, rewrite (3.99) as

$$\frac{\displaystyle\sum_{m=1}^{M} |z_m|^2 \Bigg/ \left(\frac{1}{\gamma_m} + \check{\lambda}\right)^2}{\left[\displaystyle\sum_{m=1}^{M} |z_m|^2 \Bigg/ \left(\frac{1}{\gamma_m} + \check{\lambda}\right)\right]^2} = \rho \tag{3.106}$$

by using $\mathbf{R} = \mathbf{U}\boldsymbol{\Gamma}\mathbf{U}^*$ and $\mathbf{z} = \mathbf{U}^*\bar{\mathbf{a}}$. Hence we have

$$\rho \leq \frac{\|\bar{\mathbf{a}}\|^2 \Bigg/ \left(\frac{1}{\gamma_1} + \check{\lambda}\right)^2}{\|\bar{\mathbf{a}}\|^4 \Bigg/ \left(\frac{1}{\gamma_M} + \check{\lambda}\right)^2} = \frac{\left(\frac{1}{\gamma_M} + \check{\lambda}\right)^2}{M\left(\frac{1}{\gamma_1} + \check{\lambda}\right)^2} \tag{3.107}$$

which gives the following upper bound on $\check{\lambda}$:

$$\check{\lambda} \leq \frac{\frac{1}{\gamma_M} - (M\rho)^{1/2}\left(\frac{1}{\gamma_1}\right)}{(M\rho)^{1/2} - 1}. \tag{3.108}$$

To summarize, DCRCB consists of the following steps.

Step 1. Compute the eigendecomposition of $\mathbf{R}$ (or more practically of $\hat{\mathbf{R}}$).

Step 2. If (3.90) is satisfied, solve (3.106) for $\check{\lambda}$, for example, by a Newton's method, using the knowledge that the solution is unique and it is lower bounded by $-\frac{1}{\gamma_1}$ and upper bounded by (3.108), and then continue to Step 3; otherwise, compute $\hat{\sigma}_0^2 = \gamma_1/M$ (which is obtained by using $\hat{\mathbf{a}}_0 = \tilde{\mathbf{a}}$ in (3.105)) and stop.

Step 3. Use the $\check{\lambda}$ obtained in Step 2 to get

$$\hat{\mathbf{a}}_0 = \left(M - \frac{\epsilon}{2}\right) \frac{\mathbf{U}(\mathbf{I} + \check{\lambda}\mathbf{\Gamma})^{-1}\mathbf{\Gamma}\mathbf{U}^*\bar{\mathbf{a}}}{\bar{\mathbf{a}}^*\mathbf{U}(\mathbf{I} + \check{\lambda}\mathbf{\Gamma})^{-1}\mathbf{\Gamma}\mathbf{U}^*\bar{\mathbf{a}}} \tag{3.109}$$

where the inverse of the diagonal matrix $\mathbf{I} + \check{\lambda}\mathbf{\Gamma}$ is easily computed and $\mathbf{z} = \mathbf{U}^*\bar{\mathbf{a}}$ is available from Step 2.

Step 4. Compute the SOI power estimate of DCRCB using

$$\hat{\sigma}_0^2 = \frac{1}{\left(M - \frac{\epsilon}{2}\right)^2} \frac{\left[\bar{\mathbf{a}}^*\mathbf{U}(\mathbf{I} + \check{\lambda}\mathbf{\Gamma})^{-1}\mathbf{\Gamma}\mathbf{U}^*\bar{\mathbf{a}}\right]^2}{\bar{\mathbf{a}}^*\mathbf{U}(\mathbf{I} + \check{\lambda}\mathbf{\Gamma})^{-2}\mathbf{\Gamma}\mathbf{U}^*\bar{\mathbf{a}}}. \tag{3.110}$$

Note that the steps above of DCRCB do not require that $\gamma_m > 0$ for all $m = 1, 2, \ldots, M$. Hence $\mathbf{R}$ (or $\hat{\mathbf{R}}$) can also be singular in the DCRCB, which means that we can allow $N < M$ to compute $\hat{\mathbf{R}}$.

We also note that, like for NCCB and RCB, the major computational demand of DCRCB comes from the eigendecomposition of $\mathbf{R}$ (or $\hat{\mathbf{R}}$). Therefore, the computational complexity of DCRCB is also comparable to that of SCB. Moreover, like RCB, DCRCB can be modified for recursive implementation. By using the recursive eigendecomposition updating, we can update the power and waveform estimates with $O(M^2)$ flops.

Next, we explain the relationship between the RCB and DCRCB algorithms. In fact, they both start by solving the same problem in (3.85), which is not convex due to the constant norm constraint on $\mathbf{a}$. (This constraint is required due to the scaling ambiguity problem as mentioned in Section 3.4.1.) However, if we remove the constant norm constraint in (3.85), the problem becomes convex and can be easily solved. This is the idea behind the RCB algorithm, which first finds a solution to a simplified problem, that is, (3.85) without the norm constraint on $\mathbf{a}$, then it imposes the norm constraint on the solution to eliminate the scaling ambiguity. The DCRCB algorithm, on the other hand, solves the nonconvex problem in (3.85) rigorously.

Even though RCB is an approximate solution to (3.85), it has been shown to have excellent performance in Section 3.4.3 (see also [16, 30]). Moreover, in the case of RCB, the spherical uncertainty set in (3.85) can be readily generalized to both nondegenerate and flat ellipsoidal uncertainty sets. However, it appears that DCRCB is not as easy to generalize to the case of ellipsoidal uncertainty sets as such a generalization would require a two-dimensional search to determine the Lagrange multipliers $\check{\lambda}$ and $\check{\mu}$. Numerical examples in Section 3.6.4 and [17] have demonstrated that, for a reasonably tight spherical uncertainty set of the array steering vector, DCRCB is the preferred choice for applications requiring high SINR, while RCB is the favored one for applications demanding accurate signal power estimation.

3.6.2 Smallest Possible Spherical Uncertainty Set

For both RCB and DCRCB, the choice of ϵ should be made as small as possible since when ϵ is chosen too large the ability of both RCB and DCRCB to suppress interferences that are close to the SOI will degrade. Toward this end, we note that a phase shift of $\mathbf{a}$ will not change the cost function $\mathbf{a}^*\mathbf{R}^{-1}\mathbf{a}$ or the norm constraint $\|\mathbf{a}\|^2 = M$. Hence ϵ should be chosen as small as possible but such that

$$\epsilon \geq \min_{\alpha} \|\mathbf{a}_0 e^{j\alpha} - \bar{\mathbf{a}}\|^2 \triangleq \epsilon_s \tag{3.111}$$

where $\mathbf{a}_0$ is any possible true SOI steering vector. This analysis explains why it was observed in Section 3.4.3 (see also [16, 30]) that RCB can work well even when $\epsilon < \|\mathbf{a}_0 - \bar{\mathbf{a}}\|^2$. We note that although a phase error in the estimate $\hat{\mathbf{a}}_0$ will not affect the SOI power estimate or the array output SINR, the SOI waveform estimate will contain a phase error. In applications such as communications, a training sequence can be used to estimate the phase error and then compensate it out.

3.6.3 Diagonal Loading Interpretations

In many applications, such as in communications or the global positioning system, the focus is on SOI waveform estimation. The waveform of the SOI, $s_0(n)$, as in (3.5) can be estimated as follows:

$$\hat{s}_0(n) = \hat{\mathbf{w}}^*\mathbf{y}_n \tag{3.112}$$

where $\hat{\mathbf{w}}$ is the corresponding weight vector. For NCCB, $\hat{\mathbf{w}}$ can be obtained directly as the solution to the problem. For RCB and DCRCB, we can substitute the estimated steering vector $\hat{\mathbf{a}}_0$ in lieu of $\mathbf{a}_0$ in (3.10) to obtain $\hat{\mathbf{w}}$.

Diagonal loading is a popular approach to mitigate the performance degradations of SCB in the presence of steering vector error or the small sample size problem. As the name implies, its weight vector has a diagonally loaded form:

$$\mathbf{w} = \kappa(\mathbf{R} + \delta\mathbf{I})^{-1}\bar{\mathbf{a}} \tag{3.113}$$

where δ denotes the diagonal loading level. Also, in (3.113) κ is a scaling factor, which can be important for accurate power estimation; however, it is immaterial for waveform estimation since the quality of the SOI waveform estimate is typically measured by the signal-to-interference-plus-noise ratio (SINR)

$$\text{SINR} = \frac{\sigma_0^2|\hat{\mathbf{w}}^*\mathbf{a}_0|^2}{\hat{\mathbf{w}}^*\left(\sum_{k=1}^{K}\sigma_k^2\mathbf{a}_k\mathbf{a}_k^* + \mathbf{Q}\right)\hat{\mathbf{w}}} \tag{3.114}$$

which is independent of κ.

As a matter of fact, NCCB, RCB and DCRCB can all be interpretted in the unified framework of diagonal loading based approaches. Their differences lie in the distinct choices of the diagonal loading level and the scaling factor. The following subsections contain more detailed discussions on this subject.

3.6.3.1 NCCB. The NCCB weight vector in (3.73) can be rewritten as follows:

$$\hat{\mathbf{w}}_{\text{NCCB}} = \frac{(\mathbf{R} + \lambda\mathbf{I})^{-1}\bar{\mathbf{a}}}{\bar{\mathbf{a}}^*(\mathbf{R} + \lambda\mathbf{I})^{-1}\bar{\mathbf{a}}}. \tag{3.115}$$

Note that (3.115) has the same diagonally loaded form as (3.113).

3.6.3.2 RCB

Nondegenerate Ellipsoidal Uncertainty Set. Using $\hat{\mathbf{a}}_0$ in (3.26) to replace $\mathbf{a}_0$ in (3.10), we can obtain the following RCB weight vector for the case of nondegenerate ellipsoidal uncertainty set:

$$\hat{\mathbf{w}}_{\text{RCB\_N}} = \frac{\left(\mathbf{R} + \frac{1}{\nu}\mathbf{I}\right)^{-1}\bar{\mathbf{a}}}{\bar{\mathbf{a}}^*\left(\mathbf{R} + \frac{1}{\nu}\mathbf{I}\right)^{-1}\mathbf{R}\left(\mathbf{R} + \frac{1}{\nu}\mathbf{I}\right)^{-1}\bar{\mathbf{a}}}. \tag{3.116}$$

When $\mathbf{C}$ is not a scaled identity matrix, the diagonal loading is added to the weighted matrix $\breve{\mathbf{R}}$ defined in (3.20) instead of $\mathbf{R}$ and we refer to this case as the *extended diagonal loading*.

Flat Ellipsoidal Uncertainty Set. The RCB weight vector for the case of flat ellipsoidal Uncertainty set has the form:

$$\begin{aligned}\hat{\mathbf{w}}_{\text{RCB\_F}} &= \frac{\mathbf{R}^{-1}\hat{\mathbf{a}}_0}{\hat{\mathbf{a}}_0^*\mathbf{R}^{-1}\hat{\mathbf{a}}_0} \\ &= \frac{\left(\mathbf{R} + \frac{1}{\breve{\nu}}\mathbf{B}\mathbf{B}^*\right)^{-1}\bar{\mathbf{a}}}{\bar{\mathbf{a}}^*\left(\mathbf{R} + \frac{1}{\breve{\nu}}\mathbf{B}\mathbf{B}^*\right)^{-1}\mathbf{R}\left(\mathbf{R} + \frac{1}{\breve{\nu}}\mathbf{B}\mathbf{B}^*\right)^{-1}\bar{\mathbf{a}}}.\end{aligned} \tag{3.117}$$

To obtain (3.117) we have used the fact [also using (3.51) in (3.58)] that

$$\begin{aligned}\mathbf{R}^{-1}\hat{\mathbf{a}}_0 &= -\mathbf{R}^{-1}\mathbf{B}(\breve{\mathbf{R}} + \breve{\nu}\mathbf{I})^{-1}\bar{\breve{\mathbf{a}}} + \mathbf{R}^{-1}\bar{\mathbf{a}} \\ &= -\mathbf{R}^{-1}\mathbf{B}(\mathbf{B}^*\mathbf{R}^{-1}\mathbf{B} + \breve{\nu}\mathbf{I})^{-1}\mathbf{B}^*\mathbf{R}^{-1}\bar{\mathbf{a}} + \mathbf{R}^{-1}\bar{\mathbf{a}} \\ &= \left(\mathbf{R} + \frac{1}{\breve{\nu}}\mathbf{B}\mathbf{B}^*\right)^{-1}\bar{\mathbf{a}}\end{aligned} \tag{3.118}$$

where the last equality follows from the matrix inversion lemma. We see that in this case, the RCB weight vector again has an extended diagonally loaded form.

Despite the differences in the formulation of our RCB problem and that in [15] (see also Chapter 1 of this book), we prove in Appendix 3.C that the $\hat{\mathbf{w}}_{\mathrm{RCB\_F}}$ in (3.117) and the optimal weight in [15] are identical.

3.6.3.3 DCRCB. Using $\hat{\mathbf{a}}_0$ in (3.102) to replace $\mathbf{a}_0$ in (3.10), we can obtain the following DCRCB weight vector

$$\hat{\mathbf{w}}_{\mathrm{DCRCB}} = \kappa_{\mathrm{DCRCB}}\left(\mathbf{R} + \frac{1}{\check{\lambda}}\mathbf{I}\right)^{-1}\bar{\mathbf{a}} \tag{3.119}$$

where

$$\kappa_{\mathrm{DCRCB}} = \frac{\bar{\mathbf{a}}^*\left(\mathbf{R} + \frac{1}{\check{\lambda}}\mathbf{I}\right)^{-1}\mathbf{R}\bar{\mathbf{a}}}{\left(M - \frac{\epsilon}{2}\right)\bar{\mathbf{a}}^*\left(\mathbf{R} + \frac{1}{\check{\lambda}}\mathbf{I}\right)^{-1}\mathbf{R}\left(\mathbf{R} + \frac{1}{\check{\lambda}}\mathbf{I}\right)^{-1}\bar{\mathbf{a}}}. \tag{3.120}$$

Note that like the RCB weight vector, the DCRCB weight vector also has the form associated with the diagonal loading based approach, except for the real-valued scaling factor in (3.119) as well as the fact that the diagonal loading level in (3.119) can be negative.

REMARKS. The discussions above indicate that NCCB, RCB, and DCRCB all belong to the class of (extended) diagonally loaded Capon beamforming approaches. Unlike fixed diagonal loading approaches, they can adjust their diagonal loading levels adaptively with the data. The distinction between the NCCB and our RCB and DCRCB lies in the fact that the parameter ζ in NCCB is not directly linked to the steering vector uncertainty set, while RCB and DCRCB explicitly address the steering vector uncertainty problem and can be used to determine exactly the optimal amount of diagonal loading needed for a given uncertainty set of the steering vector.

3.6.4 Numerical Examples

Next, we provide numerical examples to compare the performances of the delay-and-sum beamformer, SCB, NCCB, RCB and DCRCB. In all of the examples considered in this section, we assume a uniform linear array with $M = 10$ sensors and half-wavelength sensor spacing, and a spatially white Gaussian noise whose covariance matrix is given by $\mathbf{Q} = \mathbf{I}$. For NCCB, we set $\zeta = \beta/M$, where β ($\beta \geq 1$) is a user parameter. The larger the β, the closer NCCB is to SCB. On the other hand, the smaller the β, the closer NCCB is to the delay-and-sum beamformer. When $\beta = 1$, NCCB becomes the delay-and-sum beamformer and hence it uses the assumed array steering vector divided by M as the weight vector. Unless otherwise stated, we use the beamforming methods with the theoretical array covariance matrix $\mathbf{R}$.

Example 3.6.1: Comparison of Delay-and-Sum Beamformer, SCB, NCCB, RCB and DCRCB in the Presence of Array Calibration Errors We consider an imaging example where we wish to determine the incident signal power as a function of the signal arrival angle θ relative to the array normal. We assume that there are five incident signals with powers 30, 60, 40, 35, and 10 dB from directions -35°, -15°, 0°, 10°, and 40°, respectively. To simulate the array calibration error (the sensor amplitude and phase error as well as the sensor position error), each element of the steering vector for each incident signal is perturbed with a zero-mean circularly symmetric complex Gaussian random variable normalized so that $\epsilon_s = 1.0$. The perturbing Gaussian random variables are independent of each other. For RCB and DCRCB, we use $\epsilon = 1.0$. For NCCB, we choose $\beta = 6.0$ so that the peak widths of the NCCB and DCRCB are about the same.

Figure 3.10(a) shows the signal power estimates as functions of the arrival angle θ obtained via using the delay-and-sum beamformer, SCB, NCCB and DCRCB methods. The small circles in the figure denote the true (direction of arrival, power)-coordinates of the five incident signals. Since the power estimates of RCB and DCRCB are almost the same for this example, only the DCRCB power estimates are shown in the figure. Note that SCB can give good direction-of-arrival estimates for the incident signals based on the peak locations. However, the SCB estimates of the incident signal powers are way off. NCCB is more robust than SCB but still substantially underestimates the signal powers. On the other hand, our DCRCB provides excellent power estimates of the incident sources. As expected, the delay-and-sum beamformer has poorer resolution than the other beamformers. Moreover, the sidelobes of the former result in false peaks.

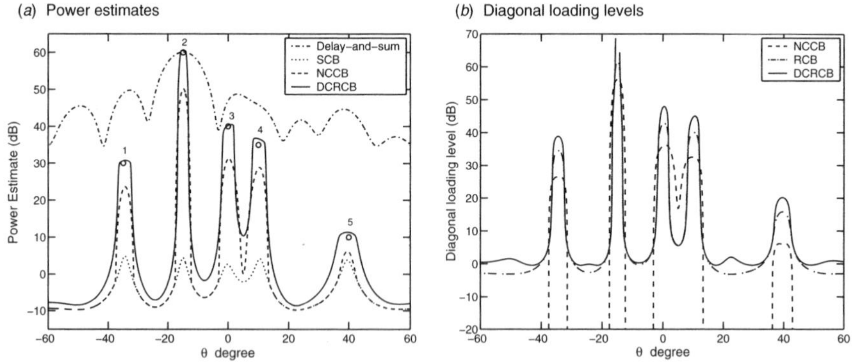

Figure 3.10 Power estimates and diagonal loading levels (using $\mathbf{R}$) versus the steering direction θ when $\epsilon = 1.0$ and $\beta = 6.0$. The true powers of the incident signals from -35°, -15°, 0°, 10°, and 40° are denoted by circles, and $\epsilon_s = 1.0$.

Figure 3.10(b) shows the diagonal loading levels of the NCCB, RCB and DCRCB approaches. Depending on whether the condition of (3.65) is satisfied or not, NCCB can have a nonzero or zero diagonal loading level. This results in the discontinuities in the NCCB diagonal loading level curve. The discontinuity in the DCRCB diagonal loading level curve is due to the fact that around the strongest signal, the condition of (3.90) is not satisfied. As a result, DCRCB is no longer a diagonal loading approach around the strongest signal.

Example 3.6.2: Making NCCB Have the Same Diagonal Loading Level as RCB in the Presence of Array Calibration Errors For each steering angle θ in Figure 3.11(a), β is chosen to make NCCB have the same diagonal loading level as RCB when $\epsilon = 1.0$ is used in RCB. We note that for NCCB and RCB to have the same diagonal level, β must be chosen in a complicated manner depending on both ϵ and the data itself. Figure 3.11(b) shows the signal power estimates as functions of θ obtained via using NCCB and RCB with the β in NCCB chosen so that NCCB and RCB have the same diagonal loading levels. We note that the RCB signal power estimates are much more accurate than those obtained using NCCB and hence the norm constraint imposed on $\hat{\mathbf{a}}_0$ in (3.37) is very helpful for accurate SOI power estimation.

Example 3.6.3: Comparison of RCB and DCRCB in the Presence of Array Calibration Errors Figures 3.12(a) and 3.12(b) show the power estimates as functions of θ obtained via using RCB and DCRCB with $\epsilon = 0.7$ and $\epsilon = 1.5$, respectively, for Example 3.6.1. Note that when $\epsilon < \epsilon_s = 1.0$, the RCB and DCRCB signal power estimates are not as accurate as when $\epsilon > \epsilon_s$, but the peaks are sharper.

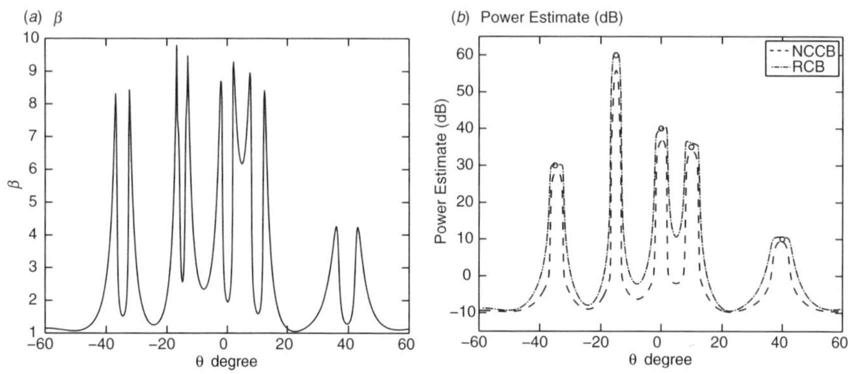

Figure 3.11 (a) For each steering direction θ, β is chosen to make NCCB have the same diagonal loading levels as RCB with $\epsilon = 1.0$. (b) Power estimates versus the steering direction θ via the RCB and NCCB approaches. For RCB, $\epsilon = 1.0$. For NCCB, β is chosen as in (a). The true powers of the incident signals from -35°, -15°, 0°, 10°, and 40° are denoted by circles, and $\epsilon_s = 1.0$.

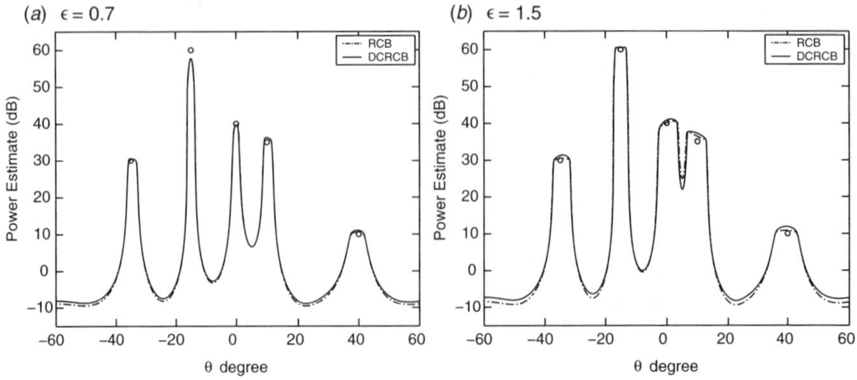

Figure 3.12 Power estimates versus the steering direction θ when (a) $\epsilon = 0.7$ and (b) $\epsilon = 1.5$. The true powers of the incident signals from $-35°$, $-15°$, $0°$, $10°$, and $40°$ are denoted by circles, and $\epsilon = 1.0$.

In Figure 3.13, we compare the SINRs and the signal power estimates for the five incident signals, as functions of ϵ, obtained via using RCB and DCRCB. Figures 3.13 (*a*), 3.13(*c*), 3.13(*e*), 3.13(*g*), and 3.13(*i*) show the SINRs of the five signals as functions of ϵ. Figures 3.13(*b*), 3.13(*d*), 3.13(*f*), 3.13(*h*), and 3.13(*j*) show the power estimates of the five signals as functions of ϵ, with the horizontal dotted lines denoting the true signal powers. Note that except for the 4th signal, the SINR of DCRCB is in general higher than that of RCB when ϵ is not too far from ϵ_s. Hence for applications requiring waveform estimation, the former may be preferred over the latter if ϵ_s is known reasonably accurately. For the 2nd signal in Figure 3.13(*c*), the condition of (3.90) is not satisfied and hence DCRCB uses the scaled principal eigenvector as the estimated steering vector. For this case, DCRCB is always better than RCB, no matter how ϵ is chosen. On the other hand, for signal power estimation RCB in general outperforms DCRCB and hence may be preferred in applications such as acoustic imaging where only the signal power distribution as a function of angle or location is of interest. We also note that the larger the ϵ, the more RCB and DCRCB will overestimate the signal power. Therefore, if possible, ϵ should not be chosen much larger than ϵ_s.

In the next examples, we concentrate on the fifth signal from 40°, which is treated as the signal-of-interest (SOI). The other four signals are considered as interferences. In the following figures for the SOI power estimates, the dotted lines correspond to the true SOI power.

Example 3.6.4: Comparison of RCB and DCRCB when the Fourth Signal Changes its Direction of Arrival We consider a scenario where the fourth signal changes its direction of arrival from 20° to 60° with the directions of arrival of the SOI and the other three interfering signals fixed. The array suffers from the same calibration error as in Figure 3.10. Note from Figure 3.14 that when the

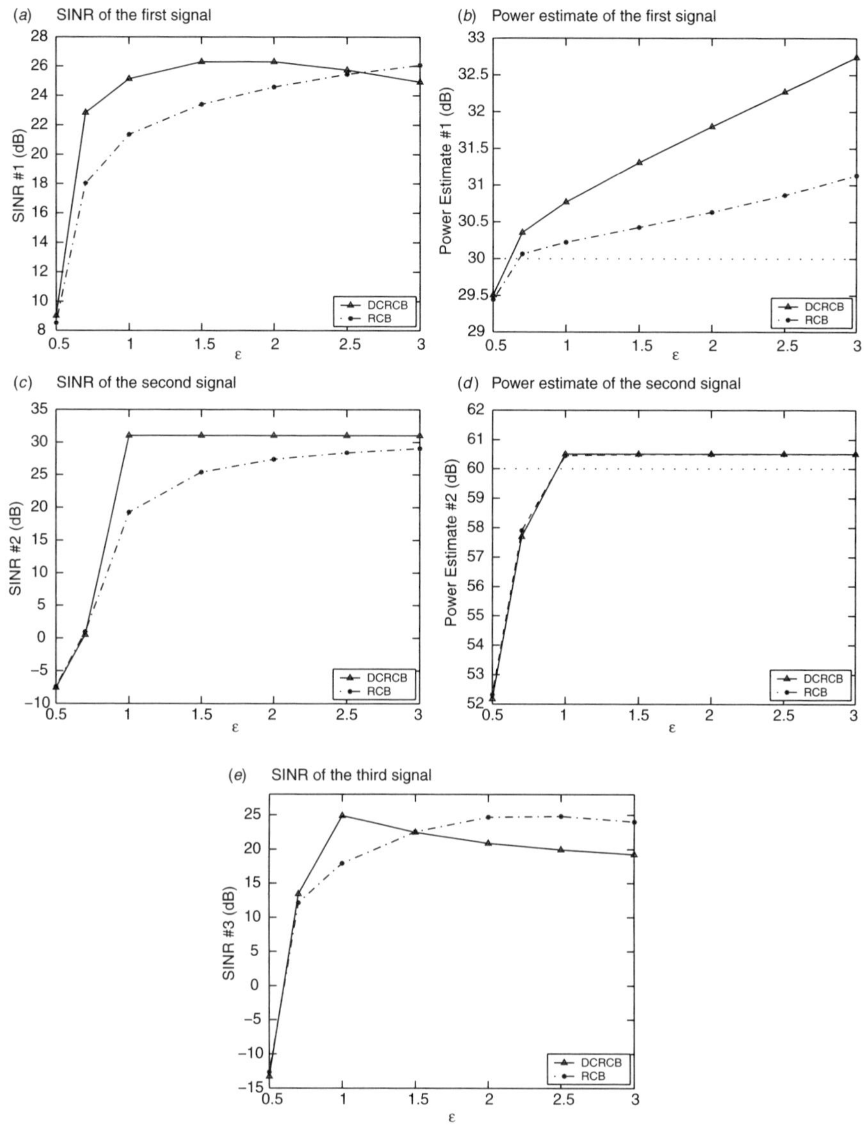

Figure 3.13 Comparison of the RCB and DCRCB approaches for each incident signal, as ϵ varies, when $\epsilon_s = 1.0$.

direction of arrival of an interference signal becomes too close to that of the SOI, both RCB and DCRCB suffer from severe performance degradations in both SINR and SOI power estimation accuracy. As expected, the larger the ϵ used the weaker the interference suppression capability of both methods when an interfering signal is nearby the SOI.

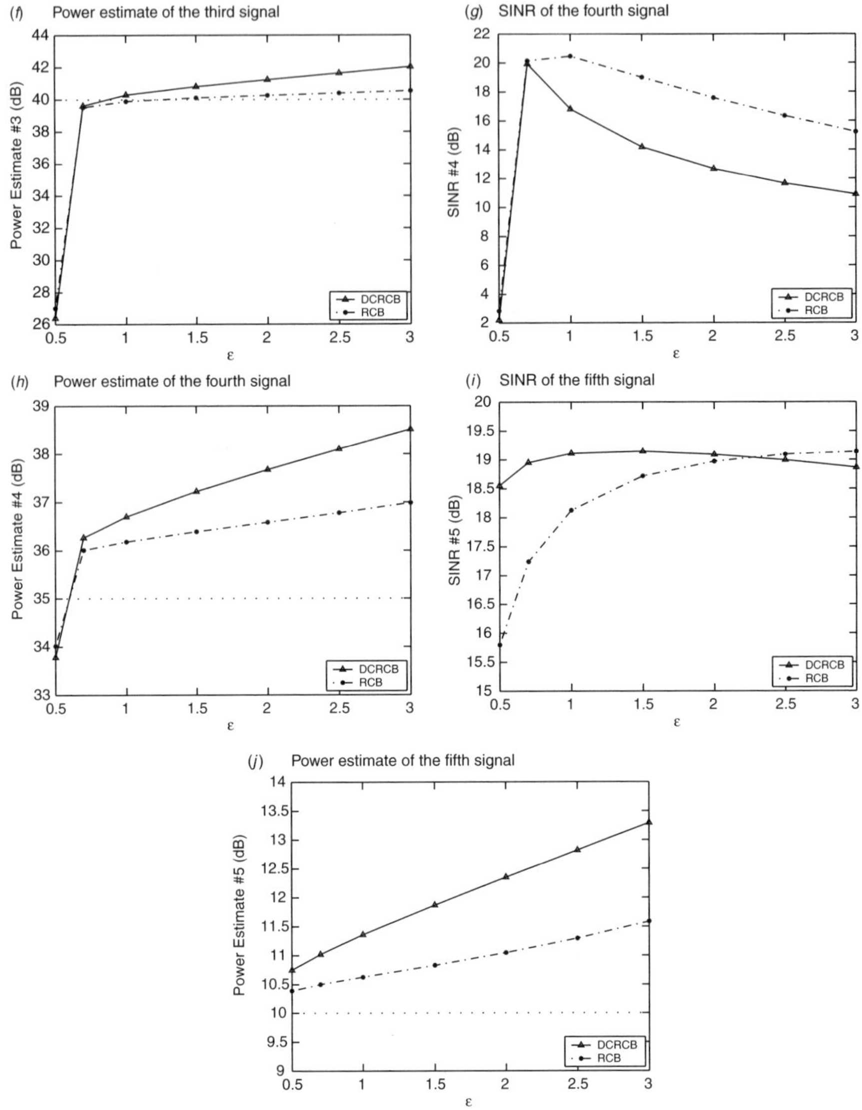

Figure 3.13 (*Continued*).

Example 3.6.5: Comparison of RCB and DCRCB for the Case of Finite Number of Snapshots in the Presence of Look Direction Errors We consider the effect of the number of snapshots N on the SINR and SOI power estimation accuracy of RCB and DCRCB when the sample covariance matrix $\hat{\mathbf{R}}$ in (3.4) is used in lieu of the theoretical array covariance matrix $\mathbf{R}$. We assume that the steering vector error is due to an error in the SOI pointing angle, which we assume to be $\theta_5 + \Delta$, where θ_5 is the true arrival angle of the SOI. In this example, $\epsilon_s = 0.5603$

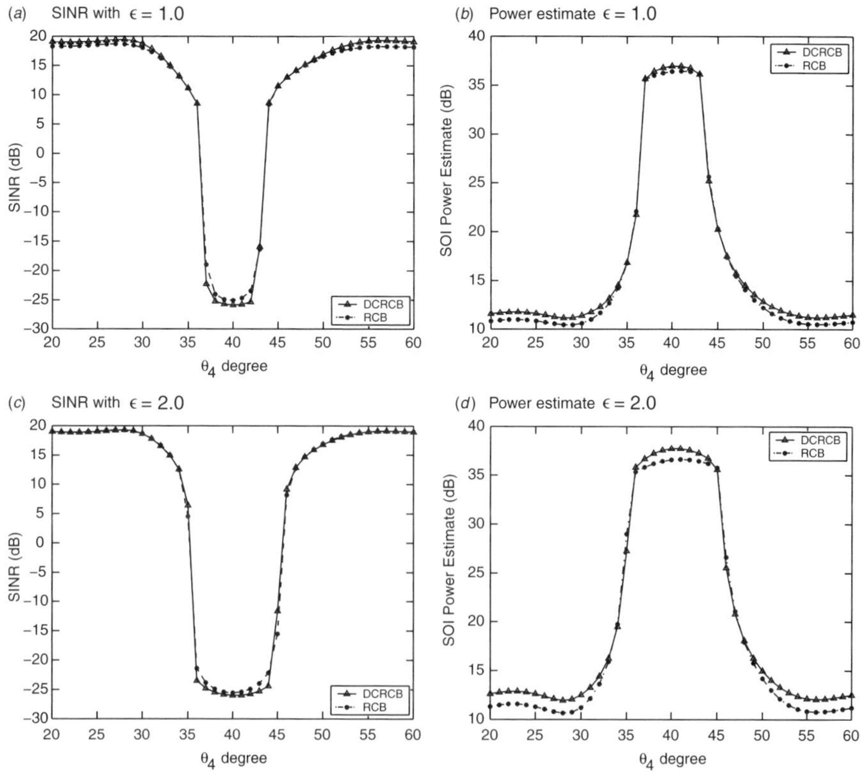

Figure 3.14 Comparison of the RCB and DCRCB approaches with (*a–b*) $\epsilon = 1.0$. and (*c–d*) $\epsilon = 2.0$ when θ_4 (the direction of arrival of the fourth signal) is changing from 20° to 60°. The SOI power is 10 dB and $\epsilon_s = 1.0$.

corresponds to $\Delta = 2.0°$. We use 100 Monte Carlo simulations to obtain the mean SINR and SOI power estimates. It is worth noting that both RCB and DCRCB allow N to be less than the number of array elements M. We use $N = 6$ and $N = 8$ for the $N < M$ case in this example. For DCRCB, when the condition of (3.90) is not satisfied, we calculate the SOI power estimate by $\hat{\sigma}_0^2 = \gamma_1/M$ (as explained in Step 2 of the DCRCB algorithm). Note from Figure 3.15 that the convergence properties of both methods are quite good and somewhat similar. Since the errors between $\hat{\mathbf{R}}$ and $\mathbf{R}$ can be viewed as equivalent steering vector errors, ϵ should be chosen larger than ϵ_s, especially for small N.

Example 3.6.6: Comparison of RCB and DCRCB when the Power of the Fourth Signal is Varying We now compare the performances of RCB and DCRCB when the power of the fourth signal is varying. As in the previous example, we have $\epsilon_s = 0.5603$ corresponding to $\Delta = 2.0°$. The INR in Figure 3.16 refers to the ratio between the 4th signal power and the noise power. Note from Figure 3.16(a) that

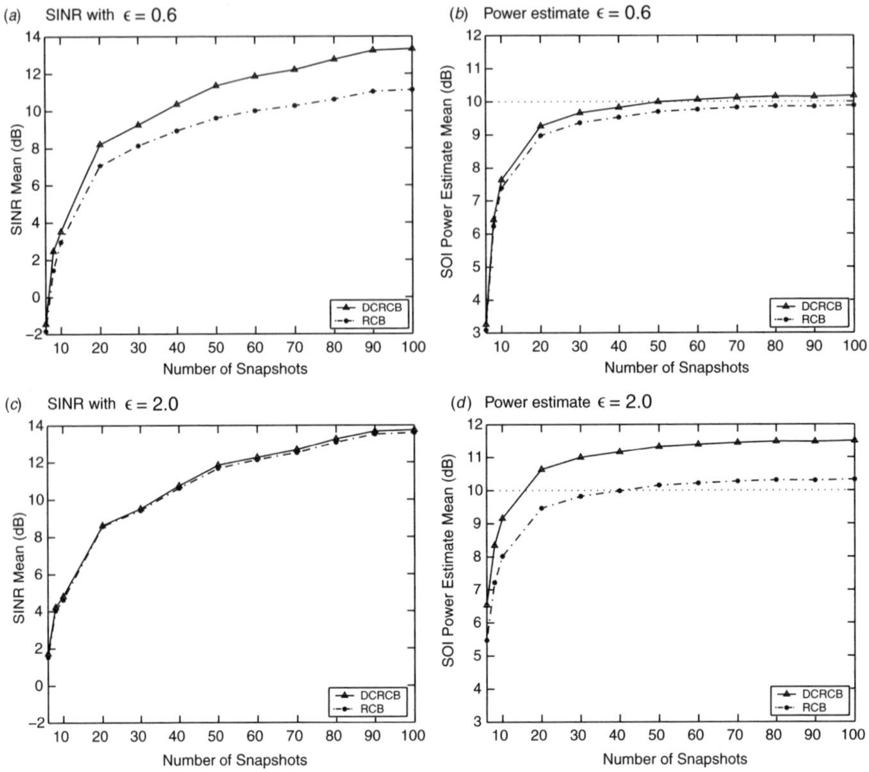

Figure 3.15 Comparison of the RCB and DCRCB approaches, as the snapshot number varies, when (*a,b*) $\epsilon = 0.6$. and (*c,d*) $\epsilon = 2.0$ The SOI power is 10 dB and $\epsilon_s = 0.5603$ (corresponding to $\Delta = 2.0°$).

the SINR of DCRCB is much better than that of RCB when $\epsilon = 0.6$. However, when ϵ is large, for example when $\epsilon = 2.0$ as in Figure 3.16(*c*), and when the INR is comparable to the SNR of the SOI, DCRCB has lower SINR than RCB. From the diagonal loading levels of the methods shown in Figures 3.16(*e*) and 3.16(*f*), it is interesting to note that the diagonal loading level of RCB when $\epsilon = 2.0$ is about the same as that of DCRCB when $\epsilon = 0.6$. As a result, the SINR and SOI power estimate of RCB when $\epsilon = 2.0$ are about the same as those of DCRCB when $\epsilon = 0.6$. Note also from Figure 3.16 that when the INR becomes close to the SNR, there is a performance drop in the array output SINR. One possible explanation is that when the INR is much smaller than the SNR, its impact on the SOI is small. As the INR increases, it causes the SINR to drop. As the INR becomes much stronger than the SNR, the adaptive beamformers start to form deep and accurate nulls on the interference and as a result, the SINR improves again and then becomes stable.

Example 3.6.7: Comparison of RCB and DCRCB when the SOI Power Varies Next, we consider the case where the SOI power varies. We choose

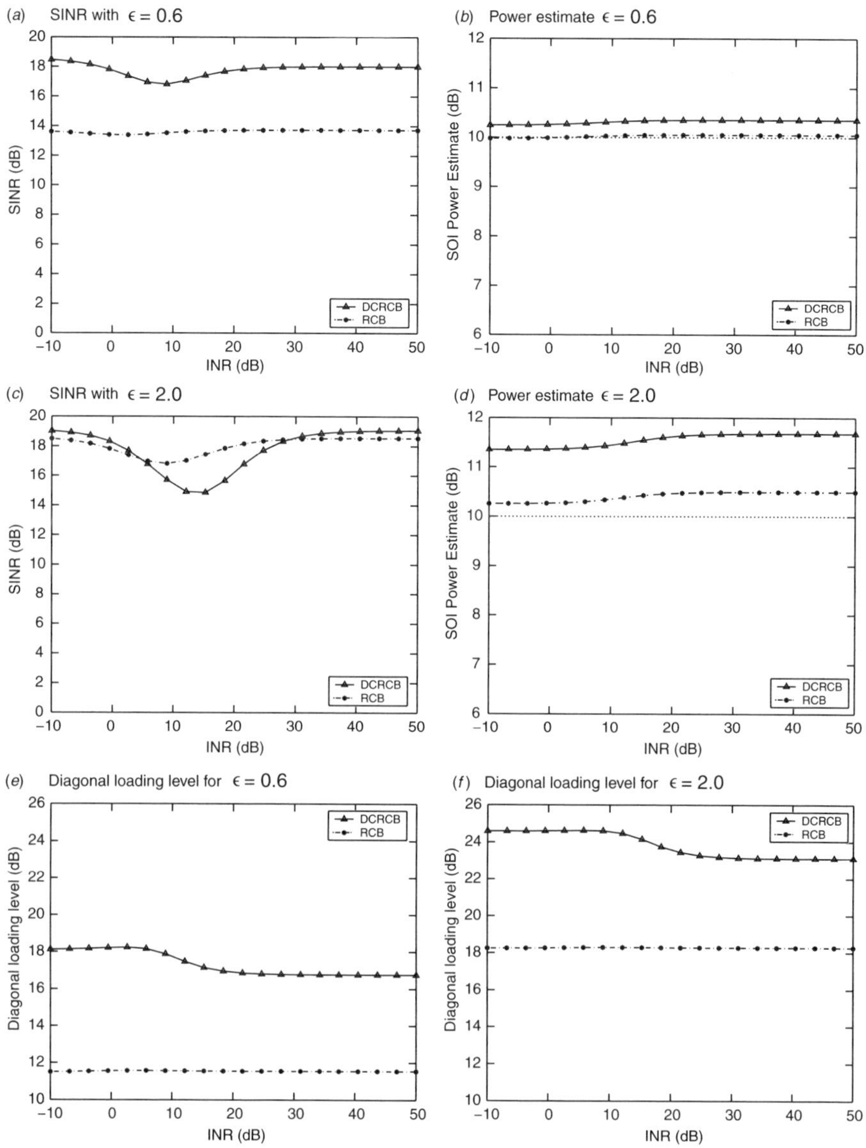

Figure 3.16 Comparison of the RCB and DCRCB approaches with (*a,b,e*) $\epsilon = 0.6$ and (*c,d,f*) $\epsilon = 2.0$. The SOI power is 10 dB and $\epsilon_s = 0.5603$ (corresponding to $\Delta = 2.0°$).

$\epsilon = 0.6$ and consider three cases: $\Delta = -1.0°$, $-2.0°$, and $3.0°$, with the corresponding ϵ_s being 0.1474, 0.5939, and 1.2289, respectively. Figure 3.17 shows that as long as ϵ is greater than ϵ_s and the SOI SNR is medium or high, the SOI power estimates of RCB and DCRCB are excellent. Their SINR curves are also quite high, but they

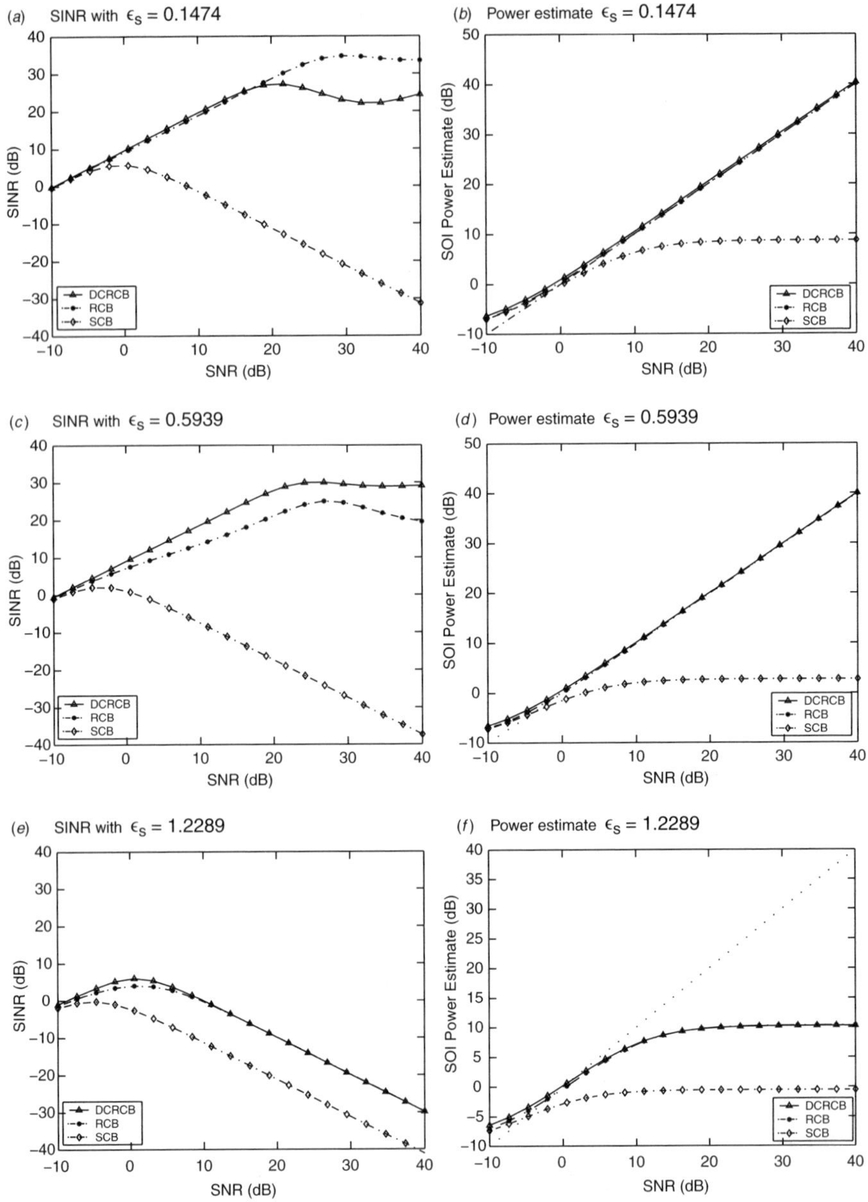

Figure 3.17 Comparison of the RCB and DCRCB approaches as SNR varies when (*a,b*) $\epsilon_S = 0.1474$, (*c,d*) $\epsilon_S = 0.5939$, and (*e,f*) $\epsilon_S = 1.2289$, corresponding to $\Delta = -1.0^\circ, -2.0^\circ$, and 3.0°, respectively. The SOI power is 10 dB and $\epsilon = 0.6$.

drop when the SOI SNR approaches the INR of one of the interfering signals. One possible explanation is given at the end of the previous paragraph. From Figures 3.17(*e*) and 3.17(*f*), we see that when ϵ is smaller than ϵ_s, the performances of both RCB and DCRCB drop drastically as the SOI SNR increases from moderate to high. This is because the SOI is suppressed as an interference for this case. Note that RCB and DCRCB significantly outperform SCB in SINR and SOI power estimates.

Next, we use 100 Monte Carlo simulations to compare the statistical performances of RCB and DCRCB.

Example 3.6.8: Comparison of the Statistical Performances of RCB and DCRCB in the Presence of Look Direction Errors We consider the case where the true arrival angle of the fifth signal is uniformly distributed between 38° and 42° but it is assumed to be 40°. Figure 3.18 compares the SINR and the SOI power estimates of RCB and DCRCB obtained in the 100 Monte Carlo trials. We note that the SINR mean of DCRCB is about the same as that of RCB but the SINR variance of DCRCB is much smaller than that of RCB (especially when $\epsilon = 0.6$, which is quite tight since $0 \leq \epsilon_s \leq 0.5939$). Hence this example shows again that DCRCB may be preferred over RCB when higher array output SINR for waveform estimation is needed. On the other hand, the bias and the variance of the SOI power estimates of RCB are smaller than those of DCRCB. This is especially so for large ϵ; note that a large ϵ is not a problem here since the interfering signals are quite far away from the SOI. Hence this example also shows that RCB may be preferred over DCRCB in applications requiring accurate SOI power estimation including radar, acoustic, and ultrasound imaging.

Example 3.6.9: Comparison of the Statistical Performances of RCB and DCRCB in the Presence of Array Calibration Errors We now consider an example of array calibration error that consists of perturbing each element of the steering vector for each incident signal with a zero-mean circularly symmetric complex Gaussian random variable with a variance equal to 0.1. The perturbing Gaussian random variables are independent of each other. The calibration error is not scaled or normalized in any way and hence $0 \leq \epsilon_s \leq \infty$. In Figure 3.19, we compare the means and variances of the SINR and SOI power estimates of RCB and DCRCB, as functions of ϵ. The figure shows once again that with a reasonable choice of ϵ, DCRCB may be preferred for applications requiring high SINR whereas RCB may be favored for applications demanding accurate SOI power estimation.

3.7 ROBUST CAPON BEAMFORMING WITH CONSTANT BEAMWIDTH AND CONSTANT POWERWIDTH

The main motivation for our work in this section comes from an acoustic imaging application in which the goal is to consistently estimate the SOI power in the presence of strong interferences as well as some uncertainty in the SOI direction of arrival. Due to its sensitivity to steering vector mismatch and small sample size, SCB has not been used very much in acoustic imaging despite its potential benefits. The various advantages of RCB, including robustness against array steering vector errors and small sample size, high resolution, and superb interference

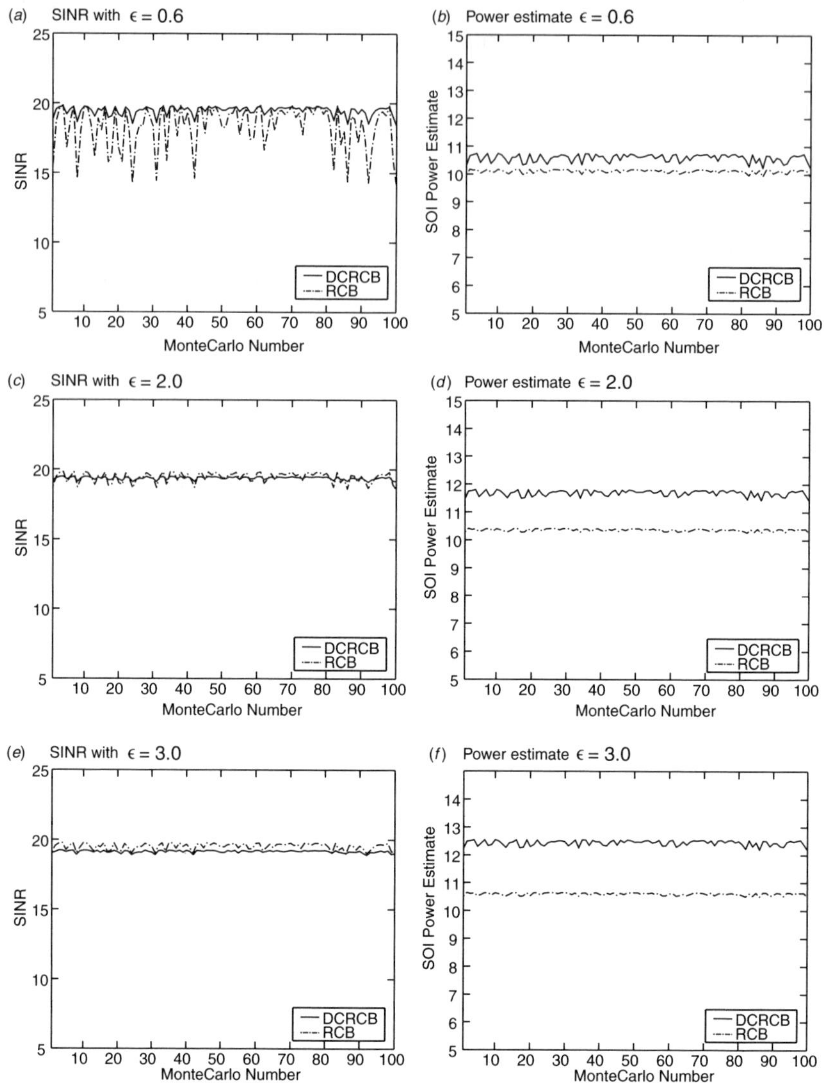

Figure 3.18 Comparison of the RCB and DCRCB approaches in 100 Monte Carlo trials when (*a,b*) $\epsilon = 0.6$, (*c,d*) $\epsilon = 2.0$, and (*e,f*) $\epsilon = 3.0$. The direction of arrival of the fifth signal is uniformly distributed between 38° and 42° and its assumed angle is 40°. The SOI power is 10 dB and $0 \leq \epsilon_s \leq 0.5939$.

suppression capability make it a very promising approach to mitigate the problem of SCB.

Although RCB was devised under the narrowband assumption, it can also deal with wideband acoustic signals by first dividing the array outputs into many narrowband frequency bins using the fast Fourier transform (FFT) and then applying the

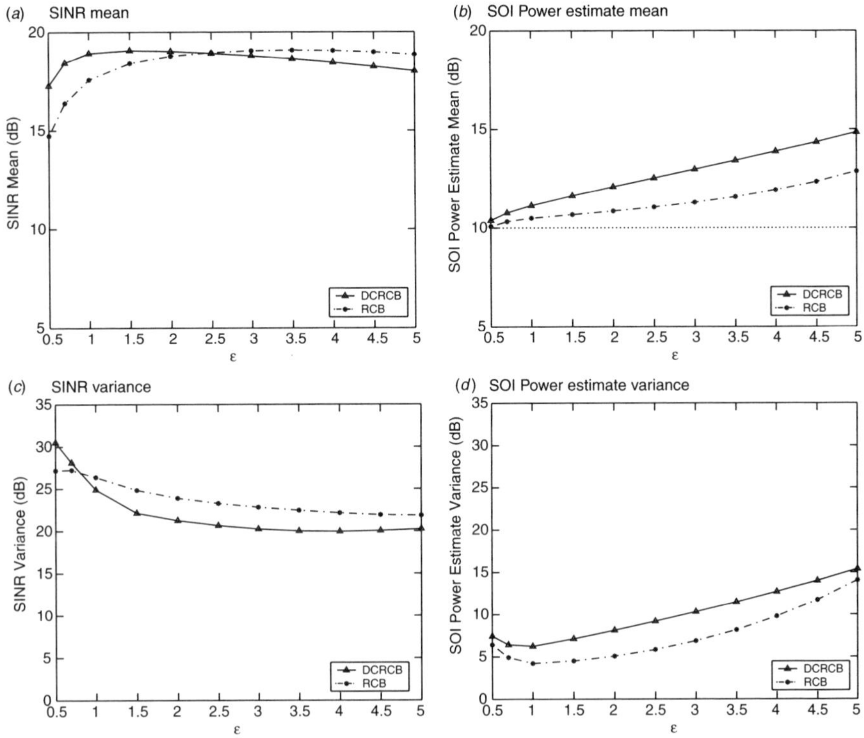

Figure 3.19 Comparison of the RCB and DCRCB approaches in the presence of potentially significant array calibration errors. The SOI power is 10 dB, and $0 \leq \epsilon_S \leq \infty$.

narrowband RCB to each bin separately. However, it is well-known that as the frequency increases, the beamwidth of both data-independent and data-adaptive beamformers decreases. This beamwidth variation as a function of frequency will subject the signals incident on the outer portions of the main beam to lowpass filtering and lead to distorted signal spectra or inaccurate SOI power estimation [46–48]. Hence it is desirable that the beamwidth of a beamformer remains approximately constant over all frequency bins of interest. In fact, a constant-beamwidth beamformer is desirable in many applications including ultrasonics, underwater acoustics, acoustic imaging and communications, and speech acquisition [46, 48, 49]. This prevents future corrections for different frequencies, and contributes to consistent sound pressure level (SPL) estimation, which means that for an acoustic wideband monopole source with a flat spectrum the acoustic image for each frequency bin stays the same.

In the past five decades, many approaches have been proposed to obtain constant-beamwidth beamformers including harmonic nesting [47, 49–51], multibeam [46, 52, 53], asymptotic theory based methods [54], and approximation of a continuously distributed sensor via a finite set of discrete sensors [55–57]. Among these approaches, harmonic nesting is commonly used for acoustic imaging via microphone arrays. For example, in [49–51] a shading scheme is used for a directional array consisting of a set of harmonically nested subarrays, each of which is designed for a particular frequency bin. For each array element, shading weights are devised as a function of the frequency. This shading scheme can provide a constant beamwidth for frequencies between 10 and 40 kHz when used with the delay-and-sum (DAS) beamformer. Hereafter, this approach will be referred to as the shaded DAS (SDAS).

In this section, we show that we can achieve a constant beamwidth across the frequency bins for an adaptive beamformer, by combining our RCB with the shading scheme devised in [49–51], provided that there are no strong interfering signals near the main beam of the array. We refer to this approach as the constant-beamwidth RCB (CBRCB) algorithm [58]. We also show that we can attain a constant powerwidth, and hence consistent power estimates across the frequency bins, by using RCB with a frequency-dependent uncertainty parameter for the array steering vector; we refer to the so-obtained beamformer as the constant-powerwidth RCB (CPRCB) [58]. CBRCB and CPRCB inherits the strength of RCB in the robustness against array steering vector errors and finite sample size problems, high resolution, and excellent interference suppression capability. Moreover, they both can be efficiently implemented at a comparable computational cost with that of SCB.

3.7.1 Data Model and Problem Formulation of Acoustic Imaging

We focus herein on forming acoustic images using a microphone array, which are obtained by determining the sound pressure estimates corresponding to the two- or three-dimensional coordinates of a grid of locations. The signal at each grid location of interest is referred to as the SOI.

First we introduce a wideband data model. Assume that a wideband SOI impinges on an array with M elements. We divide each sensor output into N nonoverlapping blocks with each block consisting of I samples. We then apply an I-point FFT to each block to obtain I narrowband frequency bins. The data vector, $\mathbf{y}_i(n)$, for the ith frequency bin and the nth snapshot can be written as

$$\mathbf{y}_i(n) = \mathbf{a}_i(\mathbf{x}_0)s_i(n) + \mathbf{e}_i(n), \quad n = 1, \ldots, N; \quad i = 1, \ldots, I \tag{3.121}$$

where $s_i(n)$ stands for the complex-valued waveform of the SOI that is present at the location coordinate $\mathbf{x}_0$ and the ith frequency bin, $\mathbf{e}_i(n)$ represents a complex noise plus interference vector for the coordinate $\mathbf{x}_0$ and the ith frequency bin, and $\mathbf{a}_i(\mathbf{x}_0)$ is the SOI array steering vector, which depends on both $\mathbf{x}_0$ and the ith frequency bin. Without loss of generality, we assume that

$$\|\mathbf{a}_i(\mathbf{x}_0)\|^2 = M. \tag{3.122}$$

The nominal or assumed array steering vector $\bar{\mathbf{a}}_i(\mathbf{x}_0)$ has the form:

$$\bar{\mathbf{a}}_i(\mathbf{x}_0) = \left[e^{j2\pi f_i \tau_1} \quad e^{j2\pi f_i \tau_2} \quad \cdots \quad e^{j2\pi f_i \tau_M}\right]^T \tag{3.123}$$

where τ_m denotes the propagation time delay between the source at $\mathbf{x}_0$ and the mth sensor, f_i is the center of the ith frequency bin, and $(\cdot)^T$ denotes the transpose. The covariance matrix for the ith frequency bin can be written as:

$$\mathbf{R}_i = E[\mathbf{y}_i(n)\mathbf{y}_i^*(n)], \quad i = 1, \ldots, I \tag{3.124}$$

where $E[\cdot]$ is the expectation operator, and $(\cdot)^*$ denotes the conjugate transpose.

In aeroacoustic measurements using arrays, the sound pressure response is normally shown. The intensity of the sound pressure response is measured on a logarithmic scale and is referred to as the sound pressure level (SPL), which is defined as [59]

$$\text{SPL} = 20 \log_{10}(p_{\text{rms}}/p_{\text{ref}})\,\text{dB} \tag{3.125}$$

where p_{rms} denotes the root-mean-squared pressure in Pa and p_{ref} stands for the reference pressure. For air, the reference pressure is 20 μPa corresponding to the hearing threshold of young people.

Next, we determine a scaling coefficient needed for calculating SPL estimates based on $\hat{\mathbf{R}}_i$. Let $y(l), l = 1, 2, \ldots, I$, represent a wideband time-domain sequence with I samples, and let $Y(i), i = 1, 2, \ldots, I$, denote the frequency-domain sequence obtained by applying an I-point FFT to $\{y(l)\}$. Let $y_i(l), l = 1, 2, \ldots, I$, be the narrowband time-domain sequence corresponding to the ith frequency bin of the above frequency-domain sequence. In other words, $\{y_i(l)\}$ is obtained by using an inverse FFT on $\{0, \ldots, 0, Y(i), 0, \ldots, 0\}$. Making use of only $Y(i)$, we can determine a root-mean-squared pressure estimate for the ith frequency bin, $\tilde{p}_{\text{rms}}$, as follows:

$$\tilde{p}_{\text{rms}} = \sqrt{\frac{1}{I}\sum_{l=1}^{I} |y_i(l)|^2} = \sqrt{\frac{1}{I^2}|Y(i)|^2}. \tag{3.126}$$

The second equality is due to the fact that $\sum_{l=1}^{I} |y_i(l)|^2 = \frac{1}{I}|Y(i)|^2$ according to the Parseval's theorem. Substituting (3.126) into (3.125), the SPL estimate for the ith narrowband frequency bin can be written as $10\log_{10}(|Y(i)|^2/(p_{\text{ref}}^2 I^2))$ in dB. Therefore, a scaling coefficient, $1/I^2$, should be used when estimating the sound pressure from the data in the frequency domain. Let σ_i denote the root-mean-squared pressure estimate for the ith narrowband frequency bin obtained using $\hat{\mathbf{R}}_i$. Then, according to the previous discussion,

$$\text{SPL} = 10\log_{10}\left[\sigma_i^2/(p_{\text{ref}}^2 I^2)\right]\text{dB}. \tag{3.127}$$

We note in passing that in many applications other than acoustics, the intensities of the sources are often measured by power estimates, which, unlike the SPL estimates, do not require scaling by a reference pressure.

In this chapter we distinguish between beampattern and powerpattern and make use of this distinction in the design of the robust constant-beamwidth and constant-powerwidth beamformers. We define the *beampattern* for the ith frequency bin as

$$\mathrm{BP}_i(\mathbf{x}) = |\mathbf{a}_i^*(\mathbf{x})\mathbf{w}(\mathbf{x}_g)|^2 \tag{3.128}$$

where $\mathbf{w}(\mathbf{x}_g)$ denotes the beamformer's weight vector corresponding to a given location, $\mathbf{x}_g$, and $\mathbf{x}$ is varied to cover each location of interest.

Next, we introduce the *powerpattern*, which at the ith frequency bin is defined as

$$\mathrm{PP}_i(\mathbf{x}) = |\mathbf{a}_i^*(\mathbf{x}_g)\mathbf{w}(\mathbf{x})|^2. \tag{3.129}$$

We remark that the beampattern shows how the beamformer will pass the SOI and interfering signals when it is steered to $\mathbf{x}_g$, whereas the powerpattern shows how the beamformer will pass the signal at $\mathbf{x}_g$ when it is steered to $\mathbf{x}$. $\mathrm{PP}_i(\mathbf{x})$ can be used to measure approximately the normalized power responses corresponding to a series of locations, and hence it is named powerpattern.

To see this, we assume that the theoretical covariance matrix for the ith frequency bin has the following form:

$$\mathbf{R}_i = \sigma_i^2 \mathbf{a}_i(\mathbf{x}_0)\mathbf{a}_i^*(\mathbf{x}_0) + \mathbf{Q}, \tag{3.130}$$

where σ_i^2 denotes the SOI power and $\mathbf{Q}$ stands for the interference-plus-noise covariance matrix. If $\mathbf{x}_g = \mathbf{x}_0$ and the signal-to-interference-plus-noise-ratio (SINR) is high, then it follows that $\mathbf{w}^*(\mathbf{x})\mathbf{R}_i\mathbf{w}(\mathbf{x}) \approx \sigma_i^2|\mathbf{w}^*(\mathbf{x})\mathbf{a}_i(\mathbf{x}_g)|^2 \propto \mathrm{PP}_i(\mathbf{x})$.

We next use an imaging example to illustrate the concept of beampattern and powerpattern. Consider a function $|\mathbf{a}_i^*(x_1)\mathbf{w}(x_2)|$ of two coordinate variables, x_1 and x_2. Then the beampattern is a slice of this function for x_2 fixed, whereas the powerpattern is a slice for x_1 fixed. Note that the beampattern and powerpattern of the DAS beamformer are identical since its weight vector and steering vector have the same functional form. However, this is not the case for an adaptive beamformer due to the fact that its weight vector depends not only on the corresponding steering vector, but also on the data.

Without loss of generality, we consider herein two-dimensional (2-D) array imaging, in which the *beamwidth* or *powerwidth* is defined as the diameter of a circle having the same area as the 3-dB contour of the main lobe of a 2-D beampattern or powerpattern. The beamwidth shows how the nearby signals impact the estimation of SOI. The powerwidth, on the other hand, shows how SOI impacts the estimation of the nearby signals. From now on, we will concentrate on the ith frequency bin.

3.7.2 Constant-Powerwidth RCB

The beamwidth of RCB decreases with the frequency but generally it does not depend on the choice of ϵ if the SOI is sufficiently separated from the interferences. On the other hand, the powerwidth of RCB depends on the signal-to-noise-ratio (SNR), the frequency, and ϵ. Since the steering vector and hence its uncertainty set are both functions of the frequency [see (3.123) and (3.7)], it is natural to consider altering the uncertainty parameter ϵ with the frequency. Intuitively, a larger ϵ will yield a larger powerwidth. By choosing a frequency-dependent parameter ϵ for RCB, we obtain the constant-powerwidth robust Capon beamformer (CPRCB), which is able to provide consistent SPL estimates across the frequency bins for the source of interest [58]. However, the beamwidth of CPRCB changes with the frequency in the same way as that of RCB.

Although it is difficult to yield an analytical formula for choosing ϵ as a function of frequency to guarantee a nearly constant powerwidth, such a choice can be readily made numerically via a contour plot of the powerwidths of RCB with respect to the frequency and ϵ. Given a desired powerwidth, we can determine ϵ as a function of frequency from the contour plot.

3.7.3 Constant-Beamwidth RCB

In [49–51] a shading scheme is used for a directional array consisting of a set of harmonically nested subarrays, each of which is designed for a particular frequency bin. For each array element, shading weights are devised as a function of the frequency. This shading scheme can provide a constant beamwidth for frequencies between 10 and 40 kHz when used with the DAS beamformer. We refer to this approach as the shaded DAS (SDAS). Our RCB can be readily combined with the shading scheme of [49, 50] to obtain a constant beamwidth for the desired frequency band, as explained below. We refer to this approach as the constant-beamwidth RCB (CBRCB) algorithm [58].

Let $\mathbf{v}$ denote the $M \times 1$ vector containing the array element shading weights for the given frequency bin. The assumed array steering vector for CBRCB can now be written as:

$$\tilde{\mathbf{a}}_i = \mathbf{v} \odot \bar{\mathbf{a}}_i \tag{3.131}$$

where $\odot$ denotes elementwise multiplication [1]. Accordingly, the covariance matrix for the CBRCB is tapered as follows:

$$\tilde{\mathbf{R}}_i = \mathbf{R}_i \odot (\mathbf{v}\mathbf{v}^T). \tag{3.132}$$

Since both $\mathbf{R}$ and $\mathbf{v}\mathbf{v}^T$ are positive semidefinite matrices, $\tilde{\mathbf{R}}$ is also positive semidefinite (see [1, 60]). Note that RCB can be viewed as a special case of CBRCB with all the elements of $\mathbf{v}$ being 1.

Similarly to (3.14) in Section 3.4.1, the CBRCB has the form:

$$\max_{\sigma^2, \mathbf{a}_i} \sigma^2 \quad \text{subject to } \tilde{\mathbf{R}}_i - \sigma^2 \mathbf{a}_i \mathbf{a}_i^* \geq 0$$
$$\|\mathbf{a}_i - \tilde{\mathbf{a}}_i\|^2 \leq \epsilon \tag{3.133}$$

which can be solved like the RCB problem.

REMARKS. The ability of CBRCB to retain a constant beamwidth across the frequencies makes it suitable for many applications such as speech acquisition. Furthermore, CBRCB can also achieve constant powerwidth, which is essential for consistent imaging. However, CBRCB is less powerful and flexible than CPRCB for applications where constant powerwidth is demanded. First, CPRCB has more degrees of freedom for interference suppression than CBRCB since at each frequency the shading scheme involved in the latter tends to deactivate some elements of the full array. In addition, CPRCB can be used with arbitrary arrays, while a special underlying array structure is required for CBRCB due to the particular shading scheme employed. As mentioned earlier, SDAS can also be utilized to yield constant beamwidth. Nevertheless, SDAS is data-independent and hence has poorer resolution and worse interference suppression capability than CBRCB and CPRCB. The features of the CBRCB and CPRCB approaches are listed in Table 3.1 in terms of constant beamwidth, constant powerwidth and loss of degree of freedom (DOF).

3.7.4 Numerical Examples

We provide several simulated examples to compare the performances of the DAS, SDAS, SCB, RCB, CPRCB, and CBRCB, approaches for acoustic imaging. We use the Small Aperture Directional Array (SADA) [49, 50], which consists of 33 microphones arranged in four circles of eight microphones each and one microphone at the array center. The diameter of each circle is twice that of the closest circle it encloses. The maximum radius of the array is 3.89 inches. Figure 3.20 shows the microphone layout of the SADA and its three subarrays used in the shading scheme (referred to as clusters in [49, 50]). Note that some microphones are shared by different clusters. Each cluster of SADA has the same directional characteristics for a given wavenumber-length product kD_n, where k is the wavenumber and D_n is the diagonal distance between the elements of the nth cluster. The wavenumber-length products at 10 kHz for Cluster 3, at 20 kHz for Cluster 2, and at

TABLE 3.1 Comparison of the CBRCB and CPRCB Approaches

Approach	Constant Beamwidth	Constant Powerwidth	Loss of DOF
CBRCB	Yes	Yes	Yes
CPRCB	No	Yes	No

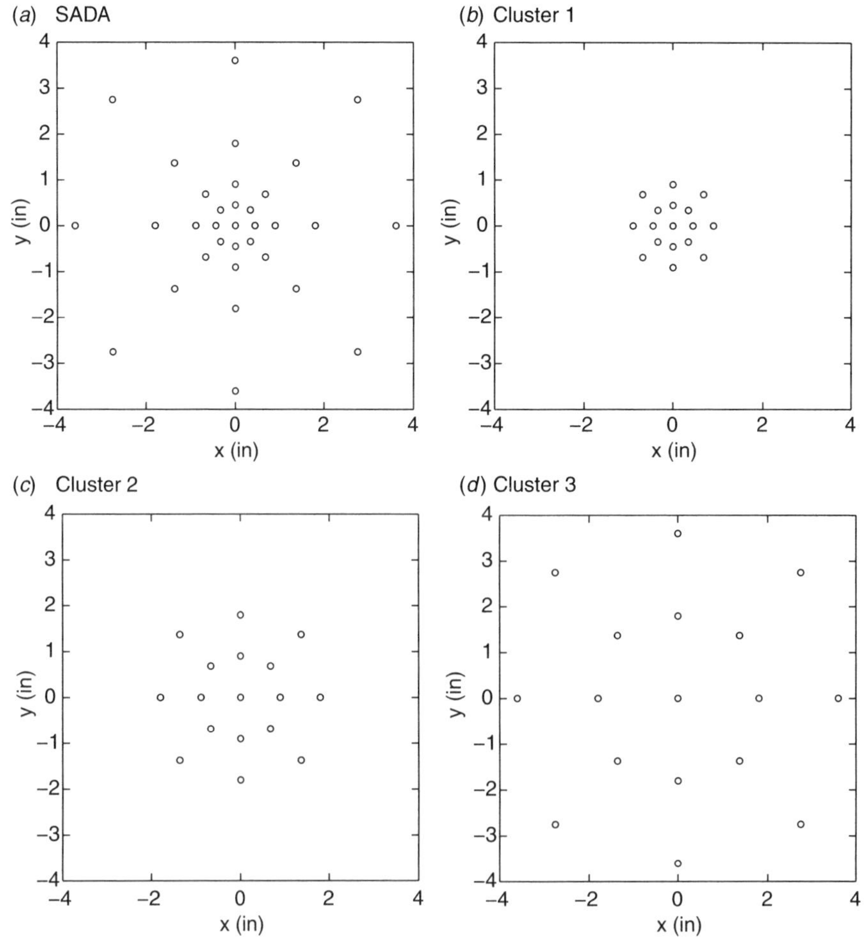

Figure 3.20 Microphone layout of the Small Aperture Directional Array (SADA) and its three clusters.

40 kHz for Cluster 1 are the same. According to the array coordinate frame, the array is located in the x-y plane, with center location at (0, 0, 0). Note that inch is used as the unit for the 3-D coordinates.

In what follows we assume that the distance between the array and the source is known and plot the 2-D images by scanning the locations on a plane parallel to the array and situated 5 feet above. In fact, even if we only had an approximate knowledge of this distance, the imaging results would still be similar. A simple explanation is as follows. Assume that the array has an aperture of A and is located at a distance of L from a narrowband point source with a wavelength of λ_0. According to [61, 62], if $L \gg A$, the array range resolution is $\lambda_0(L/A)^2$, which is much larger than $\lambda_0(L/A)$, the cross range resolution. Hence the distance is not a key issue here. We assume that

a belongs to the uncertainty set

$$\|\mathbf{a} - \bar{\mathbf{a}}\|^2 \leq \epsilon, \tag{3.134}$$

where ϵ is a user parameter chosen to account for the steering vector uncertainty. Note that this form of uncertainty set used in the CPRCB can cover all kinds of array errors, including calibration errors, look direction errors, or array covariance estimation errors due to a small snapshot number (sample size). The uncertainty set for the CBRCB is the same as above except that $\tilde{\mathbf{a}}$ is used instead of $\bar{\mathbf{a}}$. Figure 3.21 shows the SADA cluster shading weights as a function of the frequency bins, with w_1, w_2 and w_3 corresponding to Cluster 1, Cluster 2 and Cluster 3, respectively. Since some array elements are shared by different clusters, the shading weights of those elements are the sum of the corresponding cluster shading weights.

In the simulated examples below, we consider an array that is identical to SADA, with a wideband monopole source (flat spectrum from 0 Hz to 70 kHz) located at (0,0,60) (except for Figure 3.28) in the array coordinate frame and a spatially white Gaussian noise with SNR equal to 20 dB and the SPL equal to 20 dB for each frequency. We use an 8192-point FFT on the nonoverlapping blocks (each containing 8192 samples) of simulated data to convert the wideband signal into 8192 narrowband frequency bins.

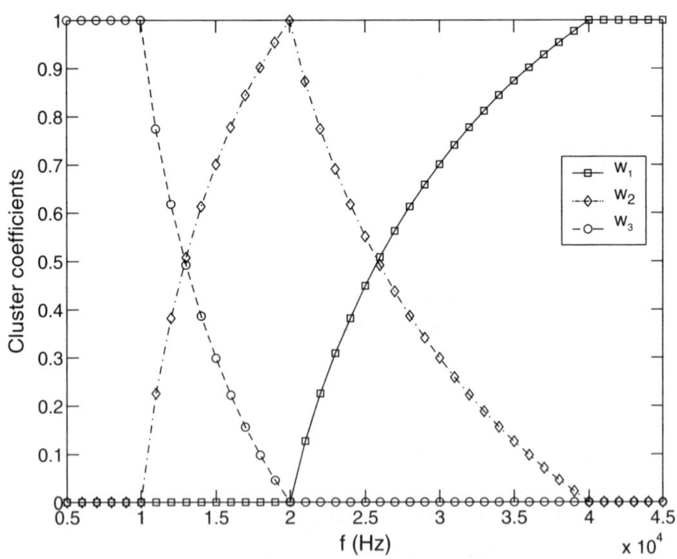

Figure 3.21 Cluster shading weights for SADA, as functions of the frequency, with w_1, w_2 and w_3 corresponding to Cluster 1, Cluster 2, and Cluster 3, respectively.

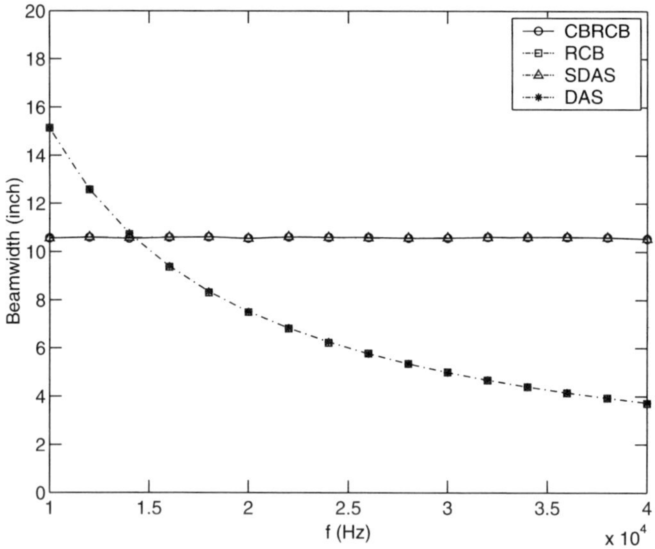

Figure 3.22 Comparison of the beamwidths for the CBRCB, RCB, SDAS and DAS methods with $N = 64$. We used $\epsilon = 2.0$ for RCB and CBRCB.

Example 3.7.1: Comparison of CBRCB, RCB, SDAS, and DAS in Terms of Beamwidth, Powerwidth, and Consistency of Acoustic Imaging Figure 3.22 compares the 3-dB beamwidths as functions of the frequency, corresponding to the CBRCB, RCB, SDAS and DAS methods when $N = 64$. Note that herein the beamwidth of RCB coincides with that of DAS and the beamwidth of CBRCB coincides with that of SDAS. CBRCB and SDAS achieve constant beamwidth over the frequency band from 10 to 40 kHz, whereas the beamwidths of RCB and DAS are frequency dependent and decrease appreciably with the frequency. We used $\epsilon = 2.0$ for RCB and CBRCB. Other choices of ϵ for RCB and CBRCB yield the same results and hence they are not shown here. We remark that one can not achieve constant beamwidth for RCB by varying ϵ. For example, it can be shown that the beampattern of RCB is independent of ϵ if $\mathbf{Q}$ in (3.130) is white Gaussian noise, that is, $\mathbf{Q} = \sigma_n^2\mathbf{I}$, where σ_n^2 denotes the noise power and $\mathbf{I}$ is an identity matrix.

Figure 3.23 compares the 3-dB powerwidths of the CBRCB, RCB, SDAS and DAS methods, as functions of the frequency, when $N = 64$. Note that the powerwidths of the DAS and RCB methods decrease drastically as the frequency increases, while SDAS and CBRCB can both achieve approximately constant powerwidth. In addition, CBRCB has much smaller powerwidth than SDAS. As can be seen from the figure, we can also adjust the powerwidth for CBRCB and RCB by choosing different values of ϵ.

Figure 3.24 compares the acoustic imaging results or sound pressure level (SPL) estimates obtained via the DAS, SDAS, RCB and CBRCB methods for the narrowband frequency bins at 10 kHz and 40 kHz, with $N = 64$. The z axes show the SPL. We used $\epsilon = 2.0$ for RCB and $\epsilon = 1.0$ for CBRCB. Note that we choose ϵ for

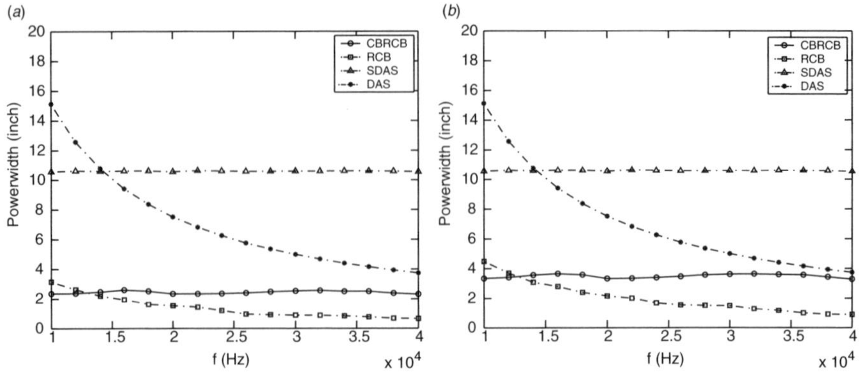

Figure 3.23 Comparison of the powerwidths for the CBRCB, RCB, SDAS, and DAS methods with $N = 64$ when (*a*) $\epsilon = 1.0$ for RCB and $\epsilon = 0.5$ for CBRCB and (*b*) $\epsilon = 2.0$ for RCB and $\epsilon = 1.0$ for CBRCB.

CBRCB to be one half of that for RCB due to the fact that the squared norm of the steering vector for CBRCB is about one half of that of RCB. As can be seen, the DAS method has poor resolution and high sidelobes and its images vary considerably with the frequency. RCB cannot be used to obtain consistent imaging results over different frequency bins, either, though it has much better resolution than DAS. It is worth noting that both SDAS and CBRCB maintain approximately the same SPL estimates across the frequency bins, but the latter has much better resolution and lower sidelobes and hence better interference rejection capability than the former. It is obvious that CBRCB significantly outperforms the other methods. According to the previous discussions and the results shown in Figures 3.22 and 3.23, it is the constant powerwidth rather than the constant beamwidth that contributes to the better performance of CBRCB as compared to SDAS.

Example 3.7.2: Comparison of CPRCB and CBRCB in Terms of Consistency of Acoustic Imaging Figure 3.25 shows the contours of the 3-dB powerwidth of RCB as ϵ and the frequency vary, when $N = 64$. As can be seen, the contours are almost linear with respect to the frequency and ϵ. Therefore, given a desired powerwidth, we can readily determine ϵ as a function of the frequency from the corresponding contour plot. Then CPRCB will have a constant powerwidth across the frequency bins.

In Figure 3.26, we show the imaging results obtained via the CPRCB approach by choosing $\epsilon = 1.3$ when $f = 10\,\text{kHz}$ and $\epsilon = 13$ when $f = 40\,\text{kHz}$ from the 3 inch powerwidth contour in Figure 3.25. The similarity of the SOI SPL estimates obtained with CPRCB at these two frequencies, especially near the powerwidth area, verifies the consistency of CPRCB in powerpattern across the frequencies.

Figure 3.27 shows the imaging results obtained via the CBRCB approach with $\epsilon = 0.65$, for $f = 10\,\text{kHz}$ and $f = 40\,\text{kHz}$. Again we note the consistency in the

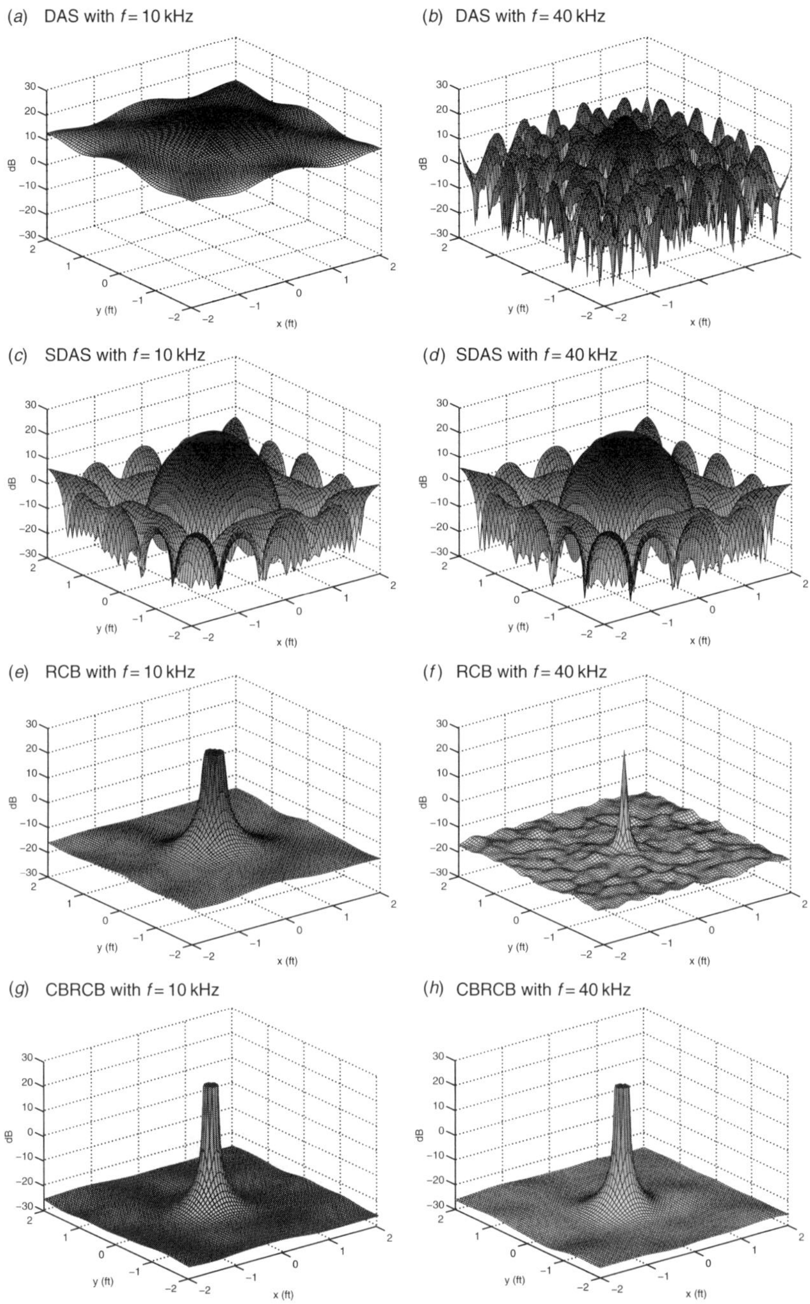

Figure 3.24 Comparison of the acoustic imaging results obtained via the DAS, SDAS, RCB, and CBRCB methods with $N = 64$, for the narrowband frequency bins at $f = 10$ kHz for (*a,c,e,g*), and $f = 40$ kHz for (*b,d,f,h*), respectively. For RCB, $\epsilon = 2.0$. For CBRCB, $\epsilon = 1.0$. The z axes show the SPL.

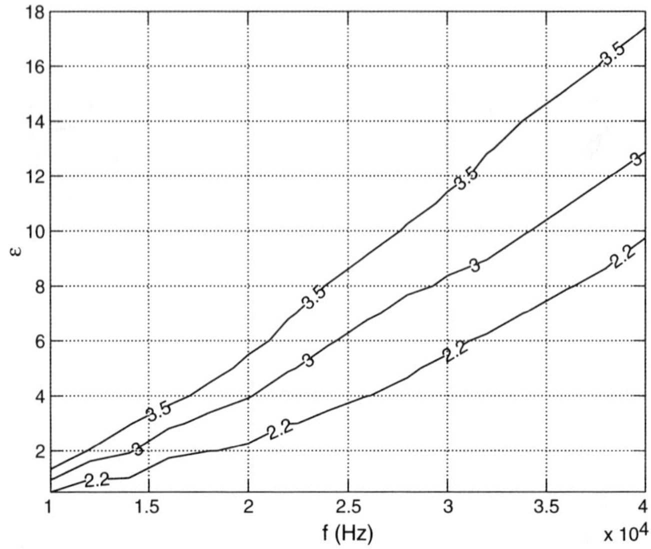

Figure 3.25 Contour plots of the powerwidth, versus ϵ and the frequency, for the RCB method. The numbers on the contours are the 3-dB powerwidths in inch.

imaging results. Therefore, both CPRCB and CBRCB are suitable for applications where consistent SPL estimates are desirable. However, the sidelobes in Figure 3.26(*b*) are higher and rougher than those in Figure 3.27(*b*). Despite this fact, CPRCB does not perform worse than CBRCB, see the next example.

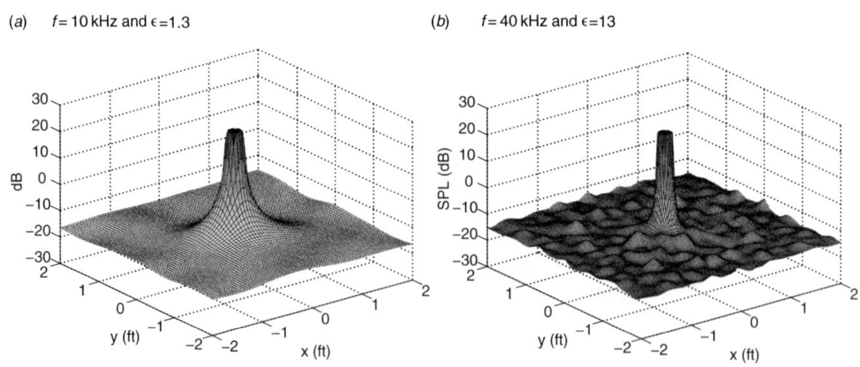

Figure 3.26 Acoustic imaging results obtained via the CPRCB method with $N = 64$ when (*a*) $f = 10$ kHz and $\epsilon = 1.3$, (*b*) $f = 40$ kHz and $\epsilon = 13$.

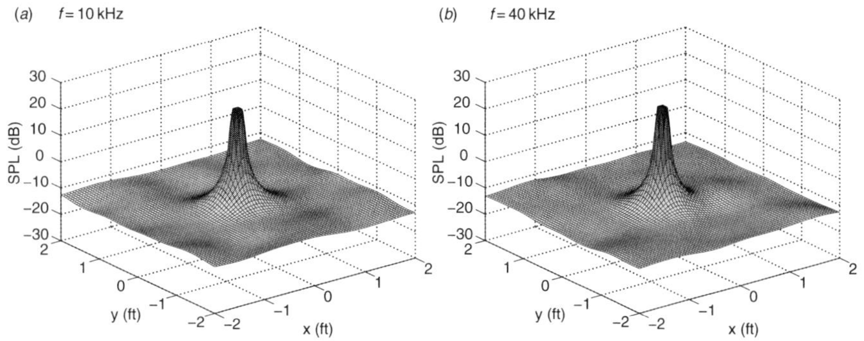

Figure 3.27 Acoustic imaging results obtained via the CBRCB method with $N = 64$ and $\epsilon = 0.65$ when (*a*) $f = 10$ kHz and (*b*) $f = 40$ kHz.

Example 3.7.3: Comparison of CPRCB, CBRCB, SCB, SDAS, and DAS in the Presence of Look Direction Errors We consider a look direction error case where the assumed source location is (0,0,60) but the actual source is located at (0.2,0.2,60) with SNR equal to 20 dB. Also we consider a varying number of interferences from $K = 0$ to $K = 20$, which are situated on a circle with a radius of 20 inches and have an INR equal to 40 dB. The circle is on a plane parallel to the array and situated 60 inches above. We assume that the theoretical covariance matrix $\mathbf{R}$ is known here.

Figure 3.28 compares the SINR and SOI SPL estimates obtained via the CBRCB, CPRCB, SCB, SDAS and DAS methods, versus the number of interferences K, for

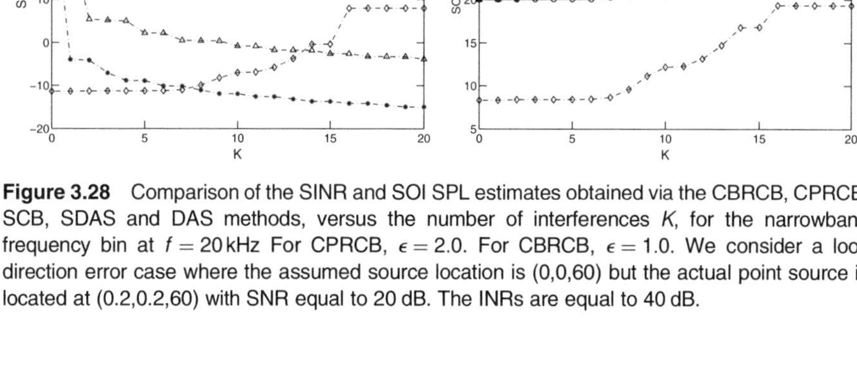

Figure 3.28 Comparison of the SINR and SOI SPL estimates obtained via the CBRCB, CPRCB, SCB, SDAS and DAS methods, versus the number of interferences K, for the narrowband frequency bin at $f = 20$ kHz For CPRCB, $\epsilon = 2.0$. For CBRCB, $\epsilon = 1.0$. We consider a look direction error case where the assumed source location is (0,0,60) but the actual point source is located at (0.2,0.2,60) with SNR equal to 20 dB. The INRs are equal to 40 dB.

the narrowband frequency bin at $f = 20\,\text{kHz}$. For CPRCB, $\epsilon = 2.0$. For CBRCB, $\epsilon = 1.0$. Note that SCB is very sensitive to the steering vector mismatch and suffers from severe performance degradation in SINR and SOI SPL estimates. Although DAS and SDAS are robust against array errors, they have poor capacity for interference suppression. Consequently, their SINRs and SOI SPL estimates are unsatisfactory. CBRCB and CPRCB outperform the other approaches due to their robustness to steering vector errors, better resolution and much better interference rejection capability than the data-independent beamformers. As can be seen, CPRCB has higher SINR than CBRCB. This is due to the fact that the former has more degrees of freedom (DOFs) for interference suppression than the latter. It might seem surprising that the performance of SCB improves as the number of interferences K increases. There is a simple explanation for this. When K is small, SCB has enough many DOFs and the SOI is suppressed as interference. As K increases, SCB focuses more on suppressing the interferences than the SOI since the INR is much higher than the SNR.

3.8 RANK-DEFICIENT ROBUST CAPON FILTER-BANK SPECTRAL ESTIMATOR

Complex spectral estimation is important to many applications including, for example, synthetic aperture radar (SAR) imaging and target feature extraction (see, e.g., [1, 63] and the references therein). The conventional nonparametric discrete Fourier transform (DFT) or fast Fourier transform (FFT) methods are data-independent approaches for spectral estimation. These methods make almost no a priori assumptions on the spectrum, and hence they possess better robustness than their parametric counterparts. However, they suffer from high sidelobes, low resolution, and poor accuracy. There are several variations of the DFT or FFT based methods that are proposed for improved statistical accuracy, which are based on smoothing the spectral estimates or windowing the data [1]. However, the improved accuracy is obtained at the cost of even poorer resolution.

Nonparametric data-adaptive finite impulse response (FIR) filtering based approaches, including Capon [3, 4] and APES [64], retain the robust nature of the nonparametric methods and at the same time improve the spectral estimates by having narrower spectral peaks and lower sidelobes than the DFT or FFT based methods. It has been shown in [65, 66] that both Capon and APES are members of the matched-filterbank (MAFI) spectral estimators.

The adaptive FIR filter-bank used in the Capon spectral estimator is obtained via the Capon beamformer [1, 2]. For complex spectral estimation, the filter length is often chosen to be quite large in order to achieve high resolution; hence the number of *snapshots* is usually small. Whenever this happens, the Capon beamformer may suppress the SOI as if it were an interference, which results in a significantly underestimated SOI power. This is, in fact, the reason why the Capon spectral estimates are generally biased downward. When the number of snapshots is so small that the sample covariance matrix is rank-deficient, the Capon spectral estimator fails completely. However, using rank-deficient sample covariance matrices for spectral estimation can yield high resolutions. It was first considered by Benitz (see, e.g., [67] and the references therein). In particular, he referred to using such spectral estimation methods for SAR image formation as high-definition imaging (HDI). It has been shown in [68] that HDI can be used to significantly improve the automatic target recognition (ATR) performance of a modern SAR, which demonstrates the importance of spectral estimation based on rank-deficient sample covariance matrices.

In this section, we consider nonparametric complex spectral estimation using an adaptive filtering based approach where the FIR filter-bank is obtained via a rank-deficient RCB. We derive the rank-deficient robust Capon filter-bank (RCF) spectral estimator in detail [69]. We show that by allowing the sample covariance matrix to be rank-deficient, we can achieve much higher resolution than existing approaches, which is useful in many applications including radar target detection and feature extraction. Numerical examples are provided to demonstrate the performance of the new approach as compared to data-adaptive and data-independent FIR filtering based spectral estimation methods.

3.8.1 Problem Formulation of Complex Spectral Estimation and Some Preliminaries

Consider the problem of estimating the amplitude spectrum of a complex-valued discrete-time 1-D signal $\{y_n\}_{n=0}^{N-1}$. (Extension to 2-D data, which will be used in one of the numerical examples in Section 3.8.3, can be done in the manner of [66].) For a frequency ω of interest, the signal y_n is modeled as

$$y_n = \alpha(\omega)e^{j\omega n} + e_n(\omega), \quad n = 0, \ldots, N-1, \quad \omega \in [0, 2\pi) \tag{3.135}$$

where $\alpha(\omega)$ denotes the complex amplitude of the sinusoidal component with frequency ω, and $e_n(\omega)$ denotes the residual term (assumed zero-mean) at frequency ω, which includes the unmodeled noise and interference from frequencies other than ω. The problem of interest is to estimate $\alpha(\omega)$ from $\{y_n\}_{n=0}^{N-1}$ for any given frequency ω.

The filter-bank approaches address the aforementioned spectral estimation problem by passing the data $\{y_n\}$ through a bank of FIR bandpass filters with varying center frequency ω, and then obtaining the amplitude spectrum estimate $\hat{\alpha}(\omega)$ for $\omega \in [0, 2\pi)$ from the filtered data. We denote an M-tap FIR bandpass filter by

$$\mathbf{h}(\omega) = [h_0 \quad h_1 \quad \cdots \quad h_{M-1}]^T \tag{3.136}$$

where $(\cdot)^T$ denotes the transpose. Let the forward data vectors

$$\bar{\mathbf{y}}_l = [y_l \quad y_{l+1} \quad \cdots \quad y_{l+M-1}]^T, \quad l = 0, \ldots, L-1 \tag{3.137}$$

be the overlapping $M \times 1$ subvectors constructed from the data vector

$$\bar{\mathbf{y}} = [y_0 \quad y_1 \quad \cdots \quad y_{N-1}]^T \tag{3.138}$$

where $L = N - M + 1$. Then, according to the data model in (3.135), the forward data vectors can be written as

$$\bar{\mathbf{y}}_l = \alpha(\omega)\mathbf{a}(\omega) \cdot e^{j\omega l} + \bar{\mathbf{e}}_l(\omega) \tag{3.139}$$

where $\mathbf{a}(\omega)$ is an $M \times 1$ vector given by

$$\mathbf{a}(\omega) = [1 \quad e^{j\omega} \quad \cdots \quad e^{j\omega(M-1)}]^T \tag{3.140}$$

and $\bar{\mathbf{e}}_l(\omega) = [e_l(\omega)\ e_{l+1}(\omega)\ \cdots\ e_{l+M-1}(\omega)]^T$. Hence the output samples obtained by passing $\bar{\mathbf{y}}_l$ through the FIR filter $\mathbf{h}(\omega)$ can be written as

$$\mathbf{h}^*(\omega)\bar{\mathbf{y}}_l = \alpha(\omega)[\mathbf{h}^*(\omega)\mathbf{a}(\omega)]e^{j\omega l} + \bar{w}_l(\omega) \tag{3.141}$$

where $(\cdot)^*$ denotes the conjugate transpose and $\bar{w}_l(\omega) = \mathbf{h}^*(\omega)\bar{\mathbf{e}}_l(\omega)$ denotes the residue term at the filter output. For an undistorted spectral estimate, we require that

$$\mathbf{h}^*(\omega)\mathbf{a}(\omega) = 1. \tag{3.142}$$

From the output of the FIR filter

$$\mathbf{h}^*(\omega)\bar{\mathbf{y}}_l = \alpha(\omega)e^{j\omega l} + \bar{w}_l(\omega) \tag{3.143}$$

we can obtain the least-squares estimate of $\alpha(\omega)$ as

$$\hat{\bar{\alpha}}(\omega) = \mathbf{h}^*(\omega)\bar{\mathbf{g}}(\omega) \tag{3.144}$$

where $\bar{\mathbf{g}}(\omega)$ is the normalized Fourier transform of the forward data vectors

$$\bar{\mathbf{g}}(\omega) = \frac{1}{L}\sum_{l=0}^{L-1} \bar{\mathbf{y}}_l e^{-j\omega l}. \tag{3.145}$$

Since a combined forward-backward approach usually yields better spectral estimates than the forward-only approach, we also consider the backward data vectors

$$\tilde{\mathbf{y}}_l = [y^c_{N-l-1} \quad y^c_{N-l-2} \quad \cdots \quad y^c_{N-l-M}]^T, \quad l = 0, \ldots, L-1 \tag{3.146}$$

where $(\cdot)^c$ denotes the complex conjugate. Note that $\{\tilde{\mathbf{y}}_l\}_{l=0}^{L-1}$ are the overlapping $M \times 1$ subvectors constructed from the data vector

$$\tilde{\mathbf{y}} = [y^c_{N-1} \quad y^c_{N-2} \quad \cdots \quad y^c_0]^T. \tag{3.147}$$

Similarly to $\bar{\mathbf{y}}_l$, $\tilde{\mathbf{y}}_l$ can be written as

$$\tilde{\mathbf{y}}_l = \alpha^c(\omega)e^{-j(N-1)\omega}\mathbf{a}(\omega) \cdot e^{j\omega l} + \tilde{\mathbf{e}}_l(\omega) \tag{3.148}$$

where $\tilde{\mathbf{e}}_l(\omega) = [e^c_{N-l-1}(\omega)\ e^c_{N-l-2}(\omega)\ \cdots\ e^c_{N-l-M}(\omega)]^T$. Passing $\tilde{\mathbf{y}}_l$ through the FIR filter $\mathbf{h}(\omega)$ yields the following output

$$\mathbf{h}^*(\omega)\tilde{\mathbf{y}}_l = e^{-j(N-1)\omega} \cdot \alpha^c(\omega)e^{j\omega l} + \tilde{w}_l(\omega) \tag{3.149}$$

where $\tilde{w}_l(\omega) = \mathbf{h}^*(\omega)\tilde{\mathbf{e}}_l(\omega)$ denotes the residue term at the filter output. From the above FIR filter output, we can obtain another least-squares estimate of $\alpha(\omega)$:

$$\hat{\tilde{\alpha}} = e^{-j(N-1)\omega}\tilde{\mathbf{g}}^*(\omega)\mathbf{h}(\omega) \tag{3.150}$$

where $\tilde{\mathbf{g}}(\omega)$ is the normalized Fourier transform of the backward data vectors:

$$\tilde{\mathbf{g}}(\omega) = \frac{1}{L}\sum_{l=0}^{L-1} \tilde{\mathbf{y}}_l e^{-j\omega l}. \tag{3.151}$$

Averaging the two least-squares estimates, $\hat{\alpha}(\omega)$ and $\hat{\tilde{\alpha}}(\omega)$, gives the forward-backward estimate of $\alpha(\omega)$:

$$\hat{\alpha}(\omega) = \tfrac{1}{2}\left[\mathbf{h}^*(\omega)\bar{\mathbf{g}}(\omega) + e^{-j(N-1)\omega}\tilde{\mathbf{g}}^*(\omega)\mathbf{h}(\omega)\right]. \tag{3.152}$$

The forward-backward approach is used in all of the adaptive filtering based spectral estimators in the sections to follow, and also in the determination of $\mathbf{h}(\omega)$ which is discussed next.

3.8.2 Rank-Deficient Robust Capon Filter-Bank for Complex Spectral Estimation

We derive the rank-deficient RCF spectral estimator in a covariance matrix fitting framework by assuming that the sample covariance matrix is singular, which happens, for example, when the FIR filter length M is large. In particular, we use the rank-deficient RCB approach to determine the data-dependent FIR filter $\mathbf{h}(\omega)$ from the sample covariance matrix. Besides introducing a high-resolution spectral estimator, our derivations will also shed more light on the properties of the RCB algorithm when the sample covariance matrix is singular.

3.8.2.1 Rank-Deficient Sample Covariance Matrix. It can be observed from Section 3.8.1 that the forward and backward data vectors are related by

$$\tilde{\mathbf{y}}_l = \mathbf{J}\bar{\mathbf{y}}^c_{L-l-1} \tag{3.153}$$

where $\mathbf{J}$ denotes the exchange matrix whose antidiagonal elements are ones and all the others are zeros. Similarly, for each frequency ω of interest, we have

$$\tilde{\mathbf{e}}_l(\omega) = \mathbf{J}\bar{\mathbf{e}}^c_{L-l-1}(\omega). \tag{3.154}$$

Let the Toeplitz noise covariance matrix $\mathbf{Q}(\omega)$ be defined by

$$\mathbf{Q}(\omega) \triangleq E[\bar{\mathbf{e}}_l(\omega)\bar{\mathbf{e}}_l^*(\omega)] = E[\tilde{\mathbf{e}}_l(\omega)\tilde{\mathbf{e}}_l^*(\omega)] \tag{3.155}$$

where the second equality follows from (3.154) and the Toeplitz structure of $\mathbf{Q}(\omega)$. The covariance matrix of $\bar{\mathbf{y}}_l$ or, equivalently, of $\tilde{\mathbf{y}}_l$ is given by

$$\mathbf{R} = |\alpha(\omega)|^2\mathbf{a}(\omega)\mathbf{a}^*(\omega) + \mathbf{Q}(\omega). \tag{3.156}$$

Let $\hat{\bar{\mathbf{R}}}$ and $\hat{\tilde{\mathbf{R}}}$ denote the sample covariance matrices estimated from $\{\bar{\mathbf{y}}_l\}$ and $\{\tilde{\mathbf{y}}_l\}$, respectively, as follows:

$$\hat{\bar{\mathbf{R}}} = \frac{1}{L}\sum_{l=0}^{L-1}\bar{\mathbf{y}}_l\bar{\mathbf{y}}_l^* \tag{3.157}$$

$$\hat{\tilde{\mathbf{R}}} = \frac{1}{L}\sum_{l=0}^{L-1}\tilde{\mathbf{y}}_l\tilde{\mathbf{y}}_l^*. \tag{3.158}$$

Then the forward-backward estimate of the covariance matrix $\mathbf{R}$ is given by

$$\hat{\mathbf{R}} = \tfrac{1}{2}(\hat{\bar{\mathbf{R}}} + \hat{\tilde{\mathbf{R}}}). \tag{3.159}$$

From (3.153), it is straightforward to show that

$$\hat{\tilde{\mathbf{R}}} = \mathbf{J}\hat{\bar{\mathbf{R}}}^T\mathbf{J} \tag{3.160}$$

and hence, the $\hat{\mathbf{R}}$ in (3.159) is persymmetric. Compared with the nonpersymmetric sample covariance matrix $\hat{\bar{\mathbf{R}}}$ estimated only from the forward data vectors $\{\bar{\mathbf{y}}_l\}$, the forward-backward $\hat{\mathbf{R}}$ is generally a better estimate of the true $\mathbf{R}$. The data-adaptive FIR filtering based spectral estimation methods we consider herein obtain the data-dependent FIR filter $\mathbf{h}(\omega)$ from the above $\hat{\mathbf{R}}$.

Let $\hat{\mathbf{R}}$ be the $M \times M$ positive semidefinite sample covariance matrix defined in (3.159). Let K denote the rank of $\hat{\mathbf{R}}$. With probability one, $K = 2L$, assuming that $M > 2L$ or, equivalently, $M > \frac{2}{3}(N+1)$. By choosing such a large M, we hope to achieve high resolution for spectral estimation. Let

$$\hat{\mathbf{R}} = \hat{\mathbf{S}}\hat{\mathbf{\Psi}}\hat{\mathbf{S}}^* \tag{3.161}$$

denote the eigendecomposition of $\hat{\mathbf{R}}$, where $\hat{\mathbf{S}}$ is an $M \times K$ semiunitary matrix with full column rank ($K < M$) and $\hat{\mathbf{\Psi}}$ is a $K \times K$ positive definite diagonal matrix whose diagonal elements are the eigenvalues of $\hat{\mathbf{R}}$. Next, we derive the rank-deficient RCF spectral estimator based on this singular $\hat{\mathbf{R}}$.

3.8.2.2 Robust Capon Filter-Bank (RCF) Approach.

Owing to the small snapshot number problem, the signal term in $\hat{\mathbf{R}}$ is not well described by

$|\alpha(\omega)|^2\mathbf{a}(\omega)\mathbf{a}^*(\omega)$, but by $|\alpha(\omega)|^2\hat{\mathbf{a}}(\omega)\hat{\mathbf{a}}^*(\omega)$ with $\hat{\mathbf{a}}(\omega)$ being some vector in the vicinity of $\mathbf{a}(\omega)$ and $\hat{\mathbf{a}}(\omega) \neq \mathbf{a}(\omega)$ [7]. Consequently, if we designed $\mathbf{h}(\omega)$ by means of the standard Capon beamformer:

$$\min_{\mathbf{h}(\omega)} \mathbf{h}^*(\omega)\hat{\mathbf{R}}\mathbf{h}(\omega) \quad \text{subject to } \mathbf{h}^*(\omega)\mathbf{a}(\omega) = 1, \tag{3.162}$$

the Euclidean norm of $\mathbf{h}(\omega)$, denoted $\|\mathbf{h}(\omega)\|$, would result rather large since $\hat{\mathbf{a}}(\omega)$ is close to $\mathbf{a}(\omega)$ and $\mathbf{h}(\omega)$ passes $\mathbf{a}(\omega)$ but attempts to suppress $\hat{\mathbf{a}}(\omega)$. A large $\|\mathbf{h}(\omega)\|$ indicates a large noise gain, which may severely degrade the estimation accuracy of $\alpha(\omega)$. It follows that we should design $\mathbf{h}(\omega)$ by

$$\min_{\mathbf{h}(\omega)} \mathbf{h}^*(\omega)\hat{\mathbf{R}}\mathbf{h}(\omega) \quad \text{subject to } \mathbf{h}^*(\omega)\hat{\mathbf{a}}(\omega) = \varphi \tag{3.163}$$

where φ is determined by the constraint $\mathbf{h}^*(\omega)\mathbf{a}(\omega) = 1$; see [2] (note that φ will be close to 1 since $\hat{\mathbf{a}}(\omega)$ is close to $\mathbf{a}(\omega)$). In summary, we design $\mathbf{h}(\omega)$ based on $\hat{\mathbf{a}}(\omega)$, instead of $\mathbf{a}(\omega)$, to avoid a large noise gain; furthermore, we choose φ based on the constraint $\mathbf{h}^*(\omega)\mathbf{a}(\omega) = 1$ to get an unbiased estimate of $\alpha(\omega)$ when we use $\mathbf{h}(\omega)$ in (3.152). The solution to (3.163) is derived as follows.

Assuming that $\hat{\mathbf{a}}(\omega)$ *is given* [*determination of* $\hat{\mathbf{a}}(\omega)$ *will be discussed later on*]. With $\tilde{\mathbf{h}}(\omega) = \mathbf{h}(\omega)/\varphi$, (3.163) can be rewritten as

$$\min_{\tilde{\mathbf{h}}(\omega)} \tilde{\mathbf{h}}^*(\omega)\tilde{\mathbf{R}}\tilde{\mathbf{h}}(\omega) \quad \text{subject to } \tilde{\mathbf{h}}^*(\omega)\hat{\mathbf{a}}(\omega) = 1 \tag{3.164}$$

where $\tilde{\mathbf{R}} \triangleq |\varphi|^2\hat{\mathbf{R}} = \hat{\mathbf{S}}\tilde{\boldsymbol{\Psi}}\hat{\mathbf{S}}^*$ and $\tilde{\boldsymbol{\Psi}} \triangleq |\varphi|^2\hat{\boldsymbol{\Psi}}$.

Let $\hat{\mathbf{G}}$ denote the $M \times (M-K)$ matrix whose columns are the eigenvectors of $\hat{\mathbf{R}}$ (or $\tilde{\mathbf{R}}$) corresponding to the zero eigenvalues. Hence $\hat{\mathbf{G}}$ spans the orthogonal complement of the subspace spanned by $\hat{\mathbf{S}}$. Let $\tilde{\mathbf{h}}(\omega)$ be written as

$$\tilde{\mathbf{h}}(\omega) = \hat{\mathbf{S}}\tilde{\mathbf{h}}_1(\omega) + \hat{\mathbf{G}}\tilde{\mathbf{h}}_2(\omega) \tag{3.165}$$

where $\tilde{\mathbf{h}}_1(\omega) = \hat{\mathbf{S}}^*\tilde{\mathbf{h}}(\omega)$ and $\tilde{\mathbf{h}}_2(\omega) = \hat{\mathbf{G}}^*\tilde{\mathbf{h}}(\omega)$.

- *Case 1*: $\hat{\mathbf{a}}(\omega)$ belongs to the range space of $\tilde{\mathbf{R}}$.
  Let $\hat{\mathbf{a}}(\omega)$ be written as $\hat{\mathbf{a}}(\omega) \triangleq \hat{\mathbf{S}}\boldsymbol{\zeta}$ for some nonzero vector $\boldsymbol{\zeta}$. Using (3.165) in (3.164) for this case yields

$$\min_{\tilde{\mathbf{h}}_1(\omega)} \tilde{\mathbf{h}}_1^*(\omega)\tilde{\boldsymbol{\Psi}}\tilde{\mathbf{h}}_1(\omega) \quad \text{subject to } \tilde{\mathbf{h}}_1^*(\omega)\boldsymbol{\zeta} = 1 \tag{3.166}$$

  which can be readily solved as

$$\tilde{\mathbf{h}}_1(\omega) = \frac{\tilde{\boldsymbol{\Psi}}^{-1}\boldsymbol{\zeta}}{\boldsymbol{\zeta}^*\tilde{\boldsymbol{\Psi}}^{-1}\boldsymbol{\zeta}}. \tag{3.167}$$

Since $\tilde{\mathbf{h}}_2(\omega)$ is irrelevant in this case, we let

$$\tilde{\mathbf{h}}_2(\omega) = 0 \tag{3.168}$$

to reduce the noise gain of $\tilde{\mathbf{h}}(\omega)(\|\tilde{\mathbf{h}}(\omega)\|^2 = \|\tilde{\mathbf{h}}_1(\omega)\|^2 + \|\tilde{\mathbf{h}}_2(\omega)\|^2)$. Then the FIR filter has the form

$$\tilde{\mathbf{h}}(\omega) = \frac{\hat{\mathbf{S}}\tilde{\boldsymbol{\Psi}}^{-1}\boldsymbol{\zeta}}{\boldsymbol{\zeta}^*\tilde{\boldsymbol{\Psi}}^{-1}\boldsymbol{\zeta}} = \frac{\tilde{\mathbf{R}}^{\dagger}\hat{\mathbf{a}}(\omega)}{\hat{\mathbf{a}}^*(\omega)\tilde{\mathbf{R}}^{\dagger}\hat{\mathbf{a}}(\omega)} \tag{3.169}$$

where $\tilde{\mathbf{R}}^{\dagger} = \hat{\mathbf{S}}\tilde{\boldsymbol{\Psi}}^{-1}\hat{\mathbf{S}}^*$ is the Moore–Penrose pseudo-inverse of $\tilde{\mathbf{R}}$. Consequently,

$$\hat{\mathbf{h}}(\omega) = \frac{\varphi \cdot \hat{\mathbf{R}}^{\dagger}\hat{\mathbf{a}}(\omega)}{\hat{\mathbf{a}}^*(\omega)\hat{\mathbf{R}}^{\dagger}\hat{\mathbf{a}}(\omega).} \tag{3.170}$$

Substituting (3.170) into $\mathbf{h}^*(\omega)\mathbf{a}(\omega) = 1$, we have

$$\varphi = \frac{\hat{\mathbf{a}}^*(\omega)\hat{\mathbf{R}}^{\dagger}\hat{\mathbf{a}}(\omega)}{\mathbf{a}^*(\omega)\hat{\mathbf{R}}^{\dagger}\hat{\mathbf{a}}(\omega).} \tag{3.171}$$

Hence $\hat{\mathbf{h}}(\omega)$ is given by

$$\hat{\mathbf{h}}(\omega) = \frac{\hat{\mathbf{R}}^{\dagger}\hat{\mathbf{a}}(\omega)}{\mathbf{a}^*(\omega)\hat{\mathbf{R}}^{\dagger}\hat{\mathbf{a}}(\omega)} \tag{3.172}$$

and we obtain the complex spectral estimator by substituting (3.172) into (3.152):

$$\hat{\alpha}(\omega) = \tfrac{1}{2}\left[\hat{\mathbf{h}}^*(\omega)\bar{\mathbf{g}}(\omega) + e^{-j(N-1)\omega}\tilde{\mathbf{g}}^*(\omega)\hat{\mathbf{h}}(\omega)\right]. \tag{3.173}$$

- *Case 2*: $\hat{\mathbf{a}}(\omega)$ does not belong to the range space of $\tilde{\mathbf{R}}$.
  Let $\hat{\mathbf{a}}(\omega)$ be written as $\hat{\mathbf{a}}(\omega) = \hat{\mathbf{S}}\boldsymbol{\zeta} + \hat{\mathbf{G}}\boldsymbol{\beta}$ for some nonzero vectors $\boldsymbol{\zeta}$ and $\boldsymbol{\beta}$. Now (3.164) becomes

$$\min_{\tilde{\mathbf{h}}_1(\omega),\,\tilde{\mathbf{h}}_2(\omega)} \tilde{\mathbf{h}}_1^*(\omega)\tilde{\boldsymbol{\Psi}}\tilde{\mathbf{h}}_1(\omega) \quad \text{subject to } \tilde{\mathbf{h}}_1^*(\omega)\boldsymbol{\zeta} + \tilde{\mathbf{h}}_2^*(\omega)\boldsymbol{\beta} = 1 \tag{3.174}$$

which admits a trivial solution:

$$\tilde{\mathbf{h}}_1(\omega) = 0 \tag{3.175}$$

and (for example)

$$\tilde{\mathbf{h}}_2(\omega) = \frac{\boldsymbol{\beta}}{\|\boldsymbol{\beta}\|^2}. \tag{3.176}$$

Consequently, since $\bar{\mathbf{g}}(\omega)$ and $\tilde{\mathbf{g}}(\omega)$ are linear transformations of $\{\bar{\mathbf{y}}_l\}$ and $\{\tilde{\mathbf{y}}_l\}$, respectively, where $\{\bar{\mathbf{y}}_l\}$ and $\{\tilde{\mathbf{y}}_l\}$ are in the range space of $\tilde{\mathbf{R}}$, we have

$$\tilde{\mathbf{h}}^*(\omega)\bar{\mathbf{g}}(\omega) = \tilde{\mathbf{h}}^*(\omega)\tilde{\mathbf{g}}(\omega) = 0 \tag{3.177}$$

which gives

$$\hat{\alpha}(\omega) = 0. \tag{3.178}$$

Combining the aforementioned two cases, the complex spectral estimate can be written as

$$\hat{\alpha}(\omega) = \begin{cases} \frac{1}{2}[\hat{\mathbf{h}}^*(\omega)\bar{\mathbf{g}}(\omega) + e^{-j(N-1)\omega}\tilde{\mathbf{g}}^*(\omega)\hat{\mathbf{h}}(\omega)], & \hat{\mathbf{a}}(\omega) \in \mathcal{R}(\hat{\mathbf{R}}) \\ 0, & \hat{\mathbf{a}}(\omega) \notin \mathcal{R}(\hat{\mathbf{R}}) \end{cases} \tag{3.179}$$

where $\hat{\mathbf{h}}(\omega)$ is given in (3.172) and $\mathcal{R}(\hat{\mathbf{R}})$ denotes the range space of $\hat{\mathbf{R}}$.

We remark that we could have obtained a signal power estimate from (3.164) as follows:

$$\begin{aligned} \hat{\hat{\sigma}}^2(\omega) &\triangleq |\hat{\alpha}(\omega)|^2 = \hat{\mathbf{h}}^*(\omega)\hat{\mathbf{R}}\hat{\mathbf{h}}(\omega) = \tilde{\mathbf{h}}^*(\omega)\tilde{\mathbf{R}}\tilde{\mathbf{h}}(\omega) \\ &= \begin{cases} \dfrac{|\varphi|^2}{\hat{\mathbf{a}}^*(\omega)\hat{\mathbf{R}}^\dagger\hat{\mathbf{a}}(\omega)}, & \hat{\mathbf{a}}(\omega) \in \mathcal{R}(\hat{\mathbf{R}}) \\ 0, & \hat{\mathbf{a}}(\omega) \notin \mathcal{R}(\hat{\mathbf{R}}) \end{cases}. \end{aligned} \tag{3.180}$$

However, $\hat{\hat{\sigma}}^2(\omega)$ is of little interest for the following two reasons. First, we also want to estimate the phase of $\alpha(\omega)$ and hence we prefer to use (3.179) instead. Second, even as an estimate of $|\alpha(\omega)|^2$, $\hat{\hat{\sigma}}^2(\omega)$ in (3.180) can be shown to be less accurate than the estimate we can obtain from (3.179) since the latter is obtained by using the waveform structures in (3.143) and (3.149) while the former is not.

Determination of $\hat{\mathbf{a}}(\omega)$. By its very definition, $\hat{\mathbf{a}}(\omega)$ is a vector in the vicinity of $\mathbf{a}(\omega)$ such that $|\alpha(\omega)|^2\hat{\mathbf{a}}(\omega)\hat{\mathbf{a}}^*(\omega)$ is a good fit to $\hat{\mathbf{R}}$. This leads to the RCB formulation directly where $\hat{\mathbf{a}}(\omega)$ is assumed to belong to a spherical uncertainty set as in Section 3.4.1 (see also [16, 30]):

$$\max_{\sigma^2(\omega),\,\hat{\mathbf{a}}(\omega)} \sigma^2(\omega) \quad \text{subject to } \hat{\mathbf{R}} - \sigma^2(\omega)\hat{\mathbf{a}}(\omega)\hat{\mathbf{a}}^*(\omega) \geq 0 \\ \|\hat{\mathbf{a}}(\omega) - \mathbf{a}(\omega)\|^2 \leq \epsilon \tag{3.181}$$

where $\sigma^2(\omega) = |\alpha(\omega)|^2$, and ϵ is a user parameter. A vector $\hat{\mathbf{a}}(\omega)$ that is in the range space of $\hat{\mathbf{S}}$, that is, $\hat{\mathbf{a}}(\omega) = \hat{\mathbf{S}}\boldsymbol{\zeta}$ for some nonzero $\boldsymbol{\zeta}$, is what we are after since otherwise we will get a spectral estimate equal to zero, according to (3.178). Hence, for each frequency ω, the problem of interest has the form:

$$\max_{\sigma^2(\omega),\,\boldsymbol{\zeta}} \sigma^2(\omega) \quad \text{subject to } \hat{\mathbf{R}} - \sigma^2(\omega)\hat{\mathbf{a}}(\omega)\hat{\mathbf{a}}^*(\omega) \geq 0$$
$$\hat{\mathbf{a}}(\omega) = \hat{\mathbf{S}}\boldsymbol{\zeta} \tag{3.182}$$
$$\|\hat{\mathbf{a}}(\omega) - \mathbf{a}(\omega)\|^2 \leq \epsilon.$$

The user parameter ϵ is used to describe the uncertainty of $\mathbf{a}(\omega)$ caused by the small snapshot number, $2L < M$, in the complex spectral estimation problem. The smaller the L, the larger the ϵ should be chosen. However, to exclude the trivial solution of $\hat{\boldsymbol{\zeta}} = 0$, we require that $\epsilon < \|\mathbf{a}(\omega)\|^2 = M$.

Next we note the following equivalences for the norm constraint:

$$\begin{aligned}
&\|\hat{\mathbf{a}}(\omega) - \mathbf{a}(\omega)\|^2 \leq \epsilon \\
&\Leftrightarrow \left\| \begin{bmatrix} \hat{\mathbf{S}}^* \\ \hat{\mathbf{G}}^* \end{bmatrix} (\hat{\mathbf{a}}(\omega) - \mathbf{a}(\omega)) \right\|^2 \leq \epsilon \\
&\Leftrightarrow \|\hat{\mathbf{S}}^*(\hat{\mathbf{a}}(\omega) - \mathbf{a}(\omega))\|^2 + \|\hat{\mathbf{G}}^*(\hat{\mathbf{a}}(\omega) - \mathbf{a}(\omega))\|^2 \leq \epsilon \\
&\Leftrightarrow \|\boldsymbol{\zeta} - \bar{\boldsymbol{\zeta}}\|^2 \leq \epsilon - \|\hat{\mathbf{G}}^*\mathbf{a}(\omega)\|^2 \\
&\Leftrightarrow \|\boldsymbol{\zeta} - \bar{\boldsymbol{\zeta}}\|^2 \leq \bar{\epsilon}
\end{aligned} \tag{3.183}$$

where we have defined $\bar{\boldsymbol{\zeta}} \triangleq \hat{\mathbf{S}}^*\mathbf{a}(\omega)$ and $\bar{\epsilon} \triangleq \epsilon - \|\hat{\mathbf{G}}^*\mathbf{a}(\omega)\|^2$.

It follows from (3.183) that if $\bar{\epsilon} < 0$, which occurs when $\mathbf{a}(\omega)$ is "far" away from the range space of $\hat{\mathbf{S}}$, the optimization problem in (3.182) is infeasible: in such a case there is no $\hat{\mathbf{a}}(\omega)$ of the form $\hat{\mathbf{a}}(\omega) = \hat{\mathbf{S}}\boldsymbol{\zeta}$ that satisfies the constraint in (3.182), or equivalently, (3.183). Consequently, the vector $\hat{\mathbf{a}}(\omega)$ cannot belong to the range space of $\hat{\mathbf{S}}$ in this case and then according to (3.174)–(3.178), we get

$$\hat{\alpha}(\omega) = 0. \tag{3.184}$$

Next, consider the case when $\bar{\epsilon} \geq 0$, which occurs when $\mathbf{a}(\omega)$ is "close" to the range space of $\hat{\mathbf{S}}$ and hence of $\hat{\mathbf{R}}$. In this case, an $\hat{\mathbf{a}}(\omega)$ belonging to the range space of $\hat{\mathbf{R}}$ can be found within the spherical uncertainty set in (3.183). To exclude the trivial solution of $\boldsymbol{\zeta} = \mathbf{0}$ (hence $\hat{\mathbf{a}}(\omega) = \mathbf{0}$), we assume that

$$\|\bar{\boldsymbol{\zeta}}\|^2 > \bar{\epsilon}. \tag{3.185}$$

It can be readily verified that the condition in (3.185) is equivalent to

$$M = \|\mathbf{a}(\omega)\|^2 > \epsilon \tag{3.186}$$

(which was assumed before). For any given $\hat{\mathbf{a}}(\omega)$ in the range of $\hat{\mathbf{S}}$, the solution $\hat{\sigma}^2(\omega)$ to (3.182) is given by (see Appendix 3.E):

$$\hat{\sigma}^2(\omega) = \frac{1}{\hat{\mathbf{a}}^*(\omega)\hat{\mathbf{R}}^\dagger\hat{\mathbf{a}}(\omega)}. \tag{3.187}$$

Making use of the fact that $\hat{\mathbf{a}}^*(\omega)\hat{\mathbf{R}}^\dagger\hat{\mathbf{a}}(\omega) = \boldsymbol{\zeta}^*\hat{\boldsymbol{\Psi}}^{-1}\boldsymbol{\zeta}$ and of the equivalences in (3.183), we can therefore reformulate (3.182) as the following minimization problem with a quadratic objective function and a quadratic inequality constraint:

$$\min_{\boldsymbol{\zeta}} \boldsymbol{\zeta}^*\hat{\boldsymbol{\Psi}}^{-1}\boldsymbol{\zeta} \quad \text{subject to } \|\boldsymbol{\zeta} - \bar{\boldsymbol{\zeta}}\|^2 \leq \bar{\epsilon}. \tag{3.188}$$

Because the solution to (3.188) (under (3.185) or (3.186)) will evidently occur on the boundary of the constraint set, we can reformulate (3.188) as the following quadratic problem with a quadratic equality constraint:

$$\min_{\boldsymbol{\zeta}} \boldsymbol{\zeta}^*\hat{\boldsymbol{\Psi}}^{-1}\boldsymbol{\zeta} \quad \text{subject to } \|\boldsymbol{\zeta} - \bar{\boldsymbol{\zeta}}\|^2 = \bar{\epsilon}. \tag{3.189}$$

This problem can be solved by using the *Lagrange multiplier methodology*, which is based on the function:

$$f(\boldsymbol{\zeta}, \lambda) = \boldsymbol{\zeta}^*\hat{\boldsymbol{\Psi}}^{-1}\boldsymbol{\zeta} + \lambda\left(\|\boldsymbol{\zeta} - \bar{\boldsymbol{\zeta}}\|^2 - \bar{\epsilon}\right) \tag{3.190}$$

where $\lambda \geq 0$ is the Lagrange multiplier. Differentiation of (3.190) with respect to $\boldsymbol{\zeta}$ gives the optimal solution $\hat{\boldsymbol{\zeta}}$:

$$\hat{\boldsymbol{\Psi}}^{-1}\hat{\boldsymbol{\zeta}} + \lambda(\hat{\boldsymbol{\zeta}} - \bar{\boldsymbol{\zeta}}) = \mathbf{0}. \tag{3.191}$$

The above equation yields

$$\hat{\boldsymbol{\zeta}} = \left(\frac{\hat{\boldsymbol{\Psi}}^{-1}}{\lambda} + \mathbf{I}\right)^{-1}\bar{\boldsymbol{\zeta}} \tag{3.192}$$

$$= \bar{\boldsymbol{\zeta}} - (\mathbf{I} + \lambda\hat{\boldsymbol{\Psi}})^{-1}\bar{\boldsymbol{\zeta}} \tag{3.193}$$

where we have used the matrix inversion lemma to obtain the second equality. Substituting (3.193) into the equality constraint of (3.189), the Lagrange multiplier $\lambda \geq 0$ is obtained as the solution to the constraint equation:

$$g(\lambda) \triangleq \left\|\left(\mathbf{I} + \lambda\hat{\boldsymbol{\Psi}}\right)^{-1}\bar{\boldsymbol{\zeta}}\right\|^2 = \bar{\epsilon}. \tag{3.194}$$

It can be shown that $g(\lambda)$ is a monotonically decreasing function of $\lambda \geq 0$ (see, e.g., Section 3.4.1 and [16]). Hence a unique solution $\lambda \geq 0$ exists which can be obtained efficiently via, for example, a Newton's method. Once λ has been determined, we use it in (3.193) to get $\hat{\boldsymbol{\zeta}}$, which gives

$$\hat{\sigma}^2(\omega) = \frac{1}{\hat{\boldsymbol{\zeta}}^* \hat{\boldsymbol{\Psi}}^{-1} \hat{\boldsymbol{\zeta}}}. \tag{3.195}$$

However, for the same reasons we did not use $\hat{\hat{\sigma}}^2(\omega)$ in (3.180), we will not use $\hat{\sigma}^2(\omega)$ as an estimate of $|\alpha(\omega)|^2$. Instead, we obtain the rank-deficient RCF $\mathbf{h}(\omega)$ by substituting $\hat{\mathbf{a}}(\omega) = \hat{\mathbf{S}}\hat{\boldsymbol{\zeta}}$, with $\hat{\boldsymbol{\zeta}}$ given by (3.192), into (3.172)

$$\begin{aligned}\hat{\mathbf{h}}(\omega) &= \frac{\hat{\mathbf{S}}\hat{\boldsymbol{\Psi}}^{-1}\hat{\boldsymbol{\zeta}}}{\mathbf{a}^*(\omega)\hat{\mathbf{S}}\hat{\boldsymbol{\Psi}}^{-1}\hat{\boldsymbol{\zeta}}} \\ &= \frac{\hat{\mathbf{S}}\left(\frac{\mathbf{I}}{\lambda} + \hat{\boldsymbol{\Psi}}\right)^{-1}\hat{\mathbf{S}}^*\mathbf{a}(\omega)}{\mathbf{a}^*(\omega)\hat{\mathbf{S}}\left(\frac{\mathbf{I}}{\lambda} + \hat{\boldsymbol{\Psi}}\right)^{-1}\hat{\mathbf{S}}^*\mathbf{a}(\omega)}.\end{aligned} \tag{3.196}$$

In Appendix 3.F, we show that the $\hat{\mathbf{h}}(\omega)$ derived above using the $\hat{\mathbf{R}}$ in (3.159) satisfies

$$\mathbf{J}\hat{\mathbf{h}}^c(\omega) = \hat{\mathbf{h}}(\omega)e^{-j(M-1)\omega}. \tag{3.197}$$

Then we have

$$\begin{aligned}e^{-j(N-1)\omega}\tilde{\mathbf{g}}^*(\omega)\hat{\mathbf{h}}(\omega) &= e^{-j(N-1)\omega} \cdot \frac{1}{L}\left(\sum_{l=0}^{L-1} \tilde{\mathbf{y}}_l e^{-j\omega l}\right)^* \cdot \hat{\mathbf{h}}(\omega) \\ &= \frac{1}{L}\left(\sum_{l=0}^{L-1} \tilde{\mathbf{y}}_l e^{-j\omega l}\right)^* \cdot \mathbf{J}\hat{\mathbf{h}}^c(\omega) \cdot e^{j(M-1)\omega}e^{-j(N-1)\omega} \\ &= \frac{1}{L}\left(\sum_{l=0}^{L-1} \mathbf{J}\bar{\mathbf{y}}^c_{L-l-1} e^{-j\omega l}\right)^* \cdot \mathbf{J}\hat{\mathbf{h}}^c(\omega) \cdot e^{-j(N-M)\omega} \\ &= \frac{1}{L}\left(\mathbf{J}\sum_{l'=0}^{L-1} \bar{\mathbf{y}}^c_{l'} e^{-j\omega(L-l'-1)}\right)^* \cdot \mathbf{J}\hat{\mathbf{h}}^c(\omega) \cdot e^{-j(N-M)\omega} \\ &= [\bar{\mathbf{g}}^c(\omega)]^* \cdot e^{j(L-1)\omega}\mathbf{J} \cdot \mathbf{J}\hat{\mathbf{h}}^c(\omega) \cdot e^{-j(N-M)\omega} \\ &= [\bar{\mathbf{g}}^c(\omega)]^*\hat{\mathbf{h}}^c(\omega) \\ &= \hat{\mathbf{h}}^*(\omega)\bar{\mathbf{g}}(\omega).\end{aligned} \tag{3.198}$$

Consequently, the forward-backward spectral estimate $\hat{\alpha}(\omega)$ in (3.179) can be simplified as

$$\hat{\alpha}(\omega) = \begin{cases} \hat{\mathbf{h}}^*(\omega)\bar{\mathbf{g}}(\omega), & \hat{\mathbf{a}}(\omega) \in \mathcal{R}(\hat{\mathbf{R}}) \\ 0, & \hat{\mathbf{a}}(\omega) \notin \mathcal{R}(\hat{\mathbf{R}}) \end{cases}. \tag{3.199}$$

Substituting (3.196) into (3.199), we get

$$\hat{\alpha}(\omega) = \begin{cases} \dfrac{\mathbf{a}^*(\omega)\hat{\mathbf{S}}\left(\dfrac{\mathbf{I}}{\lambda} + \hat{\boldsymbol{\Psi}}\right)^{-1}\hat{\mathbf{S}}^*\bar{\mathbf{g}}(\omega)}{\mathbf{a}^*(\omega)\hat{\mathbf{S}}\left(\dfrac{\mathbf{I}}{\lambda} + \hat{\boldsymbol{\Psi}}\right)^{-1}\hat{\mathbf{S}}^*\mathbf{a}(\omega)}, & \bar{\epsilon} \geq 0 \\ 0, & \bar{\epsilon} < 0 \end{cases}. \tag{3.200}$$

We remark that the above rank-deficient RCF spectral estimator requires $O(K^3)$ flops, which are mainly due to the eigendecomposition of the singular sample covariance matrix $\hat{\mathbf{R}}$, while the full rank version needs $O(M^3)$ flops.

3.8.3 Numerical Examples

We study the resolution and accuracy performance of the rank-deficient RCF complex spectral estimator by using both 1-D and 2-D numerical examples. We compare the rank-deficient RCF spectral estimator with the following spectral estimators: the windowed FFT (WFFT), Capon, APES, full-rank norm constrained Capon filter-bank (NCCF), rank-deficient NCCF, full-rank RCF, and a version of the high-definition imaging (HDI) (see Appendix 3.G for brief descriptions of the NCCF and HDI). The version of HDI we have considered includes both norm and subspace constraints [67, 70]. In the first 1-D example, we consider estimating the locations and complex amplitudes of two closely spaced spectral lines in the presence of strong interferences and additive zero-mean white Gaussian noise. In the second 1-D example, we consider a single spectral line in the presence of strong interferences and noise. In the 2-D example, we investigate the usage of complex spectral estimators for synthetic aperture radar (SAR) imaging.

Example 3.8.1: 1-D Complex Spectral Estimation for Two Closely Spaced Spectral Lines We consider two closely spaced spectral lines (sinusoids) having frequencies 0.09 and 0.1 Hz. For simplicity, we assume that they both have unit amplitude and zero phase. There are 11 strong interferences that are uniformly spaced between 0.25 and 0.27 Hz in frequency with the frequency spacing between two adjacent interferences being 0.002 Hz. The interferences also have zero phase and are of equal power, which is 32 dB stronger than that of the two weak spectral lines. The data sequence has 64 samples and is corrupted by a zero-mean additive white Gaussian noise with variance σ_n^2. For the two spectral lines of interest, we have the signal-to-noise ratios $\mathrm{SNR}_1 = \mathrm{SNR}_2 = 12$ dB, where

$$\mathrm{SNR}_k = 10\log_{10}\frac{|\alpha_k|^2}{\sigma_n^2} \text{ (dB)} \tag{3.201}$$

with α_k being the complex amplitude of the kth sinusoid. The true spectrum of the signal is given in Figure 3.29(a). We are interested in estimating the two weak spectral lines. For better visualization, the corresponding zoomed-in spectrum focusing on the weak targets is shown in the upper-left corner of the figure.

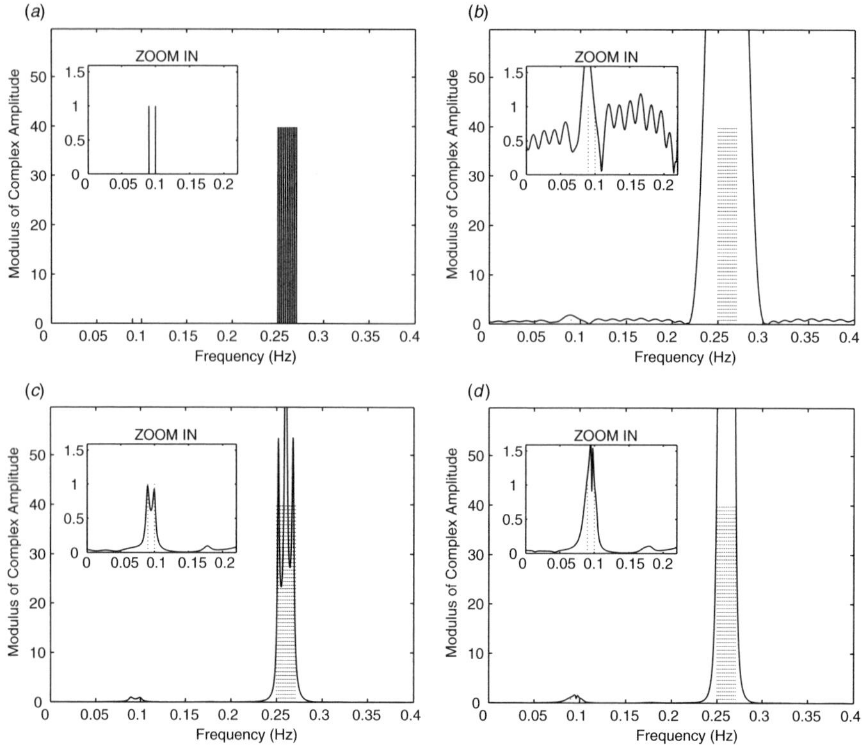

Figure 3.29 Modulus of 1-D spectral estimates ($N = 64$): (*a*) true spectrum, (*b*) WFFT, (*c*) Capon with $M = 32$, (*d*) APES with $M = 32$, (*e*) full-rank NCCF with $M = 32$ and $\eta = 0.3$, (*f*) rank-deficient NCCF with $M = 56$ and $\eta = 2$, (*g*) full-rank RCF with $M = 32$ and $\epsilon = 0.3$, and (*h*) rank-deficient RCF with $M = 56$ and $\epsilon = 0.3$.

The modulus of the spectral estimates obtained by using WFFT, Capon, APES, full-rank NCCF, rank-deficient NCCF, full-rank RCF, and rank-deficient RCF are given in Figures 3.29(*b*)–3.29(*h*), respectively. The comparison with HDI will be given later in a 2-D example. In Figure 3.29(*b*), a Taylor window with order 5 and sidelobe level -50 dB is applied to the data before the zero-padded FFT. Note that the resolution of WFFT is quite poor. In Figures 3.29(*c*) and 3.29(*d*), respectively, Capon and APES are used, both with $M = 32$. Although the Capon spectrum in Figure 3.29(*c*) gives two peaks close to the desired frequencies, they are not very well separated. The amplitude estimates of Capon are also slightly biased downward. The APES spectrum is known to give excellent amplitude estimates at the true frequency locations but suffers from biased frequency estimation [71]. As shown in Figure 3.29(*d*), APES barely resolves the two spectral lines. The full-rank NCCF spectrum obtained with $M = 32$ and a norm squared constraint on the weight vector corresponding to $\eta = 0.3$ is shown in Figure 3.29(*e*); the two closely spaced spectral lines are hardly separated. The rank-deficient NCCF

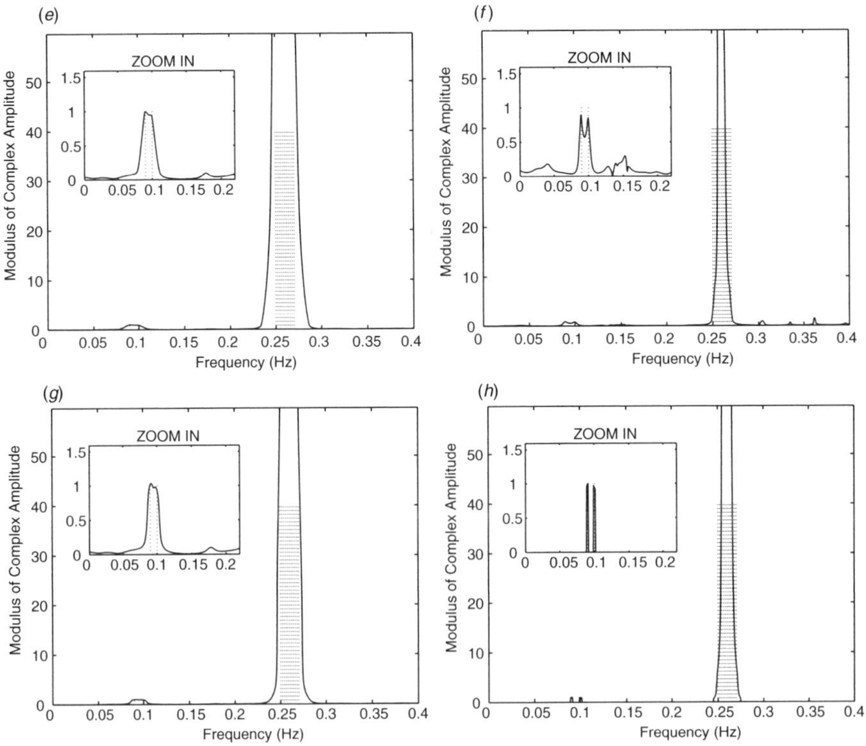

Figure 3.29 (*Continued*).

spectrum shown in Figure 3.29(f) is obtained with $M = 56$ and a norm squared constraint of $\eta = 2$. It has better resolution than its full-rank counterpart in Figure 3.29(e) but with higher sidelobes, and the amplitude estimates are biased downward. The full-rank RCF spectrum is obtained with $M = 32$ and $\epsilon = 0.3$, which is shown in Figure 3.29(g). Like the full-rank NCCF, the full-rank RCF can hardly resolve the two spectral lines. Figure 3.29(h) shows the rank-deficient RCF spectrum, which is obtained by using $M = 56$ and $\epsilon = 0.3$. Note that in Figure 3.29(h), the two closely spaced spectral lines are well resolved with no sidelobes.

Although APES can provide excellent amplitude estimates at the true frequency locations, in many cases this knowledge is not available. When this knowledge is unavailable, the frequency estimate for each of the two spectral lines can be obtained from the center frequency of the half-power (3 dB) interval of the corresponding peaks in the rank-deficient RCF spectrum. Using 100 Monte Carlo simulations (by varying the noise realizations), we obtained the root mean-squared errors (RMSEs) of the frequency estimates of the rank-deficient RCF. For the first and second lines of interest, the RMSEs of the frequency estimates obtained via the rank-deficient RCF are 6.3×10^{-4} and 5.9×10^{-4} Hz, respectively, which are quite accurate. The corresponding RMSEs of the magnitude and phase estimates obtained by using the rank-deficient RCF and APES at these estimated frequencies are listed in Table 3.2. Note that in this example, the rank-deficient RCF gives slightly more accurate magnitude estimates but worse phase estimates than APES.

Example 3.8.2: 1-D Complex Spectral Estimation for a Single Spectral Line To provide further comparisons between the more successful methods in the previous example, we next consider estimating the parameters of a single spectral line at 0.09 Hz which has unit amplitude and zero phase. The modulus of the true complex spectrum is shown in Figure 3.30(*a*). The setup of this experiment is the same as for the previous example except that we have only one spectral line instead of two. Figures 3.30(*b*)–3.30(*d*) show the spectral estimates obtained with APES, rank-deficient NCCF, and rank-deficient RCF, respectively. The APES spectrum with $M = 32$ gives a good amplitude estimate and low sidelobes, but has a relatively wide mainlobe. The rank-deficient NCCF spectrum with $M = 56$ and $\eta = 2$ shown in Figure 3.30(*c*) is clearly biased downward and has high sidelobes. The rank-deficient RCF spectrum with $M = 56$ and $\epsilon = 0.3$ shown in Figure 3.30(*d*) demonstrates a good amplitude estimate, a narrow mainlobe, and no sidelobes.

Using 100 Monte Carlo simulations, we computed the RMSEs of the frequency, magnitude, and phase estimates of the spectral line at 0.09 Hz obtained by using the rank-deficient RCF and APES. The frequency estimates of both the rank-deficient RCF and APES are obtained by using the procedure for RCF described in the previous example. The RMSEs of the frequency estimates obtained via the rank-deficient RCF and APES are listed in Table 3.3. The RMSE's of the magnitude and phase estimates of the rank-deficient RCF and APES at the frequencies estimated by the rank-deficient RCF are also listed in Table 3.3. In this example, the rank-deficient RCF gives more accurate frequency estimates than APES, but its magnitude and phase estimates are slightly worse. The RMSEs of the magnitude and phase estimates of APES at the frequencies estimated by APES are 0.028 and 0.057 (radian), respectively. These RMSEs are slightly worse than those at the frequencies determined by the rank-deficient RCF, but they are still slightly better than those of the rank-deficient RCF estimates.

Example 3.8.3: 2-D Complex Spectral Estimation for Synthetic Aperture Radar (SAR) Imaging We consider using the rank-deficient RCF for SAR imaging. The 2-D high resolution phase history data of a Slicy object at 0° azimuth

TABLE 3.2 RMSEs of the Modulus and the Phase (Radian) Estimates Obtained by the Rank-Deficient RCF and APES Spectral Estimators in the First 1-D Example

	Rank-Deficient RCF		APES	
	Modulus	Phase (Radian)	Modulus	Phase (Radian)
Signal 1	0.065	0.393	0.079	0.157
Signal 2	0.063	0.415	0.072	0.156

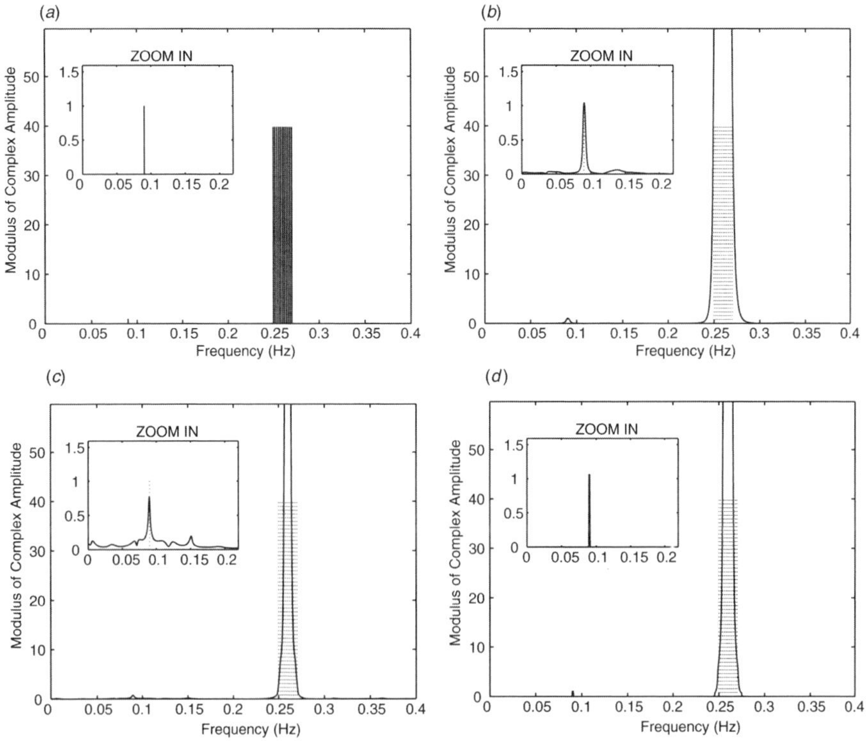

Figure 3.30 Modulus of 1-D spectral estimates ($N = 64$): (*a*) true spectrum, (*b*) APES with $M = 32$, (*c*) rank-deficient NCCF with $M = 56$ and $\eta = 2$, and (*d*) rank-deficient RCF with $M = 56$ and $\epsilon = 0.3$.

angle was generated by XPATCH [72], a high frequency electromagnetic scattering prediction code for complex 3-D objects. A photo of the Slicy object taken at 45° azimuth angle is shown in Figure 3.31(*a*). The original XPATCH data matrix has a size of $N = \bar{N} = 288$ with a resolution of 0.043 meters in both range and cross-range. Figure 3.31(*b*) shows the modulus of the 2-D WFFT image of the original data, where a Taylor window with order 5 and peak sidelobe level -35dB is applied to the data before zero-padded FFT.

TABLE 3.3 RMSEs of the Frequency, the Modulus, and the Phase (Radian) Estimates Obtained by the Rank-Deficient RCF and APES Spectral Estimators in the Second 1-D Example

	Rank-Deficient RCF	APES
Frequency (Hz)	2.72×10^{-4}	3.15×10^{-4}
Modulus	0.034	0.026
Phase (radian)	0.062	0.054

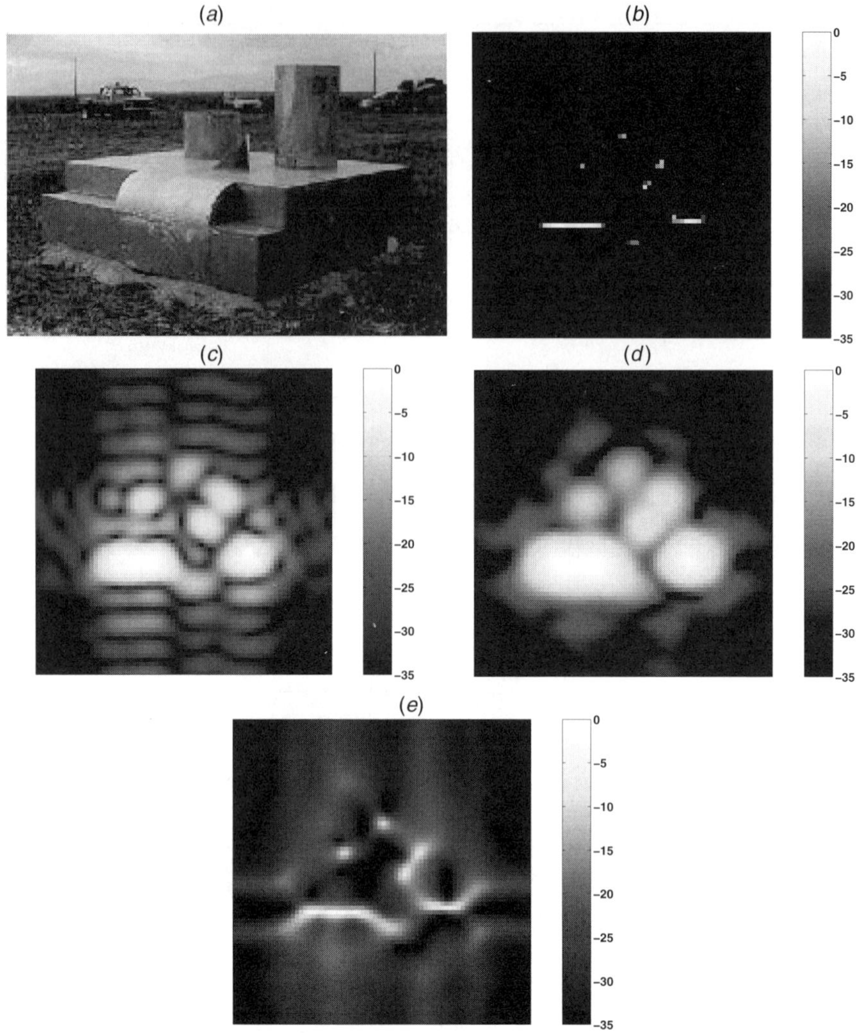

Figure 3.31 Modulus of the SAR images of the Slicy object from a 24×24 data matrix: (*a*) photograph of the object (taken at 45° azimuth angle), (*b*) 2-D WFFT with 288×288 (not 24×24) data matrix, (*c*) 2-D FFT, (*d*) 2-D WFFT, (*e*) 2-D CAPON with $M = \bar{M} = 12$, (*f*) 2-D APES with $M = \bar{M} = 12$, (*g*) 2-D full-rank RCF with $M = \bar{M} = 12$ and $\epsilon = 2$, (*h*) 2-D rank-deficient RCF with $M = \bar{M} = 16$ and $\epsilon = 2$, (*i*) 2-D rank-deficient NCCF with $M = \bar{M} = 16$ and $\eta = 0.2$, and (*j*) 2-D HDI with $M = \bar{M} = 16$ and $\varepsilon = 0.05$.

Next, we consider only a 24×24 center block of the phase history data for SAR image formation, with the purpose of using Figure 3.31(*b*) as a reference for comparison. Since some of the Slicy features, such as the spectral lines corresponding to the dihedrals, are not stationary across the cross-range, the intensity of the features relative to each other may change as the data dimension is reduced from 288×288

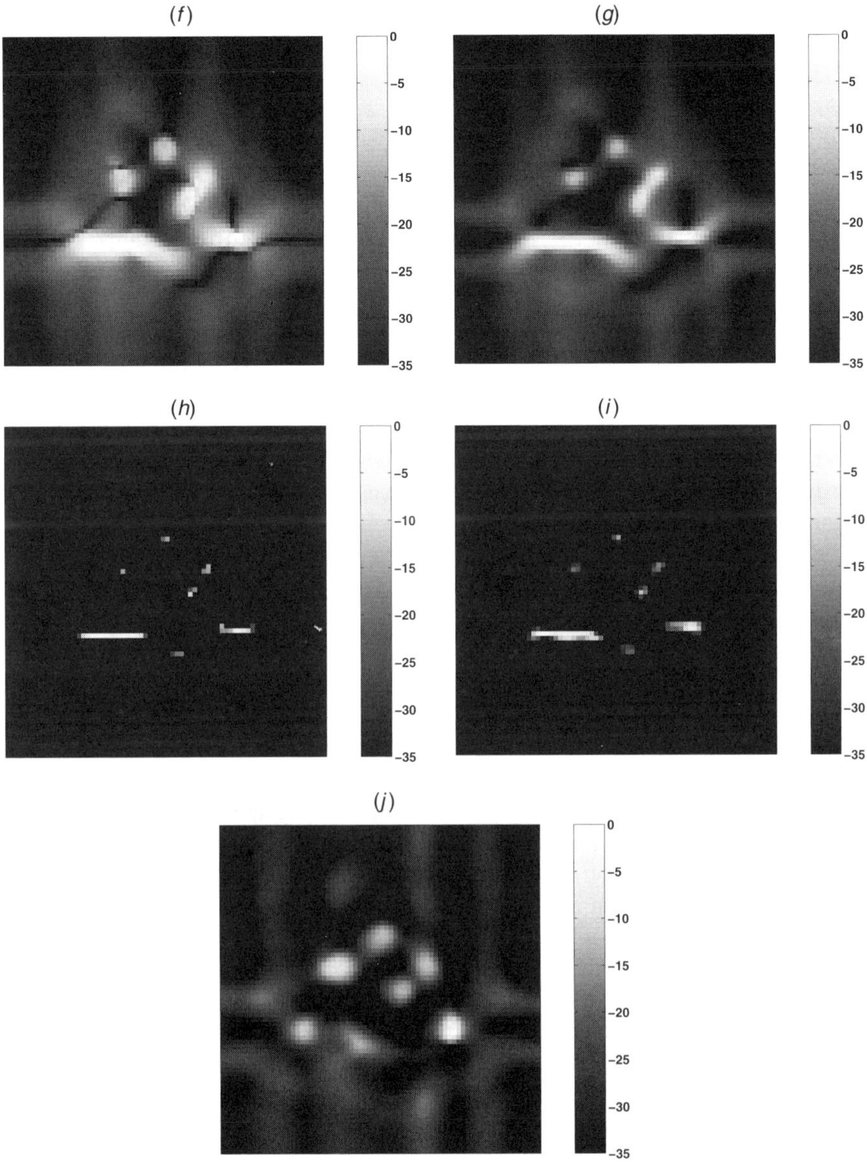

Figure 3.31 (*Continued*).

to 24 × 24. Figures 3.31(c)–3.31(f) show the modulus of 2-D FFT, 2-D WFFT [using the same type of window as for Figure 3.31(b)], 2-D Capon, and 2-D APES spectral estimates, respectively. Note the high sidelobes and smeared features in the FFT image. The WFFT image demonstrates more smeared features with some of the features not resolved. Capon gives narrow mainlobes but smaller amplitude estimates than WFFT. APES provides unbiased spectral estimates but has wider mainlobes than Capon.

The modulus of the full-rank RCF spectral estimate is shown in Figure 3.31(g) which is obtained using $M = \bar{M} = 12$ and $\epsilon = 2$. Figure 3.31(h) shows the modulus of the rank-deficient RCF spectral estimate obtained by using $M = \bar{M} = 16$ and $\epsilon = 2$. Note that the image in Figure 3.31(h) has no sidelobe problem and all important features of the Slicy object are clearly separated. Compared with Figure 3.31(b), we note that although the data size was reduced to 24×24 from 288×288, the rank-deficient RCF produces an image similar to the WFFT image using the original high-resolution data. The result for the rank-deficient NCCF is shown in Figure 3.31(i) which is obtained using $M = \bar{M} = 16$ and a norm squared constraint on the weight vector of $\eta = 0.2$. Compared with Figure 3.31(h), the features of the rank-deficient NCCF image are not as clear as those of the rank-deficient RCF image and the fidelity of the rank-deficient NCCF image is worse. Another rank-deficient spectral estimate we used in this example is the HDI with both quadratic and subspace constraints [67]. The reconstructed image using HDI is shown in Figure 3.31(j) which is obtained by using $M = \bar{M} = 16$ and a norm squared constraint of $\varepsilon = 0.05$. The subspace constraints were introduced in [67] to preserve the background. We note that the HDI image is more smeared than those obtained using the rank-deficient NCCF and rank-deficient RCF.

3.9 ADAPTIVE IMAGING FOR FORWARD-LOOKING GROUND PENETRATING RADAR

In forward-looking ground penetrating radar (FLGPR) systems, the electromagnetic (EM) wave is transmitted into the ground and the identification of targets is obtained by examining the backscattered field. Since FLGPR is able to discern the discontinuities in the electric permittivity of the propagation medium, nonmetallic objects such as plastic-cased mines can also be detected. Most FLGPRs are ultra-wideband (UWB) systems with the working frequency range from 0.5 to 3 GHz. Through the use of antenna arrays, the state-of-the-art FLGPRs can produce high resolution two-dimensional (2-D) or three-dimensional (3-D) images of buried objects for landmine detection [73–77].

Since the FLGPR system detects buried targets based on the reconstructed reflectivity image of a scene, at least for prescreening, high quality radar image formation is essential. The conventional imaging algorithm for FLGPR is the delay-and-sum (DAS) method [2], which is also known as the backprojection method [78]. However, DAS is a data-independent approach, which is known to suffer from low resolution and poor interference rejection capability. In many practical scenarios where strong clutter is present, the performance of the DAS-based algorithms degrades severely, which can result in far too many false alarms for FLGPR systems.

In this section, we present a new adaptive imaging method, referred to as the APES-RCB approach, for FLGPR image formation [79]. The new method consists of two major steps. First, the amplitude and phase estimation (APES) algorithm is used to estimate the reflection coefficients for the focal points of interest for each receiving channel. Since APES is a nonparametric data-adaptive matched-filterbank (MAFI) based algorithm, it preserves the robust nature of the nonparametric methods but at the same time it improves the spectral estimates in the sense of narrower spectral peaks and lower sidelobes than DAS, discrete Fourier transform (DFT) or fast Fourier transform (FFT) methods [64, 66]. Second, a rank-deficient robust Capon beamformer (RCB) is used to estimate the reflection coefficients for the focal points of interest from the estimates obtained via APES for all channels. By making explicit use of an uncertainty set for the array steering vector, the adaptive RCB can tolerate both array steering vector errors and low snapshot numbers (see Section 3.4 and [16]). By allowing the involved data matrix to be rank-deficient, our method can be applied to practical scenarios where the number of multilooks is smaller than the number of sensors in the array. Furthermore, by using the rank-deficient RCB, we can achieve much better interference and clutter rejection capability than most existing approaches, which is useful in many applications such as target detection and feature extraction. We apply the APES-RCB approach to experimental data collected via two recently developed FLGPR systems by PSI (Planning Systems Inc.) and SRI (Stanford Research Institute). Experimental results are used to demonstrate the excellent performance of our new imaging approach as compared with the conventional DAS-based methods.

3.9.1 Data Model and Problem Formulation of FLGPR Imaging

As shown in Figure 3.32, an FLGPR system is used to detect the buried mines in front of the vehicle. Let x, y, and z denote the cross-range, down-range, and height (also depth) axes of a coordinate system. Let $(x_{\mathrm{r},m,n}, y_{\mathrm{r},m,n}, z_{\mathrm{r},m,n})$ denote

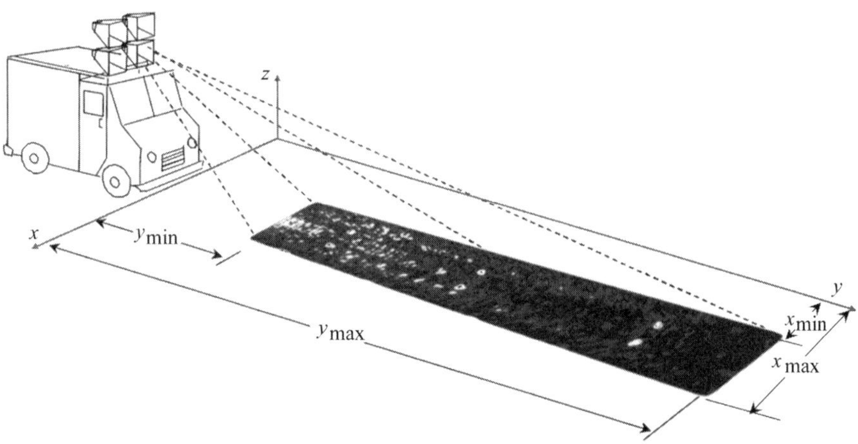

Figure 3.32 Diagram of an FLGPR system used for landmine detection.

the location of the mth receiver during the nth scan, and let $(x_{t,d,n}, y_{t,d,n}, z_{t,d,n})$ denote the location of the dth transmitter, where $m = 0, 1, \ldots, M-1$, $d = 0, 1, \ldots, D-1$, and $n = 0, 1, \ldots, N-1$ with M, D, and N denoting the total numbers of receiving antennas, transmitting antennas, and scans, respectively. The imaging region extends from $x_{\min}$ to $x_{\max}$ in the cross-range dimension and from $y_{\min}$ to $y_{\max}$ in the down-range dimension. Let $p = \{x_F, y_F, z_F\}$ represent the location of a focal point in the imaging region, where $z_F = 0$ denotes the ground surface and $z_F < 0$ denotes the underground points. For simplicity, consider a focal point on the ground ($z_F = 0$). At the nth scan, the time delay due to the system delay τ_{sys} and the EM wave propagation from the dth transmitter to the focal point p and then back to the mth receiver is

$$\begin{aligned}\tau_{d,m,n}(p) &= \frac{1}{c}[(x_{t,d,n} - x_F)^2 + (y_{t,d,n} - y_F)^2 + z_{t,d,n}^2]^{\frac{1}{2}} \\ &\quad + \frac{1}{c}[(x_{r,m,n} - x_F)^2 + (y_{r,m,n} - y_F)^2 + z_{r,m,n}^2]^{\frac{1}{2}} + \tau_{\text{sys}}\end{aligned} \tag{3.202}$$

where c is the velocity of the EM wave in the air. The stepped frequencies of FLGPR have the form:

$$f_k = f_0 + k\Delta f, \quad k = 0, 1, \ldots, K-1, \tag{3.203}$$

where f_0 denotes the initial frequency, Δf represents the frequency step, and K is the total number of stepped frequencies. Given a focal point p, the measured kth stepped-frequency response $y_{d,m,n}(k)$ corresponding to the dth transmitter and the mth receiver at the nth scan location has the form

$$\begin{aligned}y_{d,m,n}(k) &= \beta_n(p)e^{-j2\pi f_k \tau_{d,m,n}(p)} + e_{d,m,n}(k,p), \\ d &= 0, 1, \ldots, D-1, \quad m = 0, 1, \ldots, M-1, \\ n &= 0, 1, \ldots, N-1, \quad k = 0, 1, \ldots, K-1,\end{aligned} \tag{3.204}$$

where $\beta_n(p)$ denotes the reflection coefficient for the focal point p at the nth scan, and $e_{d,m,n}(k,p)$ denotes the residual term at point p, which includes the unmodeled noise and interference from scatterer responses other than p. In (3.204), we have assumed that the reflection coefficient may change from scan to scan. This is based on the fact that, in practice, as the FLGPR system moves forward, the EM wave incident angle relative to the fixed point p on the ground changes. Consequently, the reflection coefficient for point p may differ from scan to scan as the radar moves forward [80].

The problem of interest herein is to estimate $\{\beta_n(p)\}_{n=0}^{N-1}$, for each focal point of interest, from the measured data set $y_{d,m,n}(k)$ with $d = 0, 1, \ldots, D-1$, $m = 0, 1, \ldots, M-1$, $n = 0, 1, \ldots, N-1$, and $k = 0, 1, \ldots, K-1$. These estimates can then be used to form FLGPR images.

3.9.2 The Delay-and-Sum Algorithm

A brief overview of the conventional DAS method for FLGPR imaging is provided in this section. The discussion on the DAS method will be helpful for presenting our new approach later on.

The idea of DAS is to sum all measured data coherently at one focal point and repeat the process for all points of interest. The DAS-based reflection coefficient estimates for the focal point p have the form

$$\hat{\beta}_n(p) = \frac{1}{D \cdot \sum_{m=0}^{M-1} |w_r(m)|^2 \cdot \sum_{k=0}^{K-1} |w_f(k)|^2} \times \sum_{d=0}^{D-1} \sum_{m=0}^{M-1} w_r(m) \sum_{k=0}^{K-1} w_f(k) y_{d,m,n}(k) e^{j2\pi f_k \tau_{d,m,n}(p)}, \quad n = 0, 1, \ldots, N-1 \tag{3.205}$$

where $w_f(\cdot)$ and $w_r(\cdot)$ denote the weights for the frequency and receiver aperture dimensions, respectively. Based on the estimates $\{\hat{\beta}_n(p)\}_{n=0}^{N-1}$ of $\{\beta_n(p)\}_{n=0}^{N-1}$, we can obtain the radar image as

$$I_1(p) = \frac{1}{N} \sum_{n=0}^{N-1} \hat{\beta}_n(p). \tag{3.206}$$

The above method is referred to as coherent multilook processing. In practice, the phases of $\{\beta_n(p)\}_{n=0}^{N-1}$ for buried mines may vary with the scan location. Consequently, the above coherent processing tends to fail when the phase variations along the scan dimension become too large [80]. Hence, in these cases, we can take the absolute values of individual images before the multilook image is formed. This method is referred to as the noncoherent processing and can be expressed as

$$I_2(p) = \frac{1}{N} \sum_{n=0}^{N-1} \left| \hat{\beta}_n(p) \right|. \tag{3.207}$$

For stepped-frequency FLGPR systems, the above DAS-based algorithms can be efficiently implemented as follows:

1. For each channel (transmitter/receiver pair), calculate the inner sum in (3.205) via the inverse fast Fourier transform (IFFT) with zero-padding.
2. Calculate the two outer sums in (3.205) by summing up the signals corresponding to the given focal point from all channels.
3. Perform coherent or noncoherent multilook processing along the scan dimension.

Note that DAS is a data-independent approach which suffers from low resolution and poor interference and clutter rejection capability. We present next our data-adaptive imaging approach, referred to as APES-RCB, for FLGPR image formation.

3.9.3 The APES-RCB Algorithm

The APES-RCB algorithm is an adaptive imaging approach which consists of two major steps. First, instead of using the FFT-based method, APES is adopted to obtain more accurate reflection coefficient estimates for each receiving channel. Second, rank-deficient RCB is used to estimate the original reflection coefficients based on the estimates obtained via APES from all channels.

3.9.3.1 Step One: APES. Consider the data model in (3.204). Let

$$\alpha_{d,m,n}(p) = \beta_n(p)e^{-j\omega_0\tau_{d,m,n}(p)} = \beta_n(p)e^{-j2\pi f_0\tau_{d,m,n}(p)} \tag{3.208}$$

and

$$\omega_{d,m,n}(p) = 2\pi\Delta f\tau_{d,m,n}(p). \tag{3.209}$$

With these notations, (3.204) becomes

$$\begin{aligned} y_{d,m,n}(k) = \alpha_{d,m,n}(p)e^{-jk\omega_{d,m,n}(p)} &+ e_{d,m,n}(k,p), \\ d = 0, 1, \ldots, D-1, \quad & m = 0, 1, \ldots, M-1, \\ n = 0, 1, \ldots, N-1, \quad & k = 0, 1, \ldots, K-1. \end{aligned} \tag{3.210}$$

Let d, m, n, and p be fixed. Then (3.210) can be expressed as

$$y(k) = \alpha(\omega)e^{-jk\omega} + e_\omega(k), \quad k = 0, 1, \ldots, K-1. \tag{3.211}$$

(For clarity, we omit the dependence on d, m, n, and p to simplify the notation.) The problem of interest is to estimate $\alpha(\omega)$ from $\{y(k)\}_{k=0}^{K-1}$ for any given ω. This problem belongs to the classical problem of complex spectral estimation. The conventional approaches to complex spectral estimation include DFT and its variants which are typically based on smoothing the DFT spectral estimate or windowing the data [1]. These methods do not make any a priori assumptions on the data and consequently they are very robust. However, they suffer from low resolution and poor accuracy problems.

Nonparametric adaptive matched-filterbank (MAFI) methods can mitigate the low resolution and poor accuracy problems of the DFT-based methods [63, 64]. For each frequency ω of interest, a MAFI method filters the data with a normalized finite-impulse response (FIR) filter $\mathbf{h}(\omega)$. The filter is chosen according to a criterion which is different for the various spectral analysis methods, but with the common constraint that a sinusoid with frequency ω should pass the filter without any distortion. Following the filtering, a sinusoid is fitted to the filtered data in a least-squares (LS) sense, and the amplitude of the so-obtained sinusoid $\hat{\alpha}(\omega)$ is taken as the estimate of the amplitude spectrum at the frequency ω of interest. This class of estimators includes the classical Capon algorithm and the more recent APES approach. Note that it has been shown that Capon is biased downward whereas APES is unbiased. In fact, both theoretical performance analysis and numerical examples have demonstrated that APES can provide excellent accuracy for complex spectral estimation [66]. For FLGPR imaging, accurate reflection coefficient estimates for the focal points of interest for each receiving channel are essential. As we will show, APES works well for this practical problem. Additionally, APES is straightforward to use due to the fact that it requires no search over any parameter space.

Note that a computationally efficient implementation of APES can be found in [63]. After applying the fast APES of [63], which requires a uniform grid for ω, the desired estimates at different and possibly nonuniform values of ω can be obtained by using interpolation. From the APES estimate $\hat{\alpha}_{d,m,n}(p)$, we can readily obtain intermediate reflection coefficient estimates based on (3.208):

$$\hat{\bar{\beta}}_{d,m,n}(p) = e^{j\omega_0 \tau_{d,m,n}(p)} \hat{\alpha}_{d,m,n}(p), \tag{3.212}$$
$$d = 0, 1, \ldots, D-1, \quad m = 0, 1, \ldots, M-1, \quad n = 0, 1, \ldots, N-1.$$

We remark that at this stage we have obtained a total number of DMN reflection coefficient estimates for each focal point since we have overparameterized the N unknowns $\{\beta_n\}_{n=0}^{N-1}$ via DMN unknowns $\{\{\{\bar{\beta}_{d,m,n}\}_{d=0}^{D-1}\}_{m=0}^{M-1}\}_{n=0}^{N-1}$ [see (3.208)] in order to use APES in a direct manner. In the next step, we use the rank-deficient RCB to estimate the original N reflection coefficients for each focal point from the DMN estimates obtained via APES.

3.9.3.2 Step Two: Rank-Deficient RCB.

For the focal point p, the reflection coefficients estimated by APES satisfy

$$\hat{\bar{\beta}}_{d,m,n}(p) = \beta_n(p) + \mu_{d,m,n}(p), \tag{3.213}$$
$$d = 0, 1, \ldots, D-1, \quad m = 0, 1, \ldots, M-1, \quad n = 0, 1, \ldots, N-1$$

where $\{\{\{\mu_{d,m,n}(p)\}_{d=0}^{D-1}\}_{m=0}^{M-1}\}_{n=0}^{N-1}$ denote the estimation errors (such as caused by finite-sample effects and mismodelling) as well as any leftover interferences. [For each channel, the interferences from locations other than p but having the same time delay (equal to $\tau_{d,m,n}(p)$) cannot be suppressed by APES.] Let

$$\mathbf{y}_n(p) = \left[\hat{\bar{\beta}}_{0,0,n}(p) \cdots \hat{\bar{\beta}}_{0,M-1,n}(p) \cdots \hat{\bar{\beta}}_{D-1,0,n}(p) \cdots \hat{\bar{\beta}}_{D-1,M-1,n}(p)\right]^T,$$
$$n = 0, 1, \ldots, N-1 \tag{3.214}$$

and

$$\boldsymbol{\mu}_n(p) = \left[\mu_{0,0,n}(p) \cdots \mu_{0,M-1,n}(p) \cdots \mu_{D-1,0,n}(p) \cdots \mu_{D-1,M-1,n}(p)\right]^T,$$
$$n = 0, 1, \ldots, N-1. \quad (3.215)$$

Then (3.213) can be rewritten as

$$\mathbf{y}_n(p) = \beta_n(p)\mathbf{a} + \boldsymbol{\mu}_n(p), \quad n = 0, 1, \ldots, N-1 \quad (3.216)$$

where $\mathbf{a}$ is theoretically equal to $\mathbf{1}_{DM\times 1}$, with $\mathbf{1}_{DM\times 1}$ denoting a DM by 1 vector whose elements are all equal to one. Note that, in practice, the steering vector $\mathbf{a}$ in (3.216) may be imprecise, in the sense that the elements in $\mathbf{a}$ may differ slightly from 1. This may be due to many factors including array calibration errors and geo-registering errors for any given p.

We will make use of the following sample covariance matrix

$$\hat{\mathbf{R}}(p) = \frac{1}{N}\sum_{n=0}^{N-1} \mathbf{y}_n(p)\mathbf{y}_n^*(p). \quad (3.217)$$

Note that usually in applications we have $N < DM$. Hence $\hat{\mathbf{R}}(p)$ is singular. Let $\bar{N}$ denote the rank of $\hat{\mathbf{R}}(p)$ in (3.217). With probability one, $\bar{N} = N$. Let

$$\hat{\mathbf{R}}(p) = [\hat{\mathbf{S}} \quad \hat{\mathbf{G}}] \begin{bmatrix} \hat{\boldsymbol{\Lambda}} & \mathbf{0} \\ \mathbf{0} & \mathbf{0} \end{bmatrix} \begin{bmatrix} \hat{\mathbf{S}}^* \\ \hat{\mathbf{G}}^* \end{bmatrix} \quad (3.218)$$

where $\hat{\mathbf{S}}$ is a $DM \times \bar{N}(DM > \bar{N})$ full column rank matrix whose columns are the eigenvectors of $\hat{\mathbf{R}}(p)$ corresponding to the nonzero eigenvalues of $\hat{\mathbf{R}}(p)$, $\hat{\mathbf{G}}$ denotes the $DM \times (DM - \bar{N})$ orthogonal complement of $\hat{\mathbf{S}}$ with the columns of $\hat{\mathbf{G}}$ corresponding to the zero eigenvalues of $\hat{\mathbf{R}}(p)$, and $\hat{\boldsymbol{\Lambda}}$ is an $\bar{N} \times \bar{N}$ positive definite diagonal matrix whose diagonal elements are the nonzero eigenvalues of $\hat{\mathbf{R}}(p)$.

Due to the small snapshot number and the imprecise knowledge of the steering vector $\mathbf{a}$, it is natural to apply the rank-deficient robust Capon beamforming algorithm to estimate the waveform $\beta_n(p)$ from the snapshots $\{\mathbf{y}_n\}_{n=0}^{N-1}$.

Let $\bar{\mathbf{a}} = \mathbf{1}_{DM\times 1}$ denote the nominal steering vector, as discussed above. Owing to the small snapshot number and the imprecise knowledge of the true steering vector $\mathbf{a}$, the signal term in $\hat{\mathbf{R}}(p)$ is not well described by $|\beta_n(p)|^2\bar{\mathbf{a}}\bar{\mathbf{a}}^*$, but by $|\beta_n(p)|^2\hat{\mathbf{a}}\hat{\mathbf{a}}^*$ with $\hat{\mathbf{a}}$ being some vector in the vicinity of $\bar{\mathbf{a}}$ and $\hat{\mathbf{a}} \neq \bar{\mathbf{a}}$. Consequently, if we designed the weight vector $\mathbf{w}(p)$ by means of the standard Capon beamformer:

$$\min_{\mathbf{w}(p)} \mathbf{w}^*(p)\hat{\mathbf{R}}(p)\mathbf{w}(p) \quad \text{subject to } \mathbf{w}^*(p)\bar{\mathbf{a}} = 1, \quad (3.219)$$

the Euclidean norm of $\mathbf{w}(p)$, denoted as $\|\mathbf{w}(p)\|$, would be rather large since $\hat{\mathbf{a}}$ is close to $\bar{\mathbf{a}}$ and $\mathbf{w}(p)$ would pass the signal associated with $\bar{\mathbf{a}}$ undistorted [see (3.219)] but attempt to suppress the signal associated with $\hat{\mathbf{a}}$. A large $\|\mathbf{w}(p)\|$ indicates a large noise gain, which may severely degrade the estimation accuracy of $\beta_n(p)$. It follows that we should design $\mathbf{w}(p)$ by

$$\min_{\mathbf{w}(p)} \mathbf{w}^*(p)\hat{\mathbf{R}}\mathbf{w}(p) \quad \text{subject to } \mathbf{w}^*(p)\hat{\mathbf{a}} = 1 \tag{3.220}$$

where we used $\hat{\mathbf{a}}$, instead of $\bar{\mathbf{a}}$, to avoid the suppression of the signal term and a large noise gain. Before solving (3.220), which is a main step of the rank-deficient RCB estimator (see below), for convenience, we decompose the weight vector $\mathbf{w}(p)$ as

$$\mathbf{w}(p) = \hat{\mathbf{S}}\mathbf{w}_1(p) + \hat{\mathbf{G}}\mathbf{w}_2(p) \tag{3.221}$$

where $\mathbf{w}_1(p) = \hat{\mathbf{S}}^*\mathbf{w}(p)$ and $\mathbf{w}_2(p) = \hat{\mathbf{G}}^*\mathbf{w}(p)$.

First, we assume that $\hat{\mathbf{a}}$ is given (the determination of $\hat{\mathbf{a}}$ will be discussed later on in this subsection), and solve the above optimization problem in (3.220) by considering the following two cases.

- *Case 1*: $\hat{\mathbf{a}}$ belongs to the range space of $\hat{\mathbf{R}}(p)$.
  Let $\hat{\mathbf{a}}$ be written as $\hat{\mathbf{a}} \triangleq \hat{\mathbf{S}}\hat{\boldsymbol{\gamma}}$ for some nonzero vector $\hat{\boldsymbol{\gamma}}$; then we have $\hat{\boldsymbol{\gamma}} = \hat{\mathbf{S}}^*\hat{\mathbf{a}}$. Using (3.221) in (3.220) for this case yields

$$\min_{\mathbf{w}_1(p)} \mathbf{w}_1^*(p)\hat{\boldsymbol{\Lambda}}\mathbf{w}_1(p) \quad \text{subject to } \mathbf{w}_1^*(p)\hat{\boldsymbol{\gamma}} = 1 \tag{3.222}$$

  which can be readily solved as

$$\mathbf{w}_1(p) = \frac{\hat{\boldsymbol{\Lambda}}^{-1}\hat{\boldsymbol{\gamma}}}{\hat{\boldsymbol{\gamma}}^*\hat{\boldsymbol{\Lambda}}^{-1}\hat{\boldsymbol{\gamma}}}. \tag{3.223}$$

  Since $\mathbf{w}_2(p)$ is irrelevant in this case, we let

$$\mathbf{w}_2(p) = 0, \tag{3.224}$$

  to reduce the noise gain of $\mathbf{w}(p)$ ($\|\mathbf{w}(p)\|^2 = \|\mathbf{w}_1(p)\|^2 + \|\mathbf{w}_2(p)\|^2$). Then the weight vector has the form

$$\mathbf{w}(p) = \frac{\hat{\mathbf{S}}\hat{\boldsymbol{\Lambda}}^{-1}\hat{\boldsymbol{\gamma}}}{\hat{\boldsymbol{\gamma}}^*\hat{\boldsymbol{\Lambda}}^{-1}\hat{\boldsymbol{\gamma}}} = \frac{\hat{\mathbf{R}}^{\dagger}(p)\hat{\mathbf{a}}}{\hat{\mathbf{a}}^*\hat{\mathbf{R}}^{\dagger}(p)\hat{\mathbf{a}}} \tag{3.225}$$

where $\hat{\mathbf{R}}^{\dagger}(p) = \hat{\mathbf{S}}\hat{\mathbf{\Lambda}}^{-1}\hat{\mathbf{S}}^*$is the Moore–Penrose pseudo-inverse of $\hat{\mathbf{R}}(p)$. Consequently, the final estimates of the reflection coefficients can be obtained as

$$\hat{\beta}_n(p) = \mathbf{w}^*(p)\mathbf{y}_n(p) = \frac{\hat{\boldsymbol{\gamma}}^*\hat{\mathbf{\Lambda}}^{-1}\hat{\mathbf{S}}^*\mathbf{y}_n(p)}{\hat{\boldsymbol{\gamma}}^*\hat{\mathbf{\Lambda}}^{-1}\hat{\boldsymbol{\gamma}}}, \quad n = 0, 1, \ldots, N-1. \tag{3.226}$$

- *Case 2*: $\hat{\mathbf{a}}$ does not belong to the range space of $\hat{\mathbf{R}}(p)$.
Let $\hat{\mathbf{a}}$ be written as $\hat{\mathbf{a}} = \hat{\mathbf{S}}\hat{\boldsymbol{\gamma}} + \hat{\mathbf{G}}\hat{\boldsymbol{\eta}}$ for some nonzero vectors $\hat{\boldsymbol{\gamma}}$ and $\hat{\boldsymbol{\eta}}$. Now (3.220) becomes

$$\min_{\mathbf{w}_1(p), \mathbf{w}_2(p)} \quad \mathbf{w}_1^*(p)\hat{\mathbf{\Lambda}}\mathbf{w}_1(p) \quad \text{subject to } \mathbf{w}_1^*(p)\hat{\boldsymbol{\gamma}} + \mathbf{w}_2^*(p)\hat{\boldsymbol{\eta}} = 1 \tag{3.227}$$

which admits a trivial solution:

$$\mathbf{w}_1(p) = 0 \tag{3.228}$$

and (for example)

$$\mathbf{w}_2(p) = \frac{\hat{\boldsymbol{\eta}}}{\|\hat{\boldsymbol{\eta}}\|^2}. \tag{3.229}$$

Consequently, in this case the final estimate of $\beta_n(p)$ would be:

$$\hat{\beta}_n(p) = \mathbf{w}^*(p)\mathbf{y}_n(p) = 0, \quad n = 0, 1, \ldots, N-1 \tag{3.230}$$

where the last equality follows from the orthogonality between $\hat{\mathbf{G}}$ and $\mathbf{y}_n(p)$, $n = 0, 1, \ldots, N-1$.

In summary, by combining the above two cases, $\beta_n(p)$ can be estimated as

$$\hat{\beta}_n(p) = \begin{cases} \dfrac{\hat{\boldsymbol{\gamma}}^*\hat{\mathbf{\Lambda}}^{-1}\hat{\mathbf{S}}^*\mathbf{y}_n(p)}{\hat{\boldsymbol{\gamma}}^*\hat{\mathbf{\Lambda}}^{-1}\hat{\boldsymbol{\gamma}}}, & \hat{\mathbf{a}} \in \mathcal{R}(\hat{\mathbf{R}}(p)) \\ 0, & \hat{\mathbf{a}} \notin \mathcal{R}(\hat{\mathbf{R}}(p)) \end{cases}, \quad n = 0, 1, \ldots, N-1 \tag{3.231}$$

where $\mathcal{R}(\hat{\mathbf{R}}(p))$ denotes the range space of $\hat{\mathbf{R}}(p)$.

Next, we will determine $\hat{\mathbf{a}}$ via a covariance fitting approach. We assume that the only knowledge we have about $\hat{\mathbf{a}}$ is that it belongs to the following uncertainty sphere:

$$\|\hat{\mathbf{a}} - \bar{\mathbf{a}}\|^2 \leq \epsilon. \tag{3.232}$$

Furthermore, we want $\hat{\mathbf{a}}$ to be such that $|\beta_n(p)|^2\hat{\mathbf{a}}\hat{\mathbf{a}}^*$ is a good fit to $\hat{\mathbf{R}}(p)$. This leads to the following optimization problem for $\hat{\mathbf{a}}$:

$$\max_{\sigma^2(p),\,\hat{\mathbf{a}}} \sigma^2(p) \quad \text{subject to } \hat{\mathbf{R}}(p) - \sigma^2(p)\hat{\mathbf{a}}\hat{\mathbf{a}}^* \geq 0$$
$$\|\hat{\mathbf{a}} - \bar{\mathbf{a}}\|^2 \leq \epsilon \tag{3.233}$$

where $\sigma^2(p) = |\beta_n(p)|^2$. The user parameter ϵ is used to describe the uncertainty about $\hat{\mathbf{a}}$. Note that ϵ is determined by several factors such as N [7], the array calibration errors, and the system georegistering errors. Hence the smaller the N or the larger the array steering vector and system errors, the larger should the ϵ be chosen.

According to the previous discussion, a vector $\hat{\mathbf{a}}$ that is in the range space of $\hat{\mathbf{S}}$, that is, $\hat{\mathbf{a}} = \hat{\mathbf{S}}\hat{\boldsymbol{\gamma}}$ for some nonzero $\hat{\boldsymbol{\gamma}}$, is what we are after since otherwise we will get an estimate equal to zero. Observe that both the signal power $\sigma^2(p)$ and the steering vector $\hat{\mathbf{a}}$ are treated as unknowns in our covariance fitting approach [see (3.233)], hence there is a scaling ambiguity between these two unknowns (see Section 3.4.1 and [16]). To eliminate this ambiguity, we can impose the norm constraint that $\|\hat{\mathbf{a}}\|^2 = \|\hat{\boldsymbol{\gamma}}\|^2 = DM$. To determine $\hat{\boldsymbol{\gamma}}$, we first obtain $\hat{\hat{\boldsymbol{\gamma}}}$ as follows:

$$\max_{\sigma^2(p),\,\hat{\mathbf{a}}} \sigma^2(p) \quad \text{subject to } \hat{\mathbf{R}}(p) - \sigma^2(p)\hat{\mathbf{a}}\hat{\mathbf{a}}^* \geq 0$$
$$\hat{\mathbf{a}} = \hat{\mathbf{S}}\hat{\hat{\boldsymbol{\gamma}}} \tag{3.234}$$
$$\|\hat{\mathbf{a}} - \bar{\mathbf{a}}\|^2 \leq \epsilon.$$

(To exclude the trivial solution of $\hat{\hat{\boldsymbol{\gamma}}} = 0$, we require that $\epsilon < \|\bar{\mathbf{a}}\|^2 = DM$.) Then $\hat{\boldsymbol{\gamma}}$ is obtained as

$$\hat{\boldsymbol{\gamma}} = \frac{\sqrt{DM} \cdot \hat{\hat{\boldsymbol{\gamma}}}}{\|\hat{\hat{\boldsymbol{\gamma}}}\|}. \tag{3.235}$$

Consider now the solution to (3.234). Let $\bar{\boldsymbol{\gamma}} \triangleq \hat{\mathbf{S}}^*\bar{\mathbf{a}}$ and $\bar{\epsilon} \triangleq \epsilon - \|\hat{\mathbf{G}}^*\bar{\mathbf{a}}\|^2$. We consider the following two cases.

- *Case 1*: $\bar{\epsilon} < 0$, which occurs when $\bar{\mathbf{a}}$ is far from the range space of $\hat{\mathbf{S}}$. Then the optimization problem in (3.234) is infeasible [58]: in such a case there is no $\hat{\mathbf{a}}$ of the form $\hat{\mathbf{a}} = \hat{\mathbf{S}}\hat{\hat{\boldsymbol{\gamma}}}$ that satisfies the constraint in (3.234). Hence the vector $\hat{\mathbf{a}}$ cannot belong to the range space of $\hat{\mathbf{S}}$ in this case and according to (3.227)–(3.230), we get

$$\hat{\beta}_n(p) = 0, \quad n = 0, 1, \ldots, N-1. \tag{3.236}$$

- *Case 2*: $\bar{\epsilon} \geq 0$, which occurs when $\bar{\mathbf{a}}$ is close to the range space of $\hat{\mathbf{S}}$ and hence of $\hat{\mathbf{R}}$.
  In this case, an $\hat{\mathbf{a}}$ belonging to the range space of $\hat{\mathbf{R}}$ can be found within the uncertainty sphere in (3.234). By using the *Lagrange multiplier methodology* to solve (3.234) in this case, we get [58]

$$\hat{\hat{\boldsymbol{\gamma}}} = \left(\frac{\hat{\boldsymbol{\Lambda}}^{-1}}{\lambda} + \mathbf{I}\right)^{-1} \bar{\boldsymbol{\gamma}} \tag{3.237}$$

$$= \bar{\boldsymbol{\gamma}} - (\mathbf{I} + \lambda\hat{\boldsymbol{\Lambda}})^{-1}\bar{\boldsymbol{\gamma}} \tag{3.238}$$

where $\lambda \geq 0$ is the Lagrange multiplier and the second equality follows from the matrix inversion lemma; λ can be obtained as the unique solution to the constraint equation [58]:

$$g(\lambda) \triangleq \left\| \left(\mathbf{I} + \lambda\hat{\boldsymbol{\Lambda}}\right)^{-1} \bar{\boldsymbol{\gamma}} \right\|^2 = \bar{\epsilon} \tag{3.239}$$

which can be solved efficiently via, for example, a Newton's method. Once λ has been determined, we use it in (3.237) to get $\hat{\hat{\boldsymbol{\gamma}}}$, which can be used in (3.235) to compute $\hat{\boldsymbol{\gamma}}$. Then we obtain the estimate of $\beta_n(p)$ by using the $\hat{\boldsymbol{\gamma}}$ in (3.226).

Combining the two cases discussed above, the rank-deficient RCB estimate of $\beta_n(p)$ can be written as

$$\hat{\beta}_n(p) = \begin{cases} \dfrac{\hat{\boldsymbol{\gamma}}^* \hat{\boldsymbol{\Lambda}}^{-1} \hat{\mathbf{S}}^* \mathbf{y}_n(p)}{\hat{\boldsymbol{\gamma}}^* \hat{\boldsymbol{\Lambda}}^{-1} \hat{\boldsymbol{\gamma}}}, & \bar{\epsilon} \geq 0 \\ 0, & \bar{\epsilon} < 0 \end{cases}, \quad n = 0, 1, \ldots, N-1. \tag{3.240}$$

Note that the rank-deficient RCB requires $O(\bar{N}^3)$ flops, which is mainly due to the eigendecomposition of the rank-deficient (rank $\bar{N}$) matrix $\hat{\mathbf{R}}$. (See [81] for an efficient eigendecomposition of a rank-deficient matrix.) Compared with the data-independent DAS weight vector $\mathbf{w}_{\text{DAS}}(p) = \mathbf{a}(p)/\|\mathbf{a}(p)\|^2 = 1/DM \cdot \mathbf{1}_{DM\times 1}$, our rank-deficient RCB $\mathbf{w}(p)$ can provide better resolution and much better interference rejection capability.

In conclusion, the APES-RCB algorithm can be briefly summarized as follows:

Step 1. Use APES for each focal point p to estimate $\{\{\{\alpha_{d,m,n}(p)\}_{d=0}^{D-1}\}_{m=0}^{M-1}\}_{n=0}^{N-1}$. Then, obtain the intermediate estimates $\{\{\{\hat{\hat{\beta}}_{d,m,n}(p)\}_{d=0}^{D-1}\}_{m=0}^{M-1}\}_{n=0}^{N-1}$ based on (3.212).

Step 2. For each p, use the rank-deficient RCB to obtain the final estimates $\{\hat{\beta}_n(p)\}_{n=0}^{N-1}$ of $\{\beta_n(p)\}_{n=0}^{N-1}$.

Step 3. The radar image is obtained by either coherent or noncoherent multilook processing based on $\{\hat{\beta}_n(p)\}_{n=0}^{N-1}$.

3.9.4 Experimental Results

PSI and SRI have developed FLGPR systems under contracts to the U.S. Army CECOM Night Vision and Electronic Sensors Directorate [76]. These systems are designed with the goal of assessing the capability of FLGPR for detecting plastic and metallic cased surface and buried mines on roadways. We concentrate herein on the buried metal mine detection. Both of these systems are UWB stepped-frequency GPRs and can be used to form 2-D (or more precisely 3-D, but with poor resolution in depth) images of the ground. The performances of the systems have been tested on the practice mine lanes. Results obtained from experimental data collected by these systems are provided to demonstrate the performance of our new adaptive imaging approach as compared with the conventional DAS-based imaging methods. Figure 3.33(*a*) shows the data collection geometry for the FLGPR systems and the ground truth for the mine locations. In the concerned experiments, there are 12 metallic-cased mines that are buried in groups of three mines at depths of 0 (flush), 5, 10, and 15 cm, respectively. Figure 3.33(*b*) shows the photograph of a metallic-cased mine.

3.9.4.1 PSI FLGPR Experimental Results. A photograph of the PSI FLGPR phase II system is shown in Figure 3.34. This system uses a vertical three-element transmitter array precombined as a single transmitter and a receiver array consisting of two horizontal 15-element subarrays. The height of the

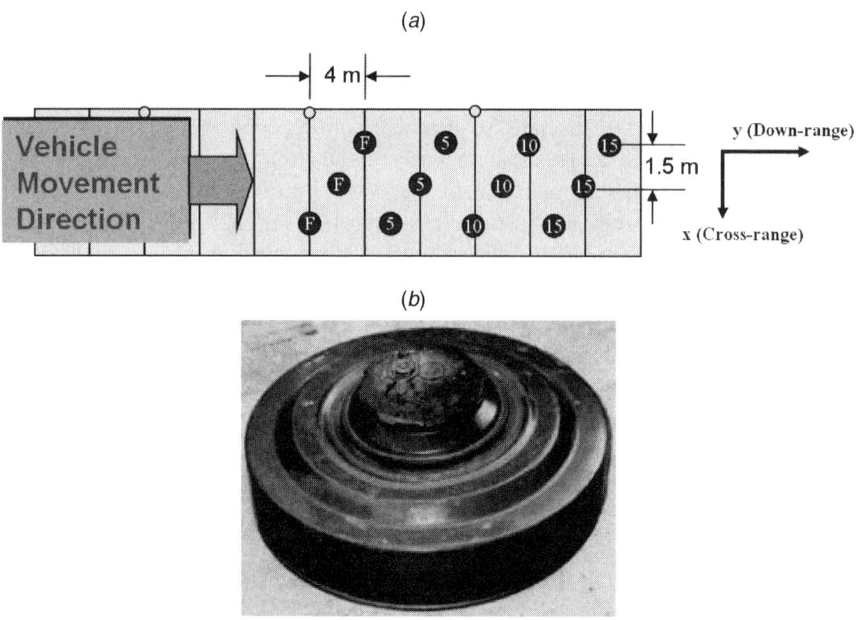

Figure 3.33 (*a*) Ground truth of the landmines on the test lane. (*b*) Photography of a metallic-cased mine.

Figure 3.34 Photograph of the PSI FLGPR phase II system.

transmitting antenna is about 2.5 m above the ground and the two receiving antenna subarrays are 1.9 m and 2.05 m above the ground. Each transmitting/receiving element uses a 14 cm Archimedean spiral antenna. The adjacent receiving antennas of each subarray are 7.62 cm apart in the aperture dimension. The stepped-frequency system operates with 201 discrete frequencies evenly spaced over a frequency range from 0.766 to 2.166 GHz. This system works in the circularly polarized mode. Data are recorded for each step of 0.1 m as the vehicle moves forward. At each scan location, the image region is 5 m (cross-range) by 3.5 m (down-range) with a 4.5 m standoff distance ahead of the vehicle. A pixel spacing of 4 cm is chosen in both the down-range and cross-range dimensions for radar imaging.

Example 3.9.1: Single-Look Imaging Results Figure 3.35 shows the single-look imaging results (the modulus is shown). In this example, evenly spaced 12 scans (each scan covering 2 m down-range) are used to form the entire image covering 24 m in the down-range. The images formed by different scans are nonoverlapping. In this figure, three different imaging methods are compared. Figure 3.35(a) is the conventional DAS imaging result where the IFFT without windowing is used. It can be observed that the imaging result is poor due to the high sidelobes and strong clutter. Figure 3.35(b) shows the DAS imaging result where the windowed IFFT is used. (We use the Kaiser window with parameter 4.) This method is referred to as the WDAS. (No weighting is used in the aperture dimension for the DAS and WDAS images.) From this figure, it is clear that the sidelobes in the down-range dimension are

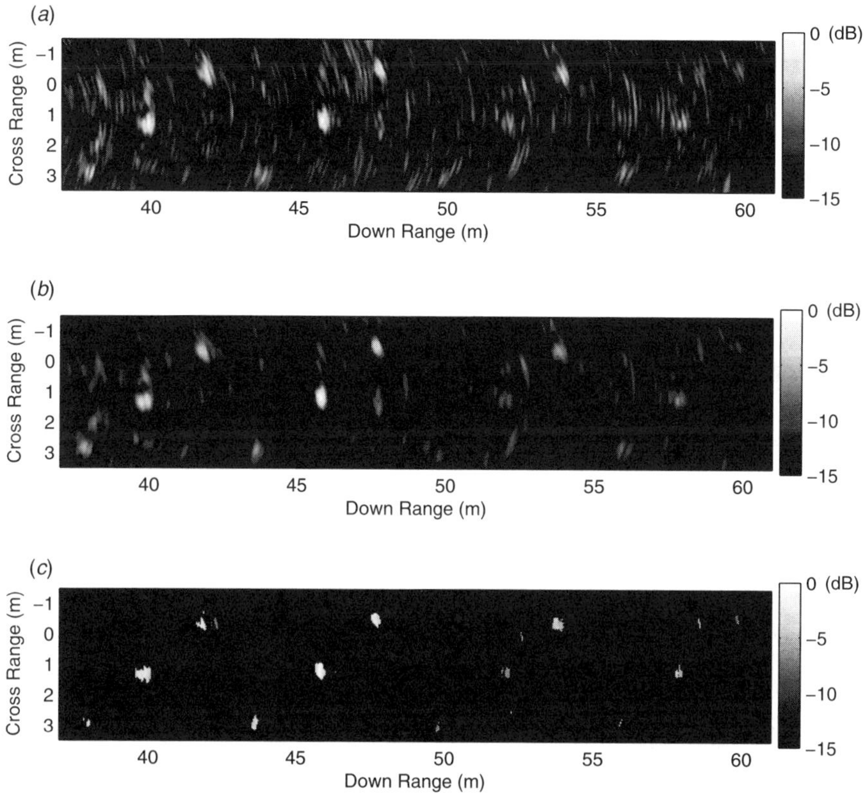

Figure 3.35 PSI single-look imaging results. (*a*) DAS imaging result. (*b*) WDAS imaging result. (*c*) APES-RCB imaging result with $\epsilon = 23$.

reduced. However, the image resolution in down-range is decreased as well. Note also that due to the poor performance of the DAS beamformer, deeply buried mines can hardly be identified from Figures 3.35(*a*) and 3.35(*b*). Figure 3.35(*c*) shows our APES-RCB imaging result. We use $\epsilon = 23$ and $\bar{\mathbf{a}} = \mathbf{1}_{DM\times 1}$ with $M = 30$ and $D = 1$ for the PSI system. From this figure, we can see that the sidelobes and clutter are effectively mitigated. Note that all 12 mines can be readily identified.

Example 3.9.2: Multilook Imaging Results The multilook imaging results based on noncoherent processing are shown in Figure 3.36. The output for each focal point in the image is obtained by 10 consecutive scans with the standoff distance from 4.5 to 5.4 m. Figures 3.36(*a*, *b*) are the DAS and the WDAS images, respectively. It is clear that, compared to their single-look counterparts, the multilook images are better. However, the strong clutter and sidelobes can still be observed in Figures 3.36(*a*) and 3.36(*b*). Figure 3.36(*c*) shows the noncoherent APES-RCB image, where $\epsilon = 14$ is used in rank-deficient RCB. Again, the adaptive imaging approach appears to

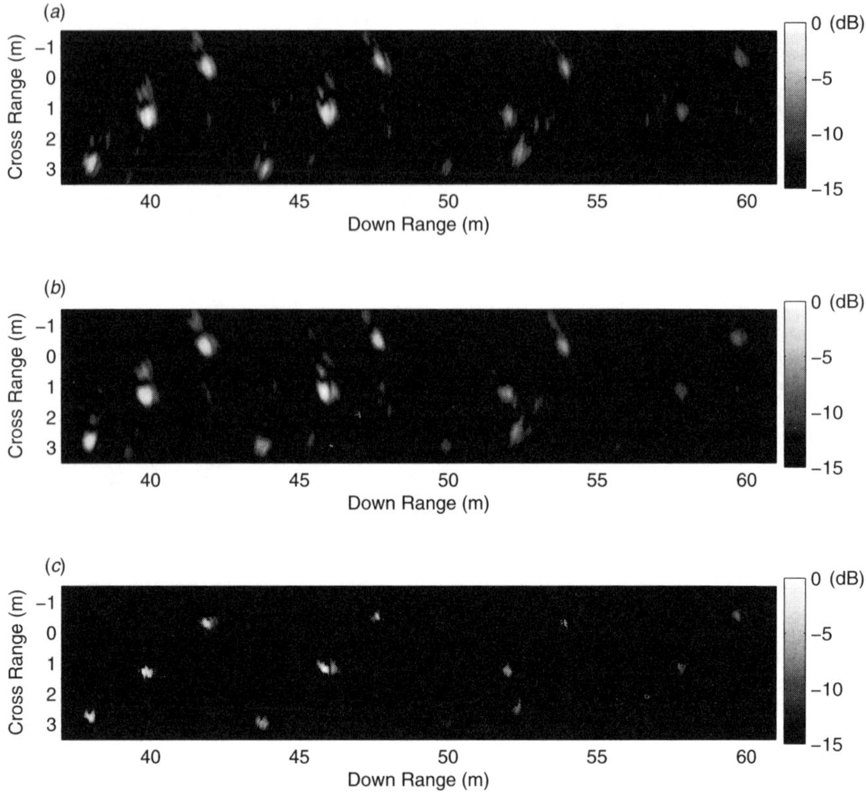

Figure 3.36 PSI noncoherent multilook processing results. (*a*) DAS imaging result. (*b*) WDAS imaging result. (*c*) APES-RCB imaging result with $\epsilon = 14$.

be the best. Note also that the multilook APES-RCB image is less sensitive to the choice of ϵ as compared with its single-look counterpart.

Example 3.9.3: Receiver Operating Characteristic Curves Figure 3.37 shows the receiver operating characteristic (ROC) curves for the PSI system based on four imaging methods, that is, single-look WDAS, multilook WDAS, single-look APES-RCB, and multilook APES-RCB. To obtain each ROC curve, each image is first segmented into connected regions by using a reasonably low threshold. For each region, the peak value and its location are retained and the rest of the pixels are set to zero. Then a simple threshold detector is used to perform the detection. The threshold increases in small steps. For each value of the threshold, we obtain a list of alarms, which is used to evaluate the probability of detection and the false alarm number. Based on the ground truth, for each mine, we define a detection circle. The center of the circle indicates the true location of the mine and the area of the circle is 1 m^2. The alarms falling within the circle are considered successful: the mine was detected. Otherwise, they are counted as false alarms.

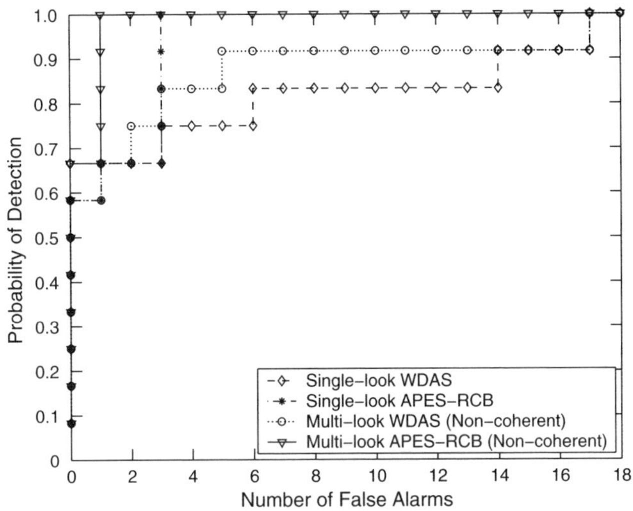

Figure 3.37 Comparison of ROC curves for the PSI FLGPR system based on the four different imaging methods.

It is clear from Figure 3.37 that, as compared with the conventional DAS-based methods, our APES-RCB imaging approach significantly improves the landmine detection capability for both single-look and multilook cases. For example, to detect all mines, the noncoherent multilook APES-RCB approach reduces the number of false alarms from 17 to 1 as compared with its noncoherent multilook WDAS counterpart. Note also that the detection results based on the multilook processing are better than those based on the single-look processing.

3.9.4.2 SRI FLGPR Experimental Results. A photograph of the SRI FLGPR system is shown in Figure 3.38. This system consists of two transmitters and 18 receivers using quad-ridged horn antennas. The height of the transmitters (two large horns) is about 3.3 m above the ground and their phase centers are 3.03 m apart. The 18 receiving antennas are horizontally equally spaced with 17 cm center to center spacing and the height for the bottom row is about 2 m above the ground. The stepped-frequency system operates at 893 discrete frequencies evenly spaced over the frequency range from 0.5 to 2.9084 GHz. The two transmitters work sequentially and all the receivers work simultaneously. Hence there is a total number of $DM = 36$ channels of received signals that can be obtained for each scan. This system can work in both VV (vertically-polarized transmitter and receiver) and HH (horizontally-polarized transmitter and receiver) modes. Data are recorded while the vehicle is moving and the distance between two adjacent scans is about 0.5 m. The GPS (global positioning system) is used to measure the location of the system for each scan. At each scan location, the image region is 5 m (cross-range) by 8 m (down-range) with an 8 m standoff distance ahead of the vehicle.

Figure 3.38 Photograph of the SRI FLGPR system.

Again, a pixel spacing of 4 cm is chosen in both the down-range and cross-range dimensions for the radar imaging. During the data collection for the SRI system, some metal cans were placed on the sides of the mine lane. To clearly illustrate the landmine imaging results, we have masked out the metal can returns in the images shown below.

Example 3.9.4: Single-Look Imaging Results Figure 3.39 shows the single-look imaging results. (Only the VV data are used here. Similar results can be obtained from the HH data.) In this example, nine evenly sampled scans (each scan covering 2.7 m in down-range) are used to form the entire image covering 24 m in the down-range. The images formed using different scans are nonoverlapping. Figures 3.39(a) and 3.39(b) show the DAS and WDAS imaging results, respectively. Note that the mines buried at the depths of 10 and 15 cm can hardly be seen in these figures due to the high sidelobes and strong clutter. Figure 3.39(c) shows the APES-RCB imaging result, where we have used $\epsilon = 28$ and $\bar{\mathbf{a}} = \mathbf{1}_{DM\times 1}$ with $M = 18$ and $D = 2$ for the SRI system. It can be noticed from this figure that the sidelobes and clutter have been effectively removed due to the excellent performance of APES-RCB, and that all 12 mines can be identified. Note also that the SRI radar images have higher resolution in the down-range dimension than the PSI radar images due to the larger system bandwidth of the SRI FLGPR system.

Example 3.9.5: Multilook Imaging Results The multilook imaging results based on noncoherent processing are shown in Figure 3.40. The output for each

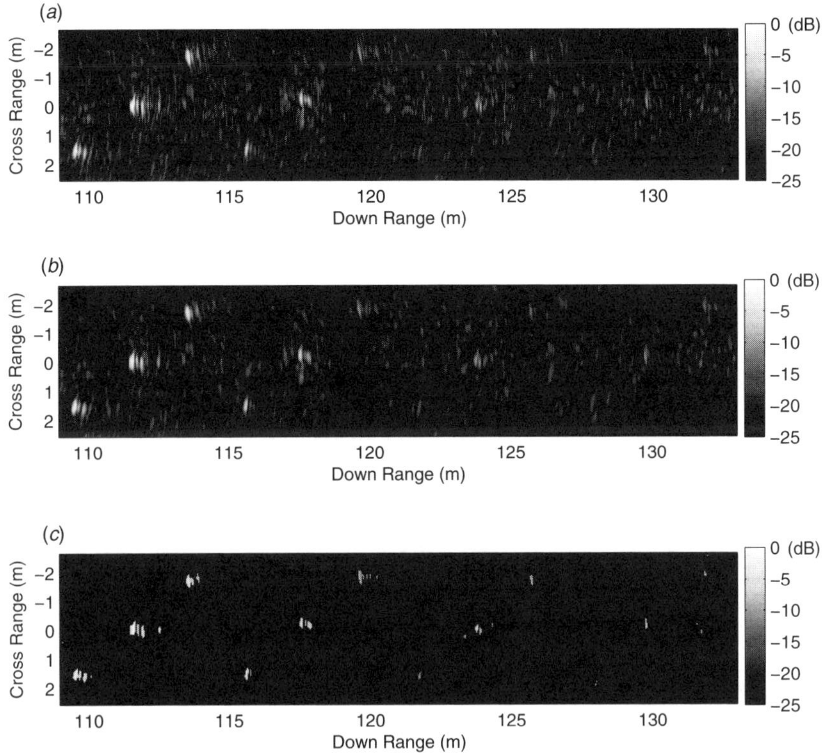

Figure 3.39 SRI single-look imaging results. (*a*) DAS imaging result. (*b*) WDAS imaging result. (*c*) APES-RCB imaging result with $\epsilon = 28$.

focal point in the image is obtained using nine consecutive scans with the standoff distance from 9 to 14 m. Figures 3.40(*a*) and 3.40(*b*) show the DAS and WDAS imaging results, respectively. Figure 3.40(*c*) shows the noncoherent APES-RCB image with $\epsilon = 14$. It can be noticed that by using APES-RCB in the multilook processing mode, high quality imaging results can be obtained.

Example 3.9.6: Receiver Operating Characteristic Curves Figure 3.41 shows the ROC curves for the SRI system based on four different methods. The same detection method as used for the PSI system is applied here. We can see from this figure that, as compared with the conventional DAS-based methods, the APES-RCB imaging approach improves the landmine detection capability for both single-look and multilook cases. In particular, to detect all mines, the noncoherent multilook APES-RCB approach reduces the number of false alarms from five to one as compared with the noncoherent multilook WDAS.

Finally, we remark that different ϵs are used in the examples above for the two different FLGPR systems due to their different array calibration errors and system

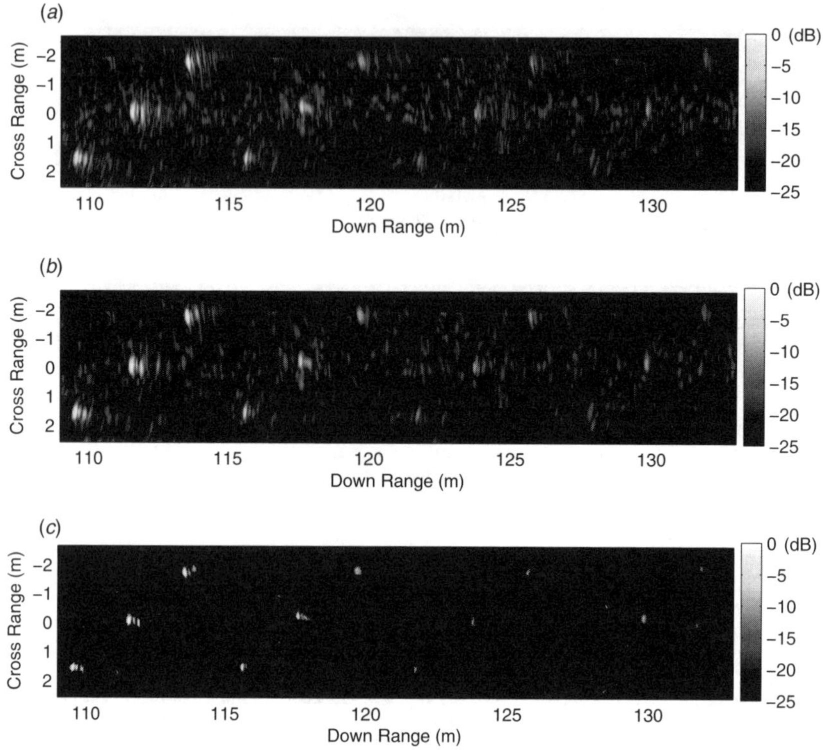

Figure 3.40 SRI noncoherent multilook processing results. (*a*) DAS imaging result. (*b*) WDAS imaging result. (*c*) APES-RCB image with $\epsilon = 14$.

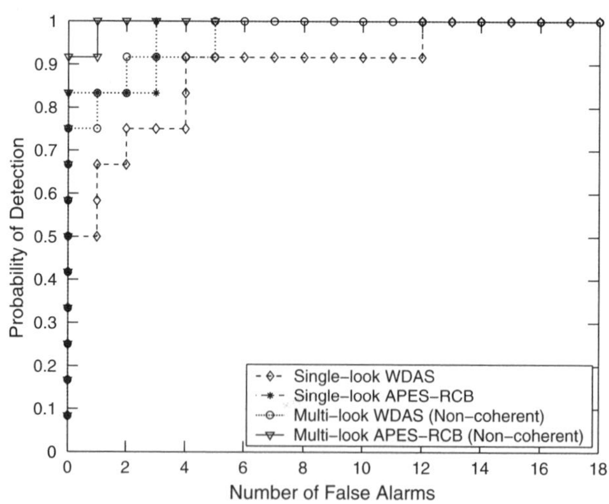

Figure 3.41 Comparison of ROC curves for the SRI FLGPR system via four different imaging methods.

georegistering errors. We also note that, in general, multilook APES-RCB images vary less with ϵ than their single-look counterparts.

3.10 SUMMARY

We have presented the robust Capon beamformer (RCB) based on an ellipsoidal uncertainty set and the doubly constrained robust Capon beamformer (DCRCB) based on a spherical uncertainty set of the array steering vector. We have provided a thorough analysis of the norm constrained Capon beamformer (NCCB) and shown that it is difficult to choose the norm constraint parameter in NCCB based on the knowledge of the array steering vector error alone. We have demonstrated that for a spherical uncertainty set, the NCCB, RCB and DCRCB are all related to the diagonal loading based approaches and they all require comparable computational costs with that associated with the SCB. However, the diagonal loading levels of these approaches are different. As a result, RCB and DCRCB can be used to obtain much more accurate signal power estimates than NCCB under comparable conditions. We have explained the relationship between RCB and DCRCB in that the former is an approximate solution while the latter is the exact solution of the same optimization problem. Our numerical examples have demonstrated that, for a reasonably tight spherical uncertainty set of the array steering vector, DCRCB is the preferred choice for applications requiring high SINR, while RCB is the favored one for applications demanding accurate signal power estimation. We have also presented several extensions and applications of RCB including constant-powerwidth RCB (CPRCB) and constant-beamwidth RCB (CBRCB) for acoustic imaging, rank-deficient robust Capon filter-bank spectral estimator for spectral estimation and radar imaging, and rank-deficient RCB for landmine detection using forward-looking ground penetrating radar (FLGPR) imaging systems. The excellent performances of RCB and DCRCB as well as those of the various extensions of RCB have been demonstrated by numerical and experimental examples.

ACKNOWLEDGMENTS

This work was supported in part by the National Science Foundation Grants CCR-0104887 and ECS-0097636, the Swedish Science Council (VR) and the Swedish Foundation for International Cooperation in Research and Higher Education (STINT). The authors also wish to thank Yanwei Wang and Yi Jiang for their helpful contributions to this book chapter.

APPENDIX 3.A: Relationship between RCB and the Approach in [14]

We repeat our optimization problem:

$$\min_{\mathbf{a}} \mathbf{a}^* \mathbf{R}^{-1} \mathbf{a} \quad \text{subject to } \|\mathbf{a} - \bar{\mathbf{a}}\|^2 = \epsilon. \tag{A.1}$$

Let $\mathbf{a}_0$ denote the optimal solution of (A.1). Let

$$\mathbf{w}_0 = \frac{\mathbf{R}^{-1}\mathbf{a}_0}{\mathbf{a}_0^*\mathbf{R}^{-1}\mathbf{a}_0}. \tag{A.2}$$

We show below that the $\mathbf{w}_0$ above is the optimal solution to the following SOCP considered in [14] (see also Chapter 2 of this book):

$$\min_{\mathbf{w}} \mathbf{w}^*\mathbf{R}\mathbf{w} \quad \text{subject to } \mathbf{w}^*\bar{\mathbf{a}} \geq \sqrt{\epsilon}\|\mathbf{w}\| + 1, \quad \text{Im}(\mathbf{w}^*\bar{\mathbf{a}}) = 0. \tag{A.3}$$

First we show that if $\|\bar{\mathbf{a}}\| \leq \sqrt{\epsilon}$, then there is no $\mathbf{w}$ that satisfies $\mathbf{w}^*\bar{\mathbf{a}} \geq \sqrt{\epsilon}\|\mathbf{w}\| + 1$. By using the Cauchy–Schwarz inequality, we have

$$\sqrt{\epsilon}\|\mathbf{w}\| + 1 \leq \mathbf{w}^*\bar{\mathbf{a}} \leq \sqrt{\epsilon}\|\mathbf{w}\| \tag{A.4}$$

which is impossible. Hence the constraint in (3.23), which is needed for our RCB to avoid the trivial solution, must also be satisfied by the approach in [14] (see also Chapter 2 of this book).

Next let

$$\mathbf{w} = \mathbf{w}_0 + \mathbf{y}. \tag{A.5}$$

We show below that the solution of (A.3) corresponds to $\mathbf{y} = \mathbf{0}$.

Insertion of (A.5) in (A.3) gives:

$$\min_{\mathbf{y}} \mathbf{y}^*\mathbf{R}\mathbf{y} + \frac{2}{\mathbf{a}_0^*\mathbf{R}^{-1}\mathbf{a}_0}\text{Re}(\mathbf{y}^*\mathbf{a}_0) + \frac{1}{\mathbf{a}_0^*\mathbf{R}^{-1}\mathbf{a}_0} \tag{A.6}$$

subject to

$$\mathbf{y}^*\bar{\mathbf{a}} + \mathbf{w}_0^*\bar{\mathbf{a}} - \sqrt{\epsilon}\|\mathbf{w}_0 + \mathbf{y}\| \geq 1 \tag{A.7}$$

and

$$\text{Im}(\mathbf{w}_0^*\bar{\mathbf{a}} + \mathbf{y}^*\bar{\mathbf{a}}) = 0. \tag{A.8}$$

Let

$$\bar{\mathbf{a}} = \mathbf{a}_0 + \boldsymbol{\mu}. \tag{A.9}$$

Then (A.7) and (A.8), respectively, become

$$\mathbf{y}^*\mathbf{a}_0 + \mathbf{y}^*\boldsymbol{\mu} + \mathbf{w}_0^*\boldsymbol{\mu} \geq \sqrt{\epsilon}\|\mathbf{w}_0 + \mathbf{y}\| \tag{A.10}$$

and

$$\text{Im}(\mathbf{w}_0^*\boldsymbol{\mu} + \mathbf{y}^*\mathbf{a}_0 + \mathbf{y}^*\boldsymbol{\mu}) = 0 \tag{A.11}$$

which implies that

$$\text{Re}(\mathbf{y}^*\mathbf{a}_0) \geq \sqrt{\epsilon}\|\mathbf{w}_0 + \mathbf{y}\| - \text{Re}(\mathbf{y}^*\boldsymbol{\mu} + \mathbf{w}_0^*\boldsymbol{\mu}). \tag{A.12}$$

Since

$$\begin{aligned} |\text{Re}[(\mathbf{y} + \mathbf{w}_0)^*\boldsymbol{\mu}]| &\leq |(\mathbf{y} + \mathbf{w}_0)^*\boldsymbol{\mu}| \\ &\leq \|\mathbf{y} + \mathbf{w}_0\|\,\|\boldsymbol{\mu}\| \\ &= \sqrt{\epsilon}\|\mathbf{w}_0 + \mathbf{y}\| \end{aligned} \tag{A.13}$$

it follows from (A.12) that

$$\text{Re}(\mathbf{y}^*\mathbf{a}_0) \geq 0. \tag{A.14}$$

This implies at once that the minimizer of (A.6) is $\mathbf{y} = \mathbf{0}$ provided that we can show that $\mathbf{y} = \mathbf{0}$ satisfies the constraints (A.7) and (A.8), or equivalently (A.10) and (A.11), that is,

$$\text{Re}\left(\mathbf{w}_0^*\boldsymbol{\mu}\right) \geq \sqrt{\epsilon}\|\mathbf{w}_0\| \tag{A.15}$$

and

$$\text{Im}\left(\mathbf{w}_0^*\boldsymbol{\mu}\right) = 0. \tag{A.16}$$

Inserting (A.2) in (A.15) yields,

$$\text{Re}(\mathbf{a}_0^*\mathbf{R}^{-1}\boldsymbol{\mu}) \geq \sqrt{\epsilon\, \mathbf{a}_0^*\mathbf{R}^{-2}\mathbf{a}_0}. \tag{A.17}$$

Using (A.2) in (A.16) gives,

$$\text{Im}(\mathbf{a}_0^*\mathbf{R}^{-1}\boldsymbol{\mu}) = 0. \tag{A.18}$$

To prove (A.17) and (A.18), we need to analyze (A.1). By using the Lagrange multiplier theory, we obtain [see (3.26)]

$$\mathbf{R}^{-1}\mathbf{a}_0 + \nu(\mathbf{a}_0 - \bar{\mathbf{a}}) = \mathbf{0} \tag{A.19}$$

where $\nu \geq 0$ is the Lagrange multiplier. Using (A.9) in (A.19) yields

$$\mathbf{R}^{-1}\mathbf{a}_0 = \nu\boldsymbol{\mu}. \tag{A.20}$$

Using (A.20) in (A.17) gives

$$\mathrm{Re}(\nu\|\boldsymbol{\mu}\|^2) = \nu\|\boldsymbol{\mu}\|^2 \geq \sqrt{\epsilon}\nu\|\boldsymbol{\mu}\| \tag{A.21}$$

However, due to the constraint in (A.1), that is, $\|\boldsymbol{\mu}\|^2 = \epsilon$, (A.21) is satisfied with equality, which proves that (A.17) is satisfied with equality. This means that the first constraint in (A.3) is satisfied with equality and hence that the optimal solution to (A.3) also occurs at the boundary of its constraint set, as expected (see also [15] or Chapter 1 of this book). Using (A.20) in (A.18) proves (A.18) since

$$\mathrm{Im}(\nu\|\boldsymbol{\mu}\|^2) = 0. \tag{A.22}$$

APPENDIX 3.B: Calculating the Steering Vector

We show how to obtain the steering vector $\mathbf{a}_0$ from the optimal solution $\mathbf{w}_0$ of the SOCP (A.3). In Appendix 3.A, we have shown that

$$\mathbf{w}_0 = \frac{\mathbf{R}^{-1}\mathbf{a}_0}{\mathbf{a}_0^*\mathbf{R}^{-1}\mathbf{a}_0}. \tag{B.1}$$

Hence

$$\mathbf{w}_0^*\mathbf{R}\mathbf{w}_0 = \frac{1}{\mathbf{a}_0^*\mathbf{R}^{-1}\mathbf{a}_0} \tag{B.2}$$

which, along with (B.1), leads to

$$\mathbf{a}_0 = \frac{\mathbf{R}\mathbf{w}_0}{\mathbf{w}_0^*\mathbf{R}\mathbf{w}_0}. \tag{B.3}$$

Hence from the optimal solution $\mathbf{w}_0$ of the SOCP (A.3), we can obtain the $\mathbf{a}_0$ as above and then correct the scaling ambiguity problem of the SOI power estimation in the same way as in our RCB approach [see (3.37)].

APPENDIX 3.C: Relationship between RCB and the Approach in [15]

Consider the SOCP with the ellipsoidal (including flat ellipsoidal) constraint on $\mathbf{w}$, not on $\mathbf{a}$ as in our formulation, considered in [15] (see also Chapter 1 of this book):

$$\min_{\mathbf{w}} \mathbf{w}^*\mathbf{R}\mathbf{w} \quad \text{subject to } \|\mathbf{B}^*\mathbf{w}\| \leq \bar{\mathbf{a}}^*\mathbf{w} - 1. \tag{C.1}$$

The Lagrange multiplier approach gives the optimal solution [15] (see also Chapter 1 of this book)

$$\begin{aligned}
\hat{\mathbf{w}} &= -\left[\frac{\mathbf{R}}{\gamma} + (\mathbf{B}\mathbf{B}^* - \bar{\mathbf{a}}\bar{\mathbf{a}}^*)\right]^{-1}\bar{\mathbf{a}} \\
&= -\left(\frac{\mathbf{R}}{\gamma} + \mathbf{B}\mathbf{B}^*\right)^{-1}\bar{\mathbf{a}} - \frac{(\mathbf{R}/\gamma + \mathbf{B}\mathbf{B}^*)^{-1}\bar{\mathbf{a}}\bar{\mathbf{a}}^*(\mathbf{R}/\gamma + \mathbf{B}\mathbf{B}^*)^{-1}\bar{\mathbf{a}}}{1 - \bar{\mathbf{a}}^*(\mathbf{R}/\gamma + \mathbf{B}\mathbf{B}^*)^{-1}\bar{\mathbf{a}}} \\
&= \frac{(\mathbf{R}/\gamma + \mathbf{B}\mathbf{B}^*)^{-1}\bar{\mathbf{a}}}{\bar{\mathbf{a}}^*(\mathbf{R}/\gamma + \mathbf{B}\mathbf{B}^*)^{-1}\bar{\mathbf{a}} - 1} \\
&= \frac{(\mathbf{R} + \gamma\mathbf{B}\mathbf{B}^*)^{-1}\bar{\mathbf{a}}}{\bar{\mathbf{a}}^*(\mathbf{R} + \gamma\mathbf{B}\mathbf{B}^*)^{-1}\bar{\mathbf{a}} - 1/\gamma}
\end{aligned} \tag{C.2}$$

where γ is the unique solution of

$$g(\gamma) = \gamma^2\bar{\mathbf{a}}^*(\mathbf{R} + \gamma\mathbf{P})^{-1}\mathbf{P}(\mathbf{R} + \gamma\mathbf{P})^{-1}\bar{\mathbf{a}} - 2\gamma\bar{\mathbf{a}}^*(\mathbf{R} + \gamma\mathbf{P})^{-1}\bar{\mathbf{a}} - 1 = 0 \tag{C.3}$$

and $\mathbf{P} = \mathbf{B}\mathbf{B}^* - \bar{\mathbf{a}}\bar{\mathbf{a}}^*$ (to obtain (C.2), we have used the matrix inversion lemma). Note that solving for the Lagrange multiplier from (C.3), as discussed in [15] (see also Chapter 1 of this book), is more complicated than solving our counterpart in (3.55).

To prove that the weight vectors in (3.117) and (C.2) are the same, we first prove that for the $\check{\nu}$ satisfying (3.52), we have

$$g\left(\frac{1}{\check{\nu}}\right) = 0. \tag{C.4}$$

To prove (C.4), note that

$$\begin{aligned}
g\left(\frac{1}{\check{\nu}}\right) &= \bar{\mathbf{a}}^*(\check{\nu}\mathbf{R} + \mathbf{P})^{-1}(\mathbf{B}\mathbf{B}^* - \bar{\mathbf{a}}\bar{\mathbf{a}}^*)(\check{\nu}\mathbf{R} + \mathbf{P})^{-1}\bar{\mathbf{a}} - 2\bar{\mathbf{a}}^*(\check{\nu}\mathbf{R} + \mathbf{P})^{-1}\bar{\mathbf{a}} - 1 \\
&= \bar{\mathbf{a}}^*(\check{\nu}\mathbf{R} + \mathbf{B}\mathbf{B}^* - \bar{\mathbf{a}}\bar{\mathbf{a}}^*)^{-1}\mathbf{B}\mathbf{B}^*(\check{\nu}\mathbf{R} + \mathbf{B}\mathbf{B}^* - \bar{\mathbf{a}}\bar{\mathbf{a}}^*)^{-1}\bar{\mathbf{a}} \\
&\quad - [\bar{\mathbf{a}}^*(\check{\nu}\mathbf{R} + \mathbf{B}\mathbf{B}^* - \bar{\mathbf{a}}\bar{\mathbf{a}}^*)^{-1}\bar{\mathbf{a}} + 1]^2.
\end{aligned} \tag{C.5}$$

Since [see (C.2)]

$$(\check{\nu}\mathbf{R} + \mathbf{B}\mathbf{B}^* - \bar{\mathbf{a}}\bar{\mathbf{a}}^*)^{-1}\bar{\mathbf{a}} = \frac{(\check{\nu}\mathbf{R} + \mathbf{B}\mathbf{B}^*)^{-1}\bar{\mathbf{a}}}{1 - \bar{\mathbf{a}}^*(\check{\nu}\mathbf{R} + \mathbf{B}\mathbf{B}^*)^{-1}\bar{\mathbf{a}}} \tag{C.6}$$

we can write $g\left(\frac{1}{\check{\nu}}\right)$ as a fraction whose numerator is:

$$\begin{aligned}\tilde{g}\left(\frac{1}{\check{\nu}}\right) &= \bar{\mathbf{a}}^*(\check{\nu}\mathbf{R} + \mathbf{BB}^*)^{-1}\mathbf{BB}^*(\check{\nu}\mathbf{R} + \mathbf{BB}^*)^{-1}\bar{\mathbf{a}} - 1 \\ &= \|\mathbf{B}^*(\check{\nu}\mathbf{R} + \mathbf{BB}^*)^{-1}\bar{\mathbf{a}}\|^2 - 1. \end{aligned} \tag{C.7}$$

Since $\check{\nu}$ satisfies (3.52), we have that

$$\begin{aligned} 1 &= \|(\check{\mathbf{R}} + \check{\nu}\mathbf{I})^{-1}\check{\bar{\mathbf{a}}}\|^2 \\ &= \|(\mathbf{B}^*\mathbf{R}^{-1}\mathbf{B} + \check{\nu}\mathbf{I})^{-1}\mathbf{B}^*\mathbf{R}^{-1}\bar{\mathbf{a}}\|^2 \\ &= \left\|\frac{1}{\check{\nu}}\mathbf{I} - \mathbf{B}^*(\check{\nu}\mathbf{R} + \mathbf{BB}^*)^{-1}\mathbf{B}]\mathbf{B}^*\mathbf{R}^{-1}\bar{\mathbf{a}}\right\|^2 \\ &= \|\mathbf{B}^*[\mathbf{I} - (\check{\nu}\mathbf{R} + \mathbf{BB}^*)^{-1}\mathbf{BB}^*](\check{\nu}\mathbf{R})^{-1}\bar{\mathbf{a}}\|^2 \\ &= \|\mathbf{B}^*(\check{\nu}\mathbf{R} + \mathbf{BB}^*)^{-1}[\check{\nu}\mathbf{R} + \mathbf{BB}^* - \mathbf{BB}^*](\check{\nu}\mathbf{R})^{-1}\bar{\mathbf{a}}\|^2 \\ &= \|\mathbf{B}^*(\check{\nu}\mathbf{R} + \mathbf{BB}^*)^{-1}\bar{\mathbf{a}}\|^2 \end{aligned} \tag{C.8}$$

which proves (C.4).

Next we prove that the denominators of (3.117) and (C.2) are the same. The denominator of (3.117) can be written as

$$\begin{aligned} &\bar{\mathbf{a}}^*\left(\mathbf{R} + \frac{1}{\check{\nu}}\mathbf{BB}^*\right)^{-1}\left(\mathbf{R} + \frac{1}{\check{\nu}}\mathbf{BB}^* - \frac{1}{\check{\nu}}\mathbf{BB}^*\right)\left(\mathbf{R} + \frac{1}{\check{\nu}}\mathbf{BB}^*\right)^{-1}\bar{\mathbf{a}} \\ &\quad = \bar{\mathbf{a}}^*\left(\mathbf{R} + \frac{1}{\check{\nu}}\mathbf{BB}^*\right)^{-1}\bar{\mathbf{a}} - \check{\nu}\left\|\mathbf{B}^*(\check{\nu}\mathbf{R} + \mathbf{BB}^*)^{-1}\bar{\mathbf{a}}\right\|^2 \\ &\quad = \bar{\mathbf{a}}^*\left(\mathbf{R} + \frac{1}{\check{\nu}}\mathbf{BB}^*\right)^{-1}\bar{\mathbf{a}} - \check{\nu} \end{aligned} \tag{C.9}$$

where we have used (C.8). Since for the $\check{\nu}$ satisfying (3.52) and γ satisfying (C.3), $\check{\nu} = \frac{1}{\gamma}$, the proof is concluded.

APPENDIX 3.D: Analysis of Equation (3.72)

Let

$$h(\lambda) = \frac{\bar{\mathbf{a}}^*(\mathbf{R} + \lambda\mathbf{I})^{-2}\bar{\mathbf{a}}}{[\bar{\mathbf{a}}^*(\mathbf{R} + \lambda\mathbf{I})^{-1}\bar{\mathbf{a}}]^2}. \tag{D.1}$$

For any matrix function $\mathbf{F}$ of λ we have:

$$(\mathbf{F}^{-1})' = -\mathbf{F}^{-1}\mathbf{F}'\mathbf{F}^{-1} \tag{D.2}$$

and

$$(\mathbf{F}^{-2})' = -\mathbf{F}^{-2}(\mathbf{F}'\mathbf{F} + \mathbf{F}\mathbf{F}')\mathbf{F}^{-2}. \tag{D.3}$$

Letting

$$\mathbf{F} = \mathbf{R} + \lambda\mathbf{I} \tag{D.4}$$

for which $\mathbf{F}' = \mathbf{I}$, we get

$$\begin{aligned} h'(\lambda) &= \frac{-(\bar{\mathbf{a}}^*\mathbf{F}^{-2}2\mathbf{F}\mathbf{F}^{-2}\bar{\mathbf{a}})(\bar{\mathbf{a}}^*\mathbf{F}^{-1}\bar{\mathbf{a}})^2 + 2(\bar{\mathbf{a}}^*\mathbf{F}^{-2}\bar{\mathbf{a}})(\bar{\mathbf{a}}^*\mathbf{F}^{-1}\bar{\mathbf{a}})(\bar{\mathbf{a}}^*\mathbf{F}^{-2}\bar{\mathbf{a}})}{(\bar{\mathbf{a}}^*\mathbf{F}^{-1}\bar{\mathbf{a}})^4} \\ &= -\frac{2(\bar{\mathbf{a}}^*\mathbf{F}^{-1}\bar{\mathbf{a}})}{(\bar{\mathbf{a}}^*\mathbf{F}^{-1}\bar{\mathbf{a}})^4}\Big[\big(\bar{\mathbf{a}}^*\mathbf{F}^{-3}\bar{\mathbf{a}}\big)\big(\bar{\mathbf{a}}^*\mathbf{F}^{-1}\bar{\mathbf{a}}\big) - \big(\bar{\mathbf{a}}^*\mathbf{F}^{-2}\bar{\mathbf{a}}\big)^2\Big]. \end{aligned} \tag{D.5}$$

For $\lambda \geq 0$, we have $\mathbf{F} > 0$. It follows that

$$\begin{aligned} (\bar{\mathbf{a}}^*\mathbf{F}^{-2}\bar{\mathbf{a}})^2 &= (\bar{\mathbf{a}}^*\mathbf{F}^{-3/2}\mathbf{F}^{-1/2}\bar{\mathbf{a}})^2 \\ &\leq (\bar{\mathbf{a}}^*\mathbf{F}^{-3}\bar{\mathbf{a}})(\bar{\mathbf{a}}^*\mathbf{F}^{-1}\bar{\mathbf{a}}) \end{aligned} \tag{D.6}$$

and therefore $h'(\lambda) \leq 0$ for $\lambda \geq 0$. Hence $h(\lambda)$ is a monotonically decreasing function of $\lambda \geq 0$. As $\lambda \to \infty$, $h(\lambda) \to 1/M < \zeta$, according to (3.77). From (3.65), $h(0) > \zeta$ since $h(0)$ is equal to the right side of (3.65). This shows that, indeed, (3.72) has a unique solution $\lambda > 0$ under (3.65) and (3.77).

APPENDIX 3.E: Rank-Deficient Capon Beamformer

In this appendix we prove (3.187) and provide some additional insights into the problem (3.182). Consider the problem (3.182) with a fixed $\hat{\mathbf{a}}(\omega)$, which can also be shown to be a covariance matrix fitting-based reformulation of the Capon beamformer (see [30] and [32]):

$$\max_{\sigma^2} \sigma^2 \quad \text{subject to } \hat{\mathbf{R}} - \sigma^2\hat{\mathbf{a}}\hat{\mathbf{a}}^* \geq 0 \tag{E.1}$$

where $\hat{\mathbf{a}}$ is a given vector, σ^2 is the signal power, and $\hat{\mathbf{R}}$ is a singular sample covariance matrix.

We solve the above optimization problem in (E.1) by first considering the case where $\hat{\mathbf{a}}$ belongs to the range space of $\hat{\mathbf{R}}$, that is,

$$\hat{\mathbf{a}} = \hat{\mathbf{S}}\boldsymbol{\zeta} \tag{E.2}$$

where $\boldsymbol{\zeta}$ is a nonzero $K \times 1$ vector. From the positive semidefinite constraint in (E.1), we have that:

$$
\begin{aligned}
&\hat{\mathbf{R}} - \sigma^2 \hat{\mathbf{a}}\hat{\mathbf{a}}^* \geq 0 \\
&\quad \Leftrightarrow \hat{\mathbf{S}}\hat{\boldsymbol{\Psi}}\hat{\mathbf{S}}^* - \sigma^2 \hat{\mathbf{S}}\boldsymbol{\zeta}\boldsymbol{\zeta}^*\hat{\mathbf{S}}^* \geq 0 \\
&\quad \Leftrightarrow \hat{\mathbf{S}}\left[\hat{\boldsymbol{\Psi}} - \sigma^2 \boldsymbol{\zeta}\boldsymbol{\zeta}^*\right]\hat{\mathbf{S}}^* \geq 0 \\
&\quad \Leftrightarrow \hat{\mathbf{S}}\hat{\boldsymbol{\Psi}}^{1/2}\left[\mathbf{I} - \sigma^2 \hat{\boldsymbol{\Psi}}^{-1/2}\boldsymbol{\zeta}\boldsymbol{\zeta}^*\hat{\boldsymbol{\Psi}}^{-1/2}\right]\hat{\boldsymbol{\Psi}}^{1/2}\hat{\mathbf{S}}^* \geq 0 \\
&\quad \Leftrightarrow \bar{\mathbf{S}}[\mathbf{I} - \sigma^2 \bar{\mathbf{v}}\bar{\mathbf{v}}^*]\bar{\mathbf{S}}^* \geq 0 \\
&\quad \Leftrightarrow \mathbf{I} - \sigma^2 \bar{\mathbf{v}}\bar{\mathbf{v}}^* \geq 0 \\
&\quad \Leftrightarrow 1 - \sigma^2 \bar{\mathbf{v}}^*\bar{\mathbf{v}} \geq 0 \\
&\quad \Leftrightarrow \sigma^2 \leq \frac{1}{\bar{\mathbf{v}}^*\bar{\mathbf{v}}}
\end{aligned}
\tag{E.3}
$$

where we have defined $\bar{\mathbf{S}} \triangleq \hat{\mathbf{S}}\hat{\boldsymbol{\Psi}}^{1/2}$ and $\bar{\mathbf{v}} \triangleq \hat{\boldsymbol{\Psi}}^{-1/2}\boldsymbol{\zeta}$. Since $\hat{\mathbf{a}} = \hat{\mathbf{S}}\boldsymbol{\zeta}$, we have $\boldsymbol{\zeta} = \hat{\mathbf{S}}^*\hat{\mathbf{a}}$. Hence

$$
\begin{aligned}
\bar{\mathbf{v}}^*\bar{\mathbf{v}} &= \boldsymbol{\zeta}^*\hat{\boldsymbol{\Psi}}^{-1/2}\hat{\boldsymbol{\Psi}}^{-1/2}\boldsymbol{\zeta} \\
&= \hat{\mathbf{a}}^*\hat{\mathbf{S}}\hat{\boldsymbol{\Psi}}^{-1}\hat{\mathbf{S}}^*\hat{\mathbf{a}} \\
&= \hat{\mathbf{a}}^*\hat{\mathbf{R}}^{\dagger}\hat{\mathbf{a}}
\end{aligned}
\tag{E.4}
$$

where $\hat{\mathbf{R}}^{\dagger} = \hat{\mathbf{S}}\hat{\boldsymbol{\Psi}}^{-1}\hat{\mathbf{S}}^*$ is the Moore–Penrose pseudo-inverse of $\hat{\mathbf{R}}$. Hence the solution to the optimization problem in (E.1) is

$$
\hat{\sigma}^2 = \frac{1}{\hat{\mathbf{a}}^*\hat{\mathbf{R}}^{\dagger}\hat{\mathbf{a}}}
\tag{E.5}
$$

which proves (3.187).

Consider now the case of an $\hat{\mathbf{a}}$ vector that does not belong to the range space of $\hat{\mathbf{R}}$. Then $\hat{\mathbf{a}}$ can be written as

$$
\hat{\mathbf{a}} = \hat{\mathbf{S}}\boldsymbol{\zeta} + \hat{\mathbf{G}}\boldsymbol{\beta} = [\hat{\mathbf{S}} \quad \hat{\mathbf{G}}]\begin{bmatrix} \boldsymbol{\zeta} \\ \boldsymbol{\beta} \end{bmatrix}.
\tag{E.6}
$$

Let

$$
\hat{\mathbf{R}} = [\hat{\mathbf{S}} \quad \hat{\mathbf{G}}]\begin{bmatrix} \hat{\boldsymbol{\Psi}} & \mathbf{0} \\ \mathbf{0} & \mathbf{0} \end{bmatrix}\begin{bmatrix} \hat{\mathbf{S}}^* \\ \hat{\mathbf{G}}^* \end{bmatrix}.
\tag{E.7}
$$

Then we have that

$$\hat{\mathbf{R}} - \sigma^2\hat{\mathbf{a}}\hat{\mathbf{a}}^* = [\hat{\mathbf{S}} \quad \hat{\mathbf{G}}]\begin{bmatrix} \hat{\boldsymbol{\Psi}} - \sigma^2\boldsymbol{\zeta}\boldsymbol{\zeta}^* & -\sigma^2\boldsymbol{\zeta}\boldsymbol{\beta}^* \\ -\sigma^2\boldsymbol{\beta}\zeta^* & -\sigma^2\boldsymbol{\beta}\boldsymbol{\beta}^* \end{bmatrix}\begin{bmatrix} \hat{\mathbf{S}}^* \\ \hat{\mathbf{G}}^* \end{bmatrix}. \tag{E.8}$$

Clearly, if $\boldsymbol{\beta} \neq \mathbf{0}$, then $\hat{\sigma}^2 = 0$ is the only solution to (E.1) for which $\hat{\mathbf{R}} - \sigma^2\hat{\mathbf{a}}\hat{\mathbf{a}}^* \geq 0$. For this case, the rank-deficient Capon beamformer will give an estimated power spectrum of zero, that is, $\hat{\sigma}^2 = 0$.

Hence the power estimate given by the rank-deficient Capon beamformer is:

$$\hat{\sigma}^2 = \begin{cases} \dfrac{1}{\hat{\mathbf{a}}^*\hat{\mathbf{R}}^\dagger\hat{\mathbf{a}}}, & \hat{\mathbf{a}} \in \mathcal{R}(\hat{\mathbf{R}}) \\ 0, & \hat{\mathbf{a}} \notin \mathcal{R}(\hat{\mathbf{R}}) \end{cases}. \tag{E.9}$$

Note that (E.9) is a scaled version of (3.180), with the scaling factor $|\varphi|^2$ being caused by the constraint $\mathbf{h}^*(\omega)\hat{\mathbf{a}}(\omega) = \varphi$ in (3.163), which indicates that the covariance matrix fitting formulation above is equivalent to the standard Capon beamforming formulation even in the case of a rank-deficient $\hat{\mathbf{R}}$. This result is an extension of the one in [32] (see also [30]), where $\hat{\mathbf{R}}$ was assumed to be full-rank.

APPENDIX 3.F: Conjugate Symmetry of the Forward-Backward FIR

For the $\mathbf{h}(\omega)$ in (3.196) we have that:

$$\begin{aligned} \mathbf{J}\mathbf{h}^c(\omega) &= \frac{\mathbf{J}\hat{\mathbf{S}}^c\left(\dfrac{\mathbf{I}}{\lambda} + \hat{\boldsymbol{\Psi}}\right)^{-1}(\hat{\mathbf{S}}^*)^c\mathbf{a}^c(\omega)}{\mathbf{a}^*(\omega)\hat{\mathbf{S}}\left(\dfrac{\mathbf{I}}{\lambda} + \hat{\boldsymbol{\Psi}}\right)^{-1}\hat{\mathbf{S}}^*\mathbf{a}(\omega)} \\ &= \frac{\mathbf{J}\hat{\mathbf{S}}^c\left(\dfrac{\mathbf{I}}{\lambda} + \hat{\boldsymbol{\Psi}}\right)^{-1}(\hat{\mathbf{S}}^*)^c\mathbf{J}\cdot\mathbf{J}\mathbf{a}^c(\omega)}{\mathbf{a}^*(\omega)\hat{\mathbf{S}}\left(\dfrac{\mathbf{I}}{\lambda} + \hat{\boldsymbol{\Psi}}\right)^{-1}\hat{\mathbf{S}}^*\mathbf{a}(\omega)} \\ &= \frac{\mathbf{J}\hat{\mathbf{S}}^c\left(\dfrac{\mathbf{I}}{\lambda} + \hat{\boldsymbol{\Psi}}\right)^{-1}(\hat{\mathbf{S}}^*)^c\mathbf{J}\cdot\mathbf{a}(\omega)\cdot e^{-j(M-1)\omega}}{\mathbf{a}^*(\omega)\hat{\mathbf{S}}\left(\dfrac{\mathbf{I}}{\lambda} + \hat{\boldsymbol{\Psi}}\right)^{-1}\hat{\mathbf{S}}^*\mathbf{a}(\omega)} \end{aligned} \tag{F.1}$$

where we have used the facts that $\mathbf{a}^*(\omega)\hat{\mathbf{S}}(\frac{\mathbf{I}}{\lambda} + \hat{\boldsymbol{\Psi}})^{-1}\hat{\mathbf{S}}^*\mathbf{a}(\omega)$ is a real-valued scalar, and that $\mathbf{J}\mathbf{J} = \mathbf{I}$.

From the persymmetry of $\hat{\mathbf{R}}$, we have that:

$$\begin{aligned}
\mathbf{J}(\hat{\mathbf{S}}\hat{\boldsymbol{\Psi}}\hat{\mathbf{S}}^*)^c\mathbf{J} &= \hat{\mathbf{S}}\hat{\boldsymbol{\Psi}}\hat{\mathbf{S}}^* \\
\Leftrightarrow \mathbf{J}\hat{\mathbf{S}}^c\hat{\boldsymbol{\Psi}}(\mathbf{J}\hat{\mathbf{S}}^c)^* &= \hat{\mathbf{S}}\hat{\boldsymbol{\Psi}}\hat{\mathbf{S}}^* \qquad \text{(F.2)} \\
\Leftrightarrow \mathbf{J}\hat{\mathbf{S}}^c &= \hat{\mathbf{S}}\boldsymbol{\Delta}, \qquad \text{(F.3)}
\end{aligned}$$

where $\boldsymbol{\Delta}$ is a $K \times K$ unitary matrix since

$$\boldsymbol{\Delta}^*\boldsymbol{\Delta} = \boldsymbol{\Delta}^*\hat{\mathbf{S}}^*\hat{\mathbf{S}}\boldsymbol{\Delta} = (\hat{\mathbf{S}}\boldsymbol{\Delta})^*(\hat{\mathbf{S}}\boldsymbol{\Delta}) = (\mathbf{J}\hat{\mathbf{S}}^c)^*(\mathbf{J}\hat{\mathbf{S}}^c) = \mathbf{I}. \qquad \text{(F.4)}$$

Note from (F.2) that we also have

$$\begin{aligned}
\hat{\mathbf{S}}\boldsymbol{\Delta}\hat{\boldsymbol{\Psi}}(\mathbf{J}\hat{\mathbf{S}}^c)^* = \hat{\mathbf{S}}\hat{\boldsymbol{\Psi}}\hat{\mathbf{S}}^* &\Rightarrow \boldsymbol{\Delta}\hat{\boldsymbol{\Psi}}(\mathbf{J}\hat{\mathbf{S}}^c)^* = \hat{\boldsymbol{\Psi}}\hat{\mathbf{S}}^* \\
&\Rightarrow \boldsymbol{\Delta}\hat{\boldsymbol{\Psi}} = \hat{\boldsymbol{\Psi}}\hat{\mathbf{S}}^*(\mathbf{J}\hat{\mathbf{S}}^c) = \hat{\boldsymbol{\Psi}}\boldsymbol{\Delta} \qquad \text{(F.5)}
\end{aligned}$$

which implies (along with (F.4)) that

$$\boldsymbol{\Delta}\hat{\boldsymbol{\Psi}}\boldsymbol{\Delta}^* = \hat{\boldsymbol{\Psi}}\boldsymbol{\Delta}\boldsymbol{\Delta}^* = \hat{\boldsymbol{\Psi}}. \qquad \text{(F.6)}$$

From (F.4) and (F.6), we get

$$\begin{aligned}
\mathbf{J}\left\{\hat{\mathbf{S}}\left(\frac{\mathbf{I}}{\lambda} + \hat{\boldsymbol{\Psi}}\right)^{-1}\hat{\mathbf{S}}^*\right\}^c\mathbf{J} &= \hat{\mathbf{S}}\boldsymbol{\Delta}\left(\frac{\mathbf{I}}{\lambda} + \hat{\boldsymbol{\Psi}}\right)^{-1}\boldsymbol{\Delta}^*\hat{\mathbf{S}}^* \\
&= \hat{\mathbf{S}}\left[\boldsymbol{\Delta}\left(\frac{\mathbf{I}}{\lambda} + \hat{\boldsymbol{\Psi}}\right)\boldsymbol{\Delta}^*\right]^{-1}\hat{\mathbf{S}}^* \qquad \text{(F.7)} \\
&= \hat{\mathbf{S}}\left(\frac{\mathbf{I}}{\lambda} + \hat{\boldsymbol{\Psi}}\right)^{-1}\hat{\mathbf{S}}^*.
\end{aligned}$$

Substituting (F.7) into (F.1), we get (3.197).

APPENDIX 3.G: Formulations of NCCF and HDI

The norm constrained Capon filter-bank (NCCF) is given by the solution to the following optimization problem:

$$\begin{aligned}
\min_{\mathbf{h}(\omega)} \mathbf{h}^*(\omega)\hat{\mathbf{R}}\mathbf{h}(\omega) \quad \text{subject to } \mathbf{h}^*(\omega)\mathbf{a}(\omega) &= 1 \\
\|\mathbf{h}(\omega)\|^2 &\leq \eta \qquad \text{(G.1)}
\end{aligned}$$

where the squared norm of the filter is constrained by η.

The subspace constrained and norm constrained HDI filter-bank is given by the solution to the following optimization problem:

$$\begin{aligned}\min_{\mathbf{h}(\omega)} \mathbf{h}^*(\omega)\hat{\mathbf{R}}\mathbf{h}(\omega) \quad \text{subject to } & \mathbf{h}(\omega) = \frac{\mathbf{a}(\omega)}{M} + \boldsymbol{v} \\ & \boldsymbol{v} = \hat{\mathbf{S}}\boldsymbol{\xi} \\ & \mathbf{a}^*(\omega)\boldsymbol{v} = 0 \\ & \|\mathbf{h}(\omega)\|^2 \leq \varepsilon \end{aligned} \tag{G.2}$$

where the squared norm of the filter is constrained by ε, the subspace constraint $\boldsymbol{v} = \hat{\mathbf{S}}\boldsymbol{\xi}$ is used to prevent $\mathbf{h}(\omega)$ from becoming orthogonal to the range space of $\hat{\mathbf{R}}$, and $\mathbf{a}^*(\omega)\boldsymbol{v} = 0$ is employed to make sure that the solution of $\boldsymbol{v}$ is unique.

APPENDIX 3.H: Notations and Abbreviations

$\|\cdot\|$	The Euclidean norm of a vector
$\|x\|$	The modulus of the (possibly complex) scalar x
$(\cdot)^*$	Conjugate transpose of a vector or matrix
$(\cdot)^c$	Complex conjugate of a vector or matrix
$(\cdot)^T$	Transpose of a vector or matrix
$(\cdot)^\dagger$	Moore–Penrose pseudo-inverse of a matrix
$\mathbf{A} > 0 (\geq 0)$	$\mathbf{A}$ is positive definite (positive semidefinite)
$\mathbf{A} < 0 (\leq 0)$	$\mathbf{A}$ is negative definite (negative semidefinite)
$\arg(x)$	The phase of the scalar x
$\max_x f(x)$	The value of x that maximizes $f(x)$
$\min_x f(x)$	The value of x that minimizes $f(x)$
$\mathcal{R}(\mathbf{A})$	The range space of a matrix $\mathbf{A}$
Re(x)	The real part of x

APES	Amplitude and phase estimation
CBRCB	Constant-beamwidth robust Capon beamformer
CPRCB	Constant-powerwidth robust Capon beamformer
DAS	Delay-and-sum
DCRCB	Doubly constrained robust Capon beamformer
DFT	Discrete Fourier transform
DOA	Direction of arrival
FFT	Fast Fourier transform
FLGPR	Forward-looking ground penetrating radar
HDI	High-definition imaging

NCCB	Norm constrained Capon beamformer
RCB	Robust Capon beamformer
RCF	Robust Capon filterbank
RMSE	Root mean-squared error
ROC	Receiver operating characteristic
SAR	Synthetic aperture radar
SCB	Standard capon beamformer
SDAS	Shaded delay-and-sum
SDP	Semidefinite program
SINR	Signal-to-interference-plus-noise ratio
SOCP	Second order cone program
SOI	Signal-of-interest
SPL	Sound pressure level

REFERENCES

1. P. Stoica and R. L. Moses, *Introduction to Spectral Analysis.* Prentice-Hall, Englewood Cliffs, NJ, 1997.
2. H. L. Van Trees, *Detection, Estimation, and Modulation Theory, Part IV: Optimum Array Processing.* John Wiley & Sons, New York, NY, 2002.
3. J. Capon, "High resolution frequency-wavenumber spectrum analysis," *Proceedings of the IEEE*, 57, 1408–1418, August 1969.
4. R. T. Lacoss, "Data adaptive spectral analysis methods," *Geophysics*, 36(4), 661–675 (1971).
5. C. D. Seligson, "Comments on high resolution frequency-wavenumber spectrum analysis," *Proceedings of the IEEE*, 58(6), 947–949 (1970).
6. H. Cox, "Resolving power and sensitivity to mismatch of optimum array processors," *Journal of the Acoustical Society of America*, 54(3), 771–785 (1973).
7. D. D. Feldman and L. J. Griffiths, "A projection approach for robust adaptive beamforming," *IEEE Transactions on Signal Processing*, 42, 867–876 (1994).
8. A. K. Steele, "Comparison of directional and derivative constraints for beamformers subject to multiple linear constraints," *IEE Proceedings, Pts. F and H*, 130, 41–45 (1983).
9. B. D. Van Veen and K. Buckley, "Beamforming: A versatile approach to spatial filtering," *IEEE ASSP Magazine*, 5, 4–24 (1988).
10. J. C. Preisig, "Robust maximum energy adaptive matched field processing," *IEEE Transactions on Signal Processing*, 42, 1585–1593 (1994).
11. A. L. Swindlehurst and M. Viberg, "Bayesian approaches in array signal processing." In T. Katayama and S. Sugimoto (Eds.), *Statistical Methods in Control and Signal Processing*, Marcel-Dekker, New York, NY, 1997.
12. A. B. Gershman, "Robust adaptive beamforming in sensor arrays," *International Journal of Electronics and Communications*, 53(6), 305–314 (1999).

13. K. L. Bell, Y. Ephraim, and H. L. Van Trees, "A Bayesian approach to robust adaptive beamforming," *IEEE Transactions on Signal Processing*, 48, 386–398 (2000).
14. S. A. Vorobyov, A. B. Gershman, and Z.-Q. Luo, "Robust adaptive beamforming using worst-case performance optimization: a solution to the signal mismatch problem," *IEEE Transactions on Signal Processing*, 51, 313–324 (2003).
15. R. G. Lorenz and S. P. Boyd, "Robust minimum variance beamforming," *IEEE Transactions on Signal Processing*, 53, 1684–1696 (2005).
16. J. Li, P. Stoica, and Z. Wang, "On robust Capon beamforming and diagonal loading," *IEEE Transactions on Signal Processing*, 51, 1702–1715 (2003).
17. J. Li, P. Stoica, and Z. Wang, "Doubly constrained robust Capon beamformer," *IEEE Transactions on Signal Processing*, 52, 2407–2423 (2004).
18. J. E. Hudson, *Adaptive Array Principles.* Peter Peregrinus, London, UK, 1981.
19. Y. I. Abramovich and A. I. Nevrev, "An analysis of effectiveness of adaptive maximisation of the signal-to-noise ratio which utilises the inversion of the estimated correlation matrix," *Radio Engineering and Electronic Physics*, 26, 67–74 (1981).
20. M. H. Er and A. Cantoni, "An alternative formulation for an optimum beamformer with robustness capability," *IEE Proceedings Part F, Communications, Radar, and Signal Processing*, 132, 447–460 (1985).
21. H. Cox, R. M. Zeskind, and M. M. Owen, "Robust adaptive beamforming," *IEEE Transactions on Acoustics, Speech, and Signal Processing*, 35, 1365–1376 (1987).
22. B. D. Carlson, "Covariance matrix estimation errors and diagonal loading in adaptive arrays," *IEEE Transactions on Aerospace and Electronic Systems*, 24, 397–401 (1988).
23. Y. I. Abramovich, V. G. Kachur, and V. N. Mikhaylyukov, "The convergence of directivity and its stabilization in fast spatial filter adaptive tuning procedures," *Soviet Journal of Communication Technology and Electronics*, 34(15), 6–11 (1989).
24. Y. I. Abramovich and V. G. Kachur, "Methods of protecting a useful signal differing from reference in adaptive procedures with unclassified teaching sample," *Soviet Journal of Communication Technology and Electronics*, 35(13), 29–35 (1990).
25. B. D. Van Veen, "Minimum variance beamforming with soft response constraints," *IEEE Transactions on Signal Processing*, 39, 1964–1972 (1991).
26. R. Wu, Z. Bao, and Y. Ma, "Control of peak sidelobe level in adaptive arrays," *IEEE Transactions on Antennas and Propagation*, 44, 1341–1347 (1996).
27. C.-C. Lee and J.-H. Lee, "Robust adaptive array beamforming under steering vector errors," *IEEE Transactions on Antennas and Propagation*, 45, 168–175 (1997).
28. Z. Tian, K. L. Bell, and H. L. Van Trees, "A recursive least squares implementation for LCMP beamforming under quadratic constraints," *IEEE Transactions on Signal Processing*, 49, 1138–1145 (2001).
29. S. Q. Wu and J. Y. Zhang, "A new robust beamforming method with antennae calibration errors," *IEEE Wireless Communications and Networking Conference*, New Orleans, LA, USA, Vol. 2, pp. 869–872, September 1999.
30. P. Stoica, Z. Wang, and J. Li, "Robust Capon beamforming," *IEEE Signal Processing Letters*, 10, 172–175 (2003).
31. S. Shahbazpanahi, A. B. Gershman, Z.-Q. Luo, and K. M. Wong, "Robust adaptive beamforming using worst-case SINR optimization: A new diagonal loading-type solution for general-rank signal models," *ICASSP*, 5, 333–336 (2003).

32. T. L. Marzetta, "A new interpretation for Capon's maximum likelihood method of frequency-wavenumber spectrum estimation," *IEEE Transactions on Acoustics, Speech, and Signal Processing*, 31, 445–449 (1983).
33. J. Ward, H. Cox, and S. Kogon, "A comparison of robust adaptive beamforming algorithms," *Proceedings of the 37th Asilomar Conference on Signals, Systems and Computers*, Pacific Grove, CA, Vol. 2, pp. 1340–1344, November 2003.
34. A. Jakobsson, F. Gini, and F. Lombardini, "Layover solution in multibaseline in SAR using robust beamforming," *IEEE International Symposium on Signal Processing and Information Technology*, December 2003.
35. T. Bowles, A. Jakobsson, and J. Chambers, "Detection of cell-cyclic elements in missampled gene expression data using a robust Capon estimator," *ICASSP*, May 2004.
36. O. Besson and F. Vincent, "Performance analysis for a class of robust adaptive beamformers," *ICASSP*, May 2004 (to appear in *IEEE Transactions on Signal Processing*).
37. R. G. Lorenz and S. P. Boyd, "Robust beamforming in GPS arrays," *Proceedings of the Institute of Navigation, National Technical Meeting*, January 2002.
38. L. Vandenberghe and S. Boyd, "Semidefinite programming," *SIAM Review*, 38, 49–95 (March 1996).
39. J. F. Sturm, "Using SeDuMi 1.02, a MATLAB toolbox for optimization over symmetric cones," *Optimization Methods and Software*, No. 11–12, pp. 625–653, 1999.
40. M. Lobo, L. Vandenberghe, S. Boyd, and H. Lebret, "Applications of second-order cone programming," *Linear Algebra and its Applications, Special Issue on Linear Algebra in Control, Signals and Image Processing*, pp. 193–228, November 1998.
41. K.-B. Yu, "Recursive updating the eigenvalue decomposition of a covariance matrix," *IEEE Transactions on Signal Processing*, 39, 1136–1145 (1991).
42. Y. Hua, M. Nikpour, and P. Stoica, "Optimal reduced-rank estimation and filtering," *IEEE Transactions on Signal Processing*, 49, 457–469 (2001).
43. D. C. Sorensen, "Newton's method with a model trust region modification," *SIAM Journal on Numerical Analysis*, 19(2), 409–426 (1982).
44. A. V. Fiacco and G. P. McCormick, *Nonlinear Programming: Sequential Unconstrained Minimization Techniques*. John Wiley & Sons, New York, NY, 1968.
45. C. A. Stutt and L. J. Spafford, "A 'best' mismatched filter response for radar clutter discrimination," *IEEE Transactions on Information Theory*, 14, 280–287 (1968).
46. M. M. Goodwin and G. W. Elko, "Constant beamwidth beamforming," *ICASSP*, Minneapolis, MN, Vol. 1, pp. 169–172, April 1993.
47. T. Chou, "Frequency-independent beamformer with low response error," *ICASSP*, Detroit, MI, Vol. 5, pp. 2995–2998, May 1995.
48. J. Lardies, "Acoustic ring array with constant beamwidth over a very wide frequency range," *Acoustic Letters*, 13(5), 77–81 (1989).
49. W. M. Humphreys, T. F. Brooks, W. W. Hunter, and K. R. Meadows, "Design and use of microphone directional arrays for aeroacoustic measurement," *AIAA Paper 98-0471, 36st Aerospace Sciences Meeting and Exhibit*, Reno, NV, January 12–15, 1998.
50. T. F. Brooks and W. M. Humphreys, "Effect of directional array size on the measurement of airframe noise components," *AIAA Paper 99-1958, 5th AIAA Aeroacoustics Conference*, Bellevue, WA, May 10–12, 1999.

51. T. F. Brooks, M. A. Marcolini, and D. S. Pope, "A directional array approach for the measurement of rotor source distributions with controlled spatial resolution," *Journal of Sound and Vibration*, 112(1), 192–197 (1987).
52. D. Tucker, "Arrays with constant beam-width over a wide frequency range," *Nature*, 180, 496–497 (1957).
53. R. Smith, "Constant beamwidth receiving arrays for broad band sonar systems," *Acustica*, 23, 21–26 (1970).
54. J. H. Doles III and F. D. Benedict, "Broad-band array design using the asymptotic theory of unequally spaced arrays," *IEEE Transactions on Antennas and Propagation*, 36, 27–33 (1988).
55. D. B. Ward, R. A. Kennedy, and R. C. Williamson, "Theory and design of broad-band sensor arrays with frequency invariant far-field beam patterns," *Journal of Acoustical Society of America*, 97, 1023–1034 (1995).
56. T. D. Abhayapala, R. A. Kennedy, and R. C. Williamson, "Nearfield broadband array design using a radially modal expansion," *Journal of Acoustical Society of America*, 107, 392–403 (2000).
57. M. Brandstein and D. Ward (Eds.), *Microphone Arrays: Signal Processing Techniques and Applications.* Springer Verlag, New York, NY, 2001.
58. Z. Wang, J. Li, P. Stoica, T. Nishida, and M. Sheplak, "Constant-beamwidth and constant-powerwidth wideband robust Capon beamformers for acoustic imaging," *Journal of the Acoustical Society of America*, 116(3), 1621–1631 (2004).
59. D. T. Blackstock, *Fundamentals of Physical Acoustics.* John Wiley & Sons, New York, NY, 2000.
60. J. R. Guerci, "Theory and application of covariance matrix tapers for robust adaptive beamforming," *IEEE Transactions on Signal Processing*, 47, 977–985 (1999).
61. M. Born and E. Wolf, *Principles of Optics.* Academic Press, New York, NY, 1970.
62. L. Borcea, G. Papanicolaou, C. Tsogka, and J. G. Berryman, "Imaging and time reversal in random media," *Inverse Problems*, 18, 1247–1279 (2002).
63. E. G. Larsson, J. Li, and P. Stoica, "High-resolution nonparametric spectral analysis: Theory and applications." In Y. Hua, A. B. Gershman, and Q. Cheng (Eds.), *High-Resolution and Robust Signal Processing.* Marcel-Dekker, Inc, New York, NY, 2003.
64. J. Li and P. Stoica, "An adaptive filtering approach to spectral estimation and SAR imaging," *IEEE Transactions on Signal Processing*, 44, 1469–1484 (1996).
65. P. Stoica, A. Jakobsson, and J. Li, "Capon, APES and matched-filterbank spectral estimation," *Signal Processing*, 66, 45–59 (1998).
66. H. Li, J. Li, and P. Stoica, "Performance analysis of forward-backward matched-filterbank spectral estimators," *IEEE Transactions on Signal Processing*, 46, 1954–1966 (1998).
67. G. R. Benitz, "High-definition vector imaging," *MIT Lincoln Laboratory Journal-Special Issue on Superresolution*, 10(2), 147–170 (1997).
68. L. M. Novak, G. J. Owirka, and A. L. Weaver, "Automatic target recognition using enhanced resolution SAR data," *IEEE Transactions on Aerospace and Electronic Systems*, 35, 157–175 (1999).
69. Y. Wang, J. Li, and P. Stoica, "Rank-deficient robust Capon filter-bank approach to complex spectral estimation," *IEEE Transactions on Signal Processing*, to appear.

70. G. R. Benitz, "High definition vector imaging for synthetic aperture radar," *Proceedings of the 31st Asilomar Conference on Signals, Systems and Computers*, Pacific Grove, CA, pp. 1204–1209, November 1997.

71. A. Jakobsson and P. Stoica, "Combining Capon and APES for estimation of spectral lines," *Circuits, Systems, and Signal Processing*, 19(2), 159–169 (2000).

72. D. J. Andersh, M. Hazlett, S. W. Lee, D. D. Reeves, D. P. Sullivan, and Y. Chu, "XPATCH: A high-frequency electromagnetic scattering prediction code and environment for complex three-dimensional objects," *IEEE Antennas and Propagation Magazine*, 36, 65–69 (1994).

73. M. Bradley, T. Witten, R. McCummins, and M. Duncan, "Mine detection with ground penetration synthetic aperture radar," *Proc. of SPIE Conference on Detection and Remediation Technologies for Mines and Minelike Target VII*, Vol. 4742, pp. 248–258, 2002.

74. K. Gu, R. Wu, J. Li, M. Bradley, J. Habersat, and G. Maksymonko, "SAR processing for GPSAR systems," *Proc. of SPIE Conference on Detection and Remediation Technologies for Mines and Minelike Target VII*, Vol. 4742, pp. 1050–1060, 2002.

75. R. Kapoor, M. Ressler, and G. Smith, "Forward-looking mine detection using an ultra-wideband radar," *Proc. of SPIE Conference on Detection and Remediation Technologies for Mines and Minelike Target V*, Vol. 4038, pp. 1067–1076, 2000.

76. J. Kositsky, R. Cosgrove, C. Amazeen, and P. Milanfar, "Results from a forward-looking GPR mine dectection system," *Proc. of SPIE Conference on Detection and Remediation Technologies for Mines and Minelike Target VII*, Vol. 4742, pp. 206–217, 2002.

77. Y. Sun and J. Li, "Time-frequency analysis for plastic landmine detection via forward-looking ground penetrating radar," *IEE Proceedings-Radar, Sonar and Navigation*, 150, 253–261 (2003).

78. W. Lertniphonphun and J. H. McClellan, "Migration of underground targets in UWB-SAR systems," *International Conference on Image Processing. Proceedings*, Vol. 1, pp. 713–716, 2000.

79. Y. Wang, X. Li, Y. Sun, and J. Li, "Adaptive imaging for forward-looking ground penetrating radar," submitted to *IEEE Transactions on Aerospace and Electronic Systems*, 2004.

80. G. Liu, Y. Wang, J. Li, and M. Bradley, "SAR imaging for a forward-looking GPR system," *Proc. of SPIE Conference on Detection and Remediation Technologies for Mines and Minelike Target VIII*, Vol. 5089, 2003.

81. Y. Wang, J. Li, G. Liu, and P. Stoica, "Polarimetric SAR target feature extraction and image formation via semi-parametric methods," *Digital Signal Processing*, 14(3), 268–293 (2004).

4

DIAGONAL LOADING FOR FINITE SAMPLE SIZE BEAMFORMING: AN ASYMPTOTIC APPROACH

Xavier Mestre and Miguel A. Lagunas
Centre Tecnològic de Telecommunicacions de Catalunya,
Barcelona, Spain

Array processing architectures have long been used as spatial filters in multiple fields, such as radar, sonar, communications, geophysics, astrophysics or biomedical applications. The main idea behind beamforming is the fact that, by conveniently weighting and combining the signals received from multiple sensors or antennas, one can synthesize an spatial response pattern that best suits the observed scenario. In general terms, the objective of the spatial filter is to optimize the response of the array so that the output contains minimal contributions from noise and signals coming from unwanted directions of arrival. In practice, the spatial characteristics of the scenario are not known to the receiver and, consequently, the filter weights have to be designed using side information and assumptions that might not completely correspond to the actual spatial characteristics of the environment. This mismatch can be caused, for instance, by environmental nonstationarity, small sample size, multipath, array manifold errors, pointing errors, fading, or other impairments (see, e.g., [1] for a more complete description of these effects). These impairments result in a general deterioration of the discrimination capabilities of the array, and have a very detrimental effect on the global performance of the spatial filter. Robust beamforming approaches emerge as specific techniques that minimize the effect of such mismatches.

From all the possible sources of impairment described above, this chapter will only concentrate on beamforming techniques that specifically try to alleviate the finite sample size effect. This type of distortion arises in many spatial filtering

Robust Adaptive Beamforming, Edited by Jian Li and Petre Stoica

applications, where the number of available observations is not sufficiently high to guarantee a proper training of the beamformer weights. Out of all the techniques proposed in the literature to mitigate the finite sample size effect, we will only concentrate on diagonal loading solutions, which can be applied to a wide range of scenarios, and are, quite probably, the most extensively used. Our approach is significantly different from previous analyses, in that we consider two-dimensional asymptotics that assume that both the number of sensors and the number of observations are high but have the same order of magnitude. It turns out that the performance of spatial filters is strongly dependent on the quotient between these two quantities and, as a consequence, two-dimensional asymptotics are an excellent approximation of the real-world nonasymptotic reality. Using this asymptotic approach, we will be able to give an accurate description of the optimum loading factor under different conditions, and then propose a consistent way of estimating it in practice.

The chapter is organized as follows. Section 4.1 gives a brief introduction to the problem of minimum variance distortionless response beamforming, the sample matrix inversion implementation, and the diagonal loading technique. The main results available in the literature are also reviewed. Section 4.2 presents a two-dimensional asymptotic analysis of the problem based on random matrix theory. It starts with a brief presentation of the mathematical tools used in random matrix theory, which are then applied to derive an asymptotic expression for the output signal to interference plus noise ratio. Most of the results available in the literature that were presented in Section 4.1 are then seen to be subsumed in the new asymptotic formulation. Section 4.3 is focused on the estimation of the optimum loading factor in a generic scenario. It proposes a new estimator that is shown to be consistent even when the number of sensors and the number of samples have the same order of magnitude. This estimator is based on the theory of general statistical analysis (or G-estimation), a novel statistical inference theory based on random matrix theory. Section 4.4 gives a more complete theoretical study of the asymptotically optimum loading factor in some simplified scenarios with one or two spatial sources, and finally, Section 4.5 summarizes and concludes the chapter. All technical proofs have been relegated to the appendices.

Notation. Boldface upper-case (resp. boldface lower-case) letters denote matrices (resp. column vectors). The superscript $(\cdot)^H$ denotes transpose complex conjugate, $\mathrm{tr}(\cdot)$ is the trace operator, and $\{\cdot\}_{i,j}$ represents the (i,j)th entry of a matrix. The real and imaginary parts of a complex number are denoted as $\mathrm{Re}(\cdot)$ and $\mathrm{Im}(\cdot)$, respectively, and j is the imaginary unit. The Euclidean norm of a vector is written $\|\cdot\|$, $\mathrm{Prob}(\cdot)$ denotes the probability of an event and $\mathbb{E}[\cdot]$ is the expectation operator. Finally, $(x)^+ = \max(x, 0)$ and $\otimes$ represents the Kronecker product.

4.1 INTRODUCTION AND HISTORICAL REVIEW

Traditionally, beamformers have been classified according to the type of reference used for the training of the filter weights [2]. In time-reference beamformers, for example, the spatial filter weights are trained using a reference signal that is known to the receiver. In spatial-reference beamformers, on the other hand, this training is carried out exploiting the information of the angle-of-arrival (or spatial signature) from which the signal of interest is assumed to come. This chapter will only focus on this second type of beamformer, that is, spatial-reference based. From the extensive range of spatial-reference architectures proposed so far, we will only concentrate on the minimum variance distortionless response beamformer and, in particular, the sample matrix inversion implementation. In this section we introduce these two concepts, along with the diagonal loading technique. We also give a brief overview of the results that have been published in the literature.

4.1.1 Signal Model and MVDR Beamforming

We consider a generic array of sensors of $M > 1$ elements from where N different spatial observations or snapshots are obtained. Let $\mathbf{y}(n)$ denote the $M \times 1$ complex snapshot obtained at the nth sample instant ($n = 1, \ldots, N$). Each of these snapshots can be modeled as

$$\mathbf{y}(n) = \beta s(n)\mathbf{s}_d + \mathbf{n}(n)$$

where $\mathbf{s}_d$ is the $M \times 1$ complex spatial signature corresponding to the signal of interest, denoted here by $s(n)$, $\mathbf{n}(n)$ is a $M \times 1$ complex vector that contains the noise plus interference contribution, and β indicates whether the desired signal is present in the observations ($\beta = 1$) or absent from them ($\beta = 0$). The presence or absence of the signal of interest is a critical issue in spatial filtering applications. In pulse radar applications, for instance, there are time intervals in which the received signal does not contain the transmitted waveform. In this situation, the spatial filter can be designed from noise plus interference samples alone, and we will see that this significantly improves the performance with respect to the case where the useful signal is present in the received data. The signal-free case is sometimes referred to as "supervised training" scenario, as opposed to the "unsupervised training" case, where the useful signal is present. Communications applications are a typical example of this last type of situation.

It is quite usual to model the signal of interest $s(n)$ and the noise plus interference vector $\mathbf{n}(n)$ as statistically independent stationary random quantities, with power and spatial correlation given respectively by

$$P_d = \mathbb{E}\left[|s(n)|^2\right], \quad \mathbf{R}_N = \mathbb{E}\left[\mathbf{n}(n)\mathbf{n}^H(n)\right].$$

From now on, and without loss of generality, we will assume that the spatial signature of the signal of interest $\mathbf{s}_d$ is normalized to have unit norm; this way, P_d can be regarded as the total received power associated with the signal of interest.

We will denote by $\mathbf{Y}$ the $M \times N$ complex matrix that contains, at each of its columns, the observations $\mathbf{y}(n)$,

$$\mathbf{Y} = [\mathbf{y}(1) \ \cdots \ \mathbf{y}(N)].$$

The signal after the spatial filter at the nth sampling instant can be mathematically described as

$$x(n) = \mathbf{w}^H \mathbf{y}(n)$$

where the beamformer weights have been gathered into an $M \times 1$ complex column vector $\mathbf{w}$. The objective of the spatial filter $\mathbf{w}$ is to suppress the noise plus interference component $\mathbf{n}(n)$ while enhancing the reception of the signal associated with the spatial signature vector $\mathbf{s}_d$.

In this chapter, we will measure the performance of the different spatial filter solutions in terms of the signal to interference plus noise ratio (SINR) at the output of the beamformer, defined as

$$\mathrm{SINR}(\mathbf{w}) = \frac{\mathbb{E}\left[|\mathbf{w}^H(s(n)\mathbf{s}_d)|^2\right]}{\mathbb{E}\left[|\mathbf{w}^H\mathbf{n}(n)|^2\right]} = \frac{P_d|\mathbf{w}^H\mathbf{s}_d|^2}{\mathbf{w}^H\mathbf{R}_N\mathbf{w}}$$

so that the optimum spatial filter will be the one maximizing this quantity, that is

$$\mathbf{w}_{\mathrm{opt}} = \arg\max_{\mathbf{w}} \mathrm{SINR}(\mathbf{w}).$$

The SINR metric is a simple indicative measure of the performance of the spatial filter, but one must keep in mind that the ultimate performance of the receiver will not depend on this quantity alone. In this sense, it would be better to have an analysis in terms of other more reliable performance metrics, such as symbol error rate in communications applications or probability of detection/false alarm in radar. However, this type of analysis complicates the formulation substantially and, for this reason, we will only concentrate on the SINR measure as far as this chapter is concerned. Note, also, that in our definition of SINR we are implicitly assuming independence between the spatial filter $\mathbf{w}$ and the received observations at which the SINR is measured. In some situations, like in communications applications, there exists an intrinsic statistical dependence between the spatial filter weights $\mathbf{w}$ and the data to which they are applied. In these cases, our performance measure is not an indicator of the actual quotient between the signal power and the noise plus interference power at the output of the spatial filter in the strict sense, but rather an indicator of such a quantity if the spatial filter was to be applied to a completely independent set of data. Even so, this definition of SINR is a good indicator of the performance of the spatial resolution capabilities of the spatial filter in terms of noise plus interference rejection and useful signal enhancement.

Now, note that output SINR of a particular spatial filter is not altered if we multiply the weight vector by a fixed constant (this constant will cancel out in the definition of SINR). Therefore, without loss of performance, we may force the spatial filter to have a fixed non-null response towards the useful signal, for instance $\mathbf{w}^H\mathbf{s}_d = 1$. In this case, the maximization of the output SINR turns out to be equivalent to the minimization of the noise variance ($\mathbf{w}^H\mathbf{R}_N\mathbf{w}$) subject to a distortionless response towards $\mathbf{s}_d$, that is

$$\mathbf{w}_{\text{opt}} = \arg\min_{\mathbf{w}} \mathbf{w}^H \mathbf{R}_N \mathbf{w}, \quad \text{subject to } \mathbf{w}^H \mathbf{s}_d = 1.$$

The solution to this optimization problem is generally referred to as minimum variance distortionless response (MVDR) beamformer [3, 4]. It turns out that the problem accepts a closed form solution, namely

$$\mathbf{w}_{\text{opt}} = \frac{1}{\mathbf{s}_d^H \mathbf{R}_N^{-1} \mathbf{s}_d} \mathbf{R}_N^{-1} \mathbf{s}_d$$

and the optimum weight vector yields a maximum output SINR given by

$$\text{SINR}_{\text{opt}} = \text{SINR}(\mathbf{w}_{\text{opt}}) = P_d \mathbf{s}_d^H \mathbf{R}_N^{-1} \mathbf{s}_d. \tag{4.1}$$

Moreover, any weight vector different from zero and proportional to $\mathbf{R}_N^{-1}\mathbf{s}_d$ will achieve the same output SINR. In particular, defining the total spatial correlation matrix of the observations as

$$\mathbf{R} = \mathbb{E}[\mathbf{y}(n)\mathbf{y}^H(n)] = \beta P_d \mathbf{s}_d \mathbf{s}_d^H + \mathbf{R}_N \tag{4.2}$$

and using the matrix inverse lemma [66], it can be seen that

$$\mathbf{w}_{\text{opt}} = \frac{1}{\mathbf{s}_d^H \mathbf{R}_N^{-1} \mathbf{s}_d} \mathbf{R}_N^{-1} \mathbf{s}_d = \left(\frac{1 + \beta P_d \mathbf{s}_d^H \mathbf{R}_N^{-1} \mathbf{s}_d}{\mathbf{s}_d^H \mathbf{R}_N^{-1} \mathbf{s}_d} \right) \mathbf{R}^{-1} \mathbf{s}_d. \tag{4.3}$$

This shows that the optimum weight vector can be formulated, up to a scalar factor, as the inverse of the spatial correlation matrix of the observations ($\mathbf{R}^{-1}$) multiplied by the spatial signature of the signal of interest ($\mathbf{s}_d$). This holds regardless of whether the useful signal is present or not in the observations, and suggests a practical way of implementing the MVDR beamformer: the sample matrix inversion (SMI) technique.

4.1.2 The Sample Matrix Inversion Technique

In practice, the spatial distribution of the noise plus interference signal is not known, and so is therefore the optimum weight vector $\mathbf{w}_{\text{opt}}$. For this reason, the weight vector $\mathbf{w}$ must be somehow "trained" from the observations in order to get as close as possible to the optimum filter $\mathbf{w}_{\text{opt}}$. In spatial-reference systems, where the only information available to the receiver is the signature vector $\mathbf{s}_d$, the form of the optimum beamformer in (4.3) naturally suggests the use of the weight vector $\hat{\mathbf{w}} = \hat{\mathbf{R}}^{-1}\mathbf{s}_d$ [3, 5], where $\hat{\mathbf{R}}$ is the sample spatial correlation matrix

$$\hat{\mathbf{R}} = \frac{1}{N}\mathbf{Y}\mathbf{Y}^H = \frac{1}{N}\sum_{n=1}^{N} \mathbf{y}(n)\mathbf{y}^H(n). \tag{4.4}$$

This direct inversion of the sample correlation matrix in order to implement the MVDR beamformer is usually referred to as the SMI technique, and it is only implementable if the number of available snapshots is higher than or equal to the number of sensors ($N \geq M$); otherwise the sample correlation matrix is not invertible. The performance of the SMI technique was first analyzed in [6] under the assumption that the signal of interest was not present in the observations ($\beta = 0$). Modeling the noise plus interference component as a sequence of independent and identically distributed (i.i.d.) circularly symmetric, zero-mean, Gaussian random vectors, the authors derived the probability density function of the output SINR normalized by $\mathrm{SINR}_{\mathrm{opt}}$, that is,

$$\rho_0 = \left. \frac{\mathrm{SINR}(\hat{\mathbf{w}})}{\mathrm{SINR}_{\mathrm{opt}}} \right|_{\beta=0}. \tag{4.5}$$

They showed that ρ_0 is beta-distributed with parameters $(N - M + 2,\ M - 1)$, that is,[1]

$$f_{\rho_0}(x) = \frac{N!}{(N - M + 1)!(M - 2)!} x^{N-M+1}(1 - x)^{M-2}.$$

In particular, the expectation of ρ_0, which can be interpreted as the average proportion of the output $\mathrm{SINR}_{\mathrm{opt}}$ that is achieved with only N samples, turns out to be equal to

$$\mathbb{E}[\rho_0] = \frac{N - M + 2}{N + 1}.$$

This shows that, in order to guarantee an average output SINR within 3 dB of the optimal one, the spatial filter weights need to be trained with approximately twice as many snapshots as the number of sensors,

$$N \geq 2M - 3 \approx 2M. \tag{4.6}$$

This useful rule of thumb has been widely used in the literature, and it gives an accurate idea of how fast the convergence rate of the SMI algorithm is whenever the signal of interest is absent from the observations.

It was early recognized that the performance of the SMI technique is severely affected when the signal of interest is present in the observed data [8–10]. This is due to the signal cancellation effect [11]. In a few words, the sample correlation matrix contains a polluted version of the useful spatial signature, which does not completely match the true one $\mathbf{s}_d$. As a consequence, the beamformer treats the desired signal as another interference source, and tries to null it out instead of

[1]The statistics of the weights $\hat{\mathbf{w}}$ themselves, which might be useful for purposes of implementation design, were derived in [7].

enhancing it. The effect is very similar to the one produced when, due to an imperfect calibration or array pointing errors, the spatial signature assumed at the receiver does not completely match the true one. The signal cancellation effect can cause a severe degradation of the performance of the SMI technique, sometimes making it even worse than the traditional phased array ($\hat{\mathbf{w}} = \mathbf{s}_d$).

Compared to the signal-free situation ($\beta = 0$), the performance analysis of the SMI beamformer when the useful signal is present in the received data ($\beta = 1$) is more difficult to tackle [9, 12, 13]. If we denote by ρ_1 the output SINR of the SMI implementation normalized by $\mathrm{SINR}_{\mathrm{opt}}$, that is,

$$\rho_1 = \left. \frac{\mathrm{SINR}(\hat{\mathbf{w}})}{\mathrm{SINR}_{\mathrm{opt}}} \right|_{\beta=1}$$

it turns out that

$$\rho_1 = \frac{\rho_1'}{\mathrm{SINR}_{\mathrm{opt}}(1 - \rho_1') + 1}$$

where

$$\rho_1' = \frac{(\mathbf{s}_d^H \hat{\mathbf{R}}^{-1} \mathbf{s}_d)^2}{\mathbf{s}_d^H \hat{\mathbf{R}}^{-1} \mathbf{R} \hat{\mathbf{R}}^{-1} \mathbf{s}_d \mathbf{s}_d^H \mathbf{R}^{-1} \mathbf{s}_d}.$$

Now, assuming that the observations form a sequence of i.i.d. circularly symmetric Gaussian random vectors with zero mean and covariance matrix $\mathbf{R}$, ρ_1' follows the same distribution as ρ_0 in (4.5), that is, a beta distribution with parameters $(N - M + 2, M - 1)$. It immediately follows that $\mathbb{E}[\rho_1] \leq \mathbb{E}[\rho_0]$, the equality holding when $\mathrm{SINR}_{\mathrm{opt}}$ goes to zero or N grows to infinity. The expected value of ρ_1 can be expressed as an infinite series [9, 12], namely

$$\mathbb{E}[\rho_1] = \frac{N!}{(N - M + 1)!} \sum_{k=0}^{\infty} \frac{1}{(N - M + k + 3) \cdots (N + k + 1)} \frac{\mathrm{SINR}_{\mathrm{opt}}^k}{(1 + \mathrm{SINR}_{\mathrm{opt}})^{k+1}}. \tag{4.7}$$

It turns out [9] that the minimum number of snapshots needed to achieve an output SINR within 3 dB of the optimum one is approximately[2]

$$N \gtrsim (2 + \mathrm{SINR}_{\mathrm{opt}})M \tag{4.8}$$

which in practice is much higher than in the signal-free case ($N \gtrsim 2M$). Interestingly enough, the convergence speed of the SMI method in the presence of useful signal ($\beta = 1$) slows down as $\mathrm{SINR}_{\mathrm{opt}}$ increases. This means that scenarios that are in principle easier to tackle (low level of interference, high signal power) might

[2]In fact, [14] proposes $N \geq \mathrm{SINR}_{\mathrm{opt}}(M - 1)$ for high $\mathrm{SINR}_{\mathrm{opt}}$. The two results coincide when the number of sensors is high.

become difficult due to the signal cancellation effect. Intuitively one may reason that, if the spatial filter does not need to devote degrees of freedom to interference nulling, they will be devoted to useful-signal cancellation. This has motivated the use of artificial noise injection methods to speed up the convergence of the beamformer weights [15].

Many authors have proposed different methods to improve the performance of the SMI technique in finite sample size situations, especially when the number of samples is lower than the number of sensors (see [1] for an interesting review of these methods). For example, the Hung–Turner projection method, originally proposed in [16], replaces the sample correlation matrix $\hat{\mathbf{R}}$ with a projection matrix onto the orthogonal complement of the columns of the observations $\mathbf{Y}$,

$$\hat{\mathbf{w}}_{HT} = (\mathbf{I}_M - \mathbf{Y}(\mathbf{Y}^H\mathbf{Y})^{-1}\mathbf{Y}^H)\mathbf{s}_d$$

where $\mathbf{I}_M$ denotes the $M \times M$ identity matrix. This method is applicable when $N \leq M$ and its performance has been evaluated in [17, 18] for the signal-free situation ($\beta = 0$) when the number of samples N is higher than the dimension of the interference subspace, denoted here by K. It was shown that, under the same the statistical assumptions as above and assuming that $N > K$, the normalized output SINR of the Hung–Turner method is distributed as the product of two independent beta-distributed random variables with parameters $(M - N, N - K)$ and $(N - K + 1, K)$ respectively, so that the mean value takes the form

$$\mathbb{E}\left[\frac{\mathrm{SINR}(\hat{\mathbf{w}}_{HT})}{\mathrm{SINR}_{\mathrm{opt}}}\right]_{\beta=0} = \frac{(M-N)(N-K+1)}{(M-K)(N+1)} \tag{4.9}$$

The best performance is achieved when the number of available samples is $N_{opt} = \sqrt{K(M+1)} + 1$.

Closely related to the Hung–Turner technique is the *eigencanceler* [19, 20, 63]. This technique is applied to the signal-free situation (otherwise, the desired signal must be filtered out in a previous stage), and it is constructed from the SMI technique by replacing the sample correlation matrix $\hat{\mathbf{R}}$ with a projection matrix onto the orthogonal complement of the interference subspace

$$\hat{\mathbf{w}}_{EC} = \left(\mathbf{I}_M - \hat{\mathbf{E}}\hat{\mathbf{E}}^H\right)\mathbf{s}_d. \tag{4.10}$$

Here, $\hat{\mathbf{E}}$ is an $M \times K$ matrix containing the K eigenvectors of $\hat{\mathbf{R}}$ corresponding to the interference subspace, that is, the ones associated with the largest eigenvalues of $\hat{\mathbf{R}}$. Note that the dimension of the interference subspace (K) must be known beforehand. As pointed out in [18], the Hung–Turner method can be seen as an eigencanceler where the projection is onto the orthogonal complement of the N (instead of K) eigenvectors associated with the largest eigenvalues of $\hat{\mathbf{R}}$. The statistical performance of the eigencanceler has been studied in [18, 21–23] for the signal-free case ($\beta = 0$) and under different asymptotic approximations. It turns out that the performance of the eigencanceler is similar to the performance of the diagonal-loading technique that we will present below. On the other hand, the method can only be applied to the signal-free scenario and presumes the knowledge of the interfering subspace dimension K. For all these reasons, we will not further consider this method in this chapter.

Diagonal loading is the most popular method to improve the convergence speed of the SMI technique. The technique is based on adding a positive[3] real number α to the diagonal entries of the sample correlation matrix,

$$\hat{\mathbf{w}}_{DL} = [\hat{\mathbf{R}} + \alpha \mathbf{I}_M]^{-1} \mathbf{s}_d. \tag{4.11}$$

The initial rationale behind diagonal loading is to extend the use of the SMI technique to the case where $\hat{\mathbf{R}}$ is singular, or otherwise facilitate its inversion by improving its condition number. The method has achieved widespread use, mainly because of its simplicity and its satisfactory performance.

Observe that there is an inherent trade-off in the choice of the loading factor α in (4.11). If the number of available samples is low, trying to steer nulls towards the interferences might result in a worse output SINR, so that a better performance could be obtained enhancing the steering capabilities of the array. The diagonally loaded solution in (4.11) provides a response between the traditional SMI spatial filter ($\alpha = 0$) and the phased array ($\alpha \to \infty$). We will show that, for each particular scenario, there will always be an optimum loading factor that maximizes the output SINR.

Note, also, that the formulation in (4.11) includes the Hung–Turner projection method as a special case. This can be shown by applying the matrix inverse lemma to the diagonally loaded spatial filter under the assumption that $N \leq M$ [24],

$$\hat{\mathbf{w}}_{DL} = [\hat{\mathbf{R}} + \alpha \mathbf{I}_M]^{-1} \mathbf{s}_d = \frac{1}{\alpha} \Big[\mathbf{I}_M - \mathbf{Y} \big[\mathbf{Y}^H \mathbf{Y} + N\alpha \mathbf{I}_N \big]^{-1} \mathbf{Y}^H \Big]^{-1} \mathbf{s}_d.$$

Dropping the term $1/\alpha$ (which does not contribute to the output SINR) and taking limits as $\alpha \to 0$ we obtain the Hung–Turner projection spatial filter.

We devote the rest of this chapter to the analysis of the diagonal loading technique. As shown above, the Hung–Turner projection approach will be a special case of this study.

4.1.3 Diagonal Loading

Diagonal loading (or diagonal incrementation, as it is sometimes referred to in the mathematics literature) has long been used as a regularization technique for different problems in statistics and engineering. The early motivation for diagonal loading

[3]Negative loading factors are sometimes used to enhance the null steering capability of the array [64, 65]. Since negative loading can sometimes cause numerical stability problems, we will restrict ourselves to positive loading factors in this chapter.

and its progressive development has been extensively discussed in [25]. It seems that the first presented use of diagonal loading dates back to 1944 is due to Kenneth Levenberg [26]. This work considered the problem of solving nonlinear least squares problems, and proposed to introduce a quadratic penalty to the linearized cost function. The result of that was a modification of the Taylor-series method that introduced an increment of the principal diagonal of the original equations. Some years later, and quite in a completely different context, James Durbin published [27], a study on the estimation of linear regression coefficients using some previous statistical information about the parameters. He showed that the usual normal equations were only slightly modified by a diagonal term that was proportional to the variance of the prior. This was later developed into the Bayes approach. Two years later, James Riley gave in [28] the first indication that diagonal loading was useful in order to improve the conditioning of the augmented matrix, thereby facilitating its numerical inversion. Later on, in the context of regularization of ill-posed problems, diagonal loading techniques were more extensively analyzed [29].

In array processing applications, the use of diagonal loading has always been regarded as the natural complement to the SMI technique. References to diagonal loading can be traced back to the work of Jack Capon [3], where he already suggests this method in order to allow the inversion of the sample correlation matrix when the number of samples is lower than the observation dimension. It was later observed [9, 24, 30, 31] that diagonal loading methods can be beneficial even in the full rank situation. It was experimentally observed that, for the signal-free case ($\beta = 0$) and assuming that the interfering sources were received with high power, the diagonal loading approach allowed a drastic reduction in the number of samples needed for convergence. An output SINR within 3 dB of the optimum one could now be achieved with only $N \approx 2K$ snapshots (with K being the dimension of the interference subspace), as opposed to the $N \approx 2M$ snapshots needed in the conventional SMI approach.

All these observations were analytically confirmed in [32], again for the signal-free case. It was shown that, under the standard statistical assumptions introduced above and assuming that (1) the interfering sources are received with high power; (2) the loading factor is chosen to be higher than the noise level but much lower than the smallest interference eigenvalue; and (3) the number of samples is higher than or equal to the dimension of the interference subspace, the output SINR of the diagonally loaded SMI solution normalized by SINR_{opt} is beta-distributed with parameters $(N - K + 1, K)$. This implies that

$$\mathbb{E}\left[\frac{\text{SINR}(\hat{\mathbf{w}}_{DL})}{\text{SINR}_{\text{opt}}}\right]_{\beta=0} \approx \frac{N-K+1}{N+1}$$

and, consequently, we see that approximately $N = 2K - 1$ snapshots are at least needed to achieve a mean output SINR within 3 dB of the optimum one. Note that, due to the different approximations made in order to derive this density, the law of the output SINR does not depend on the actual loading factor α. It is therefore unclear from this analysis whether there exists an optimum loading factor and, if so, how to calculate it.

An intuitive explanation for the performance improvement due to diagonal loading was given in [33, 34]. Still not leaving the signal-free case, let us denote by $\hat{\lambda}_{\max} = \hat{\lambda}_1 \geq \cdots \geq \hat{\lambda}_M = \hat{\lambda}_{\min}$ and $\hat{\mathbf{e}}_1, \ldots, \hat{\mathbf{e}}_M$ the eigenvalues and associated eigenvectors of the sample correlation matrix $\hat{\mathbf{R}}$. The SMI spatial filter weights can be formulated as

$$\hat{\mathbf{w}} = \left[\sum_{m=1}^{M} \frac{1}{\hat{\lambda}_m} \hat{\mathbf{e}}_m \hat{\mathbf{e}}_m^H\right] \mathbf{s}_d = \frac{1}{\hat{\lambda}_{\min}} \left[\mathbf{s}_d - \sum_{m=1}^{M} \frac{\hat{\lambda}_m - \hat{\lambda}_{\min}}{\hat{\lambda}_m} (\hat{\mathbf{e}}_m^H \mathbf{s}_d) \hat{\mathbf{e}}_m\right]. \tag{4.12}$$

This shows that the weight vector $\hat{\mathbf{w}}$ consists of two contributions: the spatial signature of the useful signal $\mathbf{s}_d$, and a weighted sum of eigenvectors $\hat{\mathbf{e}}_m$ that are subtracted from it. The factors $(\hat{\mathbf{e}}_m^H \mathbf{s}_d)$ in the second term of the above expression scale the eigenvectors so that a null is produced towards the directions associated with each of the $\hat{\mathbf{e}}_m$. Since we are dealing with the signal-free case, these directions will be associated with interfering sources or noise. The depth of the null will be proportional to the factor $(\hat{\lambda}_m - \hat{\lambda}_{\min})/\hat{\lambda}_m$, which approaches 0 for the noise eigenvalues (no nulling) and tends to 1 for eigenvalues associated with spatial interference (strong nulling). This shows that the level of nulling for an interference source represented by a single eigenvalue is proportional to $10 \log_{10} (\hat{\lambda}_m/\hat{\lambda}_{\min})$ dB with respect to the phased-array response.

Now, due to the fact that the noise eigenvalues of the sample correlation matrix are not all exactly equal, in practice the array identifies some noise eigenvectors as additional sources of spatial interference and tries to devote degrees of freedom to nulling them. In this sense, diagonal loading desensitizes the system by compressing the noise eigenvalues of the correlation matrix so that the nulling capability against small interference sources is reduced. Indeed, assume that the interference eigenvalues are high compared to the noise eigenvalues; if we choose the loading factor α to be much lower than all the directional interference eigenvalues but higher than the noise power, the effective nulling associated with noise eigenmodes will be lower, without significantly affecting the depth of the nulls towards true sources of interference. This explains the improvement of the beamformer performance due to diagonal loading in the signal-free case.

It is worth pointing out that most of the work on diagonal loading has traditionally been focused on the signal-free situation. This does not mean that the usefulness of diagonal loading is restricted to that situation. Indeed, diagonal loading was soon identified as a powerful means to improve the robustness of the spatial filter against the signal cancelation effect, caused either by mismatches in the desired spatial signature or by limited sample availability. Regarding the first type of mismatch, it is known that diagonal loading is the natural extension of classical minimum variance beamformers incorporating quadratic constraints or some degree of uncertainty of the steering vector (see, e.g., [35–39] and other chapters of this book). On the other hand, diagonal loading has also been used to alleviate the signal cancellation effect in the SMI method under finite sample size situations when the signal of interest is contaminating the observations. This application of diagonal loading has received less attention.

In [40, 41], the asymptotic output SINR of the diagonally loaded MVDR beamformer was analyzed for the case where the useful signal is present in the observations and assuming a large number of snapshots. The basic limitation of these approaches is the fact that they are asymptotic in the number of snapshots, while the actual interest of diagonal loading is in low sample size situations. Another interesting study was presented in [42], where an expression for the approximate distribution of the SINR loss due to diagonal loading was derived under the same approximations as before. This paper assumed potential mismatching in the useful signal spatial signature and studied the effect of additional linear constraints imposed on the beamformer.

Today, some important questions related to the diagonally loaded SMI beamformer remain still open. It is not clear, for instance, how to choose the best loading factor in a real scenario in order to combat the finite sample size effect. This has motivated the use of rather ad hoc methods for fixing the loading factor in low sample size situations. For the signal-free case ($\beta = 0$), the traditional choice is to fix the diagonal load to be higher than the noise power but lower than the lowest interference eigenvalue; this is more formally expressed in [32, 43]. Another formulation for the estimated loading factor was given in [44], where the authors proposed to use

$$\alpha = -\hat{\lambda}_{\min} + \sqrt{(\hat{\lambda}_K - \hat{\lambda}_{\min})(\hat{\lambda}_{K+1} - \hat{\lambda}_{\min})} \tag{4.13}$$

where $\hat{\lambda}_K$ is the minimum interference eigenvalue. This choice, which can sometimes be approximated by $\alpha \approx \sqrt{\hat{\lambda}_K \hat{\lambda}_{K+1}}$, was inspired by the expression in (4.12) and can be obtained by minimizing the following proposed cost function:

$$f(\alpha) = \left(1 - \frac{\hat{\lambda}_K - \hat{\lambda}_{\min}}{\hat{\lambda}_K + \alpha}\right) + \frac{\hat{\lambda}_{K+1} - \hat{\lambda}_{\min}}{\hat{\lambda}_{K+1} + \alpha}. \tag{4.14}$$

As shown above, the first term accounts for the nulling response towards the Kth interference eigenvector, whereas the second one is a noise-enhancement penalization. This approach has two main drawbacks. First, it is not clear whether minimization of (4.14) translates into a maximization of the output SINR. Second, the dimension of the interference subspace K is, in practice, unknown.

Other ad hoc choices have been proposed in the literature. In [36, p. 748], for instance, it is suggested that, in the signal-free case, the loading factor should be set to a value 10 dB above the minimum eigenvalue of the sample correlation matrix. On the other hand, in [45] the diagonal load is fixed equal to the standard deviation of the diagonal entries of the sample correlation matrix.

In this chapter, we give further insights into the finite sample size situation for both signal-free and signal-contaminated situations. To do that, we will consider a novel asymptotic approximation: we will assume that *both* the number sensors and the number of observations increase without bound, while their quotient remains constant. This asymptotic limit is a very appropriate model for applications with large arrays (of up to hundreds or thousands of sensors), which are quite common in sonar, radioastronomy or over-the-horizon radar. Furthermore, the asymptotic approximation will also be useful in situations with a relatively low number of sensors, because, unlike previous asymptotic analyses where only the number of observations was assumed to grow, our approach assumes that the number of sensors and the number of snapshots have the same order of magnitude, which is always the case in a practical situation. To justify the practical relevance of the asymptotic approach, we will see that most of the conclusions that have been drawn in this section could also have been obtained from the asymptotic analysis.

4.2 ASYMPTOTIC OUTPUT SINR WITH DIAGONAL LOADING

We start this section with a brief presentation of the mathematical tools used for the asymptotic study. The reader that is not interested in how the asymptotic expressions are derived may choose to skip this part and proceed directly to Section 4.2.2.

4.2.1 Introduction to Random Matrix Theory

Random matrix theory is a branch of statistics that studies the asymptotic behavior of the spectrum of random matrices when their dimensions increase without bound. The theory itself originated during the 1950s in nuclear physics, with the remarkable work of Eugene P. Wigner [46, 47]. It soon developed into a new branch of mathematical statistics which, quite recently, has gained momentum in electrical engineering applications. See, for instance, [48] for a recent tutorial on the application of random matrix theory in wireless communications.

In order to introduce the basics of random matrix theory, consider a generic $M \times M$ complex Hermitian random matrix[4] $\hat{\mathbf{R}}$, with eigenvalues $\hat{\lambda}_{\max} = \hat{\lambda}_1 \geq \cdots \geq \hat{\lambda}_M = \hat{\lambda}_{\min}$ and associated eigenvectors $\hat{\mathbf{e}}_1, \ldots, \hat{\mathbf{e}}_M$. Because the entries of $\hat{\mathbf{R}}$ are random variables, so are the eigenvalues and eigenvectors associated with $\hat{\mathbf{R}}$. We define the empirical distribution function of the eigenvalues of $\hat{\mathbf{R}}$ as the eigenvalue counting function $F_{\hat{\mathbf{R}}}(x)$ that gives, for each x, the proportion of eigenvalues that are lower than or equal to x. This can be mathematically expressed as

$$F_{\hat{\mathbf{R}}}(x) = \frac{1}{M}\sum_{m=1}^{M} \mathcal{I}(\hat{\lambda}_m \leq x) \tag{4.15}$$

where $\mathcal{I}(A)$ is an indicator function that takes the value 1 if the event A is true, and 0 otherwise. Observe that the function $F_{\hat{\mathbf{R}}}(x)$ is a random variable for each given value of x.

[4]We retain the formulation of the sample covariance matrix ($\hat{\mathbf{R}}$) because we will later concentrate on random matrices that have this particular structure. Note, however, that here $\hat{\mathbf{R}}$ denotes a generic Hermitian random matrix.

The objective of random matrix theory is to study the asymptotic behavior of the empirical distribution of eigenvalues $F_{\hat{\mathbf{R}}}(x)$ when the dimensions of the random matrix $\hat{\mathbf{R}}$ go to infinity. Interestingly enough, for some random matrix models, the empirical distribution function $F_{\hat{\mathbf{R}}}(x)$ tends, with probability one, to a nonrandom limit $F(x)$. The physical interpretation of this phenomenon is that, as the dimensions of the matrix $\hat{\mathbf{R}}$ tend to infinity and hence the number of eigenvalues increases, these eigenvalues tend to cluster around a limiting distribution described by $F(x)$.

Let us see this with an example. Consider the random matrix $\hat{\mathbf{R}} = \frac{1}{N}\mathbf{Y}\mathbf{Y}^H$ where $\mathbf{Y}$ is an $M \times N$ matrix with complex entries, whose real and imaginary parts are all i.i.d. random variables that have zero mean, variance $\frac{1}{2}$ and bounded fourth order moments. Then, as $M, N \to \infty$ while $M/N \to c, 0 < c < \infty$, the empirical distribution function of the eigenvalues of $\hat{\mathbf{R}}$ converges almost surely to a nonrandom limiting distribution $F(x)$ with density given by the so-called Marchenko–Pastur law [49].

$$f(x) = \frac{dF(x)}{dx} = \left(1 - \frac{1}{c}\right)^+ \delta(x) + \sqrt{\frac{(x-a)^+(b-x)^+}{2\pi c x}}$$

where $a = (1 - \sqrt{c})^2$, $b = (1 + \sqrt{c})^2$, $(x)^+ = \max(x, 0)$ and $\delta(x)$ being a Dirac delta centered at zero. Figure 4.1 represents the absolutely continuous part of the Marchenko–Pastur law for different values of the aspect ratio c. It should be stressed that the convergence of the empirical eigenvalue distribution does not depend on the statistical law of the entries of $\hat{\mathbf{R}}$ (only some conditions on the moments of those

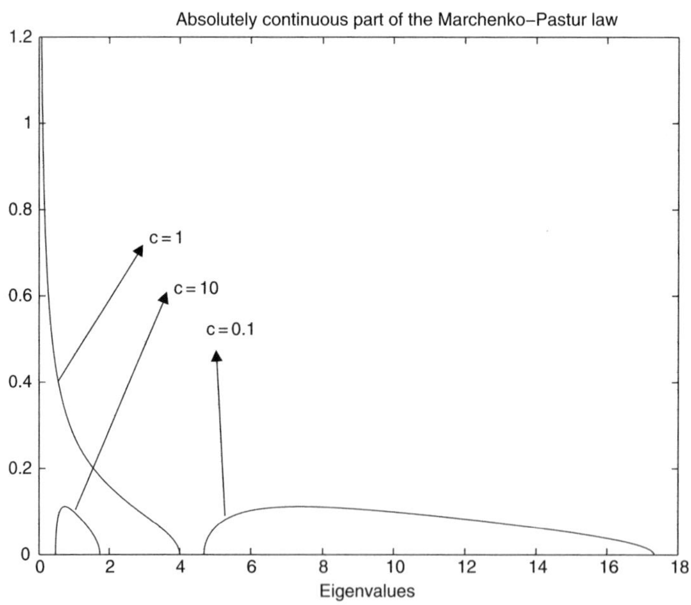

Figure 4.1 Absolutely continuous part of the Marchenko–Pastur law for different values of the ratio $c = M/N$.

entries need to be imposed). This is a general and interesting property of random matrix theory, which is sometimes referred to as the *invariance principle*.

This type of result can be generalized to more complex and useful random matrix models. In signal processing applications, for example, we are particularly interested in the sample covariance matrix model, which applies to random matrices of the form $\hat{\mathbf{R}} = \frac{1}{N}\mathbf{R}^{1/2}\mathbf{X}\mathbf{X}^H\mathbf{R}^{1/2}$, where $\mathbf{R}^{1/2}$ is the Hermitian positive definite square root of a deterministic matrix $\mathbf{R}$ (assumed also Hermitian and positive definite), and $\mathbf{X}$ is an $M \times N$ complex random matrix with the same statistical properties as the matrix $\mathbf{Y}$ above. Assume that the empirical distribution of eigenvalues of $\mathbf{R}$ converges to a limiting distribution $G(x)$ as $M \to \infty$ and that this matrix has uniformly bounded spectral norm (i.e., the maximum eigenvalue of $\mathbf{R}$ is bounded regardless of the dimension of the matrix). Then, as $M, N \to \infty$ at the same rate, the empirical distribution function of the eigenvalues of $\hat{\mathbf{R}}$ converges almost surely to a nonrandom limiting distribution $F(x)$. In this case, however, it is not possible to give a closed form expression for the generic $F(x)$ (which will depend on $G(x)$ but will in general be different from it). Instead, we can only give a relationship in terms of the so-called *Stieltjes transforms* of the eigenvalue densities of the two matrices.

The Stieltjes transform of a distribution function $F(x)$ is usually defined for complex arguments $z \in \mathbb{C}$ as

$$m(z) = \int \frac{1}{x - z} dF(x). \tag{4.16}$$

Observe that the Stieltjes transform of the empirical distribution function of the eigenvalues of a finite $M \times M$ matrix $\hat{\mathbf{R}}$ can be expressed in terms of the trace of the matrix as

$$m_{\hat{\mathbf{R}}}(z) = \int \frac{1}{x - z} dF_{\hat{\mathbf{R}}}(x) = \frac{1}{M}\sum_{m=1}^{M} \frac{1}{\hat{\lambda}_m - z} = \frac{1}{M}\mathrm{tr}\left[\left(\hat{\mathbf{R}} - z\mathbf{I}_M\right)^{-1}\right]. \tag{4.17}$$

A generic Stieltjes transform $m(z)$ contains all the information about the associated distribution function $F(x)$. In fact, if $F(x)$ is differentiable, one can easily recover the density function $f(x) = dF(x)/dx$ by means of the Stieltjes inversion formula

$$f(x) = \lim_{y \to 0^+} \frac{1}{\pi} \mathrm{Im}[m(x + jy)]. \tag{4.18}$$

It turns out that, for complicated random matrix models, working with Stieltjes transforms is substantially easier than working with distribution functions of eigenvalues.

Going back to the example of the sample covariance matrix model, we are now in a position to analyze the asymptotic density of eigenvalues of $\hat{\mathbf{R}}$. Let $m_{\hat{\mathbf{R}}}(z)$ and $m_{\mathbf{R}}(z)$ represent the Stieltjes transforms of the empirical eigenvalue distributions of $\hat{\mathbf{R}}$ and $\mathbf{R}$ respectively, namely

$$m_{\hat{\mathbf{R}}}(z) = \frac{1}{M}\mathrm{tr}[(\hat{\mathbf{R}} - z\mathbf{I}_M)^{-1}], \quad m_{\mathbf{R}}(z) = \frac{1}{M}\mathrm{tr}[(\mathbf{R} - z\mathbf{I}_M)^{-1}].$$

It turns out that, under the statistical assumptions above, and assuming that both M and N increase without bound at the same rate,

$$|m_{\hat{\mathbf{R}}}(z) - \bar{m}_{\hat{\mathbf{R}}}(z)| \to 0$$

almost surely, where, for each z, $\bar{m}_{\hat{\mathbf{R}}}(z)$ is the solution to the following equation [50, 51]

$$\bar{m}_{\hat{\mathbf{R}}}(z) = \frac{m_{\mathbf{R}}\left(z/\left(1 - c - cz\bar{m}_{\hat{\mathbf{R}}}(z)\right)\right)}{1 - c - cz\bar{m}_{\hat{\mathbf{R}}}(z)} = \frac{1}{M}\sum_{m=1}^{M}\frac{1}{\lambda_m\left(1 - c - cz\bar{m}_{\hat{\mathbf{R}}}(z)\right) - z}. \tag{4.19}$$

Observe that, when the quotient $c = M/N$ goes to zero (high number of samples in terms of the number of sensors), this equation reduces to $\bar{m}_{\hat{\mathbf{R}}}(z) = m_{\mathbf{R}}(z)$, meaning that the asymptotic distribution of the eigenvalues of $\hat{\mathbf{R}}$ is the same as the asymptotic distribution of eigenvalues of the true correlation matrix $\mathbf{R}$. For positive values of c, however, we see that the two distributions are different. In order to illustrate this point, we consider a particular example in which the eigenvalue distribution of $\mathbf{R}$ converges to a density that places mass $\{0.2, 0.25, 0.25, 0.3\}$ to the eigenvalues $\{1, 2, 3, 7\}$ respectively. Figure 4.2 represents the asymptotic distribution function

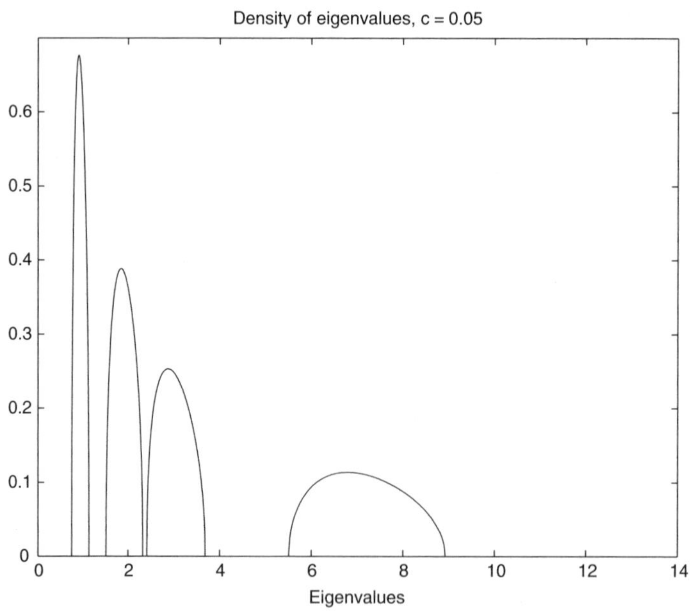

Figure 4.2 Asymptotic distribution of the eigenvalues of $\hat{\mathbf{R}}$ when $\mathbf{R}$ has four different eigenvalues $\{1, 2, 3, 7\}$ with relative multiplicity $\{0.2, 0.25, 0.25, 0.3\}$ respectively, and $c = M/N = 0.05$.

of the eigenvalues of $\hat{\mathbf{R}}$ when the quotient between the observation dimension and the number of samples was $c = 0.05$. The density was obtained from (4.19) using the inversion formula in (4.18). Observe that the eigenvalues tend to cluster around the position of the original eigenvalues of $\mathbf{R}$. On the other hand, it can be seen that the width of these clusters decreases when c goes to zero (high number of samples relative to the number of sensors) so that when $c \to 0$ the density will tend to the density of eigenvalues of $\mathbf{R}$, that is, four Dirac deltas centered at each of the four different eigenvalues. Noting that the eigenvalues of $\hat{\mathbf{R}}$ are the singular values of $\mathbf{R}\frac{1}{N}\mathbf{X}\mathbf{X}^H$, it is tempting to identify the form of the asymptotic density of the eigenvalues of $\hat{\mathbf{R}}$ with the result of convolving the asymptotic eigenvalue density of $\mathbf{R}$ and the asymptotic eigenvalue density of $\frac{1}{N}\mathbf{X}\mathbf{X}^H$ (the Marchenko–Pastur law). In fact, this is not very far from the truth, although the operation between densities is not a classical convolution, but a free multiplicative convolution [52, 53].

It turns out that a high number of random matrix models have an asymptotic eigenvalue distribution, and we now have very powerful theorems that describe the asymptotic behavior of fairly complex matrix structures (see [54] for an extensive compilation of results). It has also been noted that these limiting theorems are not restricted to transformations of the empirical distribution of eigenvalues, and more powerful results can be obtained for generic spectral functions depending on both the eigenvalues and the eigenvectors of random matrices. For instance, in [55] the following spectral functions are considered

$$H_{\hat{\mathbf{R}}}(x) = \sum_{m=1}^{M} \mathbf{a}^H \hat{\mathbf{e}}_m \hat{\mathbf{e}}_m^H \mathbf{b} \mathcal{I}(\hat{\lambda}_m \leq x) \tag{4.20}$$

where $\mathbf{a}$ and $\mathbf{b}$ are two complex $M \times 1$ column vectors and $\hat{\lambda}_m$, $\hat{\mathbf{e}}_m$ are the eigenvalues and eigenvectors of the sample correlation matrix. The Stieltjes transform associated with that spectral function is given by

$$t_{\hat{\mathbf{R}}}(z) = \int \frac{1}{x - z} dH_{\hat{\mathbf{R}}}(x) = \sum_{m=1}^{M} \frac{\mathbf{a}^H \hat{\mathbf{e}}_m \hat{\mathbf{e}}_m^H \mathbf{b}}{\hat{\lambda}_m - z} = \mathbf{a}^H (\hat{\mathbf{R}} - z\mathbf{I}_M)^{-1} \mathbf{b}. \tag{4.21}$$

Under some regularity conditions, this type of Stieltjes transform also converges to a nonrandom limit [51].

4.2.2 Asymptotic Output SINR of the SMI Technique with Diagonal Loading

In this section, we will use random matrix theory to describe the asymptotic behavior of the diagonally loaded SMI beamformer when both the number of snapshots (N) and the number of sensors (M) increase without bound at the same rate. We stress again that, since the ratio between these two quantities remains constant $M/N = c$, the results will be more representative of the finite sample size effect than those obtained under the assumption of large N with bounded M.

The following mathematical assumptions are rather technical, but are needed in order to ensure the convergence of the output SINR when both the number of sensors and the number of observations tend to infinity at the same rate. The first two assumptions describe some statistical properties of the observations that are needed to guarantee convergence. The third one is a structural mathematical condition that ensures that, when the number of sensors goes to infinity, the output SINR given by the diagonally loaded SMI beamformer converges to a fixed quantity.

(**As1**) The transmitted signal $s(n)$, $n \in \mathbb{N}$, is a sequence of independent and identically distributed (i.i.d.) complex random variables with independent real and imaginary parts that have zero mean, variance $P_d/2$ and bounded eighth order moments.

(**As2**) The noise plus interference vector $\mathbf{n}(n)$, $n \in \mathbb{N}$, is a sequence of i.i.d. complex random vectors independent of $s(n)$ for all n. The real and imaginary parts of the components of $\mathbf{n}(n)$ are independent and have zero mean, strictly positive definite covariance matrix $\mathbf{R}_N/2$ and bounded eighth order moments.

(**As3**) The eigenvalues of $\mathbf{R}$ are uniformly bounded from below and above for all M and have a limiting distribution as $M \to \infty$. Moreover,

$$\limsup_{M\to\infty} \mathbf{s}_d^H(\mathbf{R} - z\mathbf{I}_M)^{-1}\mathbf{s}_d = \liminf_{M\to\infty} \mathbf{s}_d^H(\mathbf{R} - z\mathbf{I}_M)^{-1}\mathbf{s}_d$$

for any $z \in \mathbb{C}$ outside the support of the limiting eigenvalue distribution of $\mathbf{R}$.

Some additional considerations must be made with respect to the limit when the number of sensors or antennas grows without bound. It is true that several standard array processing assumptions are violated by letting the number of antennas increase to infinity (see [56] for a more detailed discussion). Here we are only interested in the asymptotic result as an approximation of the nonasymptotic reality, and hence we will not be concerned about the actual validity of the signal model when $M \to \infty$.

The following result establishes the asymptotic behavior of the output SINR obtained with an array loading factor α, denoted from now on by SINR (we omit the argument for notational simplicity).

Proposition 1. Under (**As1–As3**) and assuming $M/N = c$ so that $0 < c < \infty$, $\alpha > 0$, we will have, for any $\varepsilon > 0$,

$$\lim_{M,N\to\infty} \text{Prob}\big(|\text{SINR} - \overline{\text{SINR}}| > \varepsilon\big) = 0$$

where $\overline{\text{SINR}} = (q(\alpha)/P_d - \beta)^{-1}$,

$$q(\alpha) = \frac{1}{1 - c\xi} \frac{\mathbf{s}_d^H(\mathbf{R} + \gamma\mathbf{I}_M)^{-1}\mathbf{R}(\mathbf{R} + \gamma\mathbf{I}_M)^{-1}\mathbf{s}_d}{(\mathbf{s}_d^H(\mathbf{R} + \gamma\mathbf{I}_M)^{-1}\mathbf{s}_d)^2}, \quad \xi = \frac{1}{M}\sum_{i=1}^{M}\left(\frac{\lambda_i}{\lambda_i + \gamma}\right)^2 \tag{4.22}$$

and $\gamma = \alpha[1 + cb]$ with b being the unique positive solution to the following equation

$$b = \frac{1}{M}\sum_{i=1}^{M} \frac{\lambda_i(1 + cb)}{\lambda_i + \alpha(1 + cb)} \tag{4.23}$$

and $\lambda_{\max} = \lambda_1 \geq \lambda_2 \ldots \geq \lambda_M = \lambda_{\min}$ denoting the eigenvalues of $\mathbf{R}$.

Proof. The asymptotic expression of the output SINR can be obtained by observing that the original quantity can be expressed in terms of a Stieltjes transform of the type defined in (4.21) and its first order derivative. Using a convergence theorem for that type of transform, we get to the result of the proposition. See Appendix 4.A for further details. □

This proposition is pointing out that, in order to analyze the asymptotic output SINR of a diagonally loaded MVDR beamformer, one can alternatively analyze the behavior of the deterministic quantity in (4.22). Because of its deterministic nature, its asymptotic behavior is much easier to characterize than the original output SINR.

It must be stressed that convergence in Proposition 1 holds regardless of the actual structure of the spatial correlation matrix $\mathbf{R}$ [provided that the requirements of (**As3**) are met]. Hence, if the useful signal is not present in the observations, $\mathbf{R}$ must be replaced with $\mathbf{R}_N$, whereas if the useful signal is present, $\mathbf{R}$ has to be fixed to $\mathbf{R} = \mathbf{R}_N + P_d\mathbf{s}_d\mathbf{s}_d^H$. The formulation can also accommodate the situation where the true spatial signature of the useful signal, denoted by $\bar{\mathbf{s}}_d$, is different from the one assumed at the receiver, $\mathbf{s}_d$. In this case, the spatial correlation matrix in Proposition 1 must be replaced with $\mathbf{R} = \mathbf{R}_N + P_d\bar{\mathbf{s}}_d\bar{\mathbf{s}}_d^H$. Also, the useful signal component of the observation does not need to have rank 1, so we can consider the case where the observation presents a spatial correlation matrix of the form $\mathbf{R} = \mathbf{R}_N + \mathbf{R}_d$ with $\mathbf{R}_d$ the spatial correlation of the signal of interest, which in general can have rank higher than 1 (this situation is further explored in Section 4.4).

Assuming perfect knowledge of the correlation matrix $\mathbf{R}$, $q(\alpha)$ in (4.22) is an objective function to minimize in terms of the loading factor α. One must stress, though, that depending on the actual correlation matrix $\mathbf{R}$, the cost function $q(\alpha)$ might not be convex in α and might even present multiple local minima. On the other hand, choosing α to minimize (4.22) does not guarantee a maximum instantaneous SINR; it only ensures an optimum asymptotic choice when both the number of sensors and the number of snapshots go to infinity at the same rate. The remarkable property of this asymptotic approximation is the fact that, as in the nonasymptotic situation, the number of sensors and snapshots have the same order of magnitude. This guarantees a very good approximation, even for moderate values of M and N. Let us now investigate the existence of the minimum of $q(\alpha)$.

REMARK 1. The function $q(\alpha)$ is well defined and positive for all $\alpha \geq 0$. Assuming that $0 < c < \infty$ and that $\mathbf{s}_d$ is not an eigenvector of $\mathbf{R}_N$, the infimum of $q(\alpha)$ over all $\alpha \geq 0$ is always attained at a finite α. If, in addition, $c < 1$, then the minimum is obtained with a strictly positive loading factor α.

Proof. The function $q(\alpha)$ is always well defined and positive, because all the quantities are positive and $c\xi < 1$. To see that, note that

$$c\xi = \frac{1}{M}\sum_{i=1}^{M}\frac{c\lambda_i^2}{\lambda_i^2 + \alpha^2(1+cb)^2 + 2\lambda_i\alpha(1+cb)}$$
$$< \frac{1}{M}\sum_{i=1}^{M}\frac{c\lambda_i}{\lambda_i + \alpha(1+cb)} = \frac{cb}{1+cb} < 1$$

where the last equality follows from (4.23). On the other hand, given the continuity and positivity of $q(\alpha)$ as a function of α, the infimum of this function will be attained at α equal to one of the following three alternatives: (1) $0 < \alpha < \infty$, (2) $\alpha = 0$, or (3) $\alpha \to +\infty$. We first show that, whenever $0 < c < \infty$, the third possibility is not valid. Indeed, assume that the optimum loading is infinitely large. Note first that, since $\|\mathbf{s}_d\| = 1$,

$$\lim_{\alpha\to\infty} q(\alpha) = \mathbf{s}_d^H\mathbf{R}\mathbf{s}_d.$$

Now, if the infimum of $q(\alpha)$ is attained with an infinitely high α, one must have

$$\lim_{\alpha\to\infty} \alpha[q(\alpha) - \mathbf{s}_d^H\mathbf{R}\mathbf{s}_d] = 2\left[\left(\mathbf{s}_d^H\mathbf{R}\mathbf{s}_d\right)^2 - \mathbf{s}_d^H\mathbf{R}^2\mathbf{s}_d\right] \geq 0. \tag{4.24}$$

On the other hand, the Cauchy–Schwarz inequality applied to the vectors $\mathbf{R}\mathbf{s}_d$ and $\mathbf{s}_d$ shows that the above inequality can never hold in the strict sense, and the equality can only occur if $\mathbf{s}_d$ is an eigenvector of $\mathbf{R}_N$. With this, we can conclude that the only possibility is that the infimum of $q(\alpha)$ is achieved with a finite $\alpha \geq 0$. It only remains to show that if $c < 1$ the minimum is achieved at a strictly positive α. Indeed, consider the case where the optimum loading is $\alpha = 0$. In this situation, the derivative of $q(\alpha)$ must be positive or zero in a sufficiently small neighborhood of $\alpha = 0$, that is,

$$\left.\frac{dq(\alpha)}{d\alpha}\right|_{\alpha=0} = \frac{-2c}{(1-c)^3}\frac{1}{\mathbf{s}_d^H\mathbf{R}^{-1}\mathbf{s}_d}\frac{1}{M}\sum_{i=1}^{M}\frac{1}{\lambda_i} \geq 0$$

and we see that $c = 0$ is the only possible option, contradicting the assumption that $c > 0$. □

Remark 1 is pointing out that the minimum of $q(\alpha)$ is always obtained with a finite loading factor. This, of course, assuming that $\mathbf{s}_d$ is not an eigenvector of $\mathbf{R}_N$, because otherwise the optimum weight vector would be proportional to $\mathbf{s}_d$ and an infinite loading factor would be the best strategy. This means that, in a practical situation, there is always a finite constant α that, when added to the diagonal of the sample correlation matrix, maximizes the output SINR (asymptotically when $M, N \to \infty$). If, in addition, the number of samples is higher than the number of sensors, we can ensure that this quantity will never be zero. In any case, note that the optimum loading factor will generally be different depending on whether the useful signal is present or not in the observations (the expression of $q(\alpha)$ is the same for both situations, but the inner structure of $\mathbf{R}$ is different).

Observe also that in practice the true value of the spatial correlation matrix $\mathbf{R}$ is unknown, and hence so is the objective function $q(\alpha)$. The traditional way of overcoming that problem would be to estimate the objective function by replacing the true correlation matrix $\mathbf{R}$ with its sample estimate, $\hat{\mathbf{R}}$. In practice, however, this gives very poor results due to the bad behavior of the direct sample estimates when M and N have the same order of magnitude. In Section 4.3 we will propose a better solution for the estimation of $q(\alpha)$. But, before that, let us investigate some of the insights that the new asymptotic formulation has to offer.

4.2.3 Relationship with Known Results

In order to justify the relevance of the double asymptotic limit, we first analyze the behavior of the asymptotic output SINR in (4.22) in some simplified scenarios. We will see that the results presented in Section 4.1 can also be obtained from the new asymptotic formulation.

4.2.3.1 Abscence of Loading ($\alpha = 0$). In the absence of diagonal loading ($\alpha = 0$), and assuming that $c \leq 1$ (more snapshots than sensors), the quantity $\overline{\text{SINR}}$ turns out to be equal to

$$\overline{\text{SINR}}|_{\alpha=0} = \left[\frac{1/(1-c)}{P_d \mathbf{s}_d^H \mathbf{R}^{-1} \mathbf{s}_d} - \beta \right]^{-1} = \frac{(1-c)\text{SINR}_{\text{opt}}}{1 + \beta c \text{SINR}_{\text{opt}}} \tag{4.25}$$

where SINR_{opt} is defined in (4.1). If $c > 1$, $\overline{\text{SINR}}|_{\alpha=0}$ is given by the same expression as in Proposition 1, with γ replaced by the positive solution of the following equation

$$\frac{1}{c} = \frac{1}{M} \sum_{i=1}^{M} \frac{\lambda_i}{\lambda_i + \gamma}. \tag{4.26}$$

Let us concentrate on the case $c \leq 1$ (number of sensors lower than or equal to the number of snapshots). The expression of $\overline{\text{SINR}}$ without diagonal loading for the case $\beta = 1$ in (4.25) can also be obtained from the expected value of the output SINR given in (4.7). Taking limits as $M, N \to \infty$ at the same rate and using Stirling's asymptotic expressions of the Gamma function, we obtain

$$\lim_{M,N \to \infty} \mathbb{E}[\text{SINR}|_{\alpha=0}] = \sum_{k=0}^{\infty} \left(\frac{(1-c)\text{SINR}_{\text{opt}}}{1 + \text{SINR}_{\text{opt}}} \right)^{k+1} = \frac{(1-c)\text{SINR}_{\text{opt}}}{1 + c\text{SINR}_{\text{opt}}}$$

which is exactly the asymptotic SINR given in (4.25). This expression is very similar to the nonasymptotic approximations given in [36, pp. 735–738].

Figure 4.3 represents the quantity $\overline{\text{SINR}}|_{\alpha=0}$ as a function of the number of samples per sensor ($N/M = 1/c$) and SINR_{opt}. In the upper plot we observe that, as expected, the SMI technique cannot attain the optimum output SINR when the number of snapshots and the number of antennas have the same order of magnitude ($c \neq 0$). If the number of snapshots is low with respect to the number of antennas ($c \to 1$), the

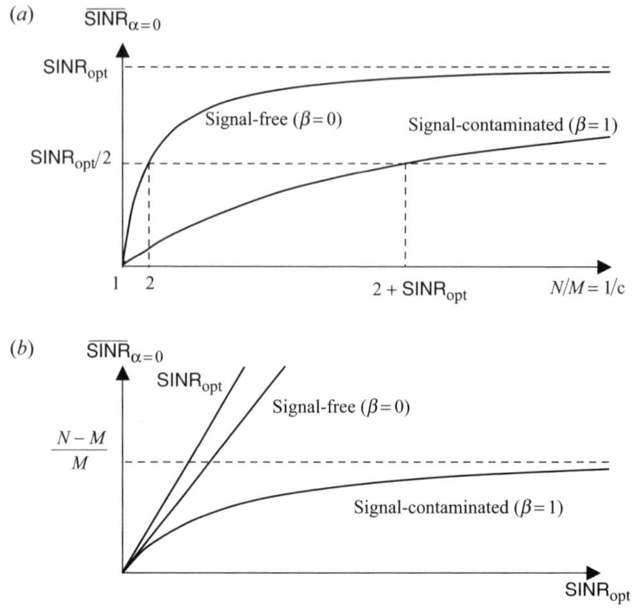

Figure 4.3 Output SINR of the SMI technique without diagonal loading for the case $c < 1$ as a function of M/N in (a) and $\mathrm{SINR}_{\mathrm{opt}}$ in (b).

asymptotic SINR tends to zero. Clearly, a simple phased array (obtained from the original solution with a large loading factor, $\alpha \to \infty$) would perform much better under these circumstances.[5] Only when the sample size increases with respect to the number of antennas ($c \to 0$) does the asymptotic output SINR approach the optimum one. Observing the form of (4.25), we can give an asymptotic approximation of the number of samples that the unloaded SMI algorithm needs to converge to an output SINR that is 3 dB within the optimum one. This can be obtained by forcing $\overline{\mathrm{SINR}}|_{\alpha=0} \geq \mathrm{SINR}_{\mathrm{opt}}/2$, resulting in the following rule of thumb:

$$N \geq (2 + \beta \mathrm{SINR}_{\mathrm{opt}})M.$$

This result was already presented in (4.6) and (4.8), for the signal-free ($\beta = 0$) and signal-contaminated ($\beta = 1$) cases respectively, in the nonasymptotic scenario.

In the lower plot of Figure 4.3 we observe that, when $\mathrm{SINR}_{\mathrm{opt}}$ tends to infinity, the asymptotic output SINR of the SMI technique has a different behavior depending on whether the useful signal is present or not in the observations. In the signal-free case, $\overline{\mathrm{SINR}}$ scales up linearly with $\mathrm{SINR}_{\mathrm{opt}}$. This is in stark contrast with the case in which the useful signal is contaminating the observations, where $\overline{\mathrm{SINR}}$ saturates to a constant value that cannot be surpassed.

[5]The asymptotic output SINR of a phased array is $P_d/\mathbf{s}_d^H \mathbf{R}_N \mathbf{s}_d > 0$ (this could be obtained taking $\alpha \to \infty$ in the original asymptotic expression and noting that we have normalized $\|\mathbf{s}_d\|^2 = 1$).

4.2.3.2 *High Directional Signal Power.* Next, we particularize the expression given in Proposition 1 to a particular scenario in which the spatial correlation of the observations can be decomposed into a directional part (useful and interfering sources) plus a diagonal matrix (background white noise). Note that the scenario with spatially white interference is not interesting, because the phased array is the optimum solution and therefore $\alpha = \infty$ will always be the best option.

Let K denote the dimension of the subspace associated with the signal plus interference directional components, so that $K \leq M$. We assume that the number of samples is higher than the signal plus interference subspace, that is, $N > K$. In this situation, the minimum eigenvalue of the spatial correlation matrix is equal to the noise power, denoted by σ^2, and has multiplicity $M - K$. Since the expression given in Proposition 1 is asymptotically valid with the number of antennas and the sample size, we will implicitly assume that the interference subspace dimension K also grows to infinity at the same rate. This way, K and M have the same magnitude as in a nonasymptotic situation. In this subsection, we analyze the case where the signal plus interference eigenvalues are much higher than σ^2, meaning that the directional sources are much stronger than the background noise.

Now, if we let the signal eigenvalues of the correlation matrix grow without bound, the expression of $\overline{\text{SINR}}$ in Proposition 1 simplifies to:

$$\lim_{\lambda_K \to \infty} \overline{\text{SINR}} = \frac{\vartheta \text{SINR}_{\text{opt}}}{1 + \beta(1 - \vartheta)\text{SINR}_{\text{opt}}}$$

where we have defined

$$\vartheta = 1 - c\frac{K}{M} - c\left(1 - \frac{K}{M}\right)\left(\frac{1}{1 + \varkappa}\right)^2$$

and

$$\varkappa = \frac{\alpha/\sigma^2 + c - 1 + \sqrt{(\alpha/\sigma^2 + c - 1)^2 + 4\alpha \Big/ \sigma^2\left(1 - c\frac{K}{M}\right)}}{2\left(1 - c\frac{K}{M}\right)}.$$

Now, if the factor α/σ^2 is chosen to be sufficiently large, the value ϑ can be approximated by $\vartheta \approx 1 - c\frac{K}{M}$, so that

$$\lim_{\lambda_K \to \infty} \overline{\text{SINR}} \approx \frac{\left(1 - c\frac{K}{M}\right)\text{SINR}_{\text{opt}}}{1 + \beta c\frac{K}{M}\text{SINR}_{\text{opt}}}.$$

Note that this is exactly the same expression that we had in (4.25), replacing M/N with K/N. In particular, we see that an output SINR equal to $\text{SINR}_{\text{opt}}/2$ is achieved with $N \geq (2 + \text{SINR}_{\text{opt}})K$ snapshots when the desired signal is present, and $N \geq 2K$ snapshots when the signal is absent. This is indicating that, in the signal-free case, the number of samples needed to achieve convergence must be, at least, twice the dimension of the signal subspace. As explained in Section 4.1.3, this was the first motivation for using diagonal loading.

Let us go back to the general case, and assume that the loading factor is fixed to zero. In this situation, the asymptotic output SINR will tend to

$$\lim_{\lambda_K \to \infty} \overline{\mathrm{SINR}}|_{\alpha=0} = \begin{cases} \dfrac{(1-c)\mathrm{SINR}_{\mathrm{opt}}}{1+\beta c \mathrm{SINR}_{\mathrm{opt}}} & \text{if } c \leq 1 \\ \dfrac{\dfrac{c-1}{c}\left(1-c\dfrac{K}{M}\right)\mathrm{SINR}_{\mathrm{opt}}}{1-\dfrac{K}{M}+\beta\left(1-\dfrac{K}{M}-\dfrac{c-1}{c}\left(1-c\dfrac{K}{M}\right)\right)\mathrm{SINR}_{\mathrm{opt}}} & \text{if } c > 1. \end{cases}$$

Figure 4.4 represents the output $\overline{\mathrm{SINR}}$ as a function of the number of samples per sensor ($N/M = 1/c$); we do not show the evolution with $\mathrm{SINR}_{\mathrm{opt}}$ because it essentially has the behavior described in Figure 4.3(*b*). Now, observe first that the curve for the case $c \leq 1$ is the exactly the one represented in Figure 4.3(*a*). The case $c > 1$, on the other hand, corresponds to a Hung–Turner projection approach as described in Section 4.1.3. For the signal-free scenario ($\beta = 0$), we have

$$\lim_{\lambda_K \to \infty} \overline{\mathrm{SINR}}\Big|_{\substack{\alpha=0\\ \beta=0}} = \frac{(M-N)(N-K)}{(M-K)N}\mathrm{SINR}_{\mathrm{opt}}$$

which confirms the expression derived in (4.9) under the same approximations for the nonasymptotic situation. Note, in particular, that the best performance is obtained with approximately $N_{opt} = \sqrt{KM}$ samples. The signal-contaminated case ($\beta = 1$) is a bit more involved, but it offers interesting insights if we assume that

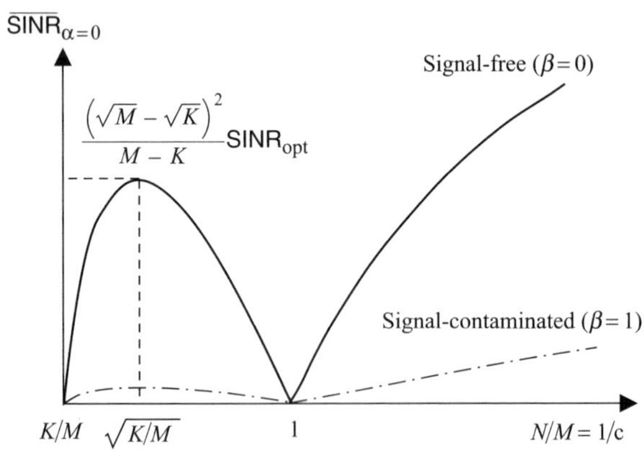

Figure 4.4 Output SINR of the SMI technique without diagonal loading (extended with the Hung–Turner projection to the case $c > 1$) as a function of the number of samples per sensor N/M. Note that the meaning of K is different depending on whether the useful signal is present or not in the observations. In the signal-free case, K is the dimension of the interference subspace. In the signal-contaminated case, K is the dimension of the signal plus interference subspace.

SINR_{opt} is sufficiently high. Indeed, in that case, $\overline{\text{SINR}}$ saturates to a constant value given by

$$\lim_{\text{SINR}_{\text{opt}}\to\infty} \lim_{\lambda_K\to\infty} \overline{\text{SINR}}\Big|_{\substack{\alpha=0\\ \beta=1}} = \frac{(M-N)(N-K)}{(M-K)N-(M-N)(N-K)}.$$

Again, the optimum number of samples in order to maximize the performance is approximately $N_{opt} = \sqrt{KM}$.

4.3 ESTIMATING THE ASYMPTOTICALLY OPTIMUM LOADING FACTOR

In this section, we propose an estimator for the asymptotically optimum loading factor that provides an excellent behavior in scenarios with limited sample size availability. The derivations are based on random matrix theory and the statistical analysis of observations with increasing dimensions, also referred to as general statistical analysis or G-analysis, introduced and developed by V. L. Girko during the 1990s (see [57, 58]). This new statistical analysis builds on random matrix theory and provides a general framework for deriving estimators that are consistent even when the number of estimated parameters increases at the same rate as the number of observations. Hence, the estimators derived from G-analysis are especially useful in situations where the number of observations has the same order of magnitude as the observation dimension. Note that this is exactly the type of situation that we are considering here, because the diagonal loading techniques are especially useful in situations where ratio between the number of observations and the number of elements of the array is not very high.

4.3.1 Introduction to General Statistical Analysis (G-Estimation)

We start by recovering the definition of the two spectral functions in (4.15) and (4.20), but now applied to the true spatial correlation matrix $\mathbf{R}$, namely

$$F_{\mathbf{R}}(x) = \frac{1}{M}\sum_{m=1}^{M} \mathcal{I}(\lambda_m \leq x), \quad H_{\mathbf{R}}(x) = \sum_{m=1}^{M} \mathbf{a}^H \mathbf{e}_m \mathbf{e}_m^H \mathbf{b} \mathcal{I}(\lambda_m \leq x) \tag{4.27}$$

where, again, $\mathbf{a}$ and $\mathbf{b}$ are two complex $M \times 1$ column vectors and $\lambda_m, \mathbf{e}_m$ are the eigenvalues and eigenvectors of $\mathbf{R}$. Instead of considering the usual definition of the Stieltjes transform of these spectral functions, we introduce here a more useful definition that is sometimes referred to as the *real-valued Stieltjes transform*. For a generic spectral function $F(\lambda)$, this transform is defined as

$$\mathcal{M}(x) = \int \frac{1}{1+\lambda x} dF(\lambda), \quad x \in \mathbb{R}, \ x \geq 0.$$

Observe that this definition and the classical one in (4.16) restricted to the negative real axis are completely equivalent, and we can obtain one from the other by a simple change of variable. The real-valued Stieltjes transforms associated with the densities in (4.27) will be denoted as

$$\mathcal{M}_{\mathbf{R}}(x) = \frac{1}{M}\mathrm{tr}[(\mathbf{I}_M + x\mathbf{R})^{-1}], \quad \mathcal{T}_{\mathbf{R}}(x) = \mathbf{a}^H(\mathbf{I}_M + x\mathbf{R})^{-1}\mathbf{b} \tag{4.28}$$

where we use calligraphic fonts to distinguish them from the usual complex Stieltjes transforms.

Now, there are a lot of quantities in communications and signal processing applications that can be expressed in terms of Stieltjes transforms of the type in (4.28). For example, the (i, j)th entry of the inverse correlation matrix $\mathbf{R}^{-1}$ can be expressed as

$$\{\mathbf{R}^{-1}\}_{i,j} = \lim_{x\to\infty} x\mathbf{u}_i^H(\mathbf{I}_M + x\mathbf{R})^{-1}\mathbf{u}_j \tag{4.29}$$

where $\mathbf{u}_i$ is an all-zeros $M \times 1$ column vector with a 1 in the ith position. On the other hand, quadratic forms such as $\mathbf{a}^H\mathbf{R}^k\mathbf{b}$, with $k \in \mathbb{Z}$ arise quite naturally in spectral estimation applications. These quantities can also be expressed in terms of the Stieltjes transform $\mathcal{T}_{\mathbf{R}}(x)$. Indeed, if $k > 0$,

$$\mathbf{a}^H\mathbf{R}^k\mathbf{b} = \frac{(-1)^k}{k!}\left[\frac{d^k}{dx^k}\mathbf{a}(\mathbf{I}_M + x\mathbf{R})^{-1}\mathbf{b}\right]_{x=0}.$$

Other relationships can be found for successive powers of the inverse correlation matrix (i.e., $k < 0$). Now, in order to obtain estimators of these quantities that are consistent when both M and N increase without bound, we only need to find a (uniformly) consistent estimator of the quantities $\mathcal{M}_{\mathbf{R}}(x)$ and $\mathcal{T}_{\mathbf{R}}(x)$. Under certain regularity conditions, the corresponding transformations of these estimators will also be consistent.

In this context, the main objective of general statistical analysis is to find estimators of Stieltjes transforms $\mathcal{M}_{\mathbf{R}}(x)$, $\mathcal{T}_{\mathbf{R}}(x)$, that are consistent when both M and N go to infinity at the same rate. Since we are concentrating on Stieltjes transforms of the correlation matrix $\mathbf{R}$, the problem is equivalent to finding a transformation of the sample correlation matrix $\hat{\mathbf{R}}$ that tends to the quantities in (4.28) when M, $N \to \infty$. Note that this is exactly the opposite of the problem considered by random matrix theory as presented in Section 4.1.3. Random matrix theory analyzes the Stieltjes transform of the sample correlation matrix $m_{\hat{\mathbf{R}}}(z)$ and finds that it is asymptotically different from the Stieltjes transform of the true correlation, $m_{\mathbf{R}}(z)$, although there is an intrinsic functional relationship between the two asymptotic densities. General statistical analysis departs from the Stieltjes transform of the true correlation matrix $\mathcal{M}_{\mathbf{R}}(x)$, and provides an answer to the following question: How does the Stieltjes transform of the sample correlation matrix $\mathcal{M}_{\hat{\mathbf{R}}}(x)$ need to be modified, so that it converges to $\mathcal{M}_{\mathbf{R}}(x)$? Following the example in Figure 4.2, the problem is equivalent to trying to deconvolve the asymptotic spectrum of $\frac{1}{N}\mathbf{X}\mathbf{X}^H$ from the spectrum of $\mathbf{R}\frac{1}{N}\mathbf{X}\mathbf{X}^H$. Note that this is theoretically possible, because the spectrum of $\frac{1}{N}\mathbf{X}\mathbf{X}^H$ is known beforehand (Marchenko–Pastur law).

In [55, 57–59], V.L. Girko proved that such an estimator exists, and is indeed given by[6]

$$\hat{\mathcal{M}}_{\mathbf{R}}(x) = \frac{1}{M}\sum_{k=1}^{M}\frac{1}{1+\theta(x)\hat{\lambda}_k}, \quad \hat{\mathcal{T}}_{\mathbf{R}}(x) = \frac{1}{M}\sum_{k=1}^{M}\frac{\mathbf{a}^H\hat{\mathbf{e}}_k\hat{\mathbf{e}}_k^H\mathbf{b}}{1+\theta(x)\hat{\lambda}_k} \tag{4.30}$$

where $\theta(x)$ denotes the positive solution to the following equation

$$\theta(x)\left[1 - c + c\frac{1}{M}\mathrm{tr}\left[\left(\mathbf{I}_M + \theta(x)\hat{\mathbf{R}}\right)^{-1}\right]\right] = x, \quad x > 0. \tag{4.31}$$

If $c < 1$, this solution always exists and is unique. If $c \geq 1$, the existence and unicity of the solution is only guaranteed for

$$x < \frac{1}{N}\sum_{m=1}^{N}\frac{1}{\hat{\lambda}_m}. \tag{4.32}$$

From this basic estimator of the real generic Stieltjes transform, one can construct estimators of more complicated quantities that are consistent even when the observation dimension increases to infinity with the sample size. For example, if $c < 1$, the G-estimator of the (i,j)th entry of $\mathbf{R}^{-1}$ can readily be found using the relationship in (4.29), replacing $\mathcal{T}_{\mathbf{R}}(x)$ with its G-estimation $\hat{\mathcal{T}}_{\mathbf{R}}(x)$ in (4.30), which gives

$$\{\hat{\mathbf{R}}^{-1}\}_{i,j} = \lim_{x\to\infty} x\mathbf{u}_i^H(\mathbf{I}_M + \theta(x)\hat{\mathbf{R}})^{-1}\mathbf{u}_j = (1-c)\mathbf{u}_i^H\hat{\mathbf{R}}^{-1}\mathbf{u}_j.$$

Hence, the G-estimator of the inverse of the correlation matrix will be given by $(1-c)\hat{\mathbf{R}}^{-1}$, which is much better that the classical estimator $\hat{\mathbf{R}}^{-1}$ whenever the number of sensors is lower than the number of samples but both quantities have the same order of magnitude.

In general, G-estimators have the following nice properties:

1. They are derived without making assumptions on the actual distribution of the observations (other than zero mean, bounded moments and circular symmetry) and are only based on the inner structure of the random matrix $\hat{\mathbf{R}}$.
2. If the sample size increases and the observation dimension remains constant ($c \to 0$), the estimator reverts to its traditional counterpart. Note that $\theta(x) \to x$ when $c = M/N \to 0$ in (4.31).
3. Under some regularity conditions, $\hat{\mathcal{M}}_{\mathbf{R}}(x)$ and $\hat{\mathcal{T}}_{\mathbf{R}}(x)$ are asymptotically $(M, N \to \infty)$ Gussian-distributed [57].

[6]The estimator $\hat{\mathcal{M}}_{\mathbf{R}}(x)$ is generally referred to as G_2-estimator, whereas $\hat{\mathcal{T}}_{\mathbf{R}}(x)$ is known as the G_{25}-estimator.

4.3.2 Consistent Estimation of the Asymptotically Optimum Loading Factor

The objective of this section is to derive an estimator of $q(\alpha)$ that is consistent when $M, N \to \infty$. The expression of the cost function $q(\alpha)$ has the same dependence on $\mathbf{R}$ regardless of whether the signal is present or not in the observations. This way, we can concentrate on deriving a generic consistent estimator of the minimum of the function $q(\alpha)$ that will ultimately depend on the sample correlation $\hat{\mathbf{R}}$, and the result will be consistent for both signal-free and signal-contaminated situations. Thus, from now on, we do not need to consider the two cases $\beta = 1$ and $\beta = 0$ separately.

A classical way of estimating the loading factor that minimizes $q(\alpha)$ would be to replace the true correlation matrix $\mathbf{R}$ with its sample estimate $\hat{\mathbf{R}}$, so that the corresponding estimator for the optimum loading factor would take the form $\hat{\alpha}_{\text{class}} = \arg\min_\alpha\{\hat{q}_{\text{class}}(\alpha)\}$, where

$$\hat{q}_{\text{class}}(\alpha) = \frac{1}{1 - c\hat{\xi}_{\text{class}}} \frac{\mathbf{s}_d^H(\hat{\mathbf{R}} + \hat{\gamma}_{\text{class}}\mathbf{I}_M)^{-1}\hat{\mathbf{R}}(\hat{\mathbf{R}} + \hat{\gamma}_{\text{class}}\mathbf{I}_M)^{-1}\mathbf{s}_d}{(\mathbf{s}_d^H(\hat{\mathbf{R}} + \hat{\gamma}_{\text{class}}\mathbf{I}_M)^{-1}\mathbf{s}_d)^2} \tag{4.33}$$

$$\hat{\xi}_{\text{class}} = \frac{1}{M}\sum_{i=1}^{M}\left(\frac{\hat{\lambda}_i}{\hat{\lambda}_i + \hat{\gamma}_{\text{class}}}\right)^2$$

and $\hat{\gamma}_{\text{class}} = \alpha[1 + c\hat{b}_{\text{class}}]$ with $\hat{b}_{\text{class}}$ being the positive solution to the following equation

$$\hat{b}_{\text{class}} = \frac{1}{M}\sum_{k=1}^{M}\frac{\hat{\lambda}_i[1 + c\hat{b}_{\text{class}}]}{\hat{\lambda}_i + \alpha[1 + c\hat{b}_{\text{class}}]} \tag{4.34}$$

and $\hat{\lambda}_1, \ldots, \hat{\lambda}_M$ the eigenvalues of the sample correlation matrix $\hat{\mathbf{R}}$. The problem with this approach is the fact that it only yields consistent estimates when the number of observations (N) increases without bound while their dimension (M) stays constant. As explained before, this is because the entries of $\hat{\mathbf{R}}$ are known to be consistent estimators of those of $\mathbf{R}$ only when $N \to \infty$ for constant M. Since the true interest of diagonal loading techniques is the situation where these two quantities have the same order of magnitude, in general the solution obtained with the classical approach will not be satisfactory. Hence, our objective now is to derive an estimator that performs well when N and M have the same order of magnitude. Since performance optimization for any value of N and M is in general difficult, our approximation will be the design of an estimator that is consistent when both N and M increase without bound at the same rate (i.e., N and M both large but with the same order of magnitude). We will use the General Statistical Analysis tools that have been introduced in the last section.

It is easy to see that $q(\alpha)$ can be expressed as an arithmetic combination of different Stieltjes transforms of the type in (4.28) and their derivatives. Replacing these transforms with the G-estimators in (4.30), one can get to the following final expression for the consistent estimation of the asymptotically optimum loading factor α (see Appendix 4.C for details of the derivation)

$$\hat{\alpha} = \arg\min_{\alpha} \{\hat{q}(\alpha)\},$$
$$\hat{q}(\alpha) = \frac{1}{(1 - c\hat{\varphi}(\alpha))^2} \frac{\mathbf{s}_d^H(\alpha\mathbf{I}_M + \hat{\mathbf{R}})^{-1}\hat{\mathbf{R}}(\alpha\mathbf{I}_M + \hat{\mathbf{R}})^{-1}\mathbf{s}_d}{(\mathbf{s}_d^H(\alpha\mathbf{I}_M + \hat{\mathbf{R}})^{-1}\mathbf{s}_d)^2} \tag{4.35}$$

where

$$\hat{\varphi}(\alpha) = \frac{1}{M}\mathrm{tr}\left[\hat{\mathbf{R}}\left(\alpha\mathbf{I}_M + \hat{\mathbf{R}}\right)^{-1}\right] = \frac{1}{M}\sum_{m=1}^{M}\frac{\hat{\lambda}_m}{\alpha + \hat{\lambda}_m}.$$

Hence, in order to obtain an appropriate estimation of the asymptotically optimum loading factor, one must first evaluate the function $\hat{q}(\alpha)$ in (4.35) and search for its global minimum.

REMARK 2. Assume that the entries of $\hat{\mathbf{R}}$ are absolutely continuous random variables and that $0 < M/N < \infty$. Then, the function $\hat{q}(\alpha)$ is well defined and positive for all $\alpha > 0$, and the infimum of $\hat{q}(\alpha)$ over all $\alpha \geq 0$ is attained at a finite α. If, in addition, $c < 1$ (more samples than sensors) then the minimum is obtained with a strictly positive loading factor α. All the assertions hold with probability one.

Proof. The proof is very similar to that of Remark 1. To see that the cost function is well defined, one must see that $c\hat{\varphi}(\alpha) < 1$. Indeed, if $\alpha > 0$,

$$c\hat{\varphi}(\alpha) = \frac{1}{N}\sum_{m=1}^{M}\frac{\hat{\lambda}_m}{\alpha + \hat{\lambda}_m} = \frac{1}{N}\sum_{m=1}^{\min(M,N)}\frac{\hat{\lambda}_m}{\alpha + \hat{\lambda}_m} < \frac{\min(M,N)}{N} \leq 1.$$

The rest of the proof is exactly as in Remark 1, replacing $\mathbf{R}$ with $\hat{\mathbf{R}}$ and noting that the probability that $\mathbf{s}_d$ is an eigenvector of $\hat{\mathbf{R}}$ is zero. □

We now formulate the main result of this section: the estimator of the loading factor obtained by minimizing $\hat{q}(\alpha)$ is consistent.

Proposition 2. Under (**As1**–**As3**), and assuming that $M, N \to \infty$ at the same rate, $\hat{\alpha}$ is a weakly consistent estimator of the asymptotically optimum loading factor.

Proof. See Appendix 4.D. □

In order to find the minimum of $\hat{q}(\alpha)$ in (4.35), one needs to avoid unnecessary matrix inversions every time a new α is explored. A computationally more efficient way of carrying out a search of the global minimum of this cost function can be proposed by using the decomposition of $\hat{\mathbf{R}}$ in terms of eigenvalues and eigenvectors (denoted as $\{\hat{\lambda}_m\}$ and $\{\hat{\mathbf{e}}_m\}$ respectively). Noting that

$$\mathbf{s}_d^H(\alpha\mathbf{I}_M + \hat{\mathbf{R}})^{-1}\hat{\mathbf{R}}(\alpha\mathbf{I}_M + \hat{\mathbf{R}})^{-1}\mathbf{s}_d = \sum_{m=1}^{M} \frac{|\mathbf{s}_d^H\hat{\mathbf{e}}_m|^2\hat{\lambda}_m}{(\alpha + \hat{\lambda}_m)^2}$$

$$\mathbf{s}_d^H(\alpha\mathbf{I}_M + \hat{\mathbf{R}})^{-1}\mathbf{s}_d = \sum_{m=1}^{M} \frac{|\mathbf{s}_d^H\hat{\mathbf{e}}_m|^2}{\alpha + \hat{\lambda}_m}$$

we see that all the matrix inversions are avoided by simply storing up the values of $|\mathbf{s}_d^H\hat{\mathbf{e}}_m|^2$ and $\hat{\lambda}_m$ for $m = 1, \ldots, M$.

It is interesting to analyze the form of the proposed estimator when the number of samples is very low. For instance, if there is only one sample available ($N = 1$), the sample correlation matrix has a dyadic form, $\hat{\mathbf{R}} = \mathbf{y}\mathbf{y}^H$, and consequently

$$\hat{q}(\alpha)|_{N=1} = \|\mathbf{y}\|^2\hat{\eta}\left[\frac{\alpha + \|\mathbf{y}\|^2}{\alpha + \|\mathbf{y}\|^2(1 - \hat{\eta})}\right]^2, \quad \hat{\eta} = \frac{\mathbf{s}_d^H\mathbf{y}\mathbf{y}^H\mathbf{s}_d}{\|\mathbf{y}\|^2}.$$

The estimated cost function for this case is a monotonically increasing function of α, so that the estimator chooses $\hat{\alpha} = 0$ as the estimated loading factor. This means that the beamformer weights are given by

$$\hat{\mathbf{w}} = \left(\mathbf{I}_M - \frac{\mathbf{y}\mathbf{y}^H}{\|\mathbf{y}\|^2}\right)\mathbf{s}_d$$

which corresponds to a Hung–Turner projection approach.

When two samples are available ($N = 2$), $\mathbf{y}(1)$ and $\mathbf{y}(2)$, the estimated cost function $q(\alpha)$ becomes rather involved,

$$\hat{q}(\alpha) = 2\left[\frac{4\alpha^2 + 2(\|\mathbf{y}(1)\|^2 + \|\mathbf{y}(2)\|^2)\alpha + \Delta_1}{4\alpha + (\|\mathbf{y}(1)\|^2 + \|\mathbf{y}(2)\|^2)}\right]^2 \times \frac{4\phi^2\alpha^2 + 4\Delta_2\alpha + (\Delta_2(\|\mathbf{y}(1)\|^2 + \|\mathbf{y}(2)\|^2) - \Delta_1\phi^2)}{[4\alpha^2 + 2(\|\mathbf{y}(1)\|^2 + \|\mathbf{y}(2)\|^2 - \phi^2)\alpha + (\Delta_1 - \Delta_2)]^2}$$

where we have defined

$$\phi_1 = \mathbf{s}_d^H\mathbf{y}(1), \quad \phi_2 = \mathbf{s}_d^H\mathbf{y}(2), \quad \phi^2 = |\phi_1|^2 + |\phi_2|^2$$

$$\Delta_1 = \|\mathbf{y}(1)\|^2\|\mathbf{y}(2)\|^2 - |\mathbf{y}(1)^H\mathbf{y}(2)|^2$$

$$\Delta_2 = |\phi_1|^2\|\mathbf{y}(2)\|^2 + |\phi_2|^2\|\mathbf{y}(1)\|^2 - 2\mathrm{Re}\left[\phi_2^*\phi_1\mathbf{y}(2)^H\mathbf{y}(1)\right].$$

Even for this simple case, numerical methods for searching the global minimum are needed.

Next, we present a numerical evaluation of the performance of the proposed estimator. The reader that is more interested in the theoretical insights of the asymptotic formulation may choose to proceed directly to Section 4.4.

4.3.3 Numerical Analysis of the Convergence Rate

The objective of this section is to study, via simulation, the convergence of the presented estimator of the loading factor. We consider here an scenario with four directional sources received with a power 20 dB above spatially white noise, impinging on a linear array of M equispaced antennas separated half a wavelength apart. We consider the case where the useful signal is contaminating the available observations; similar conclusions can be drawn from the signal-free case. In order to show the convergence towards the asymptotic limit, we fixed $M = M_0 k = 10k$ and $N = N_0 k$ with $N_0 = 5, 15, 30, 60, 120$; and k varying from 1 to 10. The dimension of the signal subspaces also scaled up with the number of antennas and was fixed equal to k for all the sources. Concerning the actual generation of the signature matrices, for the initial signature vectors ($k = 1$) we fixed a direction of arrival of -30° (desired signal), -35°, 20°, 40° (interferences) with respect to the broadside of the array. As for the case $k > 1$, we constructed the signature matrix associated with the jth source as $\mathbf{S}_j = \mathbf{F}_k \otimes \mathbf{s}_j$, where $\mathbf{s}_j$ is the signature matrix for $k = 1$ and $\mathbf{F}_k \in \mathbb{C}^{k\times k}$ is an orthogonal Fourier matrix. This is a mathematically convenient way of modeling the fact that the signal subspace dimension increases with the number of antennas while maintaining the original angular properties between the signals in the scenario. Note that the original signature $\mathbf{s}_j$ can be recovered from $\mathbf{S}_j$ by simply adding all of its columns, and therefore one can see $\mathbf{s}_j$ as belonging to the subspace of the columns of $\mathbf{S}_j$. On the other hand, we also see that $\mathbf{S}_i^H\mathbf{S}_j = \mathbf{I}_k(\mathbf{s}_i^H\mathbf{s}_j)$ so that the orthogonality properties of the original signatures are maintained in the expanded subspaces.

In Figure 4.5 we represent the evolution of different loading factors as k increases. Apart from the values of the estimated and optimal asymptotic loading factors (solid and dashed lines), which are obtained minimizing (4.35) and (4.22) respectively, we also represent the classical estimation obtained by minimizing (4.33) (dash-dotted line) and the optimal nonasymptotic loading factor (dotted line), obtained by maximizing the instantaneous output SINR at each realization. All the results were obtained averaging over 500 realizations. Observe that the proposed estimation is able to give results that are quite close to the asymptotic and nonasymptotic optimum ones even for moderate values of M, N. The classical estimation, which is obtained by minimizing the asymptotic cost function replacing the true correlation matrix with its sample estimate [cf. (4.33)] gives strongly biased estimates of the optimum loading factor. The reason for that behavior is illustrated in Figure 4.6, where we represent the empirically averaged cost functions for the cases (a) $M = 10$, $N = 15$, and (b) $M = 50$, $N = 75$. The nonasymptotic cost function is obtained by a simple transformation of the instantaneous SINR as described in

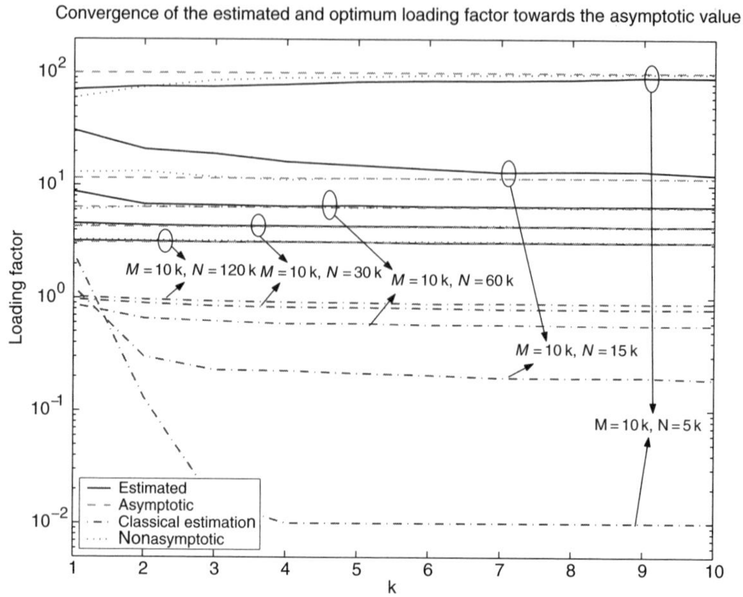

Figure 4.5 Convergence of the estimated and real loading factors towards the asymptotic limit for different values of M and N. We represent the estimated loading factor (solid line), the asymptotically optimum (one dashed line), the loading factor estimated according to the classical approach ($\hat{\alpha}_{\text{class}}$, dash-dotted line), and its real nonasymptotic optimum value (dotted line). The scenario consisted of four directional sources, received with a power equal to 20 dB above the noise floor, and spanning an associated subspace that occupied a tenth of the total dimension M (each one). The useful signal was present in the observations. The noise power was normalized to $\sigma^2 = 1$, and the results were obtained averaging over 500 realizations.

Proposition 1, that is, as

$$P_d(\text{SINR}(\hat{\mathbf{w}})^{-1} + 1)$$

where here $\text{SINR}(\hat{\mathbf{w}})$ is the instantaneous signal to interference plus noise ratio, which can be calculated because the true spatial correlation matrices are known in the simulation.

Note that, as the number of antennas grows large, the estimated and nonasymptotic cost functions tend uniformly to the corresponding asymptotic limit. On the other hand, the classical estimation of the cost function does not converge to the asymptotic one, and the corresponding minimizing loading factors are strongly biased even for high M, N.

4.3.4 Numerical Assessment of the Nonasymptotic Performance

In this section, we try to determine whether the use of the proposed estimation of the loading factor provides much better results than other ad hoc methods in realistic

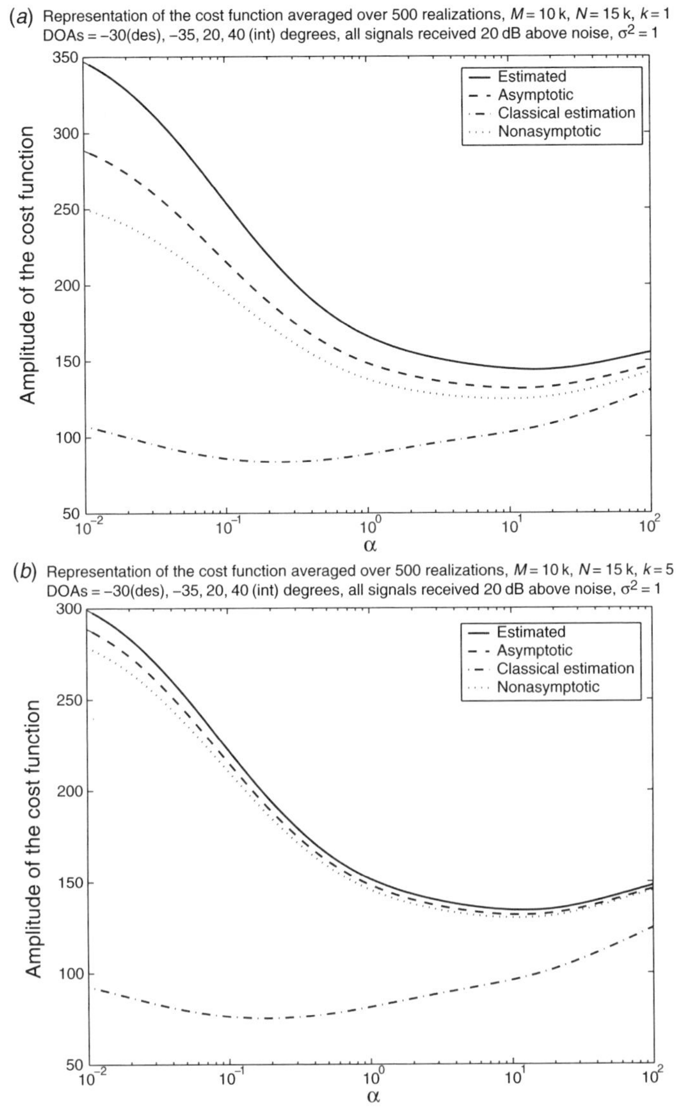

Figure 4.6 Empirical mean of the cost function $\hat{q}(\alpha)$ averaged over 500 realizations, in a scenario with four directional sources, whose received power was received 20 dB above the noise floor, and whose associated subspaces occupied (each one) a tenth of the total dimension (equal to the number of antennas). The noise power was normalized to $\sigma^2 = 1$, and two different situations were simulated: (*a*) $M = 10, N = 15$; and (*b*) $M = 50, N = 75$.

nonasymptotic scenarios. To that effect, we used the same configuration as in the last section, now fixing M, N and the number of sources. To create variability in the scenario, the direction of arrival of each source was generated as an independent random variable uniformly distributed on $[-90^\circ, 90^\circ]$, and each source contributed with a single dimension in the corresponding signal subspace. Figures 4.7 and 4.8 represent the cumulative distribution function of the SINR at the output of the SMI spatial filter for different diagonal loading methods under four different configurations of N and M, taking into account the cases $N > M$ and $N < M$. Method 1 [36, p. 748] fixes the diagonal load as 10 times the smallest eigenvalue of $\hat{\mathbf{R}}$ (if this matrix is singular, we fix the diagonal load to zero); Method 2 [45] sets the diagonal load equal to the standard deviation of the diagonal elements of $\hat{\mathbf{R}}$; finally Method 3 [44] fixes the diagonal load as in (4.13). This last method was simulated assuming a perfect knowledge of

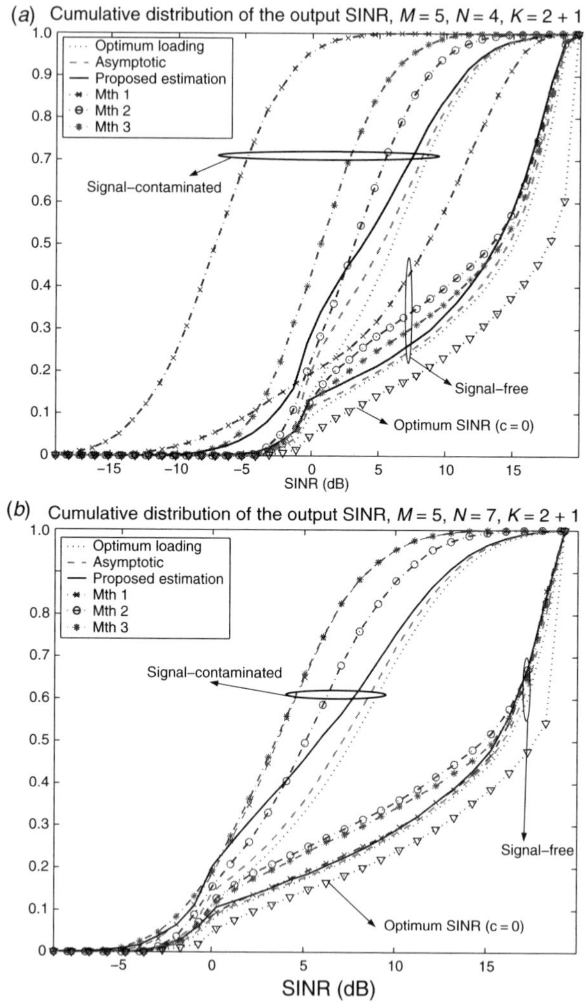

Figure 4.7 Cumulative distribution function of the output SINR of different diagonal loading methods when $M = 5$. The scenario consisted of two interfering sources.

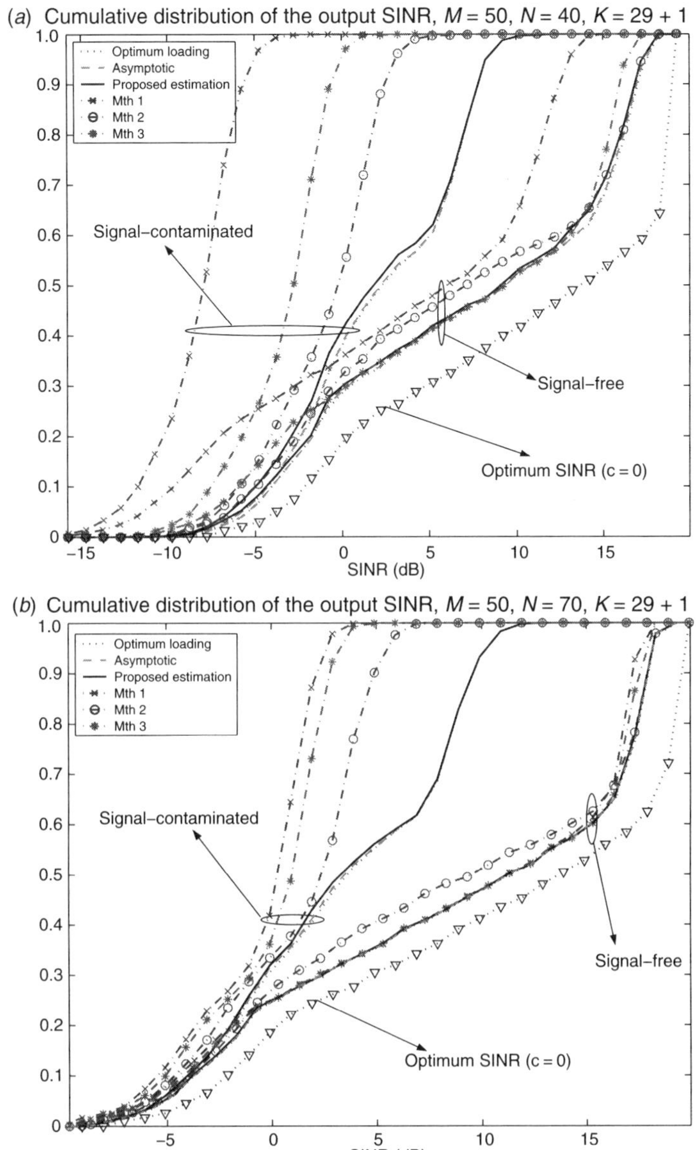

Figure 4.8 Cumulative distribution function of the output SINR of different diagonal loading methods when $M = 50$. The scenario consisted of 29 interfering sources.

the interfering subspace dimension. We compare the performance of the different methods for both the signal-free and the signal-contaminated scenarios. This is somewhat unfair, because Methods 1–3 were originally proposed for the signal-free scenario when the number of samples is higher than the number of sensors. However, we are not aware of the existence of any specific method for fixing the diagonal loading factor when the useful signal is contained in the observations.

Observe that the proposed method gives significant gains in terms of output SINR with respect to Methods 1–3 in the performance region of interest, and note that these gains are especially high in signal-contaminated scenario. For the signal-free case, the method also gives the best performance, which is only approximated by Method 3. Note, however, that this method is based on the perfect knowledge of the interfering subspace dimension, which is in practice unknown. We can therefore conclude that, contrary to previous methods, the proposed estimation of the optimum loading factor gives a good performance for both signal-free and signal-contaminated scenarios. Moreover, and in spite of being an intrinsically asymptotic approach, it has a remarkable performance even when the number of antennas is relatively low.

4.4 CHARACTERIZATION OF THE ASYMPTOTICALLY OPTIMUM LOADING FACTOR

Before drawing this chapter to a close, it seems interesting to study the behavior of the asymptotically optimum loading factor in some simplistic scenarios with only a few spatial sources. The objective is to investigate the potential influence of different parameters, such as the dimension of the interfering subspace, the desired signal power, or the degree of orthogonality between the desired and interfering sources, on the actual value of the optimum α. We will treat the signal-free and the signal-contaminated cases separately. In the first situation, we consider an scenario with only one spatial interfering source received in white noise. For the second case, we also include the influence of the desired signal in the observations.

In our signal model, the true correlation matrix can be decomposed as

$$\mathbf{R} = \beta\mathbf{R}_d + \mathbf{R}_i + \sigma^2\mathbf{I}_M, \tag{4.36}$$

where $\mathbf{R}_d \in \mathbb{C}^{M\times M}$ and $\mathbf{R}_i \in \mathbb{C}^{M\times M}$ are the spatial correlation matrices of the desired and interfering signals, σ^2 is the white noise power and β indicates whether the signal is present or not in the observations. Let us denote by $M_d \leq M$ and $M_i \leq M$ the rank of $\mathbf{R}_d$ and $\mathbf{R}_i$ respectively, and consider the eigenvalue decomposition of these two matrices, namely

$$\mathbf{R}_d = \mathbf{S}_d\boldsymbol{\Phi}_d\mathbf{S}_d^H, \quad \mathbf{S}_d^H\mathbf{S}_d = \mathbf{I}_{M_d}$$
$$\mathbf{R}_i = \mathbf{S}_i\boldsymbol{\Phi}_i\mathbf{S}_i^H, \quad \mathbf{S}_i^H\mathbf{S}_i = \mathbf{I}_{M_i}$$

where $\boldsymbol{\Phi}_d \in \mathbb{C}^{M_d\times M_d}$ and $\boldsymbol{\Phi}_i \in \mathbb{C}^{M_i\times M_i}$ are two diagonal matrices containing the nonzero eigenvalues of $\mathbf{R}_d$ and $\mathbf{R}_i$, respectively. We now make the following simplification: we assume that all the eigenvalues of $\mathbf{R}_d$ and $\mathbf{R}_i$ are concentrated around two values (P_d and P_i, i.e., desired and interfering power), so that one can take $\boldsymbol{\Phi}_d = P_d\mathbf{I}_{M_d}$ and $\boldsymbol{\Phi}_i = P_i\mathbf{I}_{M_i}$. This isotropic coherence property allows us to describe the true correlation matrix as

$$\mathbf{R} = \beta P_d \mathbf{S}_d \mathbf{S}_d^H + P_i \mathbf{S}_i \mathbf{S}_i^H + \sigma^2 \mathbf{I}_M \tag{4.37}$$

with $\mathbf{S}_d \in \mathbb{C}^{M \times M_d}$ and $\mathbf{S}_i \in \mathbb{C}^{M \times M_i}$ respectively, the desired and interfering spatial signature matrices, assumed to have full column rank. Note here that this model is slightly more general than the traditional narrowband case, where $M_d = M_i = 1$. We prefer to maintain this level of generality because this formulation retains the dimensionality of the signal plus interference subspace which, as shown in Section 4.2, plays an important part in the behavior of the diagonally loaded beamformer. We will also assume that $M_i \geq M_d$ and

$$\mathbf{S}_d^H \mathbf{S}_i \mathbf{S}_i^H \mathbf{S}_d = \eta \mathbf{I}_{M_d} \tag{4.38}$$

where η is an orthogonality parameter between the desired and interfering sources. This condition is an isotropic multidimensional generalization of the traditional definition of the scalar product of two one-dimensional spatial signature vectors. Note that the definition of $\mathbf{S}_d$ and $\mathbf{S}_i$ in (4.37) is invariant with respect to right multiplication of orthogonal matrices. This means that we can always choose $\mathbf{S}_d$ and $\mathbf{S}_i$ to ensure that $\mathbf{S}_d^H \mathbf{S}_i \mathbf{S}_i^H \mathbf{S}_d$ is a diagonal matrix (otherwise, one just needs to replace $\mathbf{S}_d$ by $\mathbf{S}_d \mathbf{E}$ with $\mathbf{E}$ the matrix of eigenvectors of $\mathbf{S}_d^H \mathbf{S}_i \mathbf{S}_i^H \mathbf{S}_d$). Thus, in (4.38) we are making the approximation that all the eigenvalues of $\mathbf{S}_d^H \mathbf{S}_i \mathbf{S}_i^H \mathbf{S}_d$ are concentrated around η.

At this point, the reader might argue that the narrowband formulation of MVDR beamformer is a suboptimal solution in a scenario where the desired signal subspace has dimension higher than one. Let us now show that optimality is not lost.

REMARK 3. Under the signal model above, the MVDR beamformer constructed with $\mathbf{s}_d$ fixed to any nonzero combination of the columns of $\mathbf{S}_d$, maximizes the output SINR.

Proof. Indeed, if $M_d \geq 1$, the optimum beamformer in terms of output SINR is known to be the maximum generalized eigenvalue eigenvector of the pencil $(P_d \mathbf{S}_d \mathbf{S}_d^H, \mathbf{R}_N)$, that is, [9]

$$(P_d \mathbf{S}_d \mathbf{S}_d^H) \mathbf{w}_{\text{opt}} = \delta_{\max} \mathbf{R}_N \mathbf{w}_{\text{opt}} \tag{4.39}$$

where $\mathbf{R}_N = \mathbf{R}_i + \sigma^2 \mathbf{I}_M$. Now, defining $\mathbf{u}_{\text{opt}} = \mathbf{R}_N^{1/2} \mathbf{w}_{\text{opt}}$ (where $\mathbf{R}_N^{1/2}$ is the Hermitian positive definite square root of $\mathbf{R}_N$) we see that

$$(P_d \mathbf{R}_N^{-1/2} \mathbf{S}_d \mathbf{S}_d^H \mathbf{R}_N^{-1/2}) \mathbf{u}_{\text{opt}} = \delta_{\max} \mathbf{u}_{\text{opt}} \tag{4.40}$$

so that $\mathbf{u}_{\text{opt}}$ is the maximum eigenvalue eigenvector of $P_d \mathbf{R}_N^{-1/2} \mathbf{S}_d \mathbf{S}_d^H \mathbf{R}_N^{-1/2}$. Now, this matrix has a single nonzero eigenvalue, with multiplicity M_d, and the associated eigenspace is spanned by the columns of $\mathbf{R}_N^{-1/2} \mathbf{S}_d$. Indeed, the matrix in question has rank M_d, and its nonzero eigenvalues are equal to the eigenvalues of $P_d \mathbf{S}_d^H \mathbf{R}_N^{-1} \mathbf{S}_d$. Now, thanks to (4.38), $\mathbf{S}_d^H \mathbf{R}_N^{-1} \mathbf{S}_d$ is proportional to an identity matrix, and consequently all the nonzero eigenvalues of $P_d \mathbf{R}_N^{-1/2} \mathbf{S}_d \mathbf{S}_d^H \mathbf{R}_N^{-1/2}$ are equal. To show that the eigenspace associated with this nonzero eigenvalue is the column space of $\mathbf{R}_N^{-1/2}\mathbf{S}_d$, observe that

$$(P_d \mathbf{R}_N^{-1/2}\mathbf{S}_d\mathbf{S}_d^H\mathbf{R}_N^{-1/2})\mathbf{R}_N^{-1/2}\mathbf{S}_d = \text{SINR}_{\text{opt}}\mathbf{R}_N^{-1/2}\mathbf{S}_d$$

where, in this scenario,

$$\text{SINR}_{\text{opt}} = \frac{P_d}{\sigma^2}\frac{(1-\eta)P_i + \sigma^2}{P_i + \sigma^2}.$$

Thus, any linear combination of the columns of $\mathbf{R}_N^{-1/2}\mathbf{S}_d$ is a maximum eigenvalue eigenvector of (4.40) and, consequently, any linear combination of the columns of $\mathbf{R}_N^{-1}\mathbf{S}_d$ will be a maximum generalized eigenvalue eigenvector of (4.39). Now, if our receiver has access to a linear combination of the columns of $\mathbf{S}_d$ (for instance, $\mathbf{s}_d = \mathbf{S}_d\mathbf{m}$, where $\|\mathbf{m}\|^2 = 1$ to preserve the normalization of $\mathbf{s}_d$), the optimum solution can be implemented as $\mathbf{w}_{\text{opt}} = \mathbf{R}_N^{-1}\mathbf{s}_d$ which, using the matrix inverse lemma, can be shown to be proportional to the MVDR solution $\mathbf{R}^{-1}\mathbf{s}_d$. Hence, we can conclude that the traditional MVDR spatial filter is the optimum solution even in this special situation. □

Next, we particularize the expressions of the asymptotic output SINR for the signal model that has just been presented.

4.4.1 One Spatial Interference in the Signal-Free Case

In this situation, the correlation matrix has two eigenvalues, $\mu_1 = P_i + \sigma^2$ and $\mu_2 = \sigma^2$, with relative multiplicity $\upsilon_i = M_i/M$ and $(1-\upsilon_i)$ respectively, associated with two orthogonal sets of eigenvectors. The output SINR of the diagonally loaded beamformer tends asymptotically to $\overline{\text{SINR}}|_{\beta=0} = P_d/q(\alpha)$, where

$$\begin{aligned}
q(\alpha) &= \frac{P_d}{1-c\xi}\frac{P_d\sigma^2}{(\sigma^2+\gamma)^2}\frac{1-\eta+\eta\left(1+\dfrac{P_i}{\sigma^2}\right)\left(1+\dfrac{P_i}{\sigma^2+\gamma}\right)^{-2}}{(\text{SINR}^*)^2}\\
\xi &= \upsilon_i\left(\frac{P_i+\sigma^2}{P_i+\sigma^2+\gamma}\right)^2 + (1-\upsilon_i)\left(\frac{\sigma^2}{\sigma^2+\gamma}\right)^2\\
\text{SINR}^* &= \frac{P_d}{(\sigma^2+\gamma)}\frac{(1-\eta)P_i + (\sigma^2+\gamma)}{P_i+(\sigma^2+\gamma)}
\end{aligned} \tag{4.41}$$

and $\gamma = \alpha[1+cb]$ with b the unique positive solution to the following cubic polynomial equation:

$$\frac{b}{1+cb} = \frac{\upsilon_i\mu_1}{\mu_1+\alpha[1+cb]} + \frac{(1-\upsilon_i)\mu_2}{\mu_2+\alpha[1+cb]}.$$

4.4.2 One Spatial Interference with the Useful Signal Present

If the desired signal is present in the observations, the spectral content of **R** can be assumed to concentrate around four eigenvalues. This is described in the following table, where we have written $v_d = M_d/M$ and $v_i = M_i/M$,

Eigenvalue	Relative multiplicity
$\mu_1 = \sigma^2 + \frac{P_d + P_i}{2} + \frac{1}{2}\sqrt{(P_d + P_i)^2 - 4P_d P_i(1-\eta)}$	v_d
$\mu_2 = \sigma^2 + P_i$	$v_i - v_d$
$\mu_3 = \sigma^2 + \frac{P_d + P_i}{2} - \frac{1}{2}\sqrt{(P_d + P_i)^2 - 4P_d P_i(1-\eta)}$	v_d
$\mu_4 = \sigma^2$	$1 - (v_d + v_i)$

Paticularizing the expression of $\overline{\text{SINR}}$ in Proposition 1 to this situation, we see that the output SINR of the diagonally loaded beamformer tends asymptotically to $\overline{\text{SINR}}|_{\beta=1} = (q(\alpha)/P_d - 1)^{-1}$, where

$$q(\alpha) = \frac{P_d}{1 - c\xi}\left[1 + \frac{P_d\sigma^2}{(\sigma^2+\gamma)^2}\frac{1 - \eta + \eta\left(1 + \frac{P_i}{\sigma^2}\right)\left(1 + \frac{P_i}{\sigma^2+\gamma}\right)^{-2}}{(\text{SINR}^*)^2}\right]$$

$$\xi = v_d\left(\frac{\mu_1}{\mu_1+\gamma}\right)^2 + (v_i - v_d)\left(\frac{\mu_2}{\mu_2+\gamma}\right)^2 + v_d\left(\frac{\mu_3}{\mu_3+\gamma}\right)^2 + (1-(v_d+v_i))\left(\frac{\mu_4}{\mu_4+\gamma}\right)^2$$

SINR^* as defined in (4.41) and $\gamma = \alpha[1 + cb]$, with b the unique positive solution to the following fifth order polynomial equation:

$$\frac{b}{1+cb} = \frac{v_d\mu_1}{\mu_1 + \alpha[1+cb]} + \frac{(v_i - v_d)\mu_2}{\mu_2 + \alpha[1+cb]} + \frac{v_d\mu_3}{\mu_3 + \alpha[1+cb]} + \frac{(1-(v_d+v_i))\mu_4}{\mu_4 + \alpha[1+cb]}.$$

4.4.3 Influence of the Scenario on the Optimum Loading Factor

Figure 4.9 represents the output SINR for the scenario under consideration as a function of the loading factor (normalized by the noise power) for different values of $c = M/N$. Observe that the loading factor that gives the best performance in this scenario is quite different depending on whether the desired signal is present or not in the observations. If the signal of interest is present, the optimum loading factor is always higher. Note also that the performance loss caused by a wrong choice of the optimum loading factor tends to be higher when $c \approx 1$, that is when the number of observations is approximately equal to the number of sensors.

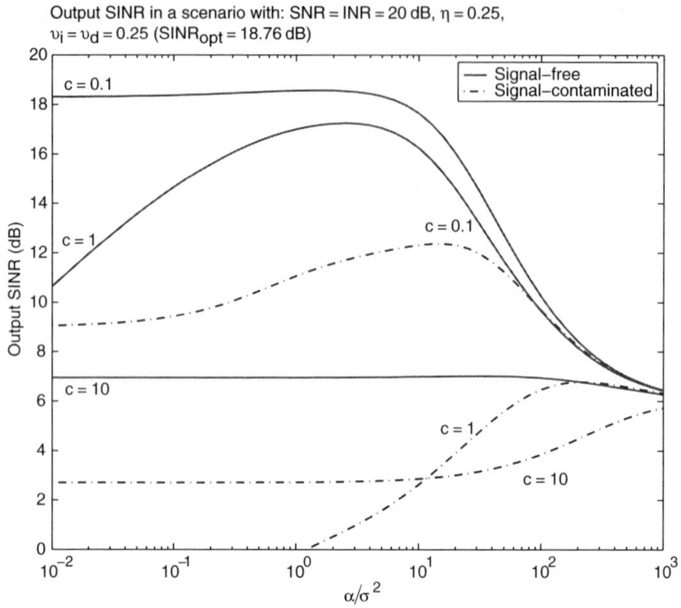

Figure 4.9 Output SINR as a function of the normalized loading factor (α/σ^2) for different values of *c*.

In Figure 4.10 we represent the influence of the orthogonality parameter η on the optimum loading factor (*a*) and the associated output SINR (*b*). Observe first that the optimum diagonal load tends to infinity as either $\eta \to 0$ or $\eta \to 1$. In fact, it can be seen that

$$\lim_{\eta\to 0} \overline{\text{SINR}} = \frac{(1 - c\xi_0)\dfrac{P_d}{\sigma^2}}{1 + \beta\dfrac{P_d}{\sigma^2}c\xi_0}, \qquad \lim_{\eta\to 1} \overline{\text{SINR}} = \frac{(1 - c\xi_1)\dfrac{P_d}{\sigma^2 + P_i}}{1 + \beta\dfrac{P_d}{\sigma^2 + P_i}c\xi_1}$$

where $\xi_0 = \lim_{\eta\to 0}\xi$ and $\xi_1 = \lim_{\eta\to 1}\xi$. Since these two quantities are monotonically decreasing with the loading factor, the optimum performance is obtained, in both cases, with an infinite α. In the first case ($\eta \to 0$), the sources tend to be increasingly orthogonal and consequently the optimum spatial filter becomes a phased array. Obviously, an infinite loading factor will achieve the best performance in this situation. In the second case, when $\eta \to 1$, the two sources become parallel (this is as if the two sources were coming from the same direction of arrival). In this situation, the use of an imperfect version of the sample correlation matrix would generate signal cancellation effects at the output of the beamformer [11]. Clearly, this situation also calls for a high loading factor, which enhances the steering capabilities of the array and avoids potential signal cancellation effects. Moreover, in the case where the two signatures are

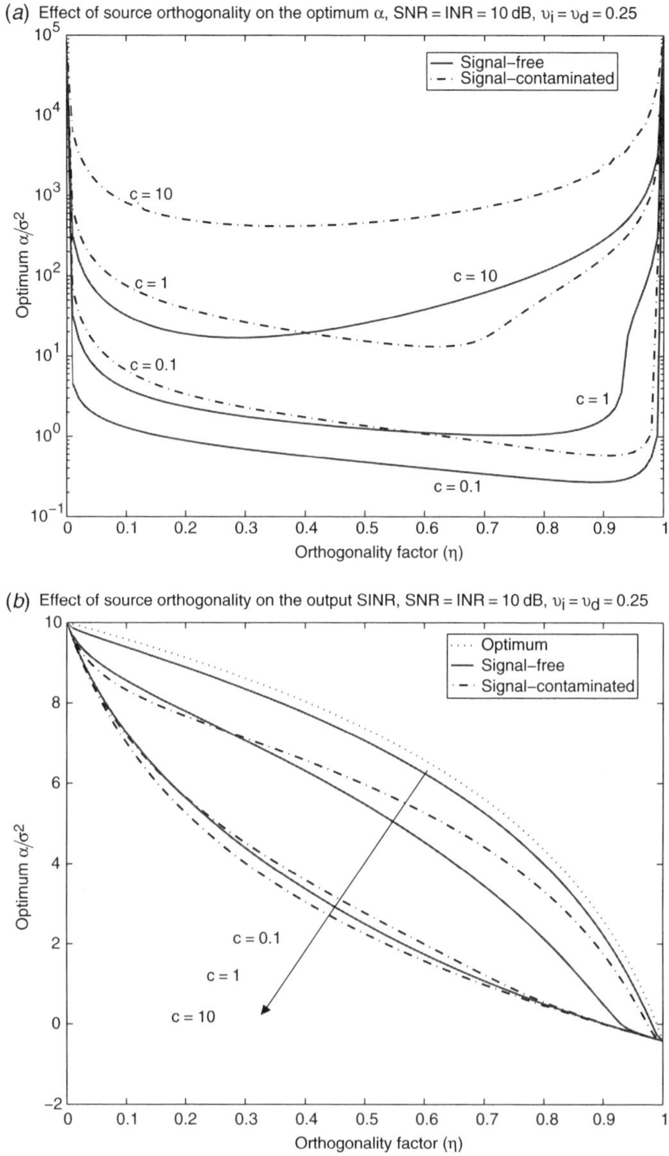

Figure 4.10 Effect of the source orthogonality parameter η on the optimum loading factor (*a*) and the corresponding output SINR (*b*). We fixed $P_d/\sigma^2 = P_i/\sigma^2 = 10\,\text{dB}$ and $\upsilon_d = \upsilon_i = 0.25$.

completely parallel ($\eta = 1$), the optimum array processor becomes the phased array, and consequently an infinite loading factor becomes the optimum alternative.

Figure 4.11 represents the influence of SNR (P_d/σ^2) and INR (P_i/σ^2) on the performance of the diagonally loaded beamformer. Note that, as expected, the SNR

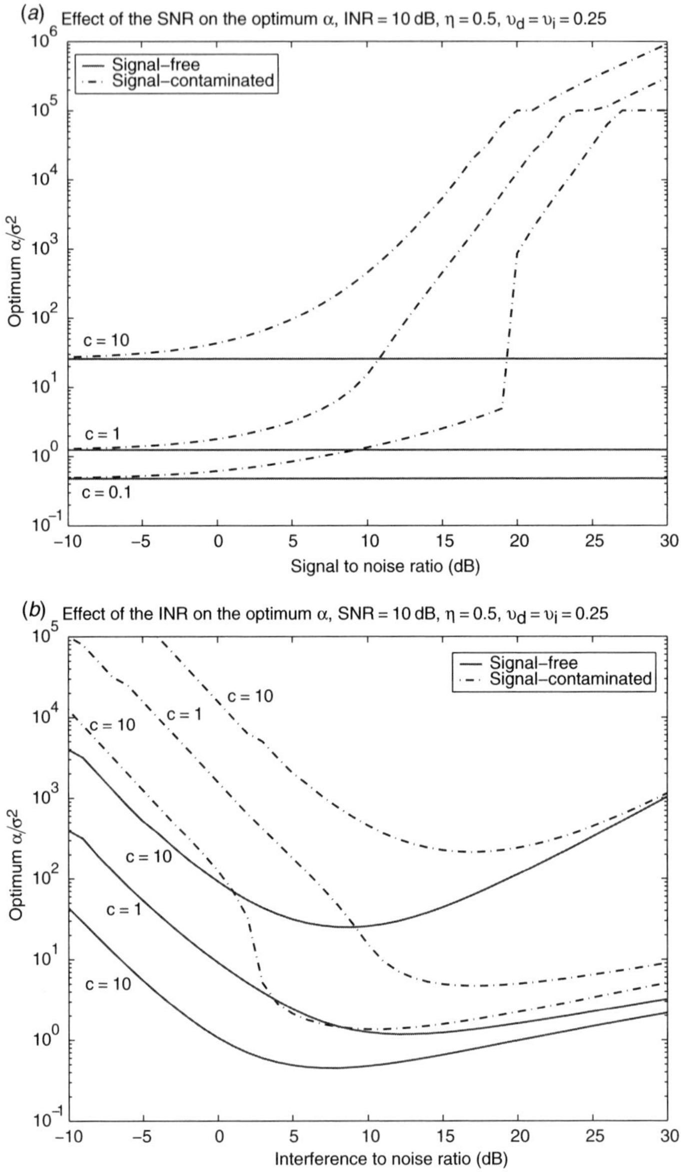

Figure 4.11 Effect of the SNR (P_d/σ^2) (a), and the INR (P_i/σ^2) (b) on the optimum loading factor. We fixed $\eta = 0.5$ and $\upsilon_d = \upsilon_i = 0.25$.

does not effect the optimum loading factor in the signal-free case. At high values of SNR and low values of INR, the optimum spatial filter tends to be a phased array, and consequently the optimum loading factor becomes increasingly high. Observe also that, when the SNR tends to zero, the optimum loading factor is the same for both signal-free and signal-contaminated scenarios. On the other hand, as the value of SNR gets higher, a higher value of the loading factor is needed in order to combat the signal cancellation effect. We do not represent the influence of the relative dimension of signal and interference subspaces because, interestingly enough, these two parameters have almost no effect on the choice of the optimum loading factor and the performance of the corresponding spatial filter.

4.5 SUMMARY AND CONCLUSIONS

This chapter has considered the use of diagonal loading to combat the finite sample size effect in the SMI implementation of the MVDR beamformer. We have reviewed the history of the application of this technique in array processing applications in order to combat the finite sample size effect. The addition of a positive constant to the diagonal entries of the spatial correlation matrix improves the behavior of the SMI beamformer in both the signal-free and the signal-contaminated scenarios. In the signal-free case, diagonal loading desensitizes the system, making it more robust against small interferences. If the useful signal is present in the observations, diagonal loading is a powerful method to avoid signal cancellation effects. Despite its widespread use, diagonal loading for these problems have seldom been considered in the literature and, in practice, the loading factor is fixed according to rather ad hoc methods. In this chapter, we have seen that random matrix theory can provide interesting insights into the performance of the diagonal loaded SMI technique and, using the theory of general statistic analysis or G-estimation, we have derived a powerful method for estimating the optimum loading factor in a completely unknown scenario.

ACKNOWLEDGMENTS

The authors would like to thank Prof. Yuri Abramovich and Prof. Alex Gershman for their valuable comments.

APPENDIX 4.A: Proof of Proposition 1

The proof of Proposition 1 is based on the use of convergence properties of spectral functions of random matrices. Observe that the original expression for the output SINR can be expressed as

$$\text{SINR} = \left[\frac{\eta_n}{P_d(\eta_d)^2} - 1\right]^{-1} \tag{A.1}$$

where we have defined the following two random quantities:

$$\eta_n = \mathbf{s}_d^H(\hat{\mathbf{R}} + \alpha\mathbf{I}_M)^{-1}\mathbf{R}(\hat{\mathbf{R}} + \alpha\mathbf{I}_M)^{-1}\mathbf{s}_d, \quad \eta_d = \mathbf{s}_d^H(\hat{\mathbf{R}} + \alpha\mathbf{I}_M)^{-1}\mathbf{s}_d.$$

Thus, convergence in probability of SINR will follow from convergence in probability of the two quantities η_n and η_d. One can readily see that under (**As1**–**As2**), the sample correlation matrix can be expressed as $\hat{\mathbf{R}} = \mathbf{R}^{1/2}\boldsymbol{\Xi}\boldsymbol{\Xi}^H\mathbf{R}^{1/2}$, where $\mathbf{R}^{1/2}$ is the positive-definite Hermitian square root of the true correlation matrix and $\boldsymbol{\Xi}$ an $M \times N$ random matrix with independent and identically distributed complex entries whose real and imaginary parts are independent and have zero mean and variance $1/(2N)$. Now, with this decomposition one can express η_d and η_n as

$$\eta_d = m(z)|_{z=0}, \quad \eta_n = \left.\frac{dm(z)}{dz}\right|_{z=0}$$

where we have defined the following complex function:

$$m(z) = \mathbf{s}_d^H\mathbf{R}^{-1/2}\left(\boldsymbol{\Xi}\boldsymbol{\Xi}^H + \alpha\mathbf{R}^{-1} - z\mathbf{I}_M\right)^{-1}\mathbf{R}^{-1/2}\mathbf{s}_d, \tag{A.2}$$

which is well-defined and holomorphic for all $z \in \mathbb{C}$ except for a segment of the positive real axis. As explained in Section 4.2.1, this function (and, in general, a function of the form $\mathbf{a}^H(\mathbf{M} - z\mathbf{I}_M)^{-1}\mathbf{a}$ where $\mathbf{M}$ is an $M \times M$ random matrix and $\mathbf{a}$ a deterministic $M \times 1$ vector) is very important in the statistical analysis of large random matrices. Denoting by $\mathbf{q}_i$ and υ_i, $i = 1, \ldots, M$ the eigenvectors and associated eigenvalues of the generic random matrix $\mathbf{M}$, we have

$$\mathbf{a}^H(\mathbf{M} - z\mathbf{I}_M)^{-1}\mathbf{a} = \sum_{i=1}^{M}\frac{\mathbf{a}^H\mathbf{q}_i\mathbf{q}_i^H\mathbf{a}}{\upsilon_i - z} \tag{A.3}$$

and this function is usually referred to as the Stieltjes transform of the spectral function

$$F_{\mathbf{M}}(x) = \sum_{i=1}^{M}\mathbf{a}^H\mathbf{q}_i\mathbf{q}_i^H\mathbf{a}\mathcal{I}(\upsilon_i \leq x), \tag{A.4}$$

where $\mathcal{I}(\upsilon_i \leq x)$ is the indicator function for the event $\{\upsilon_i \leq x\}$. The proof of Proposition 1 is based on the study of the asymptotic behavior of $m(z)$ in (A.2), which can be interpreted as a Stieltjes transform of a spectral function such as the one in (A.4) defined from the random matrix $\mathbf{M} = \boldsymbol{\Xi}\boldsymbol{\Xi}^H + \alpha\mathbf{R}^{-1}$ with $\mathbf{a} = \mathbf{R}^{-1/2}\mathbf{s}_d$. We need the following result on the asymptotic convergence of spectral functions of matrices of the type $\boldsymbol{\Xi}\boldsymbol{\Xi}^H + \alpha\mathbf{R}^{-1}$:

Lemma 1. Write $\mathbf{B} = \mathbf{A} + \boldsymbol{\Xi}\boldsymbol{\Xi}^H$ with $\mathbf{A}$ a deterministic positive definite Hermitian $M \times M$ matrix whose eigenvalues $\lambda_1(\mathbf{A}) \geq \cdots \geq \lambda_M(\mathbf{A}) > 0$ are uniformly bounded for all M and have a limiting spectral distribution, and $\boldsymbol{\Xi}$ an $M \times N$ random matrix with independent and identically distributed complex entries such that the real and imaginary parts of the entries of $\sqrt{N}\boldsymbol{\Xi}$ are independent and have zero mean, variance $1/2$ and bounded eighth moments. Let $\mathbb{D}_\epsilon$ be the open disk centered at $z = 0$ with radius $\epsilon < \lambda_M(\mathbf{A})/3$. Then

$$\lim_{M,N\to\infty} \sup_{z\in\mathbb{D}_\epsilon} \left| \frac{1}{M}\operatorname{tr}[(\mathbf{B} - z\mathbf{I}_M)^{-1}] - b(z) \right| = 0 \tag{A.5}$$

almost surely, where $b(z)$ satisfies the following equation[7]

$$b(z) = \frac{1}{M}\sum_{k=1}^{M} \frac{1 + cb(z)}{1 + (\lambda_k(\mathbf{A}) - z)(1 + cb(z))}. \tag{A.6}$$

Consider also an $M \times 1$ complex vector $\mathbf{a}$ with uniformly bounded norm and such that

$$\sup_M \sup_{z\in\mathbb{D}_{2\epsilon}} |\mathbf{a}^H(\mathbf{A} - z\mathbf{I}_M)^{-1}\mathbf{a}| < \infty. \tag{A.7}$$

Then,

$$\lim_{M,N\to\infty} \sup_{z\in\mathbb{D}_\epsilon} \left| \mathbf{a}^H[\mathbf{B} - z\mathbf{I}_M]^{-1}\mathbf{a} - \sum_{k=1}^{M} \frac{\mathbf{a}^H\mathbf{e}_k\mathbf{e}_k^H\mathbf{a}(1 + cb(z))}{1 + (\lambda_k(\mathbf{A}) - z)(1 + cb(z))} \right| = 0 \tag{A.8}$$

almost surely, where $\mathbf{e}_1, \ldots, \mathbf{e}_M$ denote the eigenvectors of $\mathbf{A}$.

Proof. The proof follows quite easily from the corresponding proof of pointwise convergence given in, for example, [60], [61] or [62]. See Appendix 4.B for further details. □

Now, with the result given in the above lemma, we readily see that

$$\lim_{M,N\to\infty} |m(z) - \bar{m}(z)| = 0$$

almost surely (a.s.) uniformly on a sufficiently small disk around the origin, where we have defined the deterministic function:

$$\bar{m}(z) = \sum_{k=1}^{M} \frac{\mathbf{s}_d^H\mathbf{R}^{-1/2}\mathbf{e}_i\mathbf{e}_i^H\mathbf{R}^{-1/2}\mathbf{s}_d(1 + cb(z))}{1 + (\alpha\lambda_k^{-1} - z)(1 + cb(z))} \tag{A.9}$$

[7]If z is restricted to the real positive axis, $b(z)$ is also real and positive, and is the unique solution to (A.6) with these properties.

with $b(z)$ the positive solution to (A.6). Now, using $\eta_d = m(z)|_{z=0}$ and hence forcing $z = 0$ in (A.9), we readily see that $\lim_{M,N\to\infty} |\eta_d - \bar{\eta}_d| = 0$ in probability, where

$$\bar{\eta}_d = \bar{m}(z)|_{z=0} = \sum_{m=1}^{M} \frac{1 + cb}{\lambda_m + \alpha(1 + cb)} |\mathbf{s}_d^H \mathbf{e}_m|^2 \tag{A.10}$$

and being $b = b(0)$ the positive solution to equation (4.23). Let us now draw our attention to the term η_n. Now, since uniform convergence of holomorphic functions implies uniform convergence of all derivatives (Weierstrass theorem), we will have $p\lim_{M,N\to\infty} |\eta_n - \bar{\eta}_n| = 0$, where now

$$\bar{\eta}_n = \frac{d}{dz}[\bar{m}(z)]_{z=0} = \sum_{k=1}^{M} |\mathbf{s}_d^H \mathbf{e}_k|^2 \lambda_k \frac{[1 + cb]^2 + cb'}{(\lambda_k + \alpha[1 + cb])^2} \tag{A.11}$$

being $b = b(0)$ the positive solution to (4.23), and

$$b' = \left.\frac{db(z)}{dz}\right|_{z=0} = \left[1 - \frac{1}{M}\sum_{k=1}^{M} \frac{c\lambda_k^2}{(\lambda_k + \alpha[1 + cb])^2}\right]^{-1} \frac{1}{M}\sum_{k=1}^{M} \frac{\lambda_k^2[1 + cb]^2}{(\lambda_k + \alpha[1 + cb])^2}.$$

Inserting (A.10) and (A.11) into (A.1) we get to the result of the proposition.

APPENDIX 4.B: Proof of Lemma 1

We will first see that, in order to prove uniform convergence, it is enough to show pointwise convergence towards the limiting function. Indeed, define

$$f_1(z) = \frac{1}{M}\mathrm{tr}[(\mathbf{B} - z\mathbf{I}_M)^{-1}], \quad f_2(z) = \mathbf{a}^H[\mathbf{B} - z\mathbf{I}_M]^{-1}\mathbf{a}$$

and note that they are both almost surely holomorphic on $z \in \mathbb{D}_{2\epsilon}$ (the open disk centered at $z = 0$ with radius $2\epsilon < 2\lambda_M(\mathbf{A})/3$). This can easily be seen by expressing these two functions as

$$f_1(z) = \frac{1}{M}\sum_{m=1}^{M} \frac{1}{\lambda_m(\mathbf{B}) - z}, \quad f_2(z) = \sum_{m=1}^{M} \frac{|\mathbf{a}^H \mathbf{f}_m|^2}{\lambda_m(\mathbf{B}) - z}, \tag{B.1}$$

where $\mathbf{f}_m$ is the eigenvector of $\mathbf{B}$ associated with the mth eigenvalue $\lambda_m(\mathbf{B})$, and noting[8] that for $z \in \mathbb{D}_{2\epsilon}$,

$$|\lambda_m(\mathbf{B}) - z| \overset{(a)}{>} |\lambda_M(\mathbf{B}) - 2\epsilon| \overset{(b)}{\geq} |\lambda_M(\mathbf{A}) - 2\epsilon| \overset{(c)}{\geq} \lambda_M(\mathbf{A})/3 > 0.$$

[8]From this point on, all the statements should be understood to hold with probability one (that is, they are valid for all realizations except for a set with probability zero).

In the last chain of inequalities, (a) follows from the fact that $|z| < 2\epsilon$, (b) by noticing that $\lambda_m(\mathbf{B}) = \lambda_m(\mathbf{A} + \boldsymbol{\Xi}\boldsymbol{\Xi}^H) \geq \lambda_m(\mathbf{A})$, and (c) from the fact that $\epsilon < \lambda_M(\mathbf{A})/3$ by hypothesis. With this, we see that the denominators in (B.1) are positive on the whole disk $\mathbb{D}_{2\epsilon}$ and, thanks to (A.7), both $f_1(z)$ and $f_2(z)$ will be bounded for all $z \in \mathbb{D}_{2\epsilon}$. In particular, they will be bounded on any compact subset within $\mathbb{D}_{2\epsilon}$. Now, if we are able to show that $f_1(z)$ and $f_2(z)$ are pointwise almost surely convergent for all $z \in \mathbb{D}_{2\epsilon}$, it follows automatically that they are almost surely uniformly convergent on any compact subset of $\mathbb{D}_{2\epsilon}$, in particular $\bar{\mathbb{D}}_\epsilon$ as we wanted to prove.

Now, it is well-known that under **(As1–As2)** $f_1(z)$ converges pointwise almost surely to $b(z)$ for every $z \in \mathbb{D}_{2\epsilon}$ (cf., for instance, [61, 62]), so that (A.5) is a direct consequence of the above line of reasoning. As for the almost sure pointwise convergence of $f_2(z)$, note that

$$\left| \mathbf{a}^H \left([\mathbf{B} - z\mathbf{I}_M]^{-1} - \sum_{k=1}^{M} \frac{\mathbf{e}_k \mathbf{e}_k^H (1 + cb(z))}{1 + (\lambda_k(\mathbf{A}) - z)(1 + cb(z))} \right) \mathbf{a} \right|$$
$$\leq \|\mathbf{a}\|^2 \max_i \left| \left\{ [\mathbf{B} - z\mathbf{I}_M]^{-1} - \sum_{k=1}^{M} \frac{\mathbf{e}_k \mathbf{e}_k^H (1 + cb(z))}{1 + (\lambda_k(\mathbf{A}) - z)(1 + cb(z))} \right\}_{ii} \right|.$$

Almost sure pointwise convergence to zero of the ith diagonal entry of the matrix above can now be proven as in [54, Vol. 1, Theorem 17.2] (see also [51, Theorem 2]).

APPENDIX 4.C: Derivation of the Consistent Estimator

In this appendix, we give a brief derivation of the consistent estimator presented in Section 4.3. The basic idea is to express the deterministic quantity $q(\alpha)$ in (4.22) in terms of real Stieltjes transforms $\mathcal{M}_{\mathbf{R}}(x)$ and $\mathcal{T}_{\mathbf{R}}(x)$ as defined in (4.28), fixing for the last one $\mathbf{a} = \mathbf{b} = \mathbf{s}_d$. Then, using the consistent estimators $\hat{\mathcal{M}}_{\mathbf{R}}(x)$ and $\hat{\mathcal{T}}_{\mathbf{R}}(x)$ presented in (4.30), we obtain a consistent estimator of the objective function to be minimized, $\hat{q}(\alpha)$.

1. Estimation of b. Observe that b is the positive solution to the equation in (4.23), which can also be written as

$$\frac{b}{1 + cb} = 1 - \mathcal{M}_{\mathbf{R}}\left(\frac{1}{\alpha(1 + cb)} \right).$$

In order to obtain a consistent estimator of b, one can readily use the consistent estimator of $\mathcal{M}_{\mathbf{R}}(x)$, so that $\hat{b}$ will be defined as the positive solution to

$$\frac{\hat{b}}{1 + c\hat{b}} = 1 - \hat{\mathcal{M}}_{\mathbf{R}}\left(\frac{1}{\alpha(1 + c\hat{b})} \right).$$

Now, using the expression of $\hat{\mathcal{M}}_{\mathbf{R}}(x)$ given in (4.30), one can see that $\hat{b}$ accepts a closed form expression, given by:

$$\hat{b} = \frac{1 - \frac{\alpha}{M}\operatorname{tr}\left[\left(\alpha\mathbf{I}_M + \hat{\mathbf{R}}\right)^{-1}\right]}{1 - c\left(1 - \frac{\alpha}{M}\operatorname{tr}\left[\left(\alpha\mathbf{I}_M + \hat{\mathbf{R}}\right)^{-1}\right]\right)}.$$

On the other hand, the consistent estimator of $\gamma = \alpha[1 + cb]$ will be given by $\hat{\gamma} = \alpha[1 + c\hat{b}]$. It is worth noting here that, if $c \geq 1$,

$$\hat{\gamma}^{-1} = \frac{1}{N}\sum_{m=1}^{N}\frac{1}{\alpha + \hat{\lambda}_m} \leq \frac{1}{N}\sum_{m=1}^{N}\frac{1}{\hat{\lambda}_m}$$

so, by virtue of (4.32), we see that the G-equation in (4.31) will always have a unique positive solution when evaluated at $x = \hat{\gamma}^{-1}$.

2. Estimation of ξ. Note that this quantity can be written as

$$\begin{aligned}\xi &= \frac{1}{M}\operatorname{tr}\left[\mathbf{R}^2(\mathbf{R} + \gamma\mathbf{I}_M)^{-2}\right] \\ &= 1 - 2\frac{1}{M}\operatorname{tr}\left[\left(\mathbf{I}_M + \gamma^{-1}\mathbf{R}\right)^{-1}\right] + \frac{1}{M}\operatorname{tr}\left[\left(\mathbf{I}_M + \gamma^{-1}\mathbf{R}\right)^{-2}\right] \\ &= 1 - \mathcal{M}_{\mathbf{R}}(x)|_{x=\gamma^{-1}} + \gamma^{-1}\frac{d}{dx}\mathcal{M}_{\mathbf{R}}(x)\bigg|_{x=\gamma^{-1}}.\end{aligned}$$

As shown above, $\hat{\gamma} = \alpha[1 + c\hat{b}]$ is a consistent estimator of γ, and consequently we can use the following consistent estimator of ξ:

$$\hat{\xi} = 1 - \hat{\mathcal{M}}_{\mathbf{R}}(x)\bigg|_{x=\hat{\gamma}^{-1}} + \hat{\gamma}^{-1}\frac{d}{dx}\hat{\mathcal{M}}_{\mathbf{R}}(x)\bigg|_{x=\hat{\gamma}^{-1}},$$

where we have implicitly used that $\hat{\mathcal{M}}_{\mathbf{R}}(x) \to \mathcal{M}_{\mathbf{R}}(x)$ uniformly in x as $M, N \to \infty$. Now, replacing $\hat{\mathcal{M}}_{\mathbf{R}}(x)$ with its corresponding expression in (4.30) and after some algebraic manipulations, we get to

$$\hat{\xi} = \frac{\frac{1}{M}\operatorname{tr}\left[\hat{\mathbf{R}}^2\left(\alpha\mathbf{I}_M + \hat{\mathbf{R}}\right)^{-2}\right] - c\left(\frac{1}{M}\operatorname{tr}\left[\hat{\mathbf{R}}\left(\alpha\mathbf{I}_M + \hat{\mathbf{R}}\right)^{-1}\right]\right)^2}{1 - c + c\alpha^2\frac{1}{M}\operatorname{tr}\left[\left(\alpha\mathbf{I}_M + \hat{\mathbf{R}}\right)^{-2}\right]}.$$

3. Estimation of $\breve{\eta}_d \stackrel{\text{def}}{=} \mathbf{s}_d^H(\gamma\mathbf{I}_M + \mathbf{R})^{-1}\mathbf{s}_d$ and $\breve{\eta}_n \stackrel{\text{def}}{=} \mathbf{s}_d^H(\gamma\mathbf{I}_M + \mathbf{R})^{-1}\mathbf{R}(\gamma\mathbf{I}_M + \mathbf{R})^{-1}\mathbf{s}_d$. Noting that $\breve{\eta}_d = \gamma^{-1}\mathcal{T}_{\mathbf{R}}(\gamma^{-1})$ and using the estimation in (4.30)

for $\mathcal{T}_{\mathbf{R}}(x)$, we obtain the following estimator for $\breve{\eta}_d$:

$$\hat{\eta}_d = \left[1 - c\left(1 - \frac{\alpha}{M}\operatorname{tr}\left[\left(\alpha\mathbf{I}_M + \hat{\mathbf{R}}\right)^{-1}\right]\right)\right]\mathbf{s}_d^H\left(\alpha\mathbf{I}_M + \hat{\mathbf{R}}\right)^{-1}\mathbf{s}_d.$$

On the other hand, one can also show that

$$\breve{\eta}_n = -\left[x^2\frac{d}{dx}\mathbf{s}_d^H(\mathbf{I}_M + x\mathbf{R})^{-1}\mathbf{s}_d\right]_{x=\gamma^{-1}} = -\left[x^2\frac{d}{dx}\mathcal{T}_{\mathbf{R}}(x)\right]_{x=\gamma^{-1}}.$$

Hence, using again the consistent estimator of $\mathcal{T}_{\mathbf{R}}(x)$ given in (4.30]),

$$\hat{\eta}_n = \frac{\left(1 - c\left(1 - \frac{\alpha}{M}\operatorname{tr}\left[\left(\alpha\mathbf{I}_M + \hat{\mathbf{R}}\right)^{-1}\right]\right)\right)^2}{1 - c + c\frac{1}{M}\operatorname{tr}\left[\left(\mathbf{I}_M + \alpha^{-1}\hat{\mathbf{R}}\right)^{-2}\right]} \times \mathbf{s}_d^H\left(\alpha\mathbf{I}_M + \hat{\mathbf{R}}\right)^{-1}\hat{\mathbf{R}}\left(\alpha\mathbf{I}_M + \hat{\mathbf{R}}\right)^{-1}\mathbf{s}_d.$$

The final estimator of $q(\alpha)$ given in Section 4.3 can be obtained inserting the all above estimations in the corresponding expression of Proposition 1.

APPENDIX 4.D: Proof of Proposition 2

To show the weak consistency of the estimated loading factor, it is sufficient to show that, for any given $\varepsilon > 0$,

$$\lim_{M,N\to\infty} \operatorname{Prob}\left(\sup_{\alpha\geq 0}\left|\hat{q}(\alpha) - q(\alpha)\right| > \varepsilon\right) = 0. \tag{D.1}$$

Let therefore ε be fixed. Let $\hat{\lambda}_{\max}$ and $\hat{\lambda}_{\min}$ denote, as usual, the maximum and minimum eigenvalues of $\hat{\mathbf{R}}$. It is well-known [51] that, as $M, N \to \infty$ at the same rate, $\hat{\lambda}_{\max}$ and $\hat{\lambda}_{\min}$ tend almost surely to two nonrandom constants, denoted by $\bar{\lambda}_{\max}$ and $\bar{\lambda}_{\min}$ respectively. Consider the following events:

$$A = \left\{\sup_{\alpha\geq 0}\left|\hat{q}(\alpha) - q(\alpha)\right| > \varepsilon\right\},$$

$$E = \left\{\left|\hat{\lambda}_{\max} - \bar{\lambda}_{\max}\right| < \varepsilon\right\}, \quad F = \left\{\left|\hat{\lambda}_{\min} - \bar{\lambda}_{\min}\right| < \varepsilon\right\},$$

and note that, because almost sure convergence implies convergence in probability,

$$\lim_{M,N\to\infty} \operatorname{Prob}(E) = \lim_{M,N\to\infty} \operatorname{Prob}(F) = 1.$$

Now, observe that one can write

$$\begin{aligned}\operatorname{Prob}(A) \leq\ & \operatorname{Prob}(A|E,F)\operatorname{Prob}(E)\operatorname{Prob}(F) \\ & + \operatorname{Prob}(A|E,F^c)\operatorname{Prob}(E)\operatorname{Prob}(F^c) \\ & + \operatorname{Prob}(A|E^c,F)\operatorname{Prob}(E^c)\operatorname{Prob}(F) \\ & + \operatorname{Prob}(A|E^c,F^c)\operatorname{Prob}(E^c)\operatorname{Prob}(F^c).\end{aligned} \tag{D.2}$$

where A^c denotes the complementary of the event A. Since

$$\lim_{M,N\to\infty} \operatorname{Prob}(E) = \lim_{M,N\to\infty} \operatorname{Prob}(F) = 1$$

(and hence $\lim_{M,N\to\infty} \operatorname{Prob}(E^c) = \lim_{M,N\to\infty} \operatorname{Prob}(F^c) = 0$), it is sufficient to show that

$$\lim_{M,N\to\infty} \operatorname{Prob}\left(\sup_{\alpha\geq 0}|\hat{q}(\alpha) - q(\alpha)| > \varepsilon \,\middle|\, E, F \right) = 0,$$

that is, we can assume from now on that all the eigenvalues of $\hat{\mathbf{R}}$ are situated on the interval $(\bar{\lambda}_{\min} - \varepsilon, \bar{\lambda}_{\max} + \varepsilon)$. We also assume that $\varepsilon < \bar{\lambda}_{\min}/2$; otherwise, we just need to take a lower ε and (D.1) will still hold.[9]

Note that one can express $\hat{q}(\alpha) - q(\alpha)$ as

$$\hat{q}(\alpha) - q(\alpha) = Q(\alpha)[\Delta_1(\alpha) - \Delta_2(\alpha)],$$

where we have introduced the following definitions:

$$\Delta_1(\alpha) = (1 - c\xi)\frac{\mathbf{s}_d^H(\alpha\mathbf{I}_M + \hat{\mathbf{R}})^{-1}\hat{\mathbf{R}}(\alpha\mathbf{I}_M + \hat{\mathbf{R}})^{-1}\mathbf{s}_d}{\mathbf{s}_d^H(\gamma\mathbf{I}_M + \mathbf{R})^{-1}\mathbf{R}(\gamma\mathbf{I}_M + \mathbf{R})^{-1}\mathbf{s}_d} - 1$$

$$\Delta_2(\alpha) = [1 - c\hat{\varphi}]^2 \frac{(\mathbf{s}_d^H(\alpha\mathbf{I}_M + \hat{\mathbf{R}})^{-1}\mathbf{s}_d)^2}{(\mathbf{s}_d^H(\gamma\mathbf{I}_M + \mathbf{R})^{-1}\mathbf{s}_d)^2} - 1$$

$$Q(\alpha) = \frac{\mathbf{s}_d^H(\gamma\mathbf{I}_M + \mathbf{R})^{-1}\mathbf{R}(\gamma\mathbf{I}_M + \mathbf{R})^{-1}\mathbf{s}_d}{[1 - c\hat{\varphi}]^2[1 - c\xi]P_d(\mathbf{s}_d^H(\alpha\mathbf{I}_M + \hat{\mathbf{R}})^{-1}\mathbf{s}_d)^2}.$$

[9]Indeed, if $\varepsilon' < \varepsilon$, $\operatorname{Prob}(\sup_{\alpha\geq 0}|\hat{q}(\alpha) - q(\alpha)| > \varepsilon) \leq \operatorname{Prob}(\sup_{\alpha\geq 0}|\hat{q}(\alpha) - q(\alpha)| > \varepsilon')$, so that if the second term goes asymptotically to zero, so does the first one.

Let us now prove that $\sup_{\alpha \geq 0} Q(\alpha) < \infty$. Consider first the case $c < 1$ and observe also that, using the definitions in Section 4.2, we can write

$$\begin{aligned} Q(\alpha) &\overset{(a)}{\leq} \frac{\mathbf{s}_d^H(\gamma \mathbf{I}_M + \mathbf{R})^{-1}\mathbf{R}(\gamma \mathbf{I}_M + \mathbf{R})^{-1}\mathbf{s}_d}{(1-c)^3 P_d(\mathbf{s}_d^H(\alpha \mathbf{I}_M + \hat{\mathbf{R}})^{-1}\mathbf{s}_d)^2} \\ &\overset{(b)}{\leq} \frac{\mathbf{s}_d^H(\gamma \mathbf{I}_M + \mathbf{R})^{-1}\mathbf{R}(\gamma \mathbf{I}_M + \mathbf{R})^{-1}\mathbf{s}_d}{(1-c)^3 P_d \|\mathbf{s}_d\|^4}(\alpha + \hat{\lambda}_{\max})^2 \\ &\overset{(c)}{\leq} \frac{\mathbf{s}_d^H(\gamma \mathbf{I}_M + \mathbf{R})^{-1}\mathbf{R}(\gamma \mathbf{I}_M + \mathbf{R})^{-1}\mathbf{s}_d}{(1-c)^3 P_d \|\mathbf{s}_d\|^4}(\alpha + \bar{\lambda}_{\max} + \varepsilon)^2. \end{aligned}$$

In the above chain of inequalities, (a) follows from the fact that $0 \leq \xi, \hat{\varphi} \leq 1$ a.s., (b) from the definition of spectral radius, and (c) from the fact that $\hat{\lambda}_{\max} < \bar{\lambda}_{\max} + \varepsilon$ by assumption. Note that $\gamma = \alpha[1 + cb]$ where b is a continuous and monotonically decreasing function of α that takes values from $(1-c)^{-1}$ when $\alpha = 0$ to 0 when $\alpha \to \infty$. This implies that $\sup_{\alpha \geq 0} \gamma/\alpha = (1-c)^{-1} < \infty$, and consequently the last term in the equation above is almost everywhere bounded uniformly over all the range of values of α under consideration.

Consider now the case $c \geq 1$. Now, for this case, one can see that, as $\alpha \to 0$, γ tends to the solution of (4.26), while as $\alpha \to \infty$, $\gamma \to \infty$ at the same rate. On the other hand,

$$1 - c\hat{\varphi} = \frac{1}{N}\sum_{m=1}^{N} \frac{\alpha}{\hat{\lambda}_m + \alpha} \geq \frac{\alpha}{\hat{\lambda}_{\max} + \alpha} \geq \frac{\alpha}{(\bar{\lambda}_{\max} + \varepsilon) + \alpha}$$

so that

$$\begin{aligned} Q(\alpha) &\leq \left[\frac{(\bar{\lambda}_{\max} + \varepsilon) + \alpha}{\alpha}\right]^2 \frac{\mathbf{s}_d^H(\gamma \mathbf{I}_M + \mathbf{R})^{-1}\mathbf{R}(\gamma \mathbf{I}_M + \mathbf{R})^{-1}\mathbf{s}_d}{(1-c\xi)P_d(\mathbf{s}_d^H(\alpha \mathbf{I}_M + \hat{\mathbf{R}})^{-1}\mathbf{s}_d)^2} \\ &= \frac{[(\bar{\lambda}_{\max} + \varepsilon) + \alpha]^2 \mathbf{s}_d^H(\gamma \mathbf{I}_M + \mathbf{R})^{-1}\mathbf{R}(\gamma \mathbf{I}_M + \mathbf{R})^{-1}\mathbf{s}_d}{(1-c\xi)P_d(\alpha \mathbf{s}_d^H(\alpha \mathbf{I}_M + \hat{\mathbf{R}})^{-1}\mathbf{s}_d)^2}. \end{aligned}$$

Both numerator and denominator are almost surely bounded uniformly in $\alpha \geq 0$.

Now, having established that $\sup_{\alpha \geq 0} Q(\alpha)$ is bounded (conditioned on E, F) and noting that, by virtue of the triangular inequality,

$$|\hat{q}(\alpha) - q(\alpha)| \leq Q(\alpha)(|\Delta_1(\alpha)| + |\Delta_2(\alpha)|),$$

we see that

$$\operatorname{Prob}\left(\sup_{\alpha \geq 0} |\hat{q}(\alpha) - q(\alpha)| > \varepsilon \middle| E, F \right) \leq \operatorname{Prob}\left(\sup_{\alpha \geq 0} |\Delta_1(\alpha)| > \varepsilon' \middle| E, F \right) + \operatorname{Prob}\left(\sup_{\alpha \geq 0} |\Delta_2(\alpha)| > \varepsilon' \middle| E, F \right)$$

with $\varepsilon' = \varepsilon / \sup_{\alpha \geq 0} Q(\alpha)$. So, we only need to show that

$$\sup_{\alpha \geq 0} |\Delta_1(\alpha)| \to 0, \quad \sup_{\alpha \geq 0} |\Delta_2(\alpha)| \to 0$$

both in probability as $M, N \to \infty$ at the same rate (conditioned, of course, on the events E, F). Let us first concentrate on $\Delta_1(\alpha)$. Consider the complex-valued function obtained replacing α with $-z$ in the expression of $\Delta_1(\alpha)$, that is,

$$\Delta_1(-z) = \frac{(1 - c\psi(z))\mathbf{s}_d^H(\hat{\mathbf{R}} - z\mathbf{I}_M)^{-1}\hat{\mathbf{R}}(\hat{\mathbf{R}} - z\mathbf{I}_M)^{-1}\mathbf{s}_d}{\mathbf{s}_d^H(\mathbf{R} - z(1 + c\beta(z))\mathbf{I}_M)^{-1}\mathbf{R}(\mathbf{R} - z(1 + c\beta(z))\mathbf{I}_M)^{-1}\mathbf{s}_d} - 1, \tag{D.3}$$

where

$$\psi(z) = \frac{1}{M}\sum_{i=1}^{M}\left(\frac{\lambda_i}{\lambda_i - z(1 + c\beta)}\right)^2$$

and with $\beta(z)$ being the solution of

$$\beta(z) = \frac{1}{M}\sum_{i=1}^{M}\left(\frac{\lambda_i(1 + c\beta(z))}{\lambda_i - z(1 + c\beta(z))}\right). \tag{D.4}$$

Now, we will show that uniform convergence is implied by pointwise convergence. Consider an open disk $\mathbb{D}_{x_0, r}$ centered on $x_0 = (\bar{\lambda}_{\max} + \bar{\lambda}_{\min})/2$ and with radius $r = (\bar{\lambda}_{\max} - \bar{\lambda}_{\min})/2 + 2\varepsilon$. Note that $\bar{\lambda}_{\min} - \epsilon \in \mathbb{D}_{x_0, r}$, $\bar{\lambda}_{\max} + \epsilon \in \mathbb{D}_{x_0, r}$ and $0 \notin \mathbb{D}_{x_0, r}$ (because $\varepsilon < \bar{\lambda}_{\min}/2$ by assumption). Now, consider the mapping T that transforms $\mathbb{D}_{x_0, r}$ into $\mathbb{C}\backslash\mathbb{D}$ (here $\mathbb{D}$ denotes the unit disk centered at zero) and vice versa, namely

$$T : z \mapsto w = \frac{r}{z - x_0}.$$

Note that this mapping is bijective and holomorphic on $z \in \mathbb{C}\backslash\mathbb{D}_{x_0, r}$ so that, in particular, T^{-1}will also be holomorphic on $\mathbb{D}$. Since $\Delta_1(-z)$ in (D.3) is holomorphic and bounded on $\mathbb{C}\backslash\mathbb{D}_{x_0, r}$, the composition $\Delta_1(-z) \circ T^{-1}$ will be holomorphic and bounded on $\mathbb{D}$ and any compact set therein. Now, assume that we can show that (D.3) converges pointwise almost surely to zero (this will be shown below). This implies a.s. pointwise convergence of $\Delta_1(-z) \circ T^{-1}$ to zero. We can therefore state that convergence of $\Delta_1(-z) \circ T^{-1}$ will be uniform on any compact subset contained in $\mathbb{D}$. In particular, it will be uniform on the closure of a disk centered at $w = 0$ with radius $0 < \rho < 1$, denoted by $\mathbb{D}_\rho$. Now, inverting the map T we see that $\Delta_1(-z)$ converges almost surely to zero uniformly over the region $\mathbb{C} \backslash T^{-1}(\mathbb{D}_\rho)$. So, if we choose $\rho > r^2/(r+\epsilon)^2$, a.s. convergence will also hold uniformly over all the negative real axis, as we wanted to prove.

In conclusion, we only need to show pointwise convergence of (D.3) to zero for all $z \in \mathbb{C} \backslash \mathbb{D}_{x_0, r}$. This can readily be shown by noting that (assuming, as before, sufficiently high M, N),

$$\begin{aligned} &\mathbf{s}_d^H \left(\hat{\mathbf{R}} - z\mathbf{I}_M\right)^{-1} \hat{\mathbf{R}} \left(\hat{\mathbf{R}} - z\mathbf{I}_M\right)^{-1} \mathbf{s}_d \\ &= \mathbf{s}_d^H \left(\hat{\mathbf{R}} - z\mathbf{I}_M\right)^{-1} \mathbf{s}_d + z\frac{d}{dz}\mathbf{s}_d^H \left(\hat{\mathbf{R}} - z\mathbf{I}_M\right)^{-1} \mathbf{s}_d. \end{aligned} \tag{D.5}$$

Now, using[10] [51, Theorem 2] and introducing a simple change of variable, we see that for $z \in \mathbb{C} \backslash \mathbb{D}_{x_0, r}$,

$$\begin{aligned} &\left| \mathbf{s}_d^H \left[\hat{\mathbf{R}} - z\mathbf{I}_M\right]^{-1} \mathbf{s}_d - \sum_{m=1}^{M} \frac{|\mathbf{s}_d^H \mathbf{e}_m|^2}{\lambda_m (1 + c\beta(z))^{-1} - z} \right| \\ &\leq \|\mathbf{s}_d\|^2 \max_i \left| \left\{ \left[\hat{\mathbf{R}} - z\mathbf{I}_M\right]^{-1} - \sum_{m=1}^{M} \frac{\mathbf{e}_m \mathbf{e}_m^H}{\lambda_m (1 + c\beta(z))^{-1} - z} \right\}_{ii} \right| \to 0 \end{aligned}$$

in probability (conditioned on E, F) as $M, N \to \infty$ at the same rate, where $\beta(z)$ is the solution to (D.4). Now, proceeding exactly as in the proof of Lemma [1] given in Appendix 4.B, one can show that convergence is uniform on sufficiently small disks[11] centered on z. Since uniform convergence of holomorphic functions implies uniform convergence of all derivatives, we also have

$$\left| \frac{d}{dz} \mathbf{s}_d^H \left[\hat{\mathbf{R}} - z\mathbf{I}_M\right]^{-1} \mathbf{s}_d - \sum_{m=1}^{M} \frac{(1 + c\beta(z))^2 + \lambda_m c\beta'(z)}{(\lambda_m - z(1 + c\beta(z)))^2} |\mathbf{s}_d^H \mathbf{e}_m|^2 \right| \to 0$$

a.s. as $M, N \to \infty$ at the same rate, where $\beta'(z)$ is the derivative of $\beta(z)$, which can be expressed as

$$\beta'(z) = \frac{1}{1 - c\psi(z)} \frac{1}{M} \sum_{i=1}^{M} \frac{\lambda_i}{\left(\lambda_i (1 + c\beta(z))^{-1} - z\right)^2}$$

[10]In [51, Theorem 2], convergence is proven without conditioning on the events E, F. However, we can see that conditioned convergence also holds reinterpreting (D.2) with $A = \left\{ \left|\Delta_1(-z)\right| > \varepsilon' \right\}$.

[11]Note that this is much easier than proving uniform convergence over a noncompact subset, as done before.

In consequence, plugging all these results into (D.5) and using (A.6), one can state that

$$\left| \mathbf{s}_d^H (\hat{\mathbf{R}} - z\mathbf{I}_M)^{-1} \hat{\mathbf{R}} (\hat{\mathbf{R}} - z\mathbf{I}_M)^{-1} \mathbf{s}_d - \frac{1}{1 - c\psi(z)} \sum_{m=1}^{M} \frac{\lambda_m |\mathbf{s}_d^H \mathbf{e}_m|^2}{(\lambda_m - z(1 + c\beta(z)))^2} \right| \to 0$$

with probability one, and pointwise for all $z \in \mathbb{C} \backslash \mathbb{D}_{x_0, r}$. Hence, we have shown that $\sup_{\alpha \geq 0} |\Delta_1(\alpha)| \to 0$ in probability.

To prove $\sup_{\alpha \geq 0} |\Delta_2(\alpha)| \to 0$ one must proceed exactly in the same way. First, with the same reasoning as before, it can be seen that pointwise convergence of $|\Delta_2(-z)|$ to zero implies uniform convergence on the negative real axis. On the other hand, using the convergence results presented above together with the fact that

$$\left| \frac{1}{M} \operatorname{tr} \left[\hat{\mathbf{R}} - z\mathbf{I}_M \right]^{-1} - \frac{1}{M} \sum_{m=1}^{M} \frac{1 + c\beta(z)}{\lambda_m - z(1 + c\beta(z))} \right| \to 0$$

almost surely (see, e.g., [51, Theorem 1]) one can readily see that $|\Delta_2(-z)| \to 0$ in probability uniformly over $\mathbb{C} \backslash \mathbb{D}_{x_0, r}$ as we want to show.

REFERENCES

1. A. B. Gershman. Robust adaptive beamforming in sensor arrays. *International Journal of Electronics and Communications (AEÜ)*, 53(3), 1–17 (1999).
2. B. D. Van Veen and K. M. Buckley. Beaforming: A versatile approach to spatial filtering. *IEEE Acoustics, Speech and Signal Processing (ASSP) Magazine*, pp. 4–24, Apr. 1988.
3. J. Capon. High resolution frequency-wavenumber spectrum analysis. *Proceedings of the IEEE*, 57, 1408–1418 (1969).
4. O. L. Frost. An algorithm for linearly constrained adaptive array processing. *Proceedings of the IEEE*, 60, 926–935 (1972).
5. J. Capon and N. R. Goodman. Probability distributions for estimators of the frequency-wavenumber spectrum. *Proceedings of the IEEE*, 58, 1785–1786 (1970).
6. I. Reed, J. Mallett, and L. Brennan. Rapid convergence rate in adaptive arrays. *IEEE Trans. on Aerospace and Electronic Systems*, 10(6), 853–863 (1974).
7. A. O. Steinhardt. The pdf of adaptive beamforming weights. *IEEE Transactions on Signal Processing*, 39(5), 1232–1235 (1991).
8. T. W. Miller. The transient response of adaptive arrays in TDMA systems. Technical Report RADC-TR-76-390, Ohio State University (Electroscience Lab.), 1976.
9. R. A. Monzingo and T. W. Miller. *Introduction to Adaptive Arrays*. John Wiley and Sons, NY, 1980.

10. D. M. Boroson. Sample size considerations for adaptive arrays. *IEEE Trans. on Aerospace and Electronic Systems*, 16(4), 446–451 (1980).

11. B. Widrow, K. M. Duvall, R. P. Gooch, and W. C. Newman. Signal cancellation phenomena in adaptive antennas: Causes and cures. *IEEE Transactions on Antennas and Propagation*, 30(3), 469–478 (1982).

12. D. D. Feldman and L. J. Griffiths. A constraint projection approach for robust adaptive beamforming. In *Proc. of the IEEE International Conference of Acoustics, Speech and Signal Processing*, Vol. 2, pp. 1381–1384, Apr. 1991.

13. B. D. Van Veen. Adaptive covergence of linearly constrained beamformers based on the sample covariance matrix. *IEEE Transactions on Signal Processing*, 39, 1470–1473 (1991).

14. D. D. Feldman and L. J. Griffiths. A projection approach for robust adaptive beamforming. *IEEE Transactions on Signal Processing*, 42, 867–876 (1994).

15. N. K. Jablon. Adaptive beamforming with the generalized sidelobe canceller in the presence of array imperfections. *IEEE Transactions on Antennas and Propagation*, 34(8), 996–1012 (1986).

16. E. K. Hung and R. M. Turner. A fast beamforming algorithm for large arrays. *IEEE Transactions on Aerospace and Electronic Systems*, 19(4), 598–607 (1983).

17. C. H. Gierull. Performance analysis of fast projections of the Hung-Turner type for adaptive beamforming. *Signal Processing (EURASIP)*, 50, 17–28 (1996).

18. M. A. Zatman. Properties of the Hung-Turner projections and their relationship to the eigencanceller. In *Proceedings of the 30th Asilomar Conference*, pp. 1176–1180, 1996.

19. T. K. Citron and T. Kailath. An improved eigenvector beamformer. In *Proceedings of the IEEE International Conference on Acoustics, Speech and Signal Processing*, pp. 33.3.1–33.3.4, San Diego, 1984.

20. A. M. Haimovich and Y. Bar-Ness. An eigenanalysis interference canceler. *IEEE Transactions on Signal Processing*, 39(1), 76–84 (1991).

21. C. H. Gierull. Statistical analysis of the eigenvector projection method for adaptive spatial filtering of interference. *IEE Proceedings of Radar, Sonar and Navigation*, 144(2), 57–63 (1997).

22. A. M. Haimovich. Asymptotic distribution of the conditional signal to noise ratio in an eigenanalysis-based adaptive array. *IEEE Transactions on Aerospace and Electronic Systems*, 33(3), 988–997 (1997).

23. C. D. Peckham, A. M. Haimovich, J. S. Goldstein, and I. S. Reed. Reduced-rank STAP performance analysis. *IEEE Transactions on Aerospace and Electronic Systems*, 36(2), 664–676 (2000).

24. Y. I. Abramovich and A. I. Nevrev. An analysis of effectiveness of adaptive maximization of the signal-to-noise ratio which utilizes the inversion of the estimated correlation matrix. *Radiotekhnika i Electronika (Radio Engineering and Electronic Physics)*, 26(12), 67–74 (1981).

25. W. W. Piegorsch and G. Casella. The early use of matrix diagonal increments in statistical problems. *SIAM Review*, 31(3), 428–434 (1989).

26. K. Levenberg. A method for the solution of certain non-linear problems in least squares. *Quarterly of Applied Mathematics*, 2, 164–168 (1944).

27. J. Durbin. A note on regression when there is extraneous information about one of the coefficients. *Journal of the American Statistical Association*, 48, 799–808 (1953).

28. J. D. Riley. Solving systems of linear equations with a positive definite, symmetric, but possibly ill-conditioned matrix. *Mathematical Tables and Other Aids to Computation*, 9, 96–101 (1955).

29. A. N. Tikhonov and V. Ya. Arsenin. *Methods for the Solution of Ill-Posed Problems*. Nauka, Moscow, 1974.

30. Y. I. Abramovich. A controlled method for adaptive optimization of filters using the criterion of maximum SNR. *Radiotekhnika i Electronika (Radio Engineering and Electronic Physics)*, 26(3), 87–95 (1981).

31. J. E. Hudson. *Adaptive Array Principles*. IEE Press, London, 1981.

32. O. P. Cheremisin. Efficiency of adaptive algorithms with regularized sample covariance matrix. *Radiotekhnika i Electronika (Radio Engineering and Electronic Physics)*, 27(10), 69–77 (1982).

33. W. F. Gabriel. Using spectral estimation techniques in adaptive processing antenna systems. *IEEE Transactions on Antennas and Propagation*, 34(3), 291–300 (1986).

34. B. D. Carlson. Covariance matrix estimation errors and diagonal loading in adaptive arrays. *IEEE Trans. on Aerospace and Electronic Systems*, 24(4), 397–401 (1988).

35. H. Cox, R. M. Zeskind, and M. M. Owen. Robust adaptive beamforming. *IEEE Transactions on Acoustics, Speech and Signal Processing, ASSP*. 35(10), 1365–1376 (1987).

36. H. L. Van Trees. *Optimum Array Processing*. John Wiley and Sons, NY, 2002.

37. S. A Vorobyov, A. B. Gershman, and Z.-Q. Luo. Robust adaptive beamforming using worst-case performance optimization: A solution to the signal mismatch problem. *IEEE Transactions on Signal Processing*, 51(2), 313–324 (2003).

38. J. Li, P. Stoica, and Z. Wang. On robust Capon beamforming and diagonal loading. *IEEE Transactions on Signal Processing*, 51, 1702–1715 (2003).

39. S. Shahbazpanahi, A. B. Gershman, Zhi-Quan Luo, and Kon Max Wong. Robust adaptive beamforming for general-rank signal models. *IEEE Transactions on Signal Processing*, 51(9), 2257–2269 (2003).

40. M. W. Ganz, R. L. Moses, and S. L. Wilson. Convergence of the SMI and the diagonally loaded SMI algorithms with weak interference. *IEEE Transactions on Antennas and Propagation*, 38(3), 394–399 (1990).

41. R. L. Dilsavor and R. L. Moses. Analysis of modified SMI method for adaptive array weight control. *IEEE Transactions on Signal Processing*, 41(2), 721–726 (1993).

42. Y. I. Abramovich. Convergence analysis of linearly constrained SMI and LSMI adaptive algorithms. In *IEEE Adaptive Systems for Signal Processing, Communications and Control Symposium*, pp. 255–259, Oct. 2000.

43. O. P. Cheremisin. The parameter selection for the controlled method of the adaptive optimization of filters. *Radiotekhnika i Electronika (Radio Engineering and Electronic Physics)*, 30(2), 2369–2377 (1985).

44. Y. L. Kim, S. U. Pillai, and J. R. Guerci. Optimal loading factor for minimal sample support space-time adaptive radar. In *Proceedings of the IEEE International Conference on Acoustics, Speech and Signal Processing*, pp. 2505–2508, Seattle, 1998.

45. N. Ma and J. T. Goh. Efficient method to determine diagonal loading value. In *Proc. of the IEEE International Conference on Acoustics, Speech and Signal Processing (Vol. V)*, pp. 341–344, 2003.

46. E. P. Wigner. On the distributions of the roots of certain symmetric matrices. *Annals of Mathematics*, 67, 325–327 (1958).

47. E. P. Wigner. Characteristic vectors of bordered matrices with infinite dimensions. *Annals of Mathematics*, 62(2), 548–564 (1965).

48. A. M. Tulino and S. Verdú. Random matrix theory and wireless communications. *Foundations and Trends in Communications and Information Theory*, 1(1), 1–182 (2004).

49. V. A. Marchenko and L. A. Pastur. The distribution of eigenvalues in certain sets of random matrices. *Math. Sbornik*, 72, 507–536 (1967).

50. J. W. Silverstein. Strong convergence of the empirical distribution of eigenvalues of large dimensional random matrices. *Journal of Multivariate Analysis*, 5, 331–339 (1995).

51. V. L. Girko. Strong law for the eigenvalues and eigenvectors of empirical covariance matrices. *Random Operators and Stochastic Equations*, 4(2), 176–204 (1996).

52. D. V. Voiculescu, K. J. Dykema, and A. Nica. *Free Random Variables*, Volume 1 of CRM Monograph Series (Université de Montréeal). American Mathematical Society, 1992.

53. D. V. Voiculescu. Lectures on free probability theory. In Pierre Bernard, Ed., *Lecture Notes in Mathematics*, pp. 280–349. Springer-Verlag, Berlin, 2000.

54. V. L. Girko. *Theory of Stochastic Canonical Equations*. Kluwer Academic Publishers, The Netherlands, 2001, 2 vols.

55. V. L. Girko. G25-estimators of principal components. *Theory of Probability and Mathematical Statistics*, 40, 1–10 (1990).

56. M. Viberg, B. Ottersten, and A. Nehorai. Performance analysis of direction finding with large arrays and finite data. *IEEE Transactions on Signal Processing*, 43(2), 469–477 (1995).

57. V. L. Girko. *Statistical Analysis of Observations of Increasing Dimension*, Volume 28 of *Mathematical and Statistical Methods*. Kluwer Academic Publishers, The Netherlands, 1995.

58. V. L. Girko. *An Introduction to Statistical Analysis of Random Arrays*. VSP, The Netherlands, 1998.

59. V. L. Girko. G2-estimators of the spectral functions of covariance matrices. *Theory of Probability and Mathematical Statistics*, 35, 27–30 (1987).

60. L. A. Pastur. Spectra of random self-adjoint operators. *Russian Mathematical Surveys*, 28(1), 4–63 (1973).

61. V. L. Girko. *Random Matrices*. Kiev University, Kiev, 1975.

62. J. W. Silverstein and Z. D. Bai. On the empirical distribution of eigenvalues of a class of large dimensional random matrices. *Journal of Multivariate Analysis*, 54(2), 175–192 (1995).

63. W. Bürhing. Adaptive orthogonal projection for rapid converging interference suppression. *IEE Electronics Letters*, 14(16), 515–516 (1978).

64. I. J. Gupta. SMI adaptive antenna arrays for weak interfering signals. *IEEE Transactions on Antennas and Propagation*, 34(10), 1237–1242 (1986).

65. M. A. Lagunas and M. Nájar. Source reference in CIR beamforming. *Signal Processing*, 29(2), 141–149 (1992).

66. J. R. Magnus and H. Neudecker. *Matrix Differential Calculus with Applications in Statistics and Econometrics*. John Wiley and Sons, NY, 1999.

5

MEAN-SQUARED ERROR BEAMFORMING FOR SIGNAL ESTIMATION: A COMPETITIVE APPROACH

Yonina C. Eldar
Department of Electrical Engineering,
Technion–Israel Institute of Technology, Haifa, Israel

Arye Nehorai
Department of Electrical Engineering and Computer Science,
University of Illinois at Chicago, Chicago, IL 60607

5.1 INTRODUCTION

Beamforming is a classical method of processing temporal sensor array measurements for signal estimation, interference cancellation, source direction, and spectrum estimation. It has ubiquitously been applied in areas such as radar, sonar, wireless communications, speech processing, and medical imaging (see, e.g., [1–8] and the references therein).

Conventional approaches for designing data dependent beamformers typically attempt to maximize the signal-to-interference-plus-noise ratio (SINR). Maximizing the SINR requires knowledge of the interference-plus-noise covariance matrix and the array steering vector. Since this covariance is unknown, it is often replaced by the sample covariance of the measurements, resulting in deterioration of performance with higher signal-to-noise ratio (SNR) when the signal is present in the training data. Some beamforming techniques are designed to mitigate this effect [9–12], whereas others are developed to also overcome uncertainty in the steering vector, for example [13–19] (see also Chapters 2 and 3 of this book and the references therein).

Robust Adaptive Beamforming, Edited by Jian Li and Petre Stoica

Despite the fact that the SINR has been used as a measure of beamforming performance and as a design criterion in many beamforming approaches, we note that maximizing SINR may not guarantee a good estimate of the signal. In an *estimation* context, where our goal is to design a beamformer in order to obtain an estimate of the signal amplitude that is close to its true value, it may make more sense to choose the weights to minimize an objective that is related to the estimation error, that is, the difference between the true signal amplitude and its estimate, rather than the SINR. Furthermore, in comparing performance of different beamforming methods, it may be more informative to consider the estimation error as a measure of performance.

In this chapter we derive five beamformers for estimating a signal in the presence of interference and noise using the mean-squared error (MSE) as the performance criterion assuming either a known steering vector or a random steering vector with known second-order statistics. Computing the MSE shows, however, that it depends explicitly on the unknown signal magnitude in the deterministic case, or the unknown signal second order moment in the stochastic case [8], hence cannot be minimized directly. Thus, we aim at designing robust beamformers whose performance in terms of MSE is good across all possible values of the unknowns. To develop beamformers with this property, we rely on a recent minimax estimation framework that has been developed for solving robust estimation problems [20, 21]. This framework considers a general linear estimation problem, and suggests two classes of linear estimators that optimize an MSE-based criterion. In the first approach, developed in [20], a linear estimator is developed to minimize the worst-case MSE over an ellipsoidal region of uncertainty on the parameter set. In the second approach, developed in [21], a linear estimator is designed whose performance is as close as possible to that of the optimal linear estimator for the case of known model parameters. Specifically, the estimator is designed to minimize the worst-case *regret*, which is the difference between the MSE of the estimator in the presence of uncertainties, and the smallest attainable MSE with a linear estimator that knows the exact model. Note that as we explain further in Section 5.2.2, even when the signal magnitude is known, we cannot achieve a zero MSE with a *linear* estimator.

Using the general minimax MSE framework [20, 21] we develop two classes of robust beamformers: Minimax MSE beamformers and minimax regret beamformers. The minimax MSE beamformers minimize the worst-case MSE over all signals whose magnitude (or variance, in the zero-mean stochastic signals case) is bounded by a constant. The minimax regret beamformers minimize the worst-case regret over all bounded signals, where this approach considers both an upper and a lower bound on the signal magnitude. We note that in practice, if bounds on the signal magnitude are not known, then they can be estimated from the data, as we demonstrate in the numerical examples.

We first consider the case in which the steering vector is known completely and develop a minimax MSE and minimax regret beamformer. In this case we show that the minimax beamformers are scaled versions of the classical SINR-based beamformers. We then consider the case in which the steering vector is not completely known or fully calibrated, for example due to errors in sensor positions, gains or phases, coherent and incoherent local scatters, receiver fluctuations due to temperature changes, quantization effects, and so on (see [14, 22] and the references therein). To model the uncertainties in the steering vector, we assume that it is a random vector with known mean and covariance. Under this model, we develop three possible beamformers: minimax MSE, minimax regret, and, following the ideas in [23], a least-squares beamformer. While the minimax beamformers require bounds on the signal magnitude, the least-squares beamformer does not require such bounds. As we show, in the case of a random steering vector, the beamformers resulting from the minimax approaches are fundamentally different than those resulting from the SINR approach, and are not just scaled versions of each other as in the known steering vector case.

To illustrate the advantages of our methods we present several numerical examples comparing the proposed methods with conventional SINR-based methods, and several recently proposed robust methods. For a known steering vector, the minimax MSE and minimax regret beamformers are shown to consistently have the best performance, particularly for negative SNR values. For random steering vectors, the minimax and least-squares beamformers are shown to have the best performance for low SNR values (−10 to 5 dB). As we show, the least-squares approach, which does not require bounds on the signal magnitude, often performs better than the recently proposed robust methods [14–16] for dealing with steering vector uncertainty. In this case, the improvement in performance resulting from the proposed methods is often quite substantial.

The chapter is organized as follows. In Section 5.2 we present the problem formulation and review existing methods. In Section 5.3 we develop the minimax MSE and minimax regret beamformers for the case in which the steering vector is known. The case of a random steering vector is considered in Section 5.4. In Sections 5.5 and 5.6 we discuss practical considerations and present numerical examples illustrating the advantages of the proposed beamformers over several existing standard and robust beamformers, for a wide range of SNR values. The chapter is summarized in Section 5.7.

5.2 BACKGROUND AND PROBLEM FORMULATION

We denote vectors in $\mathbb{C}^M$ by boldface lowercase letters and matrices in $\mathbb{C}^{N\times M}$ by boldface uppercase letters. The matrix $\mathbf{I}$ denotes the identity matrix of the appropriate dimension, $(\cdot)^*$ denotes the Hermitian conjugate of the corresponding matrix, and $\widehat{(\cdot)}$ denotes an estimated variable. The eigenvector of a matrix $\mathbf{A}$ associated with the largest eigenvalue is denoted by $\mathcal{P}\{\mathbf{A}\}$.

5.2.1 Background

Beamforming methods are used extensively in a variety of areas, where one of their goals is to estimate the source signal amplitude $s(t)$ from the array observations

$$\mathbf{y}(t) = s(t)\mathbf{a} + \mathbf{i}(t) + \mathbf{e}(t), \quad 1 \le t \le N, \tag{5.1}$$

where $\mathbf{y}(t) \in \mathbb{C}^M$ is the complex vector of array observations at time t with M being the number of array sensors, $s(t)$ is the signal amplitude, $\mathbf{a}$ is the signal steering vector, $\mathbf{i}(t)$ is the interference, $\mathbf{e}(t)$ is a Gaussian noise vector and N is the number of snapshots [4, 6, 8]. In the above model we implicitly made the common assumption of a narrow-band signal.

The source signal amplitude $s(t)$ may be a deterministic unknown signal, such as a complex sinusoid, or a stochastic stationary process with unknown signal power. For concreteness, in our development below we will treat $s(t)$ as a deterministic signal. However, as we show analytically in Section 5.2.2 and through simulations in Section 5.6, the optimality properties of the algorithms we develop are valid also in the case of stochastic signals.

In some applications, such as in the case of a fully calibrated array, the steering vector can be assumed to be known exactly. In this case, we treat $\mathbf{a}$ as a known deterministic vector. However, in practice the array response may have some uncertainties or perturbations in the steering vectors. These perturbations may be due to errors in sensor positions, gains or phases, mutual couplings between sensors, receiver fluctuations due to temperature changes, quantization effects, and coherent and incoherent local scatters [14, 22]. To account for these uncertainties, several authors have tried modeling some of their effects [24, 25]. However these perturbations often take place simultaneously, which significantly complicates the model. Instead, the uncertainty in $\mathbf{a}$ can be taken into account by treating it as a deterministic vector that lies in an ellipsoid centered at a nominal steering vector [14, 15]. An alternative approach has been to treat the steering vector as a random vector assuming knowledge of its distribution [13] or the second-order statistics [16, 26–32]. In the latter case the mean value of $\mathbf{a}$ corresponds to the nominal steering vector, and the covariance matrix captures its perturbations. In this chapter we consider the cases where $\mathbf{a}$ is known (Section 5.3) or random with known second-order statistics (Section 5.4).

Our goal is to estimate the signal amplitude $s(t)$ from the observations $\mathbf{y}(t)$ using a set of beamformer weights $\mathbf{w}(t)$, where the output of a narrowband beamformer is given by

$$\hat{s}(t) = \mathbf{w}^*(t)\mathbf{y}(t), \quad 1 \leq t \leq N. \tag{5.2}$$

To illustrate our approach, in this section we focus primarily on the case in which the steering vector $\mathbf{a}$ is assumed to be known exactly. As we show in Section 5.4, the essential ideas we outline for this case can also be applied to the setting in which $\mathbf{a}$ is random.

Traditionally, the beamformer weights $\mathbf{w}(t) = \mathbf{w}$ (where we omitted the index t for brevity) are chosen to maximize the SINR, which in the case of a known steering vector is given by

$$\text{SINR} \propto \frac{|\mathbf{w}^*\mathbf{a}|^2}{\mathbf{w}^*\mathbf{R}\mathbf{w}}, \tag{5.3}$$

where

$$\mathbf{R} = E\{(\mathbf{i}+\mathbf{n})(\mathbf{i}+\mathbf{n})^*\} \tag{5.4}$$

is the interference-plus-noise covariance matrix. The weight vector maximizing the SINR is

$$\mathbf{w}_{\text{MVDR}} = \frac{1}{\mathbf{a}^*\mathbf{R}^{-1}\mathbf{a}}\mathbf{R}^{-1}\mathbf{a}. \tag{5.5}$$

The solution (5.5) is commonly referred to as the minimum variance distortionless response (MVDR) beamformer, since it can also be obtained as the solution to

$$\min_{\mathbf{w}} \mathbf{w}^*\mathbf{R}\mathbf{w} \quad \text{subject to} \quad \mathbf{w}^*\mathbf{a} = 1. \tag{5.6}$$

In practice, the interference-plus-noise covariance matrix $\mathbf{R}$ is often not available. In such cases, the exact covariance $\mathbf{R}$ in (5.5) is replaced by an estimated covariance. Various methods exist for estimating the covariance $\mathbf{R}$. The simplest approach is to choose the estimate as the sample covariance

$$\hat{\mathbf{R}}_{sm} = \frac{1}{N}\sum_{t=1}^{N}\mathbf{y}(t)\mathbf{y}(t)^*. \tag{5.7}$$

The resulting beamformer is referred to as the sample matrix inversion (SMI) beamformer or the Capon beamformer [33, 34]. (For simplicity, we assume here that the sample covariance matrix is invertible.) If the signal is present in the training data, then it is well known that the performance of the MVDR beamformer with $\mathbf{R}$ replaced by $\hat{\mathbf{R}}_{sm}$ of (5.7) degrades considerably [11].

An alternative approach for estimating $\mathbf{R}$ is the diagonal loading approach, in which the estimate is chosen as

$$\hat{\mathbf{R}}_{dl} = \hat{\mathbf{R}}_{sm} + \xi\mathbf{I} = \frac{1}{N}\sum_{t=1}^{N}\mathbf{y}(t)\mathbf{y}(t)^* + \xi\mathbf{I}, \tag{5.8}$$

where ξ is the diagonal loading factor. The resulting beamformer is referred to as the loaded SMI or the loaded Capon beamformer [9, 10]. Various methods have been proposed for choosing the diagonal loading factor ξ; see for example, [10]. A heuristic choice for ξ, which is common in applications, is $\xi \approx 10\sigma^2$, where σ^2 is the noise power in a single sensor.

Another popular approach to estimating $\mathbf{R}$ is the eigenspace approach [11], in which the inverse of the covariance matrix is estimated as

$$(\hat{\mathbf{R}}_{eig})^{-1} = (\hat{\mathbf{R}}_{sm})^{-1}\mathbf{P}_s, \tag{5.9}$$

where $\mathbf{P}_s$ is the orthogonal projection onto the sample signal + interference subspace, that is, the subspace corresponding to the $D+1$ largest eigenvalues of $\hat{\mathbf{R}}_{sm}$, were D is the known rank of the interference subspace.

Note that the Capon, loaded Capon and eigenspace beamformers, can all be viewed as MVDR beamformers with a particular estimate of $\mathbf{R}$ (or its inverse).

The class of MVDR beamformers assumes explicitly that the steering vector $\mathbf{a}$ is known exactly. Recently, several robust beamformers have been proposed for the case in which the steering vector is not known precisely, but rather lies in some uncertainty set [14–16]. Although originally developed to deal with steering vector mismatch, the authors of the referenced papers suggest using these robust methods even in the case in which $\mathbf{a}$ is known, in order to deal with the mismatch in the interference-plus-noise covariance, namely the finite sample effects, and the fact that the signal is typically present in the training data. Each of the above robust methods is designed to maximize a measure of SINR on the uncertainty set. Specifically, in [14], the authors suggest minimizing $\mathbf{w}^*\hat{\mathbf{R}}_{sm}\mathbf{w}$ subject to the constraint that $|\mathbf{w}^*\mathbf{c}| \geq 1$ for all possible values of the steering vector $\mathbf{c}$, where $\|\mathbf{c} - \mathbf{a}\| \leq \epsilon$. The resulting beamformer is given by

$$\mathbf{w} = \frac{\lambda}{\lambda\mathbf{a}^*(\hat{\mathbf{R}}_{sm} + \lambda\epsilon^2\mathbf{I})^{-1}\mathbf{a} - 1}\left(\hat{\mathbf{R}}_{sm} + \lambda\epsilon^2\mathbf{I}\right)^{-1}\mathbf{a}, \tag{5.10}$$

where λ is chosen such that $|\mathbf{w}^*\mathbf{a} - 1|^2 = \epsilon^2\mathbf{w}^*\mathbf{w}$. In practice, the solution can be found by using a second order cone program. In [15] the authors consider a similar approach in which they first estimate the steering vector by minimizing $\tilde{\mathbf{a}}^*\hat{\mathbf{R}}_{sm}^{-1}\tilde{\mathbf{a}}$ with respect to $\tilde{\mathbf{a}}$ subject to $\|\tilde{\mathbf{a}} - \mathbf{a}\|^2 = \epsilon$, and then use $\sqrt{M}\tilde{\mathbf{a}}/\|\tilde{\mathbf{a}}\|$ in the MVDR beamformer, which results in

$$\mathbf{w} = \frac{\alpha}{\sqrt{M}} \cdot \frac{(\lambda\hat{\mathbf{R}}_{sm} + \mathbf{I})^{-1}\mathbf{a}}{\mathbf{a}^*(\lambda\hat{\mathbf{R}}_{sm} + \mathbf{I})^{-1}\hat{\mathbf{R}}_{sm}(\lambda\hat{\mathbf{R}}_{sm} + \mathbf{I})^{-1}\mathbf{a}}. \tag{5.11}$$

Here λ is chosen such that $\|(\mathbf{I} + \lambda\hat{\mathbf{R}}_{sm})^{-1}\mathbf{a}\|^2 = \epsilon$, and $\alpha = \|(\hat{\mathbf{R}}_{sm}^{-1} + \lambda\mathbf{I})^{-1}\mathbf{a}\|$. Finally, in [16] the authors consider a general-rank signal model. Adapting their results to the rank-one steering vector case, their beamformer is the solution to minimizing $\mathbf{w}^*\hat{\mathbf{R}}_{dl}\mathbf{w}$ subject to $|\mathbf{w}^*\mathbf{a}|^2 \geq 1 - \mathbf{w}^*\Delta\mathbf{w}$ for all $\|\Delta\| \leq \epsilon$, and is given by

$$\mathbf{w} = \alpha\mathcal{P}\left\{\hat{\mathbf{R}}_{dl}^{-1}(\mathbf{aa}^* - \epsilon\mathbf{I})\right\}, \tag{5.12}$$

where α is chosen such that $\mathbf{w}^*(\mathbf{aa}^* - \epsilon\mathbf{I})\mathbf{w} = 1$.

The motivation behind the class of MVDR beamformers and the robust beamformers is to maximize the SINR. However, choosing $\mathbf{w}$ to maximize the SINR does not necessarily result in an estimated signal amplitude $\hat{s}(t)$ that is close to $s(t)$. In an *estimation* context, where our goal is to design a beamformer in order to obtain an estimate $\hat{s}(t)$ that is close to $s(t)$, it would make more sense to choose the weights $\mathbf{w}$ to minimize the MSE rather than to maximize the SINR, which is not directly related to the estimation error $\hat{s}(t) - s(t)$.

5.2.2 MSE Beamforming

If $\hat{s} = \mathbf{w}^*\mathbf{y}$, then, assuming that s is deterministic, the MSE between s and $\hat{s}$ is given by

$$E\left\{|\hat{s} - s|^2\right\} = V(\hat{s}) + |B(\hat{s})|^2 = \mathbf{w}^*\mathbf{R}\mathbf{w} + |s|^2|1 - \mathbf{w}^*\mathbf{a}|^2, \tag{5.13}$$

where $V(\hat{s}) = E\{|\hat{s} - E\{\hat{s}\}|^2\}$ is the variance of the estimate $\hat{s}$ and $B(\hat{s}) = E\{\hat{s}\} - s$ is the bias. In the case in which s is a zero-mean random variable with variance σ_s^2, the MSE is given by

$$E\left\{|\hat{s} - s|^2\right\} = \mathbf{w}^*\mathbf{R}\mathbf{w} + \sigma_s^2|1 - \mathbf{w}^*\mathbf{a}|^2. \tag{5.14}$$

Comparing (5.13) and (5.14) we see that the expressions for the MSE have the same form in the deterministic and stochastic cases, where $|s|^2$ in the deterministic case is replaced by σ_s^2 in the stochastic case. For concreteness, in the discussion in the rest of the chapter we assume the deterministic model. However, all the results hold true for the stochastic model where we replace $|s|^2$ everywhere with σ_s^2. In particular, in the development of the minimax MSE and regret beamformers in the stochastic case, the bounds on $|s|^2$ are replaced by bounds on the signal variance σ_s^2.

The minimum MSE (MMSE) beamformer minimizing the MSE when $|s|$ is known is obtained by differentiating (5.13) with respect to $\mathbf{w}$ and equating to 0, which results in

$$\mathbf{w}(s) = |s|^2(\mathbf{R} + |s|^2\mathbf{a}\mathbf{a}^*)^{-1}\mathbf{a}. \tag{5.15}$$

Using the matrix inversion lemma we can express $\mathbf{w}(s)$ as

$$\mathbf{w}(s) = \frac{|s|^2}{1 + |s|^2\mathbf{a}^*\mathbf{R}^{-1}\mathbf{a}}\mathbf{R}^{-1}\mathbf{a}. \tag{5.16}$$

The MMSE beamformer can alternatively be written as

$$\mathbf{w}(s) = \frac{|s|^2\mathbf{a}^*\mathbf{R}^{-1}\mathbf{a}}{1 + |s|^2\mathbf{a}^*\mathbf{R}^{-1}\mathbf{a}} \cdot \frac{\mathbf{R}^{-1}\mathbf{a}}{\mathbf{a}^*\mathbf{R}^{-1}\mathbf{a}} = \beta(s)\mathbf{w}_{\mathrm{MVDR}}, \tag{5.17}$$

where

$$\beta(s) = \frac{|s|^2\mathbf{a}^*\mathbf{R}^{-1}\mathbf{a}}{1 + |s|^2\mathbf{a}^*\mathbf{R}^{-1}\mathbf{a}}. \tag{5.18}$$

The scaling $\beta(s)$ satisfies $0 \leq \beta(s) \leq 1$, and is monotonically increasing in $|s|^2$. Therefore, for any $|s|^2$, $\|\mathbf{w}(s)\| \leq \|\mathbf{w}_{\mathrm{MVDR}}\|$. Substituting $\mathbf{w}(s)$ back into (5.13), the smallest possible MSE, which we denote by $\mathrm{MSE}_{\mathrm{OPT}}$, is given by

$$\mathrm{MSE}_{\mathrm{OPT}} = \frac{|s|^2}{1 + |s|^2\mathbf{a}^*\mathbf{R}^{-1}\mathbf{a}}. \tag{5.19}$$

Using (5.13) we can also compute the MSE of the MVDR beamformer (5.5), which maximizes the SINR, assuming that $\mathbf{R}$ is known. Substituting $\mathbf{w} = (1/\mathbf{a}^*\mathbf{R}^{-1}\mathbf{a})\mathbf{R}^{-1}\mathbf{a}$ into (5.13), the MSE is

$$\mathrm{MSE}_{\mathrm{MVDR}} = \frac{1}{\mathbf{a}^*\mathbf{R}^{-1}\mathbf{a}}. \tag{5.20}$$

Comparing (5.19) with (5.20),

$$\mathrm{MSE}_{\mathrm{OPT}} = \frac{|s|^2}{1 + |s|^2\mathbf{a}^*\mathbf{R}^{-1}\mathbf{a}} = \frac{1}{1/|s|^2 + \mathbf{a}^*\mathbf{R}^{-1}\mathbf{a}} \leq \frac{1}{\mathbf{a}^*\mathbf{R}^{-1}\mathbf{a}} = \mathrm{MSE}_{\mathrm{MVDR}}. \tag{5.21}$$

Thus, as we expect, the MMSE beamformer always results in a smaller MSE than the MVDR beamformer. Therefore, in an estimation context where our goal is to estimate the signal amplitude, the MMSE beamformer will lead to better average performance.

From (5.17) we see that the MMSE beamformer is just a shrinkage of the MVDR beamformer (as we will show below, this is no longer true in the case of a random steering vector). Therefore, the two beamformers will result in the same SINR, so that the MMSE beamformer also maximizes the SINR. However, as (5.21) shows, this shrinkage factor impacts the MSE so that the MMSE beamformer has better MSE performance. To illustrate the advantage of the MMSE beamformer, we consider a numerical example. The scenario consists of a uniform linear array (ULA) of $M = 20$ omnidirectional sensors spaced half a wavelength apart. We choose s as a complex sinewave with varying amplitude to obtain the desired SNR in each sensor; its plane-wave has a DOA of 30° relative to the array normal. The noise $\mathbf{e}$ is a zero-mean, Gaussian, complex random vector, temporally and spatially white, with a power of 0 dB in each sensor. The interference is given by $\mathbf{i} = \mathbf{a}_i i$ where i is a zero-mean, Gaussian, complex process temporally white with interference-plus-noise ratio (INR) of 20 dB and $\mathbf{a}_i$ is the interference steering vector with DOA $= -30°$. Assuming knowledge of $|s|^2$, $\mathbf{R}$ and $\mathbf{a}$, we evaluate the square-root of the normalized MSE (NMSE) over a time-window of $N = 100$ samples, where each result is obtained by averaging 200 Monte Carlo simulations.

Figure 5.1 illustrates the NMSE of the MMSE and MVDR beamformers when estimating the complex sinewave, as a function of SNR. It can be seen that the MMSE beamformer outperforms the MVDR beamformer in all the illustrated SNR range. As the SNR increases, the MVDR beamformer converges to the MMSE beamformer, since $\beta(s)$ in (5.17) approaches to 1. The absolute value of the original signal and its estimates obtained from the MMSE and MVDR beamformers are illustrated in Figure 5.2, for an SNR of -10 dB. Clearly, the MMSE beamformer leads to a better estimate than the MVDR beamformer.

The difference between the MSE and SINR based approaches is more pronounced in the case of a random steering vector. Suppose that $\mathbf{a}$ is a random

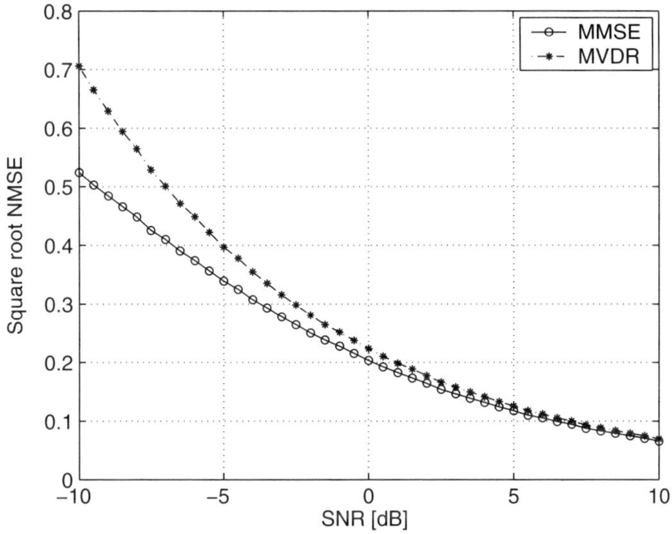

Figure 5.1 Square-root of the normalized MSE as a function of SNR using the MMSE and MVDR beamformers when estimating a complex sinewave with DOA 30°, in the presence of an interference with INR = 20 dB and DOA = −30°. We assume that $|s|^2$, the noise-plus-interference covariance matrix **R** and the steering vector **a** are known.

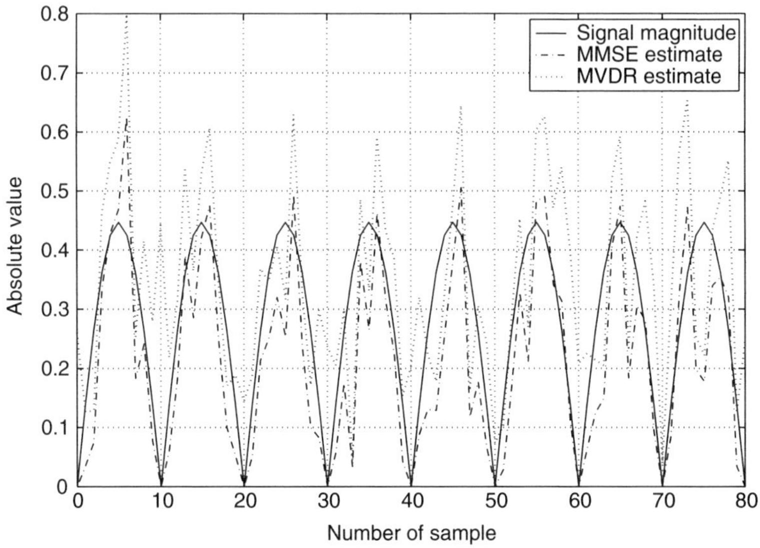

Figure 5.2 Absolute values of the true complex sinewave, and its estimates obtained by the MMSE and MVDR beamformers for SNR = −10 dB, in the presence of an interference with INR = 20 dB and DOA = −30°. We assume that $|s|^2$, the noise-plus-interference covariance matrix **R** and the steering vector **a** are known.

vector with mean $\mathbf{m}$ and covariance matrix $\mathbf{C}$. In Section 5.4 we consider this case in detail, and show that the beamformer minimizing the MSE is

$$\mathbf{w}(s) = \frac{|s|^2}{1 + |s|^2\mathbf{m}^*(\mathbf{R} + |s|^2\mathbf{C})^{-1}\mathbf{m}}(\mathbf{R} + |s|^2\mathbf{C})^{-1}\mathbf{m}. \tag{5.22}$$

On the other hand, the beamformer maximizing the SINR is given by

$$\mathbf{w} = \alpha\mathcal{P}\{\mathbf{R}^{-1}(\mathbf{C} + \mathbf{mm}^*)\}, \tag{5.23}$$

where α is chosen such that $\mathbf{w}^*(\mathbf{C} + \mathbf{mm}^*)\mathbf{w} = 1$. We refer to this beamformer as the principal eigenvector (PEIG) beamformer [35]. Comparing (5.22) and (5.23) we see that in this case the two beamformers are in general not scalar versions of each other. As we now illustrate, the MMSE beamformer can result in a much better estimate of the signal amplitude and waveform than the SINR-based PEIG beamformer.

To illustrate the advantage of the MMSE beamformer in the case of a random steering vector, we consider an example in which the DOA of the signal is random. The scenario is similar to that of the previous example, where in place of a constant signal DOA, the DOA is now given by a Gaussian random variable with mean equal to 30° and standard deviation equal to 1 (about $\pm 3^\circ$). The mean $\mathbf{m}$ and covariance matrix $\mathbf{C}$ of the steering vector are estimated from 2000 realizations of the steering vector. Assuming knowledge of $|s|^2$ and $\mathbf{R}$, we evaluate the NMSE over a time-window of $N = 100$ samples, where each result is obtained by averaging 200 Monte Carlo simulations.

Figure 5.3 illustrates the NMSE of the MMSE and PEIG beamformers as a function of SNR. It can be seen that the NMSE of the MMSE beamformer is substantially lower than that of the PEIG beamformer. The absolute value of the original signal and its estimates obtained from the MMSE and PEIG beamformers are illustrated in Figure 5.4 for an SNR of -10 dB. Clearly, the MMSE beamformer leads to a better signal estimate than the PEIG beamformer. It is also evident from this example that the signal waveform estimate obtained from both beamformers is different: the MMSE beamformer leads to a signal waveform that is much closer to the original waveform than the PEIG beamformer.

5.2.3 Robust MMSE Beamforming

Unfortunately, both in the case of known $\mathbf{a}$ and in the case of random $\mathbf{a}$ the MMSE beamformer depends explicitly on $|s|$ which is typically unknown. Therefore, in practice, we cannot implement the MMSE beamformer. The problem stems from the fact that the MSE depends explicitly on $|s|$. To illustrate the main ideas, in the remainder of this section we focus on the case of a deterministic steering vector.

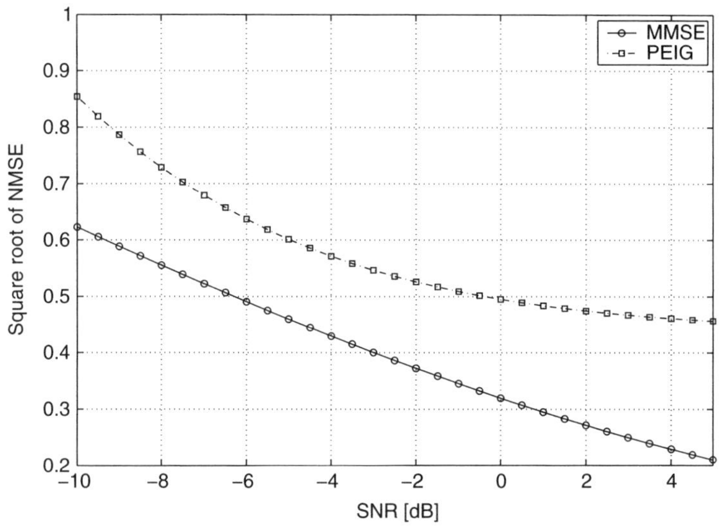

Figure 5.3 Square-root of the normalized MSE as a function of SNR using the MMSE and PEIG beamformers when estimating a complex sinewave with random DOA, in the presence of an interference with INR = 20 dB and DOA = −30°. We assume that $|s|^2$, the noise-plus-interference covariance matrix **R**, and the mean **m** and covariance **C** of the steering vector **a** are known.

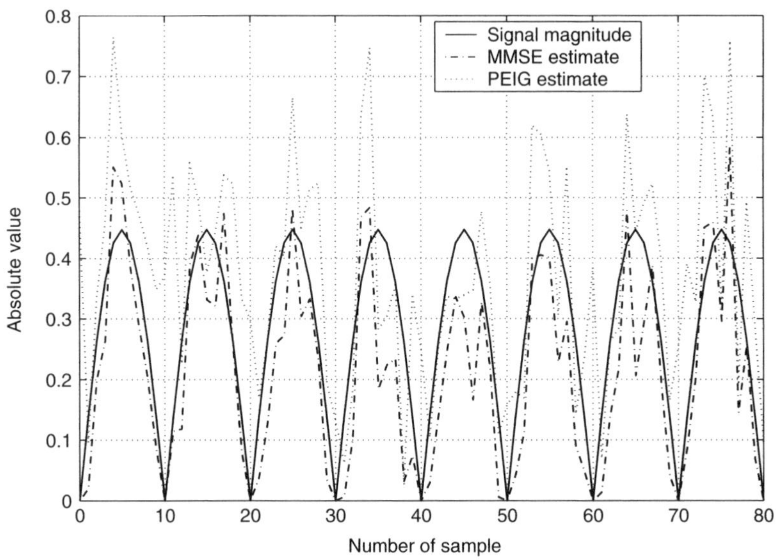

Figure 5.4 Absolute values of the true complex sinewave with random DOA, and its estimates obtained by the MMSE and PEIG beamformers for SNR = −10 dB, in the presence of an interference with INR = 20 dB and DOA = −30°. We assume that $|s|^2$, the noise-plus-interference covariance matrix **R**, and the mean **m** and covariance **C** of the steering vector **a** are known.

One approach to obtain a beamformer that does not depend on $|s|$ in this case is to force the term depending on $|s|$, namely the bias, to 0, and then minimize the MSE, that is,

$$\min_{\mathbf{w}} \mathbf{w}^* \mathbf{R} \mathbf{w} \quad \text{subject to} \quad \mathbf{w}^* \mathbf{a} = 1, \tag{5.24}$$

which leads to the class of MVDR beamformers. Thus, in addition to maximizing the SINR, the MVDR beamformer minimizes the MSE subject to the constraint that the bias in the estimator $\hat{s}$ is equal to 0. However, this does not guarantee a small MSE, so that on average, the resulting estimate of s may be far from s. Indeed, it is well known that unbiased estimators may often lead to large MSE values. The attractiveness of the SINR criterion is the fact that it is easy to solve, and leads to a beamformer that does not depend on $|s|$. Its drawback is that it does not necessarily lead to a small MSE, as can also be seen in Figures 5.1 and 5.3.

We note, that as we discuss in Section 5.4.1, the property of the MVDR beamformer that it minimizes the MSE subject to a zero bias constrain no longer holds in the random steering vector case. In fact, when $\mathbf{a}$ is random with positive covariance matrix, there is in general no choice of linear beamformer for which the MSE is independent of $|s|$.

Instead of forcing the term depending on $|s|$ to zero, it would be desirable to design a robust beamformer whose MSE is reasonably small across all possible values of $|s|$. To this end we need to define the set of possible values of $|s|$. In some practical applications, we may know a priori bounds on $|s|$, for example when the type of the source and the possible distances from the array are known, as can happen for instance in wireless communications and underwater source localization. We may have an upper bound of the form $|s| \leq U$, or we may have both an upper and (nonzero) lower bound, so that

$$L \leq |s| \leq U. \tag{5.25}$$

In our development of the minimax robust beamformers, we will assume that the bounds L and U are known. In practice, if no such bounds are given a priori, then we can estimate them from the data, as we elaborate on further in Sections 5.5 and 5.6. This is similar in spirit to the MVDR-based beamformers: In developing the MVDR beamformer it is assumed that the interference-plus-noise covariance matrix $\mathbf{R}$ is known; however, in practice, this matrix is estimated from the data.

Given an uncertainty set of the form (5.25), we may seek a beamformer that minimizes a worst-case MSE measure on this set. In the next section, we rely on ideas of [21] and [20], and propose two robust beamformers. We first assume that only an upper bound on $|s|$ is given, and develop a minimax MSE beamformer that minimizes the worst-case MSE over all $|s| \leq U$. As we show in (5.34), this beamformer is also minimax subject to (5.25). We then develop a minimax regret beamformer over the set defined by (5.25), that minimizes the worst-case difference between the MSE attainable with a beamformer that does not know $|s|$, and the optimal

MSE of the MMSE beamformer that minimizes the MSE when $|s|$ is known. In Section 5.4 we develop minimax MSE and minimax regret beamformers for the case of a random steering vector.

5.3 MINIMAX MSE BEAMFORMING FOR KNOWN STEERING VECTOR

We now treat two MSE-based criteria for developing robust beamformers when the steering vector is known: In Section 5.3.1 we consider a minimax MSE approach and in Section 5.3.2, a minimax regret approach. In our development, we assume that the covariance matrix $\mathbf{R}$ is known. In practice, as we discuss in Sections 5.5 and 5.6, the unknown $\mathbf{R}$ is replaced by an estimate $\hat{\mathbf{R}}$.

5.3.1 Minimax MSE Beamforming

The first approach we consider for developing a robust beamformer it to minimize the worst-case MSE over all bounded values of $|s|$. Thus, we seek the beamformer that is the solution to

$$\min_{\mathbf{w}} \max_{|s|\leq U} E\{|\hat{s} - s|^2\} = \min_{\mathbf{w}} \max_{|s|\leq U} \{\mathbf{w}^*\mathbf{R}\mathbf{w} + |s|^2|1 - \mathbf{w}^*\mathbf{a}|^2\}. \tag{5.26}$$

This ensures that in the worst case, the MSE of the resulting beamformer is minimized.

Now,

$$\max_{|s|\leq U} \{\mathbf{w}^*\mathbf{R}\mathbf{w} + |s|^2|1 - \mathbf{w}^*\mathbf{a}|^2\} = \mathbf{w}^*\mathbf{R}\mathbf{w} + U^2|1 - \mathbf{w}^*\mathbf{a}|^2, \tag{5.27}$$

which is equal to the MSE of (5.13) with $|s|^2 = U$. It follows that the minimax MSE beamformer, denoted $\mathbf{w}_{\mathrm{MXM}}$, is an MMSE beamformer of the form (5.16) matched to $|s| = U$:

$$\mathbf{w}_{\mathrm{MXM}} = \frac{U^2}{1 + U^2\mathbf{a}^*\mathbf{R}^{-1}\mathbf{a}}\mathbf{R}^{-1}\mathbf{a} = \beta_{\mathrm{MXM}}\mathbf{w}_{\mathrm{MVDR}}, \tag{5.28}$$

where

$$\beta_{\mathrm{MXM}} = \frac{U^2\mathbf{a}^*\mathbf{R}^{-1}\mathbf{a}}{1 + U^2\mathbf{a}^*\mathbf{R}^{-1}\mathbf{a}}. \tag{5.29}$$

The resulting MSE is

$$\mathrm{MSE}_{\mathrm{MXM}} = \frac{U^4\mathbf{a}^*\mathbf{R}^{-1}\mathbf{a} + |s|^2}{(1 + U^2\mathbf{a}^*\mathbf{R}^{-1}\mathbf{a})^2}. \tag{5.30}$$

For any $|s|^2 \leq U^2$, we have that

$$\mathrm{MSE}_{\mathrm{MXM}} \leq \frac{U^2}{1 + U^2 \mathbf{a}^* \mathbf{R}^{-1} \mathbf{a}}. \tag{5.31}$$

For comparison, we have seen in (5.20) that the MSE of the MVDR beamformer is

$$\mathrm{MSE}_{\mathrm{MVDR}} = \frac{1}{\mathbf{a}^* \mathbf{R}^{-1} \mathbf{a}}, \tag{5.32}$$

from which we have immediately that for any choice of U,

$$\mathrm{MSE}_{\mathrm{MVDR}} > \mathrm{MSE}_{\mathrm{MXM}}. \tag{5.33}$$

The inequality (5.33) is valid as long as the true covariance $\mathbf{R}$ is known. When $\mathbf{R}$ is estimated from the data, (5.33) is no longer true in general. Nonetheless, as we will see in the simulations in Section 5.6, (5.33) typically holds even when both U and $\mathbf{R}$ are estimated from the data.

Finally, we point out that the minimax MSE beamformer $\mathbf{w}_{\mathrm{MXM}}$ of (5.30) is also the solution in the case in which we have both an upper and lower bound on the norm of s. This follows from the fact that

$$\max_{L \leq |s| \leq U} |s|^2 |1 - \mathbf{w}^* \mathbf{a}|^2 = \max_{|s| \leq U} |s|^2 |1 - \mathbf{w}^* \mathbf{a}|^2, \tag{5.34}$$

since the maximum is obtained at $|s| = U$. Thus, in contrast with the minimax regret beamformer developed in the next section, the minimax MSE beamformer does not take the lower bound (if available) into account and therefore may be overconservative when such a lower bound is known.

5.3.2 Minimax Regret Beamforming

We have seen that the minimax MSE beamformer is an MMSE beamformer matched to the worst possible choice of $|s|$, namely $|s| = U$. In some practical applications, particularly when a lower bound on $|s|$ is known, this approach may be overconservative. Although it optimizes the performance in the worst case, it may lead to deteriorated performance in other cases. To overcome this possible limitation, in this section we develop a minimax regret beamformer whose performance is as close as possible to that of the MMSE beamformer that knows s, for all possible values of s in a prespecified region of uncertainty. Thus, we ensure that over a wide range of values of s, our beamformer will result in a relatively low MSE.

In [21], a minimax difference regret estimator was derived for the problem of estimating an unknown vector $\mathbf{x}$ in a linear model $\mathbf{y} = \mathbf{H}\mathbf{x} + \mathbf{n}$, where $\mathbf{H}$ is a known linear transformation, and $\mathbf{n}$ is a noise vector with known covariance matrix. The estimator was designed to minimize the worst case regret over all bounded vectors $\mathbf{x}$, namely vectors satisfying $\mathbf{x}^* \mathbf{T} \mathbf{x} \leq U^2$ for some $U > 0$ and positive definite matrix $\mathbf{T}$. It was shown that the linear minimax regret estimator can be found as a solution to a convex optimization problem that can be solved very efficiently.

In our problem, the unknown parameter $\mathbf{x} = s$ is a scalar, so that an explicit solution can be derived, as we show below (see also [36]). Furthermore, in our development we consider both lower and upper bounds on $|s|$, so that we seek the beamformer that minimizes the worst case regret over the uncertainty region (5.25).

The minimax regret beamformer $\mathbf{w}_{\text{MXR}}$ is designed to minimize the worst-case regret subject to the constraint $L \leq |s| \leq U$, where the regret, denoted $\mathcal{R}(s, \mathbf{w})$, is defined as the difference between the MSE using an estimator $\hat{s} = \mathbf{w}^*\mathbf{y}$ and the smallest possible MSE attainable with an estimator of the form $\hat{s} = \mathbf{w}^*(s)\mathbf{y}$ when s is known, so that $\mathbf{w}$ can depend explicitly on s. We have seen in (5.19) that since we are restricting ourselves to linear beamformers, even in the case in which the beamformer can depend on s, the minimal attainable MSE is not generally equal to zero. The best possible MSE is illustrated schematically in Figure 5.5. Instead of seeking an estimator to minimize the worst-case MSE, we therefore propose designing an estimator to minimize the worst-case difference between its MSE and the best possible MSE, as illustrated in Figure 5.5.

Using (5.19), we can express the regret as

$$\begin{aligned}\mathcal{R}(s, \mathbf{w}) &= E\left\{|\mathbf{w}^*\mathbf{y} - s|^2\right\} - \text{MSE}_{\text{OPT}} \\ &= \mathbf{w}^*\mathbf{R}\mathbf{w} + |s|^2|1 - \mathbf{w}^*\mathbf{a}|^2 - \frac{|s|^2}{1 + |s|^2\mathbf{a}^*\mathbf{R}^{-1}\mathbf{a}}.\end{aligned} \tag{5.35}$$

Thus $\mathbf{w}_{\text{MXR}}$ is the solution to

$$\begin{aligned}&\min_{\mathbf{w}} \max_{L \leq |s| \leq U} \mathcal{R}(s, \mathbf{w}) \\ &= \min_{\mathbf{w}} \left\{ \mathbf{w}^*\mathbf{R}\mathbf{w} + \max_{L \leq |s| \leq U} \left\{ |s|^2|1 - \mathbf{w}^*\mathbf{a}|^2 - \frac{|s|^2}{1 + |s|^2\mathbf{a}^*\mathbf{R}^{-1}\mathbf{a}} \right\} \right\}.\end{aligned} \tag{5.36}$$

The minimax regret beamformer is given by the following theorem.

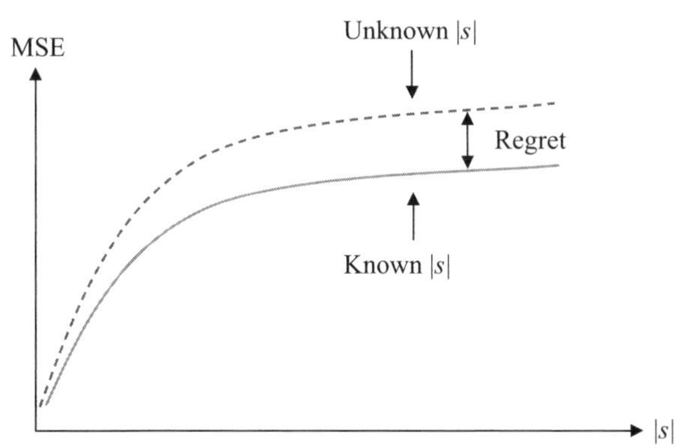

Figure 5.5 The solid line represents the best attainable MSE as a function of $|s|$ when $|s|$ is known. The dashed line represents a desirable graph of MSE with small regret as a function of $|s|$ using some linear estimator that does not depend on $|s|$.

Theorem 1. Let s denote an unknown signal amplitude in the model $\mathbf{y} = s\mathbf{a} + \mathbf{n}$, where $\mathbf{a}$ is a known length-M vector, and $\mathbf{n}$ is a zero-mean random vector with covariance $\mathbf{R}$. Then the solution to the problem

$$\min_{\hat{s}=\mathbf{w}^*\mathbf{y}} \max_{L\leq|s|\leq U} \left\{ E\{|\hat{s} - s|^2\} - \min_{\hat{s}=\mathbf{w}^*(s)\mathbf{y}} E\{|\hat{s} - s|^2\} \right\}$$

$$= \min_{\mathbf{w}} \max_{L\leq|s|\leq U} \left\{ \mathbf{w}^*\mathbf{R}\mathbf{w} + |s|^2(1 - \mathbf{w}^*\mathbf{a})^2 - \frac{|s|^2}{1 + |s|^2\mathbf{a}^*\mathbf{R}^{-1}\mathbf{a}} \right\}$$

is

$$\hat{s} = \left(1 - \frac{1}{\sqrt{(1 + L^2\mathbf{a}^*\mathbf{R}^{-1}\mathbf{a})(1 + U^2\mathbf{a}^*\mathbf{R}^{-1}\mathbf{a})}}\right) \frac{\mathbf{a}^*\mathbf{R}^{-1}\mathbf{y}}{\mathbf{a}^*\mathbf{R}^{-1}\mathbf{a}}.$$

Before proving Theorem 1, we first comment on some of the properties of the minimax regret beamformer, which, from the theorem, is given by

$$\mathbf{w}_{\text{MXR}} = \left(1 - \frac{1}{\sqrt{(1 + L^2\mathbf{a}^*\mathbf{R}^{-1}\mathbf{a})(1 + U^2\mathbf{a}^*\mathbf{R}^{-1}\mathbf{a})}}\right) \frac{1}{\mathbf{a}^*\mathbf{R}^{-1}\mathbf{a}} \mathbf{R}^{-1}\mathbf{a}. \tag{5.37}$$

Comparing (5.37) with (5.5) we see that the minimax regret beamformer is a scaled version of the MVDR beamformer, that is, $\mathbf{w}_{\text{MXR}} = \beta_{\text{MXR}}\mathbf{w}_{\text{MVDR}}$ where

$$\beta_{\text{MXR}} = 1 - \frac{1}{\sqrt{(1 + L^2\mathbf{a}^*\mathbf{R}^{-1}\mathbf{a})(1 + U^2\mathbf{a}^*\mathbf{R}^{-1}\mathbf{a})}}. \tag{5.38}$$

Clearly, the scaling β_{MXR} satisfies $0 \leq \beta_{\text{MXR}} \leq 1$, and is monotonically increasing in L and U. In addition, when $U = L$,

$$\beta_{\text{MXR}} = 1 - \frac{1}{1 + U^2\mathbf{a}^*\mathbf{R}^{-1}\mathbf{a}^2} = \frac{U^2\mathbf{a}^*\mathbf{R}^{-1}\mathbf{a}}{1 + U^2\mathbf{a}^*\mathbf{R}^{-1}\mathbf{a}}, \tag{5.39}$$

in which case the minimax regret beamformer is equal to the minimax MSE beamformer of (5.30). For $L < U$, $\|\mathbf{w}_{\text{MXR}}\| < \|\mathbf{w}_{\text{MXM}}\|$.

It is also interesting to note that the minimax regret beamformer can be viewed as an MMSE beamformer of the form (5.16), matched to

$$|s|^2 = \frac{1}{\mathbf{a}^*\mathbf{R}^{-1}\mathbf{a}} \left(\sqrt{(1 + L^2\mathbf{a}^*\mathbf{R}^{-1}\mathbf{a})(1 + U^2\mathbf{a}^*\mathbf{R}^{-1}\mathbf{a})} - 1\right), \tag{5.40}$$

for arbitrary choices of L and U. This follows immediately from substituting $|s|^2$ of (5.40) into (5.16). Since the minimax regret estimator minimizes the MSE for the signal power given by (5.40), we may view this power as the least-favorable signal power in the regret sense.

The signal power (5.40) can be viewed as an estimate of the true, unknown signal power. To gain some insight into the estimate of (5.40) we note that

$$\left(\sqrt{(1+L^2\gamma)(1+U^2\gamma)}-1\right)\left(\sqrt{1+L^2\gamma}+\sqrt{1+U^2\gamma}\right)$$
$$=U^2\gamma\sqrt{1+L^2\gamma}+L^2\gamma\sqrt{1+U^2\gamma}, \tag{5.41}$$

where for brevity, we denoted $\gamma = \mathbf{a}^*\mathbf{R}^{-1}\mathbf{a}$. Substituting (5.41) into (5.40), we have that

$$|s|^2 = \frac{U^2\sqrt{1+L^2\mathbf{a}^*\mathbf{R}^{-1}\mathbf{a}}+L^2\sqrt{1+U^2\mathbf{a}^*\mathbf{R}^{-1}\mathbf{a}}}{\sqrt{1+L^2\mathbf{a}^*\mathbf{R}^{-1}\mathbf{a}}+\sqrt{1+U^2\mathbf{a}^*\mathbf{R}^{-1}\mathbf{a}}}. \tag{5.42}$$

From (5.42) it follows that the unknown signal power is estimated as a weighted combination of the power bounds U^2 and L^2, where the weights depend explicitly on the uncertainty set and on $\mu(L)$ and $\mu(U)$, where

$$\mu(T) = T^2\mathbf{a}^*\mathbf{R}^{-1}\mathbf{a}, \tag{5.43}$$

can be viewed as the SNR in the observations, when the signal power is T^2. If $\mu(L) \gg 1$, then from (5.40),

$$|s|^2 \approx \sqrt{U^2L^2}, \tag{5.44}$$

so that in this case the unknown signal power is estimated as the geometric mean of the power bounds. If, on the other hand, $\mu(L) \ll 1$, then from (5.42),

$$|s|^2 \approx \tfrac{1}{2}(L^2+U^2). \tag{5.45}$$

Thus, in this case the unknown signal power is estimated as the algebraic mean of the power bounds.

It is interesting to note that while the minimax MSE estimator of (5.28) is matched to a signal power U^2, the minimax difference regret estimator of (5.37) is matched to a signal power $|s|^2 \leq (U^2+L^2)/2$. This follows from (5.40) by using the inequality $\sqrt{ab} \leq (a+b)/2$.

We now prove Theorem 1. Although parts of the proof are similar to the proofs in [37], we repeat the arguments for completeness.

Proof. To develop a solution to (5.36), we first consider the inner maximization problem

$$\begin{aligned} f(\mathbf{w}) &= \max_{L \le |s| \le U} \left\{ |s|^2 |1 - \mathbf{w}^* \mathbf{a}|^2 - \frac{|s|^2}{1 + |s|^2 \gamma} \right\} \\ &= \max_{L^2 \le x \le U^2} \left\{ x |1 - \mathbf{w}^* \mathbf{a}|^2 - \frac{x}{1 + x\gamma} \right\}, \end{aligned} \tag{5.46}$$

where $x = |s|^2$. To derive an explicit expression for $f(\mathbf{w})$ we note that the function

$$h(x) = ax - \frac{bx}{c + dx} \tag{5.47}$$

with $b, c, d > 0$ is convex in $x \ge 0$. Indeed,

$$\frac{d^2 h}{dx^2} = 2 \frac{bd}{(c + dx)^3} > 0, \quad x \ge 0. \tag{5.48}$$

It follows that for fixed $\mathbf{w}$,

$$g(x) = x|1 - \mathbf{w}^* \mathbf{a}|^2 - \frac{x}{1 + x\gamma} \tag{5.49}$$

is convex in $x \ge 0$, and consequently the maximum of $g(x)$ over a closed interval is obtained at one of the boundaries. Thus,

$$\begin{aligned} f(\mathbf{w}) &= \max_{L^2 \le x \le U^2} g(x) \\ &= \max\left(g(L^2), g(U^2)\right) \\ &= \max\left(L^2 |1 - \mathbf{w}^* \mathbf{a}|^2 - \frac{L^2}{1 + L^2\gamma}, U^2 |1 - \mathbf{w}^* \mathbf{a}|^2 - \frac{U^2}{1 + U^2\gamma} \right), \end{aligned} \tag{5.50}$$

and the problem (5.36) reduces to

$$\min_{\mathbf{w}} \left\{ \mathbf{w}^* \mathbf{R} \mathbf{w} + \max\left(L^2 |1 - \mathbf{w}^* \mathbf{a}|^2 - \frac{L^2}{1 + L^2\gamma}, U^2 |1 - \mathbf{w}^* \mathbf{a}|^2 - \frac{U^2}{1 + U^2\gamma} \right) \right\}. \tag{5.51}$$

We now show that the optimal value of $\mathbf{w}$ has the form

$$\mathbf{w} = d(\mathbf{a}^* \mathbf{R}^{-1} \mathbf{a})^{-1} \mathbf{R}^{-1} \mathbf{a} = \frac{d}{\gamma} \mathbf{R}^{-1} \mathbf{a}, \tag{5.52}$$

for some (possibly complex) d. To this end, we first note that the objective in (5.51) depends on $\mathbf{w}$ only through $\mathbf{w}^* \mathbf{a}$ and $\mathbf{w}^* \mathbf{R} \mathbf{w}$. Now, suppose that we are given a beamformer $\tilde{\mathbf{w}}$, and let

$$\mathbf{w} = \frac{\mathbf{a}^* \tilde{\mathbf{w}}}{\gamma} \mathbf{R}^{-1}\mathbf{a}. \tag{5.53}$$

Then

$$\mathbf{w}^*\mathbf{a} = \frac{\tilde{\mathbf{w}}^*\mathbf{a}}{\gamma} \mathbf{a}^*\mathbf{R}^{-1}\mathbf{a} = \tilde{\mathbf{w}}^*\mathbf{a}, \tag{5.54}$$

and

$$\mathbf{w}^*\mathbf{R}\mathbf{w} = \frac{|\mathbf{a}^*\tilde{\mathbf{w}}|^2}{\gamma^2} \mathbf{a}^*\mathbf{R}^{-1}\mathbf{a} = \frac{|\mathbf{a}^*\tilde{\mathbf{w}}|^2}{\gamma}. \tag{5.55}$$

From the Cauchy-Schwarz inequality we have that for any vector $\mathbf{x}$,

$$|\mathbf{a}^*\mathbf{x}|^2 = |(\mathbf{R}^{-1/2}\mathbf{a})^*\mathbf{R}^{1/2}\mathbf{x}|^2 \leq \mathbf{a}^*\mathbf{R}^{-1}\mathbf{a}\mathbf{x}^*\mathbf{R}\mathbf{x} = \gamma\mathbf{x}^*\mathbf{R}\mathbf{x}. \tag{5.56}$$

Substituting (5.56) with $\mathbf{x} = \tilde{\mathbf{w}}$ into (5.55), we have that

$$\mathbf{w}^*\mathbf{R}\mathbf{w} \leq \frac{|\mathbf{a}^*\tilde{\mathbf{w}}|^2}{\gamma} \leq \tilde{\mathbf{w}}^*\mathbf{R}\tilde{\mathbf{w}}. \tag{5.57}$$

It follows from (5.54) and (5.57) that $\mathbf{w}$ is at least as good as $\tilde{\mathbf{w}}$ in the sense of minimizing (5.51). Therefore, the optimal value of $\mathbf{w}$ satisfies

$$\mathbf{w} = \frac{\mathbf{a}^*\mathbf{w}}{\gamma} \mathbf{R}^{-1}\mathbf{a}, \tag{5.58}$$

which implies that

$$\mathbf{w} = \frac{d}{\gamma} \mathbf{R}^{-1}\mathbf{a}, \tag{5.59}$$

for some d.

Combining (5.59) and (5.51), our problem reduces to

$$\min_{d} \left\{ \frac{|d|^2}{\gamma} + \max\left(L^2|1-d|^2 - \frac{L^2}{1+\gamma L^2}, U^2|1-d|^2 - \frac{U^2}{1+\gamma U^2} \right) \right\}. \tag{5.60}$$

Since d is in general complex, we can write $d = |d|e^{j\phi}$ for some $0 \leq \phi \leq 2\pi$. Using the fact that $|1-d|^2 = 1 + |d|^2 - 2\cos(\phi)$, it is clear that at the optimal solution, $\phi = 0$. Therefore, without loss of generality, we assume in the sequel that $d \geq 0$. We can then express the problem of (5.60) as

$$\min_{t,d} t \tag{5.61}$$

subject to

$$\begin{aligned} &\frac{d^2}{\gamma} + L^2(1-d)^2 - \frac{L^2}{1+\gamma L^2} \le t; \\ &\frac{d^2}{\gamma} + U^2(1-d)^2 - \frac{U^2}{1+\gamma U^2} \le t. \end{aligned} \tag{5.62}$$

The constraints (5.62) can be equivalently written as

$$\begin{aligned} f_L(d) &\triangleq \left(\frac{1}{\gamma} + L^2\right)\left(d - \frac{\gamma L^2}{1+\gamma L^2}\right)^2 \le t; \\ f_U(d) &\triangleq \left(\frac{1}{\gamma} + U^2\right)\left(d - \frac{\gamma U^2}{1+\gamma U^2}\right)^2 \le t. \end{aligned} \tag{5.63}$$

To develop a solution to (5.61) subject to (5.63), we note that both $f_L(d)$ and $f_U(d)$ are quadratic functions in d, that obtain a minimum at d_L and d_U respectively, where

$$\begin{aligned} d_L &= \frac{\gamma L^2}{1+\gamma L^2}; \\ d_U &= \frac{\gamma U^2}{1+\gamma U^2}. \end{aligned} \tag{5.64}$$

It therefore follows, that the optimal value of d, denoted d_0, satisfies

$$d_L \le d_0 \le d_U. \tag{5.65}$$

Indeed, let $t(d) = \max[f_L(d), f_U(d)]$, and let $t_0 = t(d_0)$ be the optimal value of (5.61) subject to (5.63). Since both $f_L(d)$ and $f_U(d)$ are monotonically decreasing for $d < d_L$, $t(d) > t(d_L) \ge t_0$ for $d < d_L$ so that $d_0 \ge d_L$. Similarly, since both $f_L(d)$ and $f_U(d)$ are monotonically increasing for $d > d_U$, $t(d) > t(d_U) \ge t_0$ for $d > d_U$ so that $d \le d_U$.

Since $f_L(d)$ and $f_U(d)$ are both quadratic, they intersect at most at two points. If $f_L(d) = f_U(d)$, then

$$(1-d)^2 = \frac{1}{(1+\gamma L^2)(1+\gamma U^2)}, \tag{5.66}$$

so that $f_L(d) = f_U(d)$ for $d = d_+$ and $d = d_-$, where

$$d_\pm = 1 \pm \frac{1}{\sqrt{(1+\gamma L^2)(1+\gamma U^2)}}. \tag{5.67}$$

Denoting by $\mathcal{I}$ the interval $\mathcal{I} = [d_L, d_U]$, since $d_+ > 1$, clearly $d_+ \notin \mathcal{I}$. Using the fact that

$$\frac{1}{1+\gamma U^2} \le \frac{1}{\sqrt{(1+\gamma L^2)(1+\gamma U^2)}} \le \frac{1}{1+\gamma L^2}, \tag{5.68}$$

we have that $d_- \in \mathcal{I}$.

In Figure 5.6, we illustrate schematically the functions $f_L(d)$ and $f_U(d)$, where

$$d_e = d_- = 1 - \frac{1}{\sqrt{(1+\gamma L^2)(1+\gamma U^2)}} \tag{5.69}$$

is the unique intersection point of $f_L(d)$ and $f_U(d)$ in $\mathcal{I}$. For the specific choices of $f_L(d)$ and $f_U(d)$ drawn in the figure, it can be seen that the optimal value of d is $d_0 = d_e$. We now show that this conclusion holds true for any choice of the parameters. Indeed, if $L = U$, then $d_e = d_L = d_U$ so that from (5.65), $d_0 = d_e$. Next, assume that $L < U$. In this case, for $d \in \mathcal{I}, f_L(d)$ is monotonically increasing and $f_U(d)$ is monotonically decreasing. Denoting $t_e = t(d_e)$ and noting that $t_e = f_L(d_e) = f_U(d_e)$, we conclude that for $d_e < d \le d_L$, $f_U(d) > t_e$, and for $d_U \le d < d_e$, $f_L(d) > t_e$ so that $t(d) > t_e$ for any $d \in \mathcal{I}$ such that $d \ne d_e$, and therefore $d_0 = d_e$. □

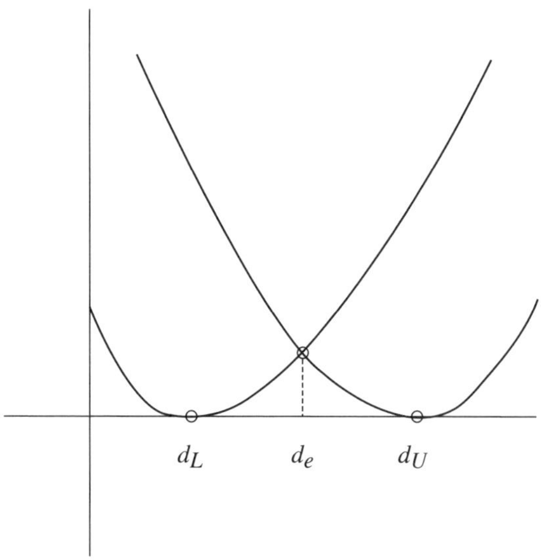

Figure 5.6 Illustration of the functions $f_L(d)$ and $f_U(d)$ of (5.63).

One advantage of the minimax regret beamformer is that it explicitly accounts for both the upper and the lower bounds on $|s|$, while the minimax MSE beamformer depends only on the upper bound. Therefore, in applications in which both bounds are available, the minimax regret beamformer can lead to better performance. As an example, in Figure 5.7 we compare the square-root of the NMSE of the minimax regret (MXR) and minimax MSE (MXM) beamformers, where for comparison we also plot the NMSE of the MVDR beamformer, and the MMSE beamformer. Note that the MMSE beamformer cannot be implemented if $|s|$ is not known, however it serves as a bound on the NMSE. Each result was obtained by averaging 200 Monte Carlo simulations. The scenario we consider consists of a ULA of $M = 20$ omnidirectional sensors spaced half a wavelength apart. The signal of interest s is a complex random process, temporally white, whose amplitude has a uniform distribution between the values 3 and 6 and its plane-wave has a DOA of 30° relative to the array normal. The noise $\mathbf{e}$ consists of a zero-mean complex Gaussian random vector, spatially and temporally white, with a varying power to obtain the desired SNR. The interference is given by $\mathbf{i} = \mathbf{a}_i i$ where i is a zero-mean complex Gaussian random process with INR $= 20$ dB and $\mathbf{a}_i$ its steering vector with DOA $= -30^\circ$. We assume that $\mathbf{R}$, $\mathbf{a}$ and the bounds $U = 6$ and $L = 3$ are known.

As we expect, the minimax regret beamformer outperforms the minimax MSE and MVDR beamformers in all the illustrated SNR range, and approaches the performance of the MMSE beamformer. In Example 5.6.1, we show the advantages

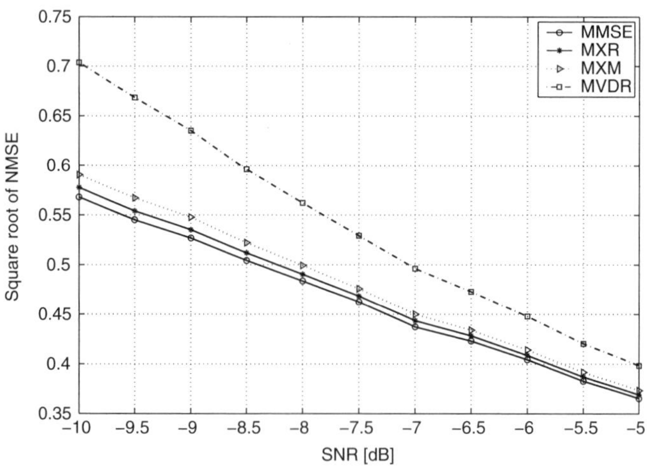

Figure 5.7 Square-root of the normalized MSE as a function of SNR using the MMSE, MXR, MXM and MVDR beamformers in estimating a complex random process with amplitude uniformly distributed between 3 and 6 and DOA $= 30^\circ$, in the presence of an interference with INR $= 20$ dB and DOA $= -30^\circ$.

of the minimax MSE and minimax regret beamformers over several other beamformers for the case where **a** is known but **R**, U and L are estimated from training data containing s.

5.4 RANDOM STEERING VECTOR

In Section 5.3, we explicitly assumed that the steering vector was deterministic and known. However, it is well known that the performance of adaptive beamformers degrades due to uncertainties or errors in the assumed array steering vectors [28, 38–40]. Several authors have considered this problem by modeling the steering vector as a random vector with known distribution [13] or known second-order statistics [16, 26–32]. In this section, we consider the case in which the steering vector is a random vector with mean **m** and covariance matrix **C**: the mean **m** corresponds to a perfectly calibrated array, that is, the perturbation-free steering vector, and **C** represents the perturbations to the steering vector.

In the case of a random steering vector, the SINR is given by [16, 35]

$$\text{SINR} \propto \frac{\mathbf{w}^*\mathbf{R}_s\mathbf{w}}{\mathbf{w}^*\mathbf{R}\mathbf{w}}, \tag{5.70}$$

where

$$\mathbf{R}_s = E\{\mathbf{a}\mathbf{a}^*\} = \mathbf{C} + \mathbf{m}\mathbf{m}^* \tag{5.71}$$

is the signal correlation matrix, whose rank can be between 1 and M. If **a** is deterministic so that $\mathbf{a} = \mathbf{m}$, then $\mathbf{R}_s = \mathbf{a}\mathbf{a}^*$ and the SINR of (5.70) reduces to the SINR of (5.3).

As in the case of deterministic **a**, the most common approach to designing a beamformer is to maximize the SINR, which results in the principle eigenvector beamformer

$$\mathbf{w} = \alpha\mathcal{P}\{\mathbf{R}^{-1}\mathbf{R}_s\}, \tag{5.72}$$

where α is chosen such that $\mathbf{w}^*\mathbf{R}_s\mathbf{w} = 1$. The beamformer (5.72) can also be obtained as the solution to

$$\min_{\mathbf{w}} \mathbf{w}^*\mathbf{R}\mathbf{w} \quad \text{subject to} \quad \mathbf{w}^*\mathbf{R}_s\mathbf{w} = 1. \tag{5.73}$$

In practice, the interference-plus-noise matrix **R** is replaced by an estimate.

However, choosing **w** to maximize the SINR does not necessarily result in an estimated signal amplitude $\hat{s}$ that is close to s. Instead, it would be desirable to minimize the MSE. In the case of a random steering vector, the MSE between s and $\hat{s}$ is given by the expectation of the MSE of (5.13) with respect to **a**, so that

$$\begin{aligned} E\{|\hat{s}-s|^2\} &= \mathbf{w}^*\mathbf{R}\mathbf{w} + |s|^2 E\{|1-\mathbf{w}^*\mathbf{a}|^2\} \\ &= \mathbf{w}^*\mathbf{R}\mathbf{w} + |s|^2 E\{|1-\mathbf{w}^*\mathbf{m}-\mathbf{w}^*(\mathbf{a}-\mathbf{m})|^2\} \\ &= \mathbf{w}^*\mathbf{R}\mathbf{w} + |s|^2(|1-\mathbf{w}^*\mathbf{m}|^2 + \mathbf{w}^*\mathbf{C}\mathbf{w}). \end{aligned} \tag{5.74}$$

The MMSE beamformer minimizing the MSE when $|s|$ is known is obtained by differentiating (5.74) with respect to **w** and equating to 0, which results in [23]

$$\begin{aligned} \mathbf{w}(s) &= |s|^2(\mathbf{R} + |s|^2\mathbf{C} + |s|^2\mathbf{m}\mathbf{m}^*)^{-1}\mathbf{m} \\ &= \frac{|s|^2}{1 + |s|^2\mathbf{m}^*(\mathbf{R} + |s|^2\mathbf{C})^{-1}\mathbf{m}}(\mathbf{R} + |s|^2\mathbf{C})^{-1}\mathbf{m}. \end{aligned} \tag{5.75}$$

Note, that if $\mathbf{C} = \mathbf{0}$, so that $\mathbf{a} = \mathbf{m}$ (with probability one), then (5.75) reduces to

$$\mathbf{w}(s) = \frac{|s|^2}{1 + |s|^2\mathbf{m}^*\mathbf{R}^{-1}\mathbf{m}}\mathbf{R}^{-1}\mathbf{m}, \tag{5.76}$$

which is equal to the MMSE beamformer of (5.16) with $\mathbf{a} = \mathbf{m}$. Comparing (5.75) with (5.72) we see that in general the MMSE beamformer and the SINR beamformer are not scaled versions of each other, as in the known steering vector case. Substituting $\mathbf{w}(s)$ back into (5.74), the smallest possible MSE, which we denote by $\mathrm{MSE}_{\mathrm{OPT}}$, is given by

$$\mathrm{MSE}_{\mathrm{OPT}} = |s|^2 - |s|^4\mathbf{m}^*(\mathbf{R} + |s|^2\mathbf{C})^{-1}\mathbf{m}. \tag{5.77}$$

Since the optimal beamformer (5.76) depends explicitly on $|s|$, it cannot be implemented if $|s|$ is not known. Following the same framework as in the deterministic steering vector case, we consider two alternative approaches for designing a beamformer when $|s|$ is not known: minimax MSE that minimizes the worst-case MSE over all $|s| \le U$, and minimax MSE regret that minimizes the worst-case regret, which in the case of a random steering vector is given by

$$\begin{aligned} \mathcal{R}(s, \mathbf{w}) = \mathbf{w}^*\mathbf{R}\mathbf{w} &+ |s|^2(-\mathbf{w}^*\mathbf{m} - \mathbf{m}^*\mathbf{w} + \mathbf{w}^*(\mathbf{C} + \mathbf{m}\mathbf{m}^*)\mathbf{w}) \\ &+ |s|^4\mathbf{m}^*(\mathbf{R} + |s|^2\mathbf{C})^{-1}\mathbf{m}. \end{aligned} \tag{5.78}$$

The minimax MSE and minimax regret estimators for the linear estimation problem of estimating s in the model $\mathbf{y} = \mathbf{h}s + \mathbf{n}$, where **h** is a random vector with mean **m** and covariance **C**, and **n** is a noise vector with covariance **R**, have been considered in

[23]. From the results in [23], the minimax MSE beamformer is

$$\mathbf{w}_{\text{MXMR}} = \frac{U^2}{1 + U^2\mathbf{m}^*(\mathbf{R} + U^2\mathbf{C})^{-1}\mathbf{m}}(\mathbf{R} + U^2\mathbf{C})^{-1}\mathbf{m}, \tag{5.79}$$

which is just an MMSE estimator matched to $|s| = U$. The minimax regret beamformer for this problem is

$$\mathbf{w}_{\text{MXRR}} = \frac{\gamma}{1 + \gamma\mathbf{m}^*(\mathbf{R} + \gamma\mathbf{C})^{-1}\mathbf{m}}(\mathbf{R} + \gamma\mathbf{C})^{-1}\mathbf{m}, \tag{5.80}$$

where $L^2 \leq \gamma \leq U^2$ is the unique root of $\mathcal{G}(\gamma)$ defined by

$$\mathcal{G}(\gamma) = \sum_{i=1}^{n} \frac{|y_i|^2(U^2 - L^2)}{(1 + U^2\lambda_i)(1 + L^2\lambda_i)\lambda_i}\left(\frac{(1 + U^2\lambda_i)(1 + L^2\lambda_i)}{(1 + \gamma\lambda_i)^2} - 1\right). \tag{5.81}$$

Here y_i is the ith component of $\mathbf{y} = \mathbf{V}^*\mathbf{R}^{-1/2}\mathbf{m}$, $\mathbf{V}$ is the unitary matrix in the eigendecomposition of $\mathbf{A} = \mathbf{R}^{-1/2}(\mathbf{C} + \mathbf{m}\mathbf{m}^*)\mathbf{R}^{-1/2}$, and λ_i is the ith eigenvalue of $\mathbf{A}$.

We see that in the case of a random steering vector, the minimax beamformers are in general no longer scaled versions of the SINR-based principle eigenvector beamformer (5.72), but rather point in different directions. Through several numerical examples (see Section 5.6, Example 5.6.2) we demonstrate the advantages of the minimax MSE and minimax regret beamformers over the principal eigenvector solution as well as over some alternative robust beamformers [14–16] for a wide range of SNR values.

5.4.1 Least-Squares Approach

In the far-field point source case, we have seen that the MVDR approach, which consists of maximizing the SINR, is equivalent to minimizing the MSE subject to the constraint that the beamformer is unbiased. The MVDR beamformer is also the least-squares beamformer, which minimizes the weighted least-squares error

$$\epsilon_{\text{LS}} = (\mathbf{y} - \mathbf{a}\hat{s})\mathbf{R}^{-1}(\mathbf{y} - \mathbf{a}\hat{s}). \tag{5.82}$$

As we now show, these equivalences no longer hold in the case of a random steering vector. First, we note that in this case of a random steering vector, the variance also depends on the unknown power $|s|^2$. Therefore, in this case, the unbiased beamformer that minimizes the variance depends explicitly on s and therefore cannot be implemented. Furthermore, for a full-rank model in which $\mathbf{C}$ is positive definite, there is no choice of beamformer that will result in an MSE that is independent of s (unless, of course, $s = 0$). This follows from the fact that the term depending on s in the MSE is

$$|s|^2(|1 - \mathbf{w}^*\mathbf{m}|^2 + \mathbf{w}^*\mathbf{C}\mathbf{w}). \tag{5.83}$$

Since $\mathbf{w}^*\mathbf{C}\mathbf{w} > 0$ for any nonzero $\mathbf{w}$, this term cannot be equal to zero.

Following the ideas in [23], we now consider the least-squares beamformer for the case of a random steering vector. In this case, the least-squares error (5.82) depends on $\mathbf{a}$, which is random. Thus, instead of minimizing the error $\epsilon_{\rm LS}$ directly, we may minimize the expectation of $\epsilon_{\rm LS}$ with respect to $\mathbf{a}$, which is given by

$$\begin{aligned} E\{\epsilon_{\rm LS}\} &= E\{(\mathbf{y} - \mathbf{m}\hat{s} - (\mathbf{a} - \mathbf{m})\hat{s})^* \mathbf{R}^{-1}(\mathbf{y} - \mathbf{m}\hat{s} - (\mathbf{a} - \mathbf{m})\hat{s})\} \\ &= (\mathbf{y} - \mathbf{m}\hat{s})^* \mathbf{R}^{-1}(\mathbf{y} - \mathbf{m}\hat{s}) + \hat{s}^2 E\{(\mathbf{a} - \mathbf{m})^* \mathbf{R}^{-1}(\mathbf{a} - \mathbf{m})\} \\ &= (\mathbf{y} - \mathbf{m}\hat{s})^* \mathbf{R}^{-1}(\mathbf{y} - \mathbf{m}\hat{s}) + \hat{s}^2 \mathrm{Tr}\,(\mathbf{R}^{-1}\mathbf{C}). \end{aligned} \tag{5.84}$$

Differentiating (5.84) with respect to $\hat{s}$ and equating to 0, we have that

$$\hat{s} = \frac{1}{\mathrm{Tr}(\mathbf{R}^{-1}\mathbf{C}) + \mathbf{m}^*\mathbf{R}^{-1}\mathbf{m}} \mathbf{m}^*\mathbf{R}^{-1}\mathbf{y}. \tag{5.85}$$

Thus, the least-squares beamformer is

$$\mathbf{w}_{\rm LS} = \frac{1}{\mathrm{Tr}(\mathbf{R}^{-1}\mathbf{C}) + \mathbf{m}^*\mathbf{R}^{-1}\mathbf{m}} \mathbf{R}^{-1}\mathbf{m}, \tag{5.86}$$

which, in general, is different than the MVDR beamformer of (5.72). It is interesting to note that the beamformer of (5.86) is a scaled version of the MVDR beamformer for known $\mathbf{a} = \mathbf{m}$.

The advantage of the least-squares approach is that it does not require bounds on the signal magnitude; in fact, it assumes the same knowledge as the principle eigenvector beamformer which maximizes the SINR. In Example 5.6.2 we illustrate through numerical examples that for a wide range of SNR values the least-squares beamformer has smaller NMSE than the principal eigenvector beamformer (5.72) as well as the robust solutions [14–16]. These observations are true even when $\mathbf{m}$ and $\mathbf{C}$ are not known exactly, but are rather chosen in an ad-hoc manner. Therefore, in terms of NMSE, the least-squares approach appears to often be preferable over standard and robust methods, while requiring the same prior knowledge. As we show, the NMSE performance can be improved further by using the minimax MSE and minimax regret methods; however, these methods require prior estimates of the signal magnitude bounds.

5.5 PRACTICAL CONSIDERATIONS

In our development of the minimax MSE and minimax regret beamformers, we assumed that there exists an upper bound U on the magnitude of the signal to be estimated, as well as a lower bound L for the minimax regret beamformer. For random steering vectors we assumed also the knowledge of their mean $\mathbf{m}$ and covariance $\mathbf{C}$.

In some applications, the bounds U and L may be known, for example based on a priori knowledge on the type of the source and its possible range of distances from the array. If no such bounds are available, then we may estimate them from the data using one of the conventional beamformers. Specifically, let $\mathbf{w}_c$ denote one of the conventional beamformers. Then, using this beamformer we can estimate $s(t)$ as $\hat{s}(t) = \mathbf{w}_c^*\mathbf{y}(t)$ (the dependence on the time index t is presented for clarity). We may then use this estimate to obtain approximate values for U and L. In the simulations below, we use $U = (1+\beta)^2\|\mathbf{w}_c^*\mathbf{Y}\|$ and $L = (1-\beta)^2\|\mathbf{w}_c^*\mathbf{Y}\|$ for some β, where $\mathbf{Y}$ is the training data matrix of dimension $M \times N$ and $\|\cdot\|$ is the average norm over the training interval.

Since in most applications the true covariance $\mathbf{R}$ is not available we have to estimate it, for example, using (5.7). However, as we discussed in Section 5.2, if $s(t)$ is present in the training data, then a diagonal loading (5.8) may perform better than (5.7). Therefore, in the simulations, the true $\mathbf{R}$ is replaced by (5.8).

In the case of a random steering vector, our beamformers rely on knowledge of the steering vector $\mathbf{m}$ and covariance $\mathbf{C}$. These parameters can either be estimated from observations of the steering vector, or, they can be approximated by choosing $\mathbf{m}$ to be equal to a nominal steering vector, and choosing $\mathbf{C} = \nu\mathbf{I}$, where ν reflects the uncertainty size around the nominal steering vector.

5.6 NUMERICAL EXAMPLES

To evaluate and compare the performance of our methods with other techniques, we conducted numerical examples using scenarios similar to [16]. Specifically, we consider a ULA of $M = 20$ omnidirectional sensors spaced half a wavelength apart. In all the examples below, we choose $s(t)$ to be either a complex sinewave or a zero-mean complex Gaussian random process, temporally white, with varying amplitude or variance respectively, to obtain the desired SNR in each sensor. The signal $s(t)$ is continuously present throughout the training data. The noise $\mathbf{e}(t)$ is a zero-mean, Gaussian, complex random vector, temporally and spatially white, with a power of 0 dB in each sensor. The interference is given by $\mathbf{i}(t) = \mathbf{a}_i i(t)$ where $i(t)$ is a zero-mean, Gaussian, complex process temporally white and $\mathbf{a}_i$ is the interference steering vector. To illustrate the performance of the beamformers we use the square-root of the NMSE, which is obtained by averaging 200 Monte Carlo simulations. Unless otherwise stated, we use a signal with SNR $= -5$ dB, and an interference with DOA $= -30°$, INR $= 20$ dB and $N = 50$ training snapshots. For brevity, in the remainder of this section we use the following notation for the different beamformers: MXM (minimax MSE), MXR (minimax regret), MXMR (minimax MSE for random $\mathbf{a}$), MXRR (minimax regret for random $\mathbf{a}$), LSR (least-squares for random $\mathbf{a}$); see also Table 5.1.

We consider two examples: Example 5.6.1 evaluates the performance of the MXM and MXR beamformers using the exact knowledge of the steering vector. Example 5.6.2 evaluates the MXMR, MXRR and LSR beamformers for a mismatch in the signal DOA. We focus on low SNR values (important e.g., in sonar) and

TABLE 5.1 Beamformers Used in the Numerical Examples

Beamformer	Expression	Parameters	Ref.
L-CAPON	$\dfrac{1}{\mathbf{a}^*\hat{\mathbf{R}}_{dl}^{-1}\mathbf{a}}\hat{\mathbf{R}}_{dl}^{-1}\mathbf{a}$	$\xi = 10$	[10]
EIG	$\dfrac{1}{\mathbf{a}^*\hat{\mathbf{R}}_{eig}^{-1}\mathbf{a}}\hat{\mathbf{R}}_{eig}^{-1}\mathbf{a}$	$D = 1$	[11]
ROB1	$\dfrac{\lambda}{\lambda\mathbf{a}^*(\hat{\mathbf{R}}_{sm} + \lambda\epsilon^2\mathbf{I})^{-1}\mathbf{a} - 1}\left(\hat{\mathbf{R}}_{sm} + \lambda\epsilon^2\mathbf{I}\right)^{-1}\mathbf{a}$	$\epsilon = 3$	[14]
ROB2	$\dfrac{\alpha_1(\lambda\hat{\mathbf{R}}_{sm} + \mathbf{I})^{-1}\mathbf{a}}{\sqrt{M}\mathbf{a}^*(\lambda\hat{\mathbf{R}}_{sm} + \mathbf{I})^{-1}\hat{\mathbf{R}}_{sm}(\lambda\hat{\mathbf{R}}_{sm} + \mathbf{I})^{-1}\mathbf{a}}$	$\epsilon = 3.5$	[15]
ROB3	$\alpha_2\mathcal{P}\{\hat{\mathbf{R}}_{dl}^{-1}(\mathbf{aa}^* - \epsilon\mathbf{I})\}$	$\xi = 30$, $\epsilon = 9$	[16]
PEIG	$\alpha_3\mathcal{P}\{\mathbf{R}^{-1}\mathbf{R}_s\}$		[35]
MXM	$\dfrac{U^2}{1 + U^2\mathbf{a}^*\mathbf{R}^{-1}\mathbf{a}}\mathbf{R}^{-1}\mathbf{a}$	U	[41]
MXR	$\dfrac{\alpha_4}{\mathbf{a}^*\mathbf{R}^{-1}\mathbf{a}}\mathbf{R}^{-1}\mathbf{a}$	U, L	[36]
MXMR	$\dfrac{U^2}{1 + U^2\mathbf{m}^*(\mathbf{R} + U^2\mathbf{C})^{-1}\mathbf{m}}(\mathbf{R} + U^2\mathbf{C})^{-1}\mathbf{m}$	U, $\mathbf{m}$, $\mathbf{C}$	
MXRR	$\dfrac{\gamma}{1 + \gamma\mathbf{m}^*(\mathbf{R} + \gamma\mathbf{C})^{-1}\mathbf{m}}(\mathbf{R} + \gamma\mathbf{C})^{-1}\mathbf{m}$	U, L, $\mathbf{m}$, $\mathbf{C}$	
LSR	$\dfrac{1}{\mathrm{Tr}(\mathbf{R}^{-1}\mathbf{C}) + \mathbf{m}^*\mathbf{R}^{-1}\mathbf{m}}\mathbf{R}^{-1}\mathbf{m}$	$\mathbf{m}$, $\mathbf{C}$	

compare the performance of the proposed methods against seven alternative methods: the Capon beamformer (CAPON) [33, 34], loading Capon beamformer (L-CAPON) [9, 10], eigenspace-based beamformer (EIG) [11, 12], and the robust beamformers of (5.10), (5.11) and (5.12) which we refer to, respectively, as ROB1, ROB2, and ROB3 [14–16]. In Example 5.6.2 we compare our methods against the principal eigenvector beamformer [35], which we refer to as PEIG, with $\mathbf{R}$ given by (5.8) and $\mathbf{R}_s$ given by exact knowledge or an ad hoc estimate (see Example 5.6.2 for more details.) The parameters of each of the compared methods were chosen as suggested in the literature. Namely, for L-CAPON (5.8) and PEIG (5.72) the diagonal loading was set as $\xi = 10\sigma^2$ [14, 15] with σ^2 being the variance of the noise in each sensor, assumed to be known ($\sigma^2 = 1$ in these examples); for the EIG beamformer (5.9) it was assumed that the low-rank condition and number of interferers are known. For the alternative robust methods, the parameters were chosen as follows: For ROB1 (5.10) the upper bound on the steering vector uncertainty was set as $\epsilon = 3$ [14], for ROB2 (5.11) $\epsilon = 3.5$, and for ROB3 (5.12) $\epsilon = 9$, and the diagonal loading was chosen as $\xi = 30$. Table 5.1 summarizes the beamformers implemented in the simulations, where $\alpha_1 = \|(\hat{\mathbf{R}}_{sm}^{-1} + \lambda\mathbf{I})^{-1}\mathbf{a}\|$, α_2 is chosen such that the corresponding beamformer satisfies $\mathbf{w}^*(\mathbf{aa}^* - \epsilon\mathbf{I})\mathbf{w} = 1$, α_3 is chosen such that $\mathbf{w}^*\mathbf{R}_s\mathbf{w} = 1$, and $\alpha_4 = 1 - 1/\sqrt{(1 + L^2\mathbf{a}^*\mathbf{R}^{-1}\mathbf{a})(1 + U^2\mathbf{a}^*\mathbf{R}^{-1}\mathbf{a})}$.

Example 5.6.1: Known Steering Vector In this example we assume that the steering vector $\mathbf{a}$ is known. We first choose $s(t)$ as a complex sinewave with DOA

TABLE 5.2 Specification of the MXM and MXR Beamformers in Example 1

Beamformer	**R**	$\mathbf{w}_c$	β
MXM	$\mathbf{R}_{dl}$, $\xi = 10$	L-CAPON, $\xi = 10$	9
MXR	$\mathbf{R}_{dl}$, $\xi = 10$	L-CAPON, $\xi = 10$	6

of its plane-wave equal to 30° relative to the array normal. We implemented the MXR and MXM beamformers with the sample covariance matrix estimated using a loading factor $\xi = 10$ [14, 15], $\mathbf{w}_c$ given by the L-CAPON beamformer with $\xi = 10$ and β set as 6 and 9, for these beamformers, respectively. The values of β selected for MXM and MXR were those that gave the best performance over a wide range of negative SNR values. Table 5.2 summaries the parameters chosen for MXM and MXR in this example.

In Figure 5.8 we plot the square-root of the NMSE as a function of the SNR using the MXR, MXM, EIG, L-CAPON, ROB2, and ROB3 beamformers. Since in all the scenarios considered in this example the NMSE of the CAPON and ROB1 beamformers was out of the illustrated scales, we do not plot the performance of these methods. It can be seen in Figure 5.8 that the MXM beamformer has the best performance for SNR values between -10 to -3 dB, and the MXR beamformer has the best performance for SNR values between -2 to 2 dB. In Figure 5.9 we plot the square-root of the NMSE as a function of the number of training data with

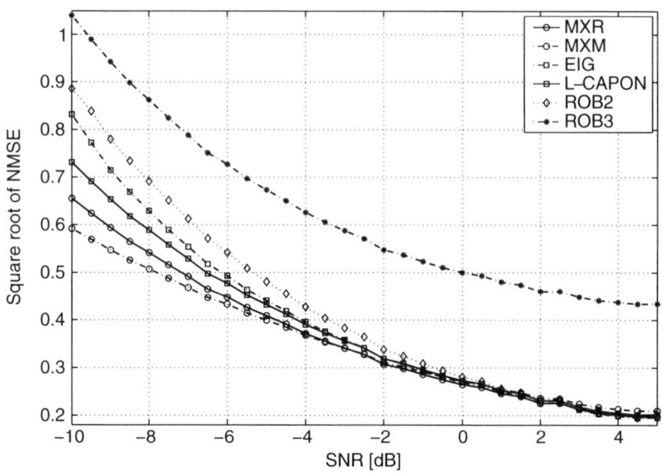

Figure 5.8 Square-root of the normalized MSE as a function of SNR when estimating a complex sinewave with known **a** and with DOA = 30°, using the MXR, MXM, EIG, L-CAPON, ROB2, and ROB3 beamformers.

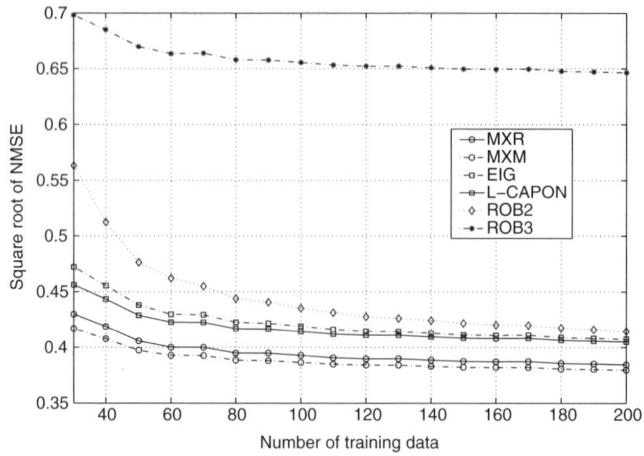

Figure 5.9 Square-root of the normalized MSE as a function of the number of training snapshots when estimating a complex sinewave with known **a** and DOA $= 30^\circ$, using the MXR, MXM, EIG, L-CAPON, ROB2, and ROB3 beamformers.

SNR $= -5$ dB. The performance of the proposed methods as a function of the difference between the signal and interference DOAs is illustrated in Figure 5.10. The NMSE of all the methods remains constant for DOA differences between 10° and 90°, but deteriorates for DOA differences close to 0°. Despite this, MXR and MXM continue to outperform the other methods. The performance of ROB2 and ROB3 is out of the illustrated scale for DOA differences less than 3°, because

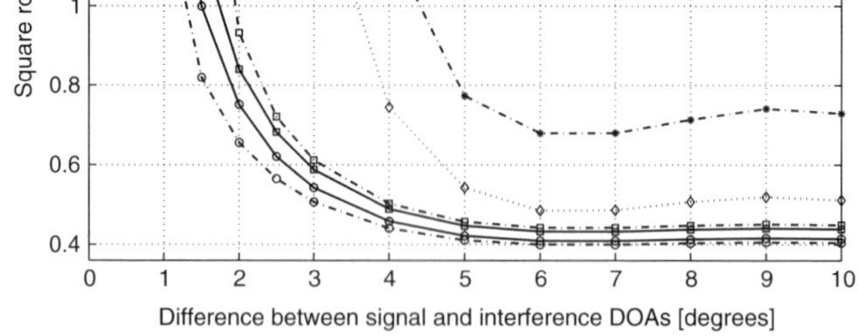

Figure 5.10 Square-root of the normalized MSE as a function of the difference between the signal and interference DOAs when estimating a complex sinewave with known **a** and DOA $= 30^\circ$, using the MXR, MXM, EIG, L-CAPON, ROB2, and ROB3 beamformers.

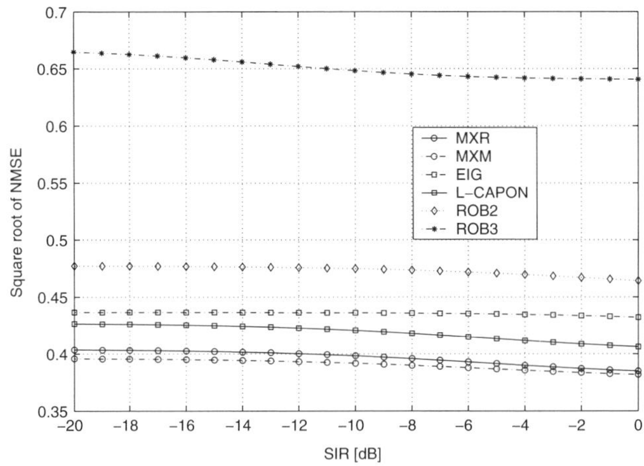

Figure 5.11 Square-root of the normalized MSE as a function of SIR when estimating a complex sinewave with known **a** and DOA $= 30^\circ$, using the MXR, MXM, EIG, L-CAPON, ROB2, and ROB3 beamformers.

the uncertainty region of their steering vectors overlaps with that of the interference. Figure 5.11 illustrates the performance as a function of the signal-to-interference-ratio (SIR) for an SNR of -5 dB. As expected, the NMSE of all the beamformers decreases as the SIR increases, however the minimax methods still outperform the alternative methods for all SIR values shown.

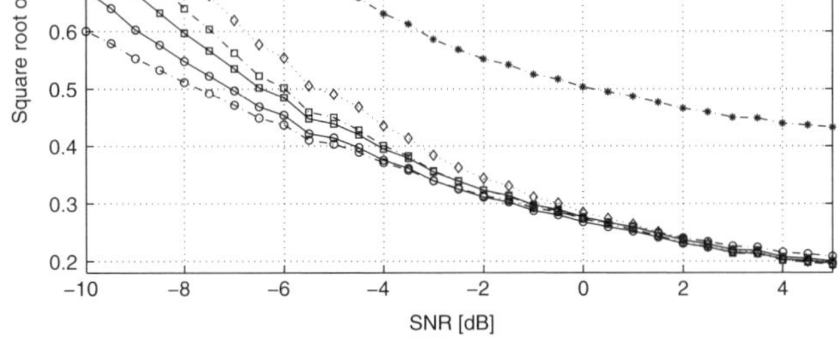

Figure 5.12 Square-root of the normalized MSE as a function of SNR when estimating a zero-mean complex Gaussian random signal with known **a** and DOA $= 30^\circ$, using the MXR, MXM, EIG, L-CAPON, ROB2, and ROB3 beamformers.

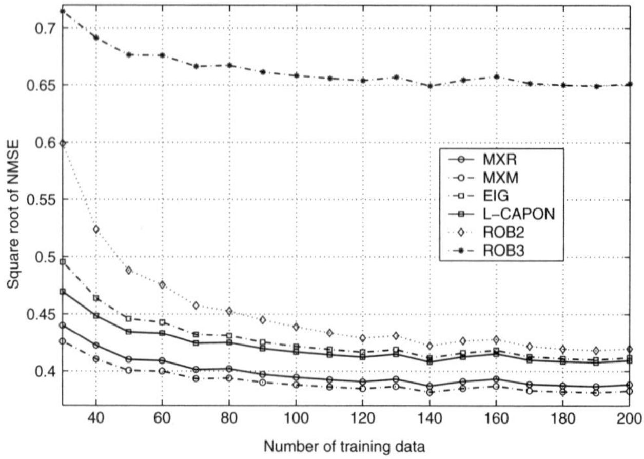

Figure 5.13 Square-root of the normalized MSE as a function of the number of training snapshots when estimating a zero-mean complex Gaussian random signal with known **a** and DOA $= 30^\circ$, using the MXR, MXM, EIG, L-CAPON, ROB2, and ROB3 beamformers.

We next repeat the simulations for $s(t)$ chosen to be a zero-mean complex Gaussian random signal, temporally white. The square-root of the NMSE as a function of the SNR, the number of training data, the difference between signal and interference DOAs, and the SIR is depicted in Figures 5.12–5.15, respectively.

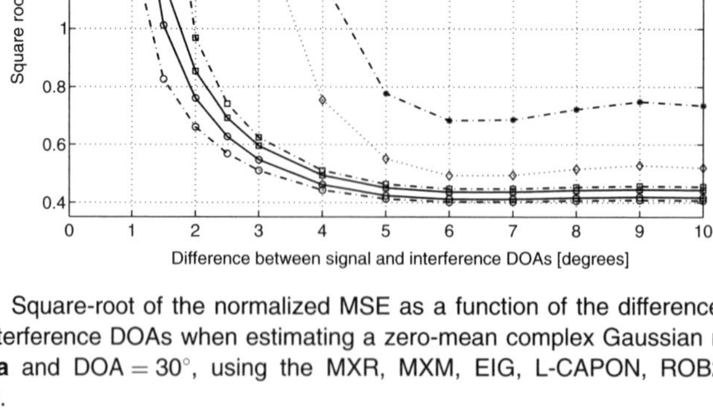

Figure 5.14 Square-root of the normalized MSE as a function of the difference between the signal and interference DOAs when estimating a zero-mean complex Gaussian random signal with known **a** and DOA $= 30^\circ$, using the MXR, MXM, EIG, L-CAPON, ROB2, and ROB3 beamformers.

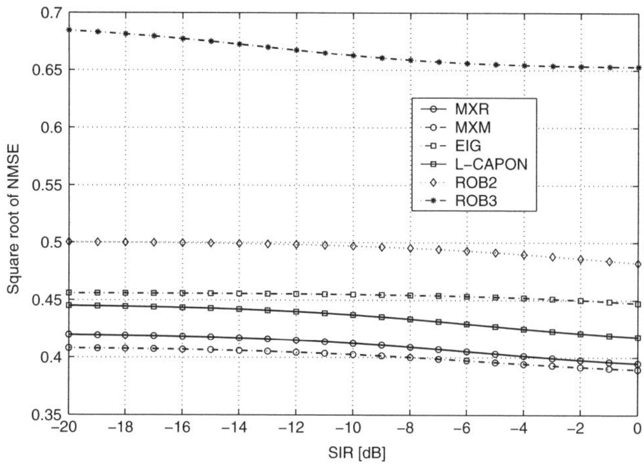

Figure 5.15 Square-root of the normalized MSE as a function of SIR when estimating a zero-mean complex Gaussian random signal with known **a** and DOA $= 30^\circ$, using the MXR, MXM, EIG, L-CAPON, ROB2, and ROB3 beamformers.

It can be seen in Figure 5.12 that the MXM has the best performance for SNR values between -10 to -4 dB and the MXR has the best performance for SNR values between -3 to 1.5 dB. The performance in Figures 5.13–5.15 is similar to the case of a deterministic sinewave.

Example 5.6.2: Steering Vector with Signal DOA Uncertainties We illustrate the robustness of the algorithms for random steering vectors developed in Section 1.4 against a mismatch in the assumed signal DOA. Specifically, the DOA of the signal is given by a Gaussian random variable with mean equal to 30° and standard deviation equal to 1 (about $\pm 3^\circ$). This DOA was independently drawn in each simulation run. To estimate the signal in this case, we implemented the MXRR, MXMR, LSR, and PEIG beamformers with parameters given by Table 5.3, for two choices of $\mathbf{m}$ and $\mathbf{C}$: the true values (estimated from 2000 realizations of the steering vector), and ad hoc values. The ad hoc value of $\mathbf{m}$ was chosen as the steering vector with DOA $= 30^\circ$ and for the covariance we chose $\mathbf{C} = \nu\mathbf{I}$ with $\nu = \epsilon/M$, where $\epsilon = 3.5$ is the norm value of the steering vector error (size of the uncertainty) used by

TABLE 5.3 Specification of the Beamformers Used in Example 2

Beamformer	R	$\mathbf{w}_c$	β
MXMR	$\mathbf{R}_{dl}$, $\xi = 10$	L-CAPON, $\xi = 10$	9
MXRR	$\mathbf{R}_{dl}$, $\xi = 10$	L-CAPON, $\xi = 10$	4
LSR	$\mathbf{R}_{dl}$, $\xi = 10$	—	—
PEIG	$\mathbf{R}_{dl}$, $\xi = 10$	—	—

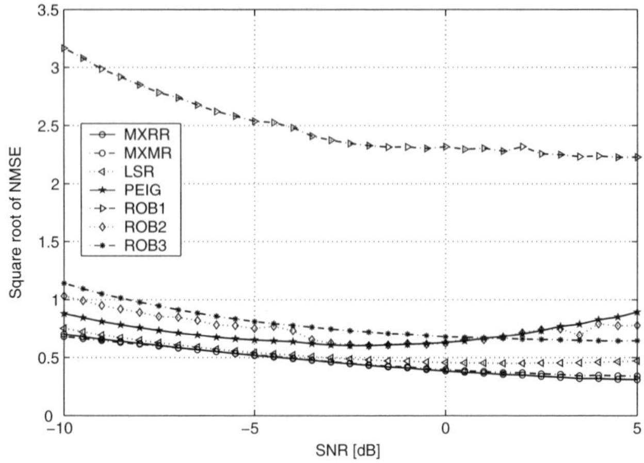

Figure 5.16 Square-root of the normalized MSE as a function of SNR when estimating a complex sinewave with random DOA and known **m** and **C**, using the MXRR, MXMR, LSR, PEIG, ROB1, ROB2, and ROB3 beamformers.

ROB2 [15] and $M = 20$ is the number of sensors. For the ROB1, ROB2 and ROB3 beamformers $\mathbf{a}$ is the steering vector for a signal DOA $= 30^\circ$.

In Figures 5.16 and 5.17 we depict the NMSE when estimating a complex sinewave as a function of SNR, using the MXRR, MXMR, LSR, PEIG, ROB1, ROB2, and ROB3 beamformers, with known and ad hoc values of **m** and **C**, respectively. As can be seen from the figures, the MXMR, MXRR and LSR methods perform better

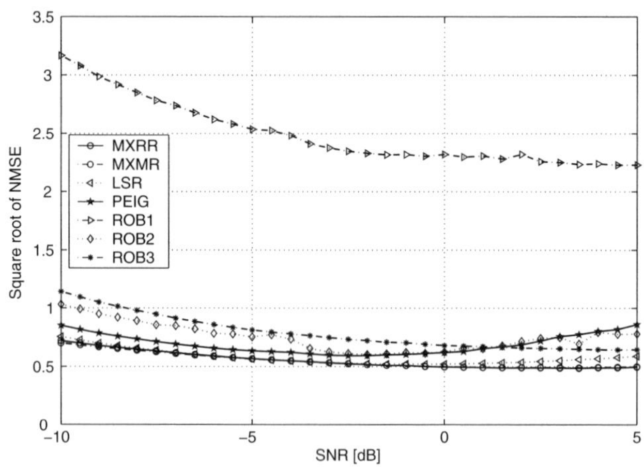

Figure 5.17 Square-root of the normalized MSE as a function of SNR when estimating a complex sinewave with random DOA and ad hoc values of **m** and **C**, using the MXRR, MXMR, LSR, PEIG, ROB1, ROB2, and ROB3 beamformers.

than all the other methods both in the case of known **m** and **C** and when ad hoc values are chosen. Although the performance of the MXMR, MXRR and LSR methods deteriorates when the ad hoc values are used in place of the true values, the difference in performance is minor. A surprising observation from the figures is that the standard PEIG method performs better, in terms of NMSE, than ROB1, ROB2 and ROB3. We also note that the LSR method, which does not use any additional parameters such as magnitude bounds, outperforms all previously proposed methods.

The performance of the MXRR, MXMR, LSR, PEIG, ROB1, ROB2, and ROB3 beamformers, assuming known **m** and **C**, as a function of the number of training snapshots and the difference between the signal and interference DOAs is illustrated in Figures 5.18 and 5.19, respectively. The NMSE of all methods remain almost constant for differences above 6°. Below this range all the beamformers decrease their performance significantly, except for the PEIG and LSR beamformers. As in the case of deterministic **a**, the performance of all the methods improves slightly as a function of negative SIR, and is therefore not shown.

In Figures 5.20 and 5.21 we plot the NMSE as a function of the SNR when estimating a complex Gaussian random signal, temporally white, with random DOA, as a function of SNR. As can be seen by comparing these figures with Figures 5.16 and 5.17, the performance of all of the methods is similar to the case of a deterministic sinewave. The NMSE as a function of the number of training snapshots and the difference between the signal and interference DOAs is also similar to the deterministic sinewave case, and therefore not shown.

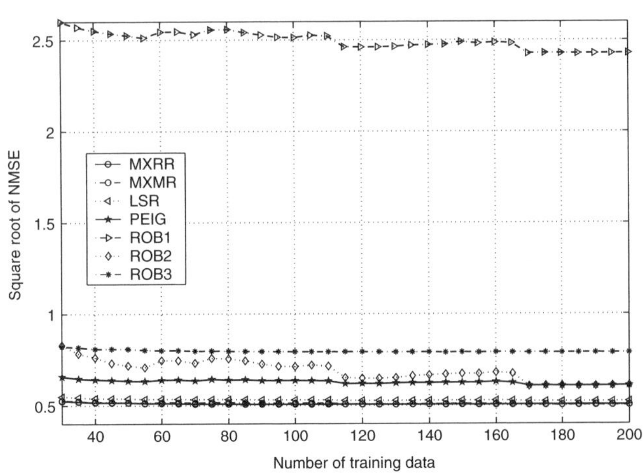

Figure 5.18 Square-root of the normalized MSE as a function of the number of training snapshots when estimating a complex sinewave with random DOA and known **m** and **C** using the MXRR, MXMR, LSR, PEIG, ROB1, ROB2, and ROB3 beamformers.

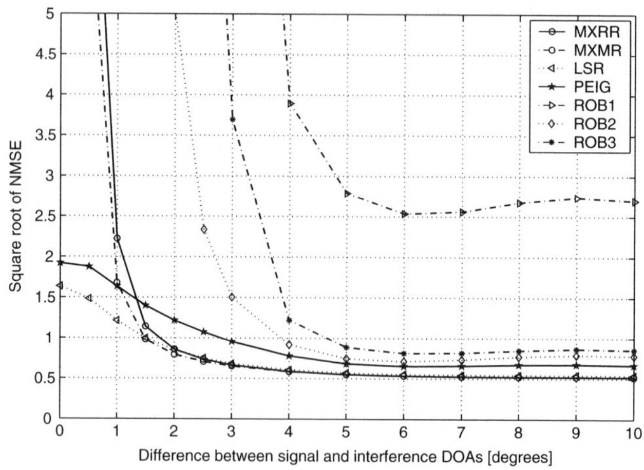

Figure 5.19 Square-root of the normalized MSE as a function of the difference between the signal and interference DOAs when estimating a complex sinewave with random DOA and known **m** and **C** using the MXRR, MXMR, LSR, PEIG, ROB1, ROB2, and ROB3 beamformers.

5.7 SUMMARY

We treated the problem of designing linear beamformers to estimate a source signal $s(t)$ from sensor array observations, where the goal is to obtain an estimate $\hat{s}(t)$ that is close to $s(t)$. Although standard beamforming approaches are aimed at maximizing the SINR, maximizing SINR does not necessarily guarantee a small MSE, hence on average a signal estimate maximizing the SINR can be far from $s(t)$. To ensure that $\hat{s}(t)$ is close to $s(t)$, we proposed using the more appropriate design criterion of

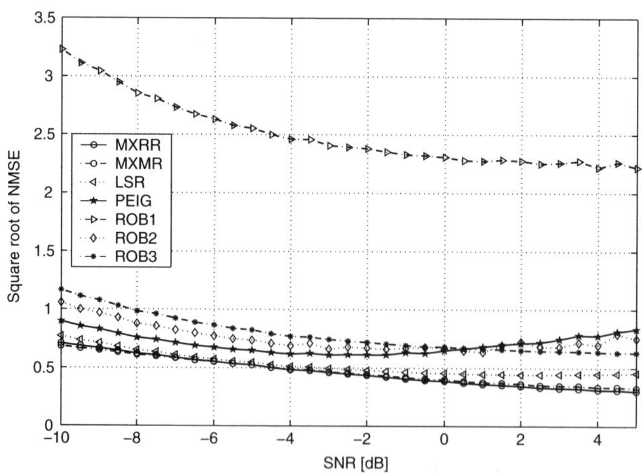

Figure 5.20 Square-root of the normalized MSE as a function of SNR when estimating a zero-mean complex Gaussian random signal with random DOA and known **m** and **C**, using the MXRR, MXMR, LSR, PEIG, ROB1, ROB2, and ROB3 beamformers.

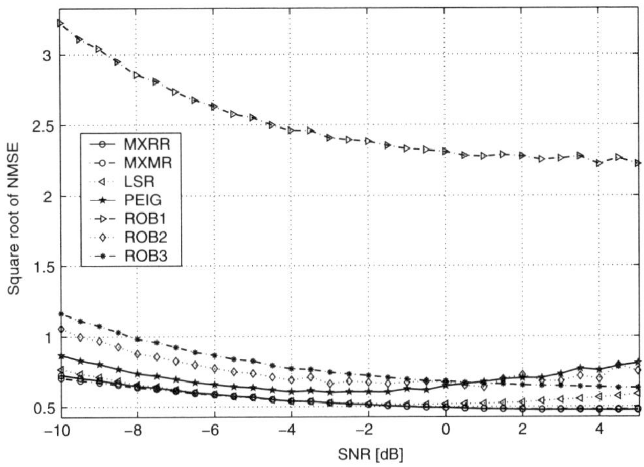

Figure 5.21 Square-root of the normalized MSE as a function of SNR when estimating a zero-mean complex Gaussian random signal with random DOA and ad hoc values of **m** and **C**, using the MXRR, MXMR, LSR, PEIG, ROB1, ROB2, and ROB3 beamformers.

MSE. Since the MSE depends in general on $s(t)$ which is unknown, it cannot be minimized directly. Instead, we suggested beamforming methods that minimize a worst-case measure of MSE assuming known and random steering vectors with known second-order statistics. We first developed a minimax MSE beamformer that minimizes the worst-case MSE. We then considered a minimax regret beamformer that minimizes the worst-case difference between the MSE using a beamformer ignorant of $s(t)$ and the smallest possible MSE attainable with a beamformer that knows $s(t)$. As we showed, even if $s(t)$ is known, we cannot achieve a zero MSE with a linear estimator. In the case of a random steering vector we also proposed a least-squares beamformer that does not require bounds on the signal magnitude.

In the numerical examples, we clearly illustrated the advantages of our methods in terms of the MSE. For both known and random steering vectors, the minimax beamformers consistently have the best performance, particularly for negative SNR values. Quite surprisingly, it was observed that the least-squares beamformer, which does not require bounds on the signal magnitude, performs better than the recently proposed robust methods in the case of random DOAs. It was also observed that for a small difference between signal and interference directions of arrival, all our methods show better performance. The behavior of our methods was similar when the signal was chosen as a deterministic sinewave or a zero-mean complex Gaussian random signal.

ACKNOWLEDGMENTS

The authors are grateful to Mr. Patricio S. La Rosa for conducting the numerical examples and for carefully reviewing several earlier versions of this chapter. The work of A. Nehorai was supported by the Air Force Office of Scientific

Research Grant F49620-02-1-0339, and the National Science Foundation Grants CCR-0105334 and CCR-0330342.

REFERENCES

1. R. Mozingo and T. Miller, *Introduction to Adaptive Arrays*. Wiley and Sons, New York, 1980.
2. J. E. Hudson, *Adaptive Array Principles*. Peter Peregrinus Ltd, London, 1981.
3. S. Haykin, *Array Signal Processing*. Prentice-Hall, Englewood Cliffs, New Jersey, 1985.
4. B. D. van Veen and K. M. Buckley, "Beamforming: A versatile approach to spatial filtering," *IEEE Signal Proc. Magazine*, Vol. 5, pp. 4–24, Apr. 1988.
5. S. Haykin and A. Steinhardt, *Adaptive Radar Detection and Estimation*. Wiley, New York, 1992.
6. H. Krim and M. Viberg, "Two decades of array signal processing research: The parametric approach," *IEEE Signal Proc. Magazine*, Vol. 13, pp. 67–94, July 1996.
7. M. Hawkes and A. Nehorai, "Acoustic vector-sensor beamforming and capon direction estimation," *IEEE Trans. on Signal Proc.*, Vol. SP-46, pp. 2291–2304, Sept. 1998.
8. H. L. van Trees, *Optimal Array Processing (Detection, Estimation, and Modulation Theory, Part IV)*. Wiley-Interscience, New York, 2002.
9. B. D. Carlson, "Covariance matrix estimation errors and diagonal loading in adaptive arrays," *IEEE Trans. Aerosp. Electron. Syst.*, Vol. 24, pp. 397–401, July 1988.
10. H. Cox, R. M. Zeskind, and M. M. Owen, "Robust adaptive beamforming," *IEEE Trans. Acoust., Speech, Signal Processing*, Vol. ASSP-35, pp. 1365–1376, Oct. 1987.
11. D. D. Feldman and L. J. Griffiths, "A projection approach for robust adaptive beamforming," *IEEE Trans. Signal Processing*, Vol. 42, pp. 867–876, Apr. 1994.
12. N. L. Owsley, "Enhanced minimum variance beamforming," in Y. T. Chan, Ed., *Underwater Acoustic Data Processing*, pp. 285–292, Kluwer, 1989.
13. K. L. Bell, Y. Ephraim, and H. L. van Trees, "A Bayesian approach to robust adaptive beamforming," *IEEE Trans. Signal Processing*, Vol. 48, pp. 386–398, Feb. 2000.
14. S. A. Vorobyov, A. B. Gershman, and Z.-Q. Luo, "Robust adaptive beamforming using worst case performance optimization," *IEEE Trans. Signal Proc.*, Vol. 51, pp. 313–324, Feb. 2003.
15. J. Li, P. Stoica and Z. Wang, "On robust capon beamforming and diagonal loading," *IEEE Trans. Signal Proc.*, Vol. 51, pp. 1702–1715, July 2003.
16. S. Shahbazpanahi, A. B. Gershman, Z.-Q. Luo, and K. M. Wong, "Robust adaptive beamforming for general-rank signal models," *IEEE Trans. Signal Proc.*, Vol. 51, pp. 2257–2269, Sep. 2003.
17. C. J. Lam and A. C. Singer, "Performance analysis of the Bayesian beamforming," *Proc. IEEE Int. Conf. Acoust., Speech, Signal Process.*, Vol. II, pp. 197–200, 2004.
18. C. D. Richmond, "The Capon-MVDR algorithm: threshold SNR prediction and the probability of resolution," *Proc. IEEE Int. Conf. Acoust., Speech, Signal Process.*, Vol. II, pp. 217–220, 2004.
19. O. Besson and F. Vincent, "Performance analysis of the Bayesian beamforming," in *Proc. IEEE Int. Conf. Acoust., Speech, Signal Process.*, Vol. II, pp. 169–172, 2004.

20. Y. C. Eldar, A. Ben-Tal, and A. Nemirovski, "Robust mean-squared error estimation in the presence of model uncertainties," *IEEE Trans. Signal Processing*, Vol. 53, pp. 168–181, Jan. 2005.
21. Y. C. Eldar, A. Ben-Tal, and A. Nemirovski, "Linear minimax regret estimation of deterministic parameters with bounded data uncertainties," *IEEE Trans. Signal Processing*, Vol. 52, pp. 2177–2188, Aug. 2004.
22. J. Yang and A. Swindlehurst, "Signal copy with array calibration errors," *Signals, Systems and Computers, 1993. Conference Record of The Twenty-Seventh Asilomar Conference on, 1–3 Nov.*, Vol. 2, pp. 405–413, 1993.
23. Y. C. Eldar, "Robust estimation in linear models with a random model matrix," to appear in *IEEE Trans. Signal Processing*.
24. L. Seymour, C. Cowan, and P. Grant, "Bearing estimation in the presence of sensors positioning errors," in *Proc. IEEE Int. Conf. Acoust., Speech, Signal Process.*, pp. 2264–2267, 1987.
25. B. Friendlander and A. Weiss, "Direction finding in the presence of mutual coupling," *IEEE Trans. Antennas and Propagation*, Vol. AP-39, pp. 273–284, 1991.
26. A. B. Gershman, V. I. Turchin, and V. A. Zverev, "Experimental results of localization of moving underwater signal by adaptive beamforming," *IEEE Trans. Signal Processing*, Vol. 43, pp. 2249–2257, Oct. 1995.
27. E. Y. Gorodetskaya, A. I. Malekhanov, A. G. Sazontov, and N. K. Vdovicheva, "Deep-water acoustic coherence at long ranges: Theoretical prediction and effects on large-array signal processing," *IEEE J. Ocean. Eng.*, Vol. 24, pp. 156–171, Apr. 1999.
28. H. Cox, "Effects of random steering vector errors in the applebaum array," *J. Acoust. Soc. Amer.*, Vol. 54, pp. 771–785, Sept. 1973.
29. D. R. Morgan and T. M. Smith, "Coherence effects on the detection performance of quadratic array processors with application to largearray matched-field beamforming," *J. Acoust. Soc. Amer.*, Vol. 87, pp. 737–747, Feb. 1988.
30. A. Paulraj and T. Kailath, "Direction of arrival estimation by eigenstructure methods with imperfect spatial coherence ofwave fronts," *J. Acoust. Soc. Amer.*, Vol. 83, pp. 1034–1040, Mar. 1988.
31. A. B. Gershman, C. F. Mecklenbräuker, and J. F. Böhme, "Matrix fitting approach to direction of arrival estimation with imperfect spatial coherence of wavefronts," *IEEE Trans. Signal Processing*, Vol. 45, pp. 1894–1899, July 1997.
32. M. Wax and Y. Anu, "Performance analysis of the minimum variance beamformer in the presence of steering vector errors," *IEEE Trans. Signal Processing*, Vol. 44, pp. 938–947, Apr. 1996.
33. J. Capon, "High resolution frequency-wavenumber spectrum analysis," *Proc. IEEE*, Vol. 57, pp. 1408–1418, Aug. 1969.
34. R. T. Lacos, "Data adaptive spectral analysis methods," *Geophys.*, Vol. 36, pp. 661–675, Aug. 1971.
35. A. B. Gershman, "Robust adaptive beamforming in sensor arrays," *Int. J. Electron. Commun.*, Vol. 53, pp. 305–324, Dec. 1999.
36. Y. C. Eldar and A. Nehorai, "Competitive mean-squared error beamforming," in *Proc. 12th Annu. Workshop Adaptive Sensor Array Processing*, Lincoln Laboratory, MIT, Lexington, MA, Mar. 2004.

37. Y. C. Eldar, "Robust competitive estimation with signal and noise covariance uncertainties," submitted to *IEEE Trans. Inform. Theory.*

38. C. L. Zahm, "Effects of errors in the direction of incidence on the performance of an adaptive array," *Proc. IEEE*, Vol. 60, pp. 1068–1069, Aug. 1972.

39. R. T. Compton, "Effects of random steering vector errors in the applebaum array," *IEEE Trans. Aerosp. Electron. Syst.*, Vol. AES-18, pp. 392–400, Sept. 1982.

40. L. C. Godara, "Error analysis of the optimal antenna array processors," *IEEE Trans. Aerosp. Electron. Syst.*, Vol. AES-22, pp. 395–409, Jul. 1986.

41. Y. C. Eldar and A. Nehorai, "Uniformly robust mean-squared error beamforming," in *Proc. 3rd IEEE Sensor Array and Multichannel Signal Processing Workshop*, Barcelona, Spain, Jul. 2004.

6

CONSTANT MODULUS BEAMFORMING

Alle-Jan van der Veen
Department of Electrical Engineering, Delft University of Technology, Delft, The Netherlands

Amir Leshem
School of Engineering, Bar-Ilan University, Ramat-Gan, Israel

Algorithms for blind source separation aim to compute beamformers that select a desired source while suppressing interfering sources, without specific knowledge of the sources or the channel. The preceding chapters have described algorithms based on direction finding: sources are separated based on differences in spatial signature vectors (array response vectors). Such algorithms need to know the parametric structure of the array response, therefore they rely on calibrated arrays. A complementary class of algorithms uses the structural properties of the source modulation, and try to reconstruct, at the output of the beamformer, a signal that has this structure. A widely used property for this is based on the fact that many sources are phase modulated, therefore have a constant modulus. The related Constant Modulus Algorithms (CMAs) are studied in this chapter.

6.1 INTRODUCTION

In wireless communications, an elementary beamforming problem arises when a number of sources at distinct locations transmit signals at nominally the same carrier frequency and in the same time slot. The signals are received by the base station, which is assumed here to contain an array of antennas. By linearly combining the antenna outputs, the objective is to separate the signals and remove the interference

Robust Adaptive Beamforming, Edited by Jian Li and Petre Stoica

from the other signals. In many cases of channel estimation and source separation, training sequences are available: a segment of the signal of interest which is known. In this chapter, we consider 'blind' algorithms: a blind beamformer is to compute the proper weight vectors $\mathbf{w}_i$ from the measured data only, *without* detailed knowledge of the signals and the channel. It can do so by comparing properties of the signal at the output of the beamformer to properties that the desired source signal would have at this point.

For example, if we know that the desired source has an amplitude $|s_k|$ constant to 1 for every sample index k (such a signal is called *constant modulus* or CM), then we can test this property for the output signal y_k of the beamformer, and define an error equal to the modulus difference $|y_k|^2 - 1$, as in Figure 6.1(a). Alternatively, we can estimate the best signal that has this property based on the output of the beamformer, that is, $\hat{s}_k = y_k/|y_k|$, and give an error equal to $\hat{s}_k - y_k$. Here, $\hat{s}_k$ is regarded as a good

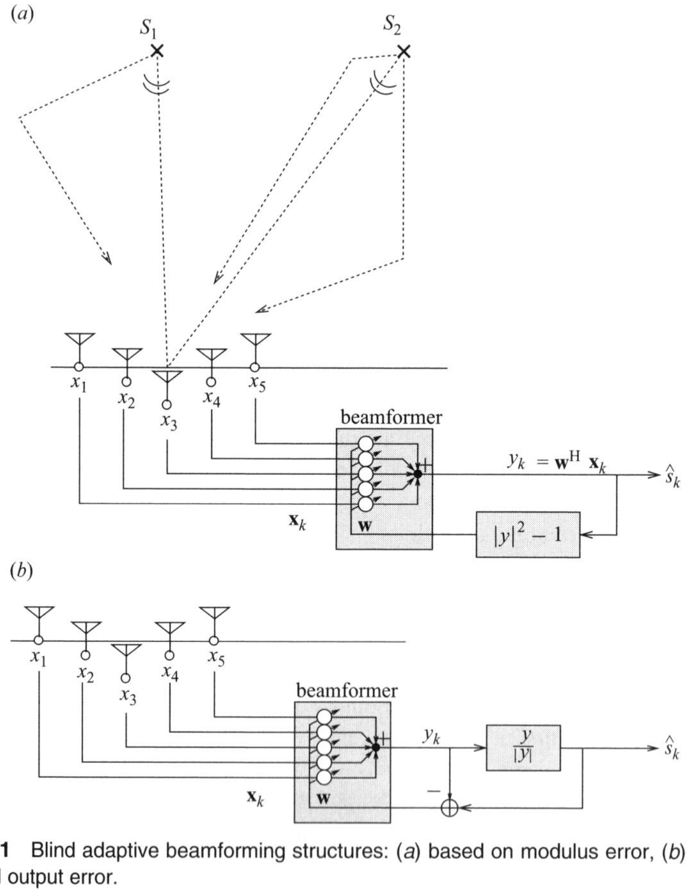

Figure 6.1 Blind adaptive beamforming structures: (a) based on modulus error, (b) based on estimated output error.

estimate of the source signal, and it is used as a reference signal instead of s_k. This is an elementary form of *decision feedback*, which could be refined further if we know that the source belongs to a certain alphabet, for example, $\{\pm 1\}$ for BPSK or $\{\pm 1, \pm j\}$ for QPSK. See Figure 6.1(*b*).

Throughout most of this chapter we assume a stationary situation with essentially no delay spread (as compared to the inverse of the signal bandwidths), so that no equalization is required. With d sources and M receive antennas, the situation is described by the simple data model

$$\mathbf{x}_k = \mathbf{A}\mathbf{s}_k + \mathbf{n}_k \tag{6.1}$$

where the vector $\mathbf{x}_k$ is a stacking of the M antenna outputs $(x_i)_k$ at discrete time k, $\mathbf{s}_k$ is a stacking of the d source signals $(s_i)_k$, and $\mathbf{A}$ is the $M \times d$ array response matrix which describes the linear combinations of the signals as received by the antennas. The M-dimensional vector $\mathbf{n}_k$ is the additive noise. A beamformer $\mathbf{w}$ takes a linear combination of the antenna outputs, which is written as an inner product $y_k = \mathbf{w}^{\mathrm{H}}\mathbf{x}_k$, where ^{H} denotes the complex conjugate transpose. The beamforming problem is to find weight vectors, one for each source, such that $\mathbf{w}_i^{\mathrm{H}}\mathbf{x}_k = (s_i)_k$ is equal to one of the original sources, without interference from the others. For this to be possible, we need to make (at least) the following assumptions, which are valid throughout the chapter:

1. $M \geq d$: more antennas than sources.
2. $\mathbf{A}$ has full column rank: its columns are linearly independent.
3. The power of the sources is absorbed in $\mathbf{A}$ (or a separate diagonal factor $\mathbf{B}$), therefore we may assume that all sources have equal, unit power.
4. $\mathbf{n}_k$ is white Gaussian noise, with covariance matrix $\mathrm{E}(\mathbf{n}_k\mathbf{n}_k^{\mathrm{H}}) = \sigma^2\mathbf{I}$, where E denotes the expectation operator.

Constant modulus algorithms have been widely studied. The specific aim of this chapter is to look at algorithms that find the complete set of *all* beamformers (one for each impinging signal). The original Constant Modulus Algorithm (CMA) [1] can find only a single signal. A 1992 extension (the CM Array [2, 3]) is a multistage algorithm based on successive cancellation; it has a poor performance. Currently there are two types of algorithms that can find all signals:

1. The Algebraic CMA (ACMA) [4] and similar algorithms such as JADE [5]: this is a noniterative block algorithm acting on a batch of data which computes jointly beamforming vectors for all constant modulus sources as the solution of a joint diagonalization problem;
2. The Multiuser Kurtosis (MUK) Algorithm [6]: an adaptive algorithm which can be viewed as a bank of CMAs with an orthogonality constraint between the beamformers.

Both algorithms are based on similar cost functions leading to fourth order equations for the beamformer coefficients, and in both algorithms, prewhitening plays an important role. Our aim is to put the two algorithms in a common framework so that they can be compared. To this end, we derive an block-iterative version and an adaptive version of ACMA.

In a second part of the chapter, we study the application of CMA algorithms to *direction finding*. In these approaches, we use both the structure of the source $\mathbf{s}_k$ and the parametric structure of $\mathbf{A}$, namely each column of $\mathbf{A}$ is a vector on the array manifold associated to a certain direction-of-arrival (DOA). By combining both properties, increased estimation accuracy and robustness of the beamformers is obtained.

Notation

We adopt the following notation:

$\bar{\cdot}$	Complex conjugation
^{T}	Matrix or vector transpose
^{H}	Matrix or vector complex conjugate transpose
†	Matrix pseudo-inverse (Moore–Penrose inverse)
$\underline{\cdot}$	Prewhitened data
$\mathbf{0}$	Vector of all 0s
$\mathbf{1}$	Vector of all 1s
$\mathrm{E}(\cdot)$	Mathematical expectation operator
$\hat{\cdot}$	Estimated value of a variable
$\mathrm{diag}(\mathbf{a})$	A diagonal matrix constructed from the vector $\mathbf{a}$
$\mathrm{vec}(\mathbf{A})$	Stacking of the columns of $\mathbf{A}$ into a vector
$\odot$	Schur–Hadamard product (entrywise multiplication)
$\otimes$	Kronecker product
$\circ$	Khatri–Rao product (column-wise Kronecker product), that is

$$\mathbf{A} \circ \mathbf{B} := [\mathbf{a}_1 \otimes \mathbf{b}_1 \quad \mathbf{a}_2 \otimes \mathbf{b}_2 \quad \cdots]. \tag{6.2}$$

Notable properties are, for matrices $\mathbf{A}, \mathbf{B}, \ldots$ and vectors $\mathbf{a}, \mathbf{b}$ of compatible sizes,

$$\mathrm{vec}(\mathbf{a}\mathbf{b}^{\mathrm{H}}) = \bar{\mathbf{b}} \otimes \mathbf{a} \tag{6.3}$$

$$(\mathbf{A} \otimes \mathbf{B})(\mathbf{C} \otimes \mathbf{D}) = \mathbf{AC} \otimes \mathbf{BD} \tag{6.4}$$

$$\mathrm{vec}(\mathbf{ABC}) = (\mathbf{C}^{\mathrm{T}} \otimes \mathbf{A})\mathrm{vec}(\mathbf{B}) \tag{6.5}$$

$$\mathrm{vec}(\mathbf{A}\ \mathrm{diag}(\mathbf{b})\mathbf{C}) = (\mathbf{C}^{\mathrm{T}} \circ \mathbf{A})\mathbf{b}. \tag{6.6}$$

6.2 THE CONSTANT MODULUS ALGORITHM

As mentioned, many communication signals have a constant modulus property. For such signals, the amplitude $|s_k|$ is a constant, typically normalized to 1, and all information is carried in the phase. If we have a single source s_k, and plot the (complex) samples in the complex plane, then all samples will lie on the unit circle, see Figure 6.2. On the other hand, if we have the sum of two sources, $(s_1)_k + \alpha(s_2)_k$, then the samples will in general not lie on a circle, unless $\alpha = 0$ (or if there are very special relations between the two sources—this is not possible if the two sources are independent). If $\alpha \neq 0$, then the received samples will be on a donut-shaped annulus.

The idea of modulus restoral is to play with the weights of a beamformer $\mathbf{w}$ until the output $y_k = \hat{s}_k = \mathbf{w}^{\mathrm{H}}\mathbf{x}_k$ has the same property, $|\hat{s}_k| = 1$, for all k. If that is the case, the output signal will be equal to one of the original sources [1], up to an unknown phase factor which cannot be established blindly.

6.2.1 CMA Cost Function

Popular implementations of such a property restoral algorithm are found by writing down a suitable cost function and minimizing it using stochastic gradient-descent techniques. For example, for a sample vector $\mathbf{x}_k$ we can consider as cost function the expected deviation of the squared modulus of the output signal $y_k = \mathbf{w}^{\mathrm{H}}\mathbf{x}_k$ to a constant, say 1:

$$J(\mathbf{w}) = \mathrm{E}(|y_k|^2 - 1)^2 = \mathrm{E}(|\mathbf{w}^{\mathrm{H}}\mathbf{x}_k|^2 - 1)^2. \tag{6.7}$$

This so-called CMA(2,2) cost function is simply a positive measure of the average amount that the beamformer output y_k deviates from the unit modulus condition. The objective in choosing $\mathbf{w}$ is to minimize J and hence to make y_k as close to a constant modulus signal as possible. Without additive noise, if we manage to achieve $J(\mathbf{w}) = 0$ then $\mathbf{w}$ reconstructs one of the sources.

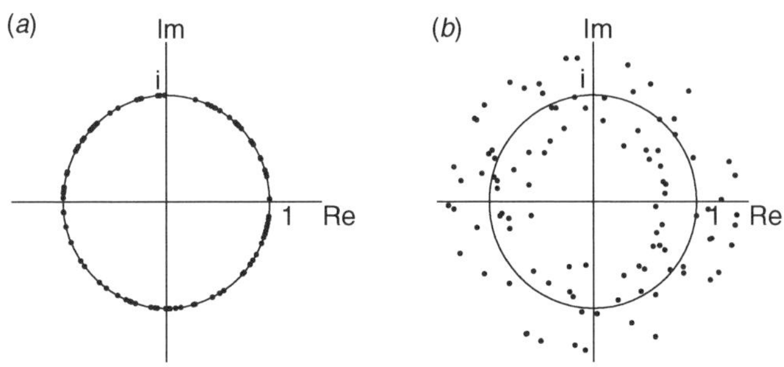

Figure 6.2 (*a*) Single CM signal, (*b*) sum of two CM signals.

A closed-form solution which will minimize the CM cost function (6.7) appears to be impossible because it is a fourth-order function with a more complicated structure. However, there are many ways in which we can iteratively search for the minimum of J. The simplest algorithm follows from a stochastic gradient-descent, similar to the derivation of the LMS algorithm by Widrow [7]. In this case, we update $\mathbf{w}$ iteratively, with small steps into the direction of the negative gradient,

$$\mathbf{w}^{(k+1)} = \mathbf{w}^{(k)} - \mu \, \nabla(J_k)$$

where μ is a small step size, and $\nabla(J_k) \equiv \nabla_{\bar{\mathbf{w}}} J(\mathbf{w}^{(k)})$ is the gradient vector of $J(\mathbf{w})$ with respect to the entries of $\bar{\mathbf{w}}$ (treated independently from $\mathbf{w}$), evaluated at the current value of $\mathbf{w}$. Using complex calculus and the fact that $|y_k|^2 = y_k \bar{y}_k = \mathbf{w}^{\mathrm{H}} \mathbf{x}_k \mathbf{x}_k^{\mathrm{H}} \mathbf{w}$, it can be verified that the gradient is given by

$$\begin{aligned}
\nabla(J_k) &= \mathrm{E}\{(|y_k|^2 - 1) \cdot \nabla(\mathbf{w}^{\mathrm{H}} \mathbf{x}_k \mathbf{x}_k^{\mathrm{H}} \mathbf{w})\} \\
&= 2\mathrm{E}\{(|y_k|^2 - 1) \cdot \mathbf{x}_k \mathbf{x}_k^{\mathrm{H}} \mathbf{w}\} \\
&= 2\mathrm{E}\{(|y_k|^2 - 1) \cdot \bar{y}_k \mathbf{x}_k\}.
\end{aligned}$$

Replacing the expectation by an instantaneous estimate, as in LMS, shows that we can find a minimizer $\mathbf{w}$ iteratively via

$$\mathbf{w}^{(k+1)} = \mathbf{w}^{(k)} - \mu \mathbf{x}_k \bar{z}_k, \quad y_k := \mathbf{w}^{(k)H} \mathbf{x}_k, \quad z_k := (|y_k|^2 - 1) y_k \tag{6.8}$$

(absorbing the factor 2 in μ). This iteration is called the Constant Modulus Algorithm (CMA, Treichler, Agee, and Larimore 1983 [1, 3, 8]) and was first introduced for the case of blind equalization. It has its roots in the work of Sato [9] and Godard [10]. See [11–13] for overviews and a historical perspective.

In comparison to the LMS algorithm, we see that the role of the update error (in the LMS equal to the output error $\epsilon_k = y_k - s_k$) is here played by z_k. In the LMS, we need a reference signal s_k to which we want to have the output converge. In CMA, however, the reference signal is not necessary, we use the a priori information that $|y_k| = 1$ in the absence of interfering signals.

We need to select a suitable step size μ and an initial point $\mathbf{w}^{(0)}$ for the iteration. Unlike LMS, we cannot choose $\mathbf{w}^{(0)} = \mathbf{0}$ since this precisely selects a local maximum of the cost function, but any other random vector will do. The maximal step size has not been theoretically derived. Because the cost function involves fourth order moments, the gradient is much steeper away from the optimum in comparison to LMS, and the maximal μ that still guarantees stability is smaller.

The constant modulus property holds for all phase-modulated and frequency modulated signals, and for several types of signals in the digital domain, such as frequency-shift keying (FSK), phase-shift keying (PSK), binary-shift keying (BPSK) and 4-QAM. For digital signals, the fact that the source symbols are selected from a finite alphabet is an even stronger property that can very well be exploited.

Even for sources that are not constant modulus, such as multilevel constellations (higher order QAM), the CMA can be successfully applied. The algorithm has been widely used in modem equalization.

6.2.2 Variants of the Adaptive CMA

Instead of the cost function in (6.7), a slightly different cost function is also often considered, namely the CMA(1,2) cost function

$$J(\mathbf{w}) = \mathrm{E}(|y_k| - 1)^2 = \mathrm{E}(|\mathbf{w}^{\mathrm{H}}\mathbf{x}_k| - 1)^2. \tag{6.9}$$

The associated update rule is given by [2]

$$\mathbf{w}^{(k+1)} := \mathbf{w}^{(k)} - \mu \mathbf{x}_k \bar{z}_k, \qquad \text{where } y_k := \mathbf{w}^{(k)H}\mathbf{x}_k, \quad z_k = y_k - \frac{y_k}{|y_k|}. \tag{6.10}$$

In this case, the update error that controls the iteration is $\left(y - \frac{y}{|y|}\right)$. Compared to the LMS, we see that $y/|y|$ plays the role of desired signal, see also Figure 6.1(*b*). Ideally, y_k is constant modulus and the error is zero. An advantage of this iteration is that the role of μ is more closely related to that of LMS, facilitating its analysis close to convergence. It also allows us to pose a normalized CMA [14] similar to the normalized LMS by Goodwin [15],

$$\mathbf{w}^{(k+1)} := \mathbf{w}^{(k)} - \frac{\mu}{\|\mathbf{x}_k\|^2}\mathbf{x}_k \bar{z}_k, \tag{6.11}$$

where μ is made data scaling independent by dividing by the instantaneous input power. For the NLMS, this modification was obtained by computing the optimal stepsize which would minimize the instantaneous error; it is known that $0 < \mu < 2$ is required for stability, although one would take $\mu \ll 1$ to obtain a sufficiently smooth performance. The same range holds for NCMA. A better but slightly more complicated *orthogonalization* version of CMA was considered in [2] and became known as orthogonal CMA (OCMA, see also [16]),

$$\mathbf{w}^{(k+1)} := \mathbf{w}^{(k)} - \mu \hat{\mathbf{R}}_k^{-1}\mathbf{x}_k \bar{z}_k, \tag{6.12}$$

where $\hat{\mathbf{R}}_k$ is an estimate of the data covariance matrix $\mathbf{R} = \mathrm{E}(\mathbf{x}_k\mathbf{x}_k^{\mathrm{H}})$. Usually $\hat{\mathbf{R}}_k$ is estimated by a sliding window estimate

$$\hat{\mathbf{R}}_k = \lambda \hat{\mathbf{R}}_{k-1} + (1 - \lambda)\mathbf{x}_k\mathbf{x}_k^{\mathrm{H}}, \tag{6.13}$$

where $0 < \lambda < 1$ determines the exponential window size (the effective window size is defined as $1/(1-\lambda)$). The inverse of $\hat{\mathbf{R}}_k$ can be efficiently updated using techniques known from the RLS algorithm [15].

There are a few other CMAs, in particular the LS-CMA, which will be discussed in Section 6.4.2.

6.2.3 The CM Array

The CMA gives only a single beamformer vector. This is sufficient for blind equalization applications, where the received signal consists of several temporal shifts of the same CM signal, and we do not have to recover all of them. In contrast, the beamforming problem frequently asks for all possible weight vectors that give back linearly independent CM signals, which is usually much harder.

If we initialize the CMA with a random vector $\mathbf{w}^{(0)}$, then the CMA tends to converge to the strongest signal. However, this cannot be guaranteed: for an initialization close enough to a weak signal, the algorithm converges to that weaker signal.[1] This gives one way to find all signals: use various initializations and determine if independent signals have been found.

A somewhat more robust algorithm is the so-called multistage CMA, also called the CM array. It was introduced in [2, 3], with analysis appearing in [17–19]. The output of a first CMA stage results in the detection of the first CM signal, and gives an estimate $\hat{s}_1(k)$. This signal can be used as a reference signal for an LMS algorithm to estimate the corresponding array response vector $\hat{\mathbf{a}}_1$, an estimate of the first column of $\mathbf{A}$. The resulting update rule is

$$\hat{\mathbf{a}}_1^{(k+1)} = \hat{\mathbf{a}}_1^{(k)} + \mu_{lms}\left[\mathbf{x}(k) - \hat{\mathbf{a}}_1^{(k)}\hat{s}_1(k)\right]\bar{\hat{s}}_1(k).$$

We can then subtract the estimated source signal from the original data sequence,

$$\mathbf{x}_1(k) = \mathbf{x}(k) - \hat{\mathbf{a}}_1^{(k)}\hat{s}_1(k)$$

and feed the resulting filtered data to a second CMA stage in order to detect a possible second CM signal. This can be repeated until all signals have been found. See Figure 6.3.[2]

A problem with this scheme is that the LMS algorithm can converge only after the first CMA has sufficiently converged. In the mean time, the second CMA may have converged to the same signal as the first CMA (especially if it is strong), and if the first LMS is not completely removing this signal, the second stage will stay at this signal. Thus, it may happen that the same signal is found twice, and/or that not all signals are found. A related problem is that the CMA converges to a point close to the Wiener solution. Hence, the estimate $\hat{s}_1(k)$ will always contain

[1] During the early days of CMA, only the reception of the strongest signal was desired, and convergence to another signal was regarded as misconvergence.

[2] The algorithm used in the CM array in [17–19] is in fact based on the CMA(1,2) as in equation (6.10).

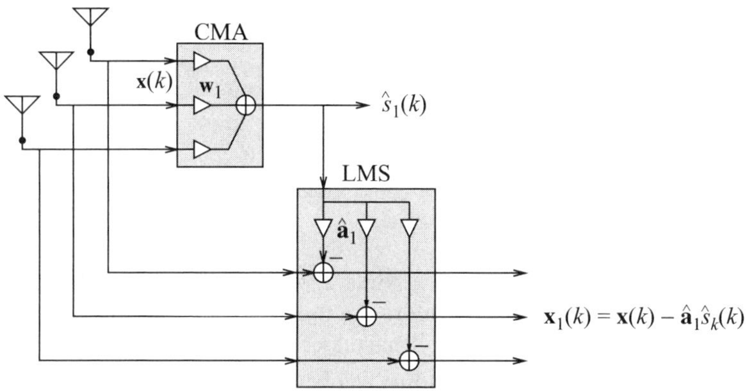

Figure 6.3 The CM Array.

components of the other signals as well, causing misadjustment in later stages. An analysis of the situation is given in [17, 19].

Another problem is that the convergence speed may be slow (several hundreds of samples), since we have a cascade of adaptive algorithms. It has been proposed to use direction-finding algorithms such as MUSIC first to initialize the cascade. An alternative approach is to augment the cost function with additional terms that express the independence of the output signals, for example, by putting a constraint on the cross-correlation of the recovered signals [20, 21]. Algorithms for this are discussed in Section 6.4. Some of the problems are alleviated by prewhitening, which is discussed in the next section.

6.3 PREWHITENING AND RANK REDUCTION

6.3.1 Rank Reduction

At this point, let us first make a small extension to our notation. Starting from the data model $\mathbf{x}_k = \mathbf{A}\mathbf{s}_k + \mathbf{n}_k$, we assume that we have received N sample vectors, $k = 1, \ldots, N$. It is often convenient to collect the data in matrices:

$$\mathbf{X} = [\mathbf{x}_1, \mathbf{x}_2, \ldots, \mathbf{x}_N], \quad \mathbf{S} = [\mathbf{s}_1, \mathbf{s}_2, \ldots, \mathbf{s}_N],$$

and likewise for a noise matrix $\mathbf{N}$, so that the model becomes

$$\mathbf{X} = \mathbf{A}\mathbf{S} + \mathbf{N}.$$

The objective is to recover all beamformers $\mathbf{w}_i$, $i = 1, \ldots, d$, one for each source.

These can also be collected into a matrix $\mathbf{W}$,

$$\mathbf{W} = [\mathbf{w}_1, \ldots, \mathbf{w}_d], \quad M \times d.$$

In the noise-free case, we would like to achieve $\mathbf{W}^{\mathrm{H}}\mathbf{X} = \mathbf{S}$. However, with no other knowledge about $\mathbf{S}$ but the constant modulus property, there are two causes of nonuniqueness:

1. The ordering of sources is arbitrary: we do not know which is "source number 1," and so on. Therefore the ordering of beamformers (columns of $\mathbf{W}$) is arbitrary: $\mathbf{W}$ will always have a permutation ambiguity.
2. The solution for each beamformer can be found only up to an arbitrary phase, since the CM cost function is phase-blind.

This kind of nonuniqueness is common to all blind source separation problems. In the estimation of a beamformer, there may be one other cause of nonuniqueness. Namely, additional nullspace solutions exist if the received data matrix $\mathbf{X}$ is rank deficient: in this case there are beamformers $\mathbf{w}_0$ such that $\mathbf{w}_0^{\mathrm{H}}\mathbf{X} = \mathbf{0}^{\mathrm{H}}$. Such solutions can be added to any beamformer $\mathbf{w}_i$ and cause nonuniqueness: two linearly independent beamformers ($\mathbf{w}_i$ and $\mathbf{w}_i + \mathbf{w}_0$) may still reconstruct the same signal. This is clearly undesired if our aim is to reconstruct all independent signals, because the simplest way to detect if independent signals have been obtained is to verify the linear independence of the corresponding beamformers.

Nullspace solutions exist if the number of sensors is larger than the number of sources ($\mathbf{A}$ tall), and if $\mathbf{A}$ is not full column rank. The former is simply treated by a prefiltering operation that reduces the number of rows of $\mathbf{X}$ from M to d, as we discuss here, whereas the latter case is hopeless, at least for linear receivers.

We will use the underscore (_) to denote prefiltered variables. Thus, let $\underline{\mathbf{X}} := \mathbf{F}^{\mathrm{H}}\mathbf{X}$ where $\mathbf{F}: M \times d$ is the prefilter. Then

$$\underline{\mathbf{X}} = \underline{\mathbf{A}}\mathbf{S} + \underline{\mathbf{N}}, \quad \text{where } \underline{\mathbf{A}} := \mathbf{F}^{\mathrm{H}}\mathbf{A}, \;\; \underline{\mathbf{N}} := \mathbf{F}^{\mathrm{H}}\mathbf{N}.$$

This is essentially the same model as before, except $\underline{\mathbf{X}}$ has only d channels and $\underline{\mathbf{A}}: d \times d$ is square. The blind beamforming problem is now replaced by finding a

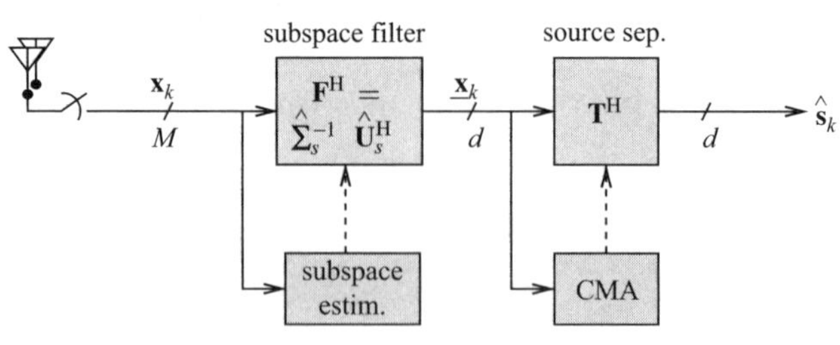

Figure 6.4 Blind beamforming prefiltering structure.

separating beamforming matrix $\mathbf{T}: d \times d$ with columns $\mathbf{t}_i$, acting on $\underline{\mathbf{X}}$. After $\mathbf{T}$ has been found, the beamforming matrix on the original data will be $\mathbf{W} = \mathbf{FT}$. The associated processing structure is illustrated in Figure 6.4. For the purpose of dimension reduction, there are many suitable $\mathbf{F}$: the only requirement is that $\mathbf{F}^{\mathrm{H}}\mathbf{A}$ should be full rank. To avoid noise enhancement, we also want it to be well conditioned. This leads to the choice of $\mathbf{F}$ to have orthogonal columns that together span the column span of $\mathbf{A}$.

6.3.2 Whitening

Assume that the noise is white i.i.d. with covariance matrix $\mathbf{R_n} = \mathrm{E}(\mathbf{nn}^{\mathrm{H}}) = \sigma^2\mathbf{I}$. We can choose $\mathbf{F}$ such that the resulting data matrix $\underline{\mathbf{X}}$ is white, as follows. Let $\hat{\mathbf{R}}_{\mathbf{x}} = \frac{1}{N}\sum_{k=1}^{N} \mathbf{x}_k\mathbf{x}_k^{\mathrm{H}} = (\frac{1}{N})\mathbf{XX}^{\mathrm{H}}$ be the noisy sample data covariance matrix, with eigenvalue decomposition

$$\hat{\mathbf{R}}_{\mathbf{x}} = \hat{\mathbf{U}}\hat{\mathbf{\Sigma}}^2\hat{\mathbf{U}}^{\mathrm{H}} = \begin{bmatrix} \hat{\mathbf{U}}_s & \hat{\mathbf{U}}_n \end{bmatrix} \begin{bmatrix} \hat{\mathbf{\Sigma}}_s^2 & \\ & \hat{\mathbf{\Sigma}}_n^2 \end{bmatrix} \begin{bmatrix} \hat{\mathbf{U}}_s^{\mathrm{H}} \\ \hat{\mathbf{U}}_n^{\mathrm{H}} \end{bmatrix}. \tag{6.14}$$

Here, $\hat{\mathbf{U}} = [\hat{\mathbf{U}}_s \; \hat{\mathbf{U}}_n]$ is $M \times M$ unitary, and $\hat{\mathbf{\Sigma}}^2$ is $M \times M$ diagonal ($\hat{\mathbf{\Sigma}}$ contains the singular values of $\mathbf{X}/\sqrt{N}$). The d largest eigenvalues are collected into a diagonal matrix $\hat{\mathbf{\Sigma}}_s^2$ and the corresponding d eigenvectors into $\hat{\mathbf{U}}_s$ (they span the signal subspace). In this notation, define $\mathbf{F}$ as

$$\mathbf{F} = \hat{\mathbf{U}}_s\hat{\mathbf{\Sigma}}_s^{-1}. \tag{6.15}$$

This prewhitening is such that $\hat{\underline{\mathbf{R}}}_{\mathbf{x}} := (\frac{1}{N})\underline{\mathbf{X}}\underline{\mathbf{X}}^{\mathrm{H}}$ is unity: $\hat{\underline{\mathbf{R}}}_{\mathbf{x}} = \mathbf{I}$, and at the same time it reduces the dimension of $\mathbf{X}$ from M rows to d rows.

An important reason to choose a prefilter that whitens the output is that the resulting $\underline{\mathbf{A}}$ in the whitened domain is approximately unitary. Indeed, let $\mathbf{A} = \mathbf{U}_A\mathbf{\Sigma}_A\mathbf{V}_A^{\mathrm{H}}$ be an economy-size SVD of $\mathbf{A}$ ($\mathbf{U}_A$ is a submatrix of a unitary matrix and has size $M \times d$, $\mathbf{V}_A$ is size $d \times d$ and is unitary, and $\mathbf{\Sigma}_A$ is $d \times d$ diagonal and contains the singular values), then for a large number of samples, $\hat{\mathbf{R}}_{\mathbf{x}} \approx \mathbf{R}_{\mathbf{x}}$, where

$$\mathbf{R}_{\mathbf{x}} = \mathbf{AA}^{\mathrm{H}} + \sigma^2\mathbf{I} = [\mathbf{U}_A \quad \mathbf{U}_A^{\perp}] \begin{bmatrix} \mathbf{\Sigma}_A^2 + \sigma^2\mathbf{I} & \\ & \sigma^2\mathbf{I} \end{bmatrix} \begin{bmatrix} \mathbf{U}_A^{\mathrm{H}} \\ (\mathbf{U}_A^{\perp})^{\mathrm{H}} \end{bmatrix}$$

so that we can identify $\mathbf{U}_s = \mathbf{U}_A$ and $\mathbf{\Sigma}_s^2 = \mathbf{\Sigma}_A^2 + \sigma^2\mathbf{I}$. Therefore,

$$\underline{\mathbf{A}} = \mathbf{F}^{\mathrm{H}}\mathbf{A} = \mathbf{\Sigma}_s^{-1}\mathbf{U}_s^{\mathrm{H}}\mathbf{U}_A\mathbf{\Sigma}_A\mathbf{V}_A^{\mathrm{H}} = (\mathbf{\Sigma}_s^{-1}\mathbf{\Sigma}_A)\mathbf{V}_A^{\mathrm{H}}.$$

This shows that $\underline{\mathbf{A}}$ is unitary if $(\mathbf{\Sigma}_A^2 + \sigma^2\mathbf{I})^{-1/2}\mathbf{\Sigma}_A = \mathbf{I}$, or a scalar multiple of $\mathbf{I}$, which is the case if there is no noise or if the singular values of $\mathbf{A}$ are all the same (then $\mathbf{A}$ has orthonormal columns, which corresponds to well-separated equal-powered sources). If this is not the case, then $\underline{\mathbf{A}}$ is only approximately unitary, but always better conditioned than $\mathbf{A}$ itself (the conditioning of a matrix is the ratio of its largest and smallest singular value, preferably this is a number close to 1).

It has sometimes been suggested to use a slightly different prewhitening filter, $\mathbf{F} = \hat{\mathbf{U}}_s \hat{\boldsymbol{\Sigma}}_A^{-1}$, where $\hat{\boldsymbol{\Sigma}}_A = (\hat{\boldsymbol{\Sigma}}_s^2 - \sigma^2 \mathbf{I})^{1/2}$ is an estimate of $\boldsymbol{\Sigma}_A$. This would yield a unitary $\underline{\mathbf{A}}$. Care has to be taken to ensure that $\hat{\boldsymbol{\Sigma}}_s^2 - \sigma^2 \mathbf{I}$ is positive, for example, by replacing σ^2 by a data-dependent estimate.

In the noise-free case, the optimal beamformer is $\mathbf{T} = \underline{\mathbf{A}}^{-H}$. The importance of having a unitary $\underline{\mathbf{A}}$-matrix is that the corresponding beamformer is $\mathbf{T} = \underline{\mathbf{A}}$ is also unitary, hence has maximally independent columns. If our aim is to find all independent sources in the data, it is very convenient to know that the beamformers are orthogonal: if we have found one solution $\mathbf{t}_1$, we can constrain the other solutions to be orthogonal to it, which is a simpler constraint than requiring them to be linearly independent. If $\underline{\mathbf{A}}$ is only approximately unitary, for example, in the presence of noise, then the orthogonality condition on $\mathbf{T}$ has to be relaxed to being well-conditioned.

It is well recognized in adaptive filtering that moving to the whitened domain improves the convergence speed: this is often limited by the conditioning of the data covariance matrix, which becomes optimal after whitening. (A disadvantage is that the noise is not longer white nor spatially independent.) This is the motivation of introducing the factor $\mathbf{R}_{\mathbf{x}}^{-1}$ in the OCMA in equation (6.12).

6.3.3 Convergence to the Wiener Beamformer

In a stochastic context, the Wiener beamformer is defined as the solution to the linear minimum mean square error (LMMSE) problem

$$\mathbf{w} = \arg\min_{\mathbf{w}} \mathrm{E}\left|\mathbf{w}^{\mathrm{H}}\mathbf{x}_k - s_k\right|^2,$$

where s_k is known to the receiver. The solution is straightforward to derive,

$$\begin{aligned}
\mathrm{E}\left|\mathbf{w}^{\mathrm{H}}\mathbf{x}_k - s_k\right|^2 &= \mathbf{w}^{\mathrm{H}}\mathrm{E}(\mathbf{x}_k\mathbf{x}_k)\mathbf{w}^{\mathrm{H}} - \mathbf{w}^{\mathrm{H}}\mathrm{E}(\mathbf{x}_k\bar{s}_k) - \mathrm{E}(s_k\mathbf{x}_k^{\mathrm{H}})\mathbf{w} + \mathrm{E}(|s_k|^2) \\
&= \mathbf{w}^{\mathrm{H}}\mathbf{R}_{\mathbf{x}}\mathbf{w} - \mathbf{w}^{\mathrm{H}}\mathbf{r}_{\mathbf{x}s} - \mathbf{r}_{\mathbf{x}s}^{\mathrm{H}}\mathbf{w} + r_s \\
&= (\mathbf{w} - \mathbf{R}_{\mathbf{x}}^{-1}\mathbf{r}_{\mathbf{x}s})^{\mathrm{H}}\mathbf{R}_{\mathbf{x}}(\mathbf{w} - \mathbf{R}_{\mathbf{x}}^{-1}\mathbf{r}_{\mathbf{x}s}) + r_s - \mathbf{r}_{\mathbf{x}s}^{\mathrm{H}}\mathbf{R}_{\mathbf{x}}^{-1}\mathbf{r}_{\mathbf{x}s},
\end{aligned}$$

where $\mathbf{R}_{\mathbf{x}}$ is the data covariance matrix, and $\mathbf{r}_{\mathbf{x}s}$ is the correlation of the received data with the transmitted signal. Therefore, the optimal beamformer is given by $\mathbf{w} = \mathbf{R}_{\mathbf{x}}^{-1}\mathbf{r}_{\mathbf{x}s}$. Similarly, if a vector $\mathbf{s}_k$ of d sources is specified, the collection of beamformers is $\mathbf{W} = \mathbf{R}_{\mathbf{x}}^{-1}\mathbf{R}_{\mathbf{xs}}$, where $\mathbf{R}_{\mathbf{xs}} = \mathrm{E}(\mathbf{x}_k\mathbf{s}_k^{\mathrm{H}})$. Assuming the noise is independent from the signals, and the sources are i.i.d., we obtain $\mathbf{R}_{\mathbf{xs}} = \mathbf{A}$ and

$$\mathbf{W} = \mathbf{R}_{\mathbf{x}}^{-1}\mathbf{A}. \tag{6.16}$$

In a deterministic context, the Wiener beamformer based on sample data is derived similarly as the solution to the least squares problem

$$\hat{\mathbf{W}} = \arg\min_{\mathbf{W}} \|\mathbf{W}^{\mathrm{H}}\mathbf{X} - \mathbf{S}\|_{\mathrm{F}}^{2} = (\mathbf{S}\mathbf{X}^{\dagger})^{\mathrm{H}} = \left(\frac{1}{N}\mathbf{X}\mathbf{X}^{\mathrm{H}}\right)^{-1}\frac{1}{N}\mathbf{X}\mathbf{S}^{\mathrm{H}}. \tag{6.17}$$

(The subscript $_{\mathrm{F}}$ indicates the Frobenius norm.) As $N \to \infty$, the deterministic Wiener beamformer converges to (6.16). The importance of the Wiener beamformer is that it can be shown to maximize the signal to interference and noise ratio (SINR) among linear receivers.

Suppose we act in the whitened domain: $\underline{\mathbf{R}}_{\mathbf{x}} = \mathbf{I}$. The Wiener solution is then $\mathbf{T} = \underline{\mathbf{A}}$: each column $\mathbf{t}_i$ of the beamformer is equal to the whitened direction vector (a matched spatial filter). If we go back to the resulting beamformer $\mathbf{w}_i$ acting on the original (unwhitened) data matrix $\mathbf{X}$, we find (for $i = 1, \ldots, d$)

$$\mathbf{t}_i = \underline{\mathbf{a}}_i = \mathbf{F}^{\mathrm{H}}\mathbf{a}_i \implies \mathbf{w}_i = \mathbf{F}\mathbf{t}_i = \mathbf{F}\mathbf{F}^{\mathrm{H}}\mathbf{a}_i = \mathbf{R}_{\mathbf{x}}^{-1}\mathbf{a}_i.$$

If the whitening did not involve a dimension reduction, the last step is clear. But even if $\mathbf{F}$ is defined to reduce the dimension to the signal subspace the result holds (assuming the noise is white i.i.d.), since

$$\mathbf{R}_{\mathbf{x}}^{-1} = \mathbf{U}\mathbf{\Sigma}^{-2}\mathbf{U}^{\mathrm{H}} = \mathbf{U}_s\mathbf{\Sigma}_s^{-2}\mathbf{U}_s^{\mathrm{H}} + \sigma^{-2}\mathbf{U}_n\mathbf{U}_n^{\mathrm{H}} = \mathbf{F}\mathbf{F}^{\mathrm{H}} + \sigma^{-2}\mathbf{U}_n\mathbf{U}_n^{\mathrm{H}}$$

whereas $\mathbf{U}_n^{\mathrm{H}}\mathbf{a}_i = \mathbf{0}$. Therefore, the beamformer in the original domain is equal to the Wiener beamformer (6.16). In general, this is a very attractive property.

Several algorithms can be shown to converge to the true $\mathbf{a}$-vector or $\mathbf{A}$-matrix asymptotically, in case the number of samples $N \to \infty$ or in case the signal to noise ratio SNR $\to \infty$. Such algorithms, when applied in the whitened domain, converge to Wiener beamformers. A caveat in the proof is that, in the whitened domain, the noise covariance $\sigma^2\mathbf{F}\mathbf{F}^{\mathrm{H}}$ is in general not white anymore whereas many algorithms are based on this assumption. (For well separated and equal powered sources, the resulting noise covariance is still white.)

In summary, the prewhitening serves two purposes:

1. The dimension reduction (from number of antennas M to number of sources d) avoids the existence of additional nullspace solutions, which facilitates the reconstruction of all independent signals.
2. The prewhitening will improve the conditioning of the problem: after prewhitening, $\underline{\mathbf{A}}$ is approximately unitary, therefore the beamformers in the whitened domain are approximately orthogonal. This greatly facilitates the convergence of iterative algorithms to independent solutions, and sometimes also guarantees the convergence to beamformers that are close to the Wiener beamformer.

6.4 MULTIUSER CMA TECHNIQUES

6.4.1 OCMA and MUK

The CMA(2,2) cost function was shown in equation (6.7). The corresponding adaptive algorithm (stochastic gradient) is

$$\mathbf{w}_{k+1} = \mathbf{w}_k - \mu \mathbf{x}_k \bar{z}_k, \quad z_k = (|y_k|^2 - 1)y_k$$

where $y_k = \mathbf{w}_k^{\mathrm{H}} \mathbf{x}_k$ is the output of the beamformer using its current estimate, and μ is a small step size. We also introduced the OCMA in equation (6.12) [2], which premultiplied the update term by $\hat{\mathbf{R}}_{\mathbf{x}}^{-1}$ to make the algorithm independent of the scaling of the data:

$$\mathbf{w}_{k+1} = \mathbf{w}_k - \mu \hat{\mathbf{R}}_{\mathbf{x}}^{-1} \mathbf{x}_k \bar{z}_k, \quad z_k = (|y_k|^2 - 1)y_k. \tag{6.18}$$

We can directly interpret this algorithm in terms of our prewhitening step. Indeed, define $\mathbf{t} = \hat{\mathbf{R}}_{\mathbf{x}}^{1/2} \mathbf{w}$ and $\underline{\mathbf{x}} = \hat{\mathbf{R}}_{\mathbf{x}}^{-1/2} \mathbf{x}$. Premultiplying (6.18) with $\hat{\mathbf{R}}_{\mathbf{x}}^{1/2}$ leads to

$$\mathbf{t}_{k+1} = \mathbf{t}_k - \mu \underline{\mathbf{x}}_k \bar{z}_k, \quad z_k = (|y_k|^2 - 1)y_k$$

where $y_k = \mathbf{w}_k^{\mathrm{H}} \mathbf{x}_k = \mathbf{t}_k^{\mathrm{H}} \underline{\mathbf{x}}_k$. Therefore, OCMA is equal to the ordinary CMA, but in the whitened domain. The algorithm is easily modified to update d beamformers in parallel:

$$\mathbf{T}_{k+1} = \mathbf{T}_k - \mu \underline{\mathbf{x}}_k \mathbf{z}_k^{\mathrm{H}}, \quad \mathbf{z}_k = (\mathbf{y}_k \odot \bar{\mathbf{y}}_k - 1) \odot \mathbf{y}_k$$

where $\mathbf{y}_k = \mathbf{W}_k^{\mathrm{H}} \mathbf{x}_k = \mathbf{T}_k^{\mathrm{H}} \underline{\mathbf{x}}_k$ and $\odot$ denotes the Schur–Hadamard product (entrywise multiplication). In spite of its appearance, the beamformers are updated independently, and there is no guarantee that they converge to independent solutions. However, since $\mathbf{T}$ is supposed to be almost orthogonal and therefore well-conditioned (linearly independent columns), it is straightforward to recondition $\mathbf{T}$ after each update.

A simple technique to restore the linear independence of the solutions is to compute a singular value decomposition of $\mathbf{T}$ as $\mathbf{T} = \sum \sigma_j \mathbf{u}_j \mathbf{v}_j^{\mathrm{H}}$, and to replace the singular values of $\mathbf{T}$ that are below some threshold (e.g., smaller than 0.5) by 1:

$$\mathbf{T}' = \mathrm{recond}(\mathbf{T}) := \sum \sigma_j' \mathbf{u}_j \mathbf{v}_j^{\mathrm{H}} \text{ where } \sigma_j' = \begin{cases} 1, & \sigma_j < 0.5 \\ \sigma_j, & \sigma_j \geq 0.5 \end{cases} \tag{6.19}$$

Experience by simulations shows that (1) this reconditioning step is very effective, and (2) with good SNR it is rarely needed, even if sources are closely spaced. The reasons for this have to do with the good conditioning of the problem after prewhitening: the columns of the desired solution are almost orthogonal.

A similar algorithm was proposed more recently, called the Multiuser Kurtosis Algorithm (MUK) [6]. MUK is not specifically targeted for CM signals, but aims to separate statistically independent non-Gaussian signals by maximizing the absolute value of the kurtosis $K(y)$ of the output, where $K(y) = \mathrm{E}(|y|^4) - 2[\mathrm{E}(|y|^2)]^2$. This is a Shalvi–Weinstein cost function [22] (in this seminal paper, criteria were introduced which involve only the computation of the second- and fourth-order moments and do not pose restrictions on the probability distribution of the input sequence). For sources with a negative kurtosis, this leads to [6]

$$\mathbf{T}_{k+1} = \mathbf{T}_k - \mu \underline{\mathbf{x}}_k \mathbf{z}_k^{\mathrm{H}}, \quad \mathbf{z}_k = \mathbf{y}_k \odot \bar{\mathbf{y}}_k \odot \mathbf{y}_k.$$

In [6], a condition that $\mathbf{T}$ should be orthogonal (rather than well-conditioned) is maintained. This can be formulated as an orthogonal Procrustes problem,

$$\mathbf{T}' = \arg\min_{\mathbf{T}'^{\mathrm{H}}\mathbf{T}'=\mathbf{I}} \|\mathbf{T} - \mathbf{T}'\|_{\mathrm{F}}^2$$

of which the solution is [23]

$$\mathbf{T} =: \sum_j \sigma_j \mathbf{u}_j \mathbf{v}_j^{\mathrm{H}} \implies \mathbf{T}' = \mathrm{reorth}(\mathbf{T}) := \sum_j \mathbf{u}_j \mathbf{v}_j^{\mathrm{H}}. \tag{6.20}$$

In [6], this solution is approximated by a QR factorization (this implements a sequential reorthogonalization where each column of $\mathbf{T}'$ is made orthogonal to the preceding columns). This does not optimize (6.20) and therefore may be less effective, but is computationally simpler. Alternatively, orthogonality of weight vectors may be enforced during adaptive processing by perturbation analysis of small rotations, as illustrated, for example, in [24] and in Section 6.6.

Both algorithms are summarized in Figure 6.5.

Given $\mathbf{X} = [\mathbf{x}_1, \mathbf{x}_2, \ldots]$ and a stepsize μ, compute beamformers $\mathbf{W}_k = \mathbf{F}_k\mathbf{T}_k$, output vector $\mathbf{s}_k = \mathbf{W}_k^{\mathrm{H}}\mathbf{x}_k$:
Initialize prewhitening filter $\mathbf{F}$ using m prior input samples; $\mathbf{T} = \mathbf{I}_{d\times d}$.
For $k = 1, 2, \ldots$ do
1. Update $\mathbf{F}$, the prewhitening filter, using $\mathbf{x}_k$ (Fig. 6.11) — $(4dM + 3d^2)$
$\underline{\mathbf{x}} = \mathbf{F}^{\mathrm{H}}\mathbf{x}_k$, the prewhitened input vector
2. $\mathbf{y} = \mathbf{T}^{\mathrm{H}}\underline{\mathbf{x}}$, the current beamformer output vector, or — (d^2)
$\mathbf{z} = \mathbf{y} \odot \bar{\mathbf{y}} \odot \mathbf{y}$ (MUK-update)
$\mathbf{z} = (\mathbf{y} \odot \bar{\mathbf{y}} - \mathbf{1}) \odot \mathbf{y}$ (OCMA-update) — $(3d)$
$\mathbf{T} = \mathbf{T} - \mu\underline{\mathbf{x}}\mathbf{z}^{\mathrm{H}}$ — (d^2)
$\mathbf{T} = \mathrm{reorth}(\mathbf{T})$, equation (6.20), or — (d^3)
$\mathbf{T} = \mathrm{recond}(\mathbf{T})$, equation (6.19)
3. $\hat{\mathbf{s}}_k = \mathbf{y}$
Total: $4dM + d^3 + 5d^2$

Figure 6.5 Multiuser CMAs: MUK and OCMA algorithm (in brackets the complexity of a single update step).

6.4.2 Least Squares CMA

In the context of block-iterative algorithms, many other solutions are possible. We may formulate the multiuser constant modulus problem as a least squares problem

$$\hat{\mathbf{S}} = \arg\min_{\mathbf{S},\mathbf{W}} \|\mathbf{W}^{\mathrm{H}}\mathbf{X} - \mathbf{S}\|^2, \quad \mathbf{S} \in \mathcal{CM}, \mathbf{W} \text{ full rank}$$

where $\mathcal{CM}$ denotes the set of constant modulus signals,

$$\mathcal{CM} = \{\mathbf{S} | |s_{ij}| = 1, \text{ all } i,j\}.$$

With an initial value for $\mathbf{W}$, we can set up an alternating least squares procedure as before: (1) find the best matching $\mathbf{S} \in \mathcal{CM}$, that is, project $\mathbf{W}^{\mathrm{H}}\mathbf{X}$ onto the set of CM signals, (2) find the best matching $\mathbf{W}$, that is $\mathbf{W}^{\mathrm{H}} = \mathbf{S}\mathbf{X}^{\dagger}$. Note that $\mathbf{W}^{\mathrm{H}}\mathbf{X} = \mathbf{S}\mathbf{X}^{\dagger}\mathbf{X}$, so that the algorithm alternates between projecting $\mathbf{S}$ onto the row span of $\mathbf{X}$, and projecting it onto the set $\mathcal{CM}$. The latter projection is simply a scaling of each entry of $\mathbf{S}$ to unit-norm:

$$\hat{\mathbf{S}} = \left[\frac{s_{ij}}{|s_{ij}|}\right]_{ij}.$$

This algorithm was derived from the CMA by Agee [25], and called the LS-CMA. A similar algorithm is well-known in the field of optics for solving the phase-retrieval problem, where it is called the Gerchberg–Saxton Algorithm (GSA) [26]. The relation was pointed out in [27].

As stated in Section 6.2, CMAs have significant problems in finding all independent sources. For example, LS-CMA has no capability to ensure that $\mathbf{W}$ has full rank. As in the preceding section, this problem can be largely avoided by

Given data $\mathbf{X}$, compute beamformers $\mathbf{W}$ and output $\mathbf{S} = \mathbf{W}^{\mathrm{H}}\mathbf{X}$:

1. SVD: $\mathbf{X} =: \hat{\mathbf{U}}\hat{\mathbf{\Sigma}}\hat{\mathbf{V}}$ $\quad (M^2N)$
   Prewhitening filter: $\mathbf{F} = \hat{\mathbf{U}}_s\hat{\mathbf{\Sigma}}_s^{-1}/\sqrt{N}$
   Prefiltering: $\underline{\mathbf{X}} := \mathbf{F}^{\mathrm{H}}\mathbf{X} = \hat{\mathbf{V}}_s\sqrt{N}$
   Data matrix inverse: $\underline{\mathbf{X}}^{\dagger} = \hat{\mathbf{V}}_s^{\mathrm{H}}/\sqrt{N}$
   Initialize: $\mathbf{T} = \mathbf{I}_{d\times d}$
2. Iterate until convergence:
   $\mathbf{S} = \mathbf{T}^{\mathrm{H}}\underline{\mathbf{X}}$ $\quad (d^2N)$
   $\mathbf{S} = \left[\frac{s_{ij}}{|s_{ij}|}\right]_{ij}$ $\quad (dN)$
   $\mathbf{T} = (\mathbf{S}\underline{\mathbf{X}}^{\dagger})^{\mathrm{H}}$ $\quad (d^2N)$
   $\mathbf{T} = \text{recond}(\mathbf{T})$, see equation (6.19) $\quad (d^3)$
3. $\mathbf{W} = \mathbf{FT}$

Figure 6.6 LS-CMA in the whitened domain.

working in the whitened domain, for which **T** is nearly orthogonal. In the whitened domain, after **T** is computed, we need to verify its conditioning and possibly improve it, as in equation (6.19). This leads to the algorithm stated in Figure 6.6. Although as a block algorithm it does not need many samples, the convergence can be slow: typically 20 iterations or more are needed, especially for small N. The reconditioning escape is typically needed only about once during the iterations. A performance simulation will follow later in Section 6.5.5.

6.5 THE ANALYTICAL CMA[4]

In the preceding sections, we have discussed some of the adaptive and iterative CMAs that have appeared in the literature. They have been derived as solutions of certain optimization problems, for example, the CMA(2,2) or CMA(1,2) cost function. It is interesting to note that the CMA(2,2) also admits an approximate solution that can be computed in closed form. In fact, given a block of $N > d^2$ samples, the complete collection of beamformers for all sources can be computed as the solution of a generalized eigenvalue problem. The algorithm is called the Analytical CMA (ACMA) [4].

6.5.1 Reformulation of the CMA(2,2) Optimization Problem

The CMA(2,2) cost function (6.7) was expressed in a stochastic framework. Given a finite batch of N data samples, it is more convenient to pose the cost function as a least squares problem:

$$\mathbf{w} = \arg\min_{\mathbf{w}} \frac{1}{N}\sum_k \left(|\hat{s}_k|^2 - 1\right)^2, \quad \hat{s}_k = \mathbf{w}^{\mathrm{H}}\mathbf{x}_k. \tag{6.21}$$

The solution of this problem coincides with that of (6.7) as $N \to \infty$.

Using the Kronecker product properties (6.3) and (6.5), we can write $|\hat{s}_k|^2$ as

$$|\hat{s}_k|^2 = \mathbf{w}^{\mathrm{H}}(\mathbf{x}_k\mathbf{x}_k^{\mathrm{H}})\mathbf{w} = (\bar{\mathbf{x}}_k \otimes \mathbf{x}_k)^{\mathrm{H}}(\bar{\mathbf{w}} \otimes \mathbf{w}).$$

We subsequently stack the rows $(\bar{\mathbf{x}}_k \otimes \mathbf{x}_k)^{\mathrm{H}}$ of the data into a matrix $\mathbf{P}$ (size $N \times M^2$). Referring to the definition of the Khatri–Rao product in (6.2), we see that $\mathbf{P} = [\bar{\mathbf{X}} \circ \mathbf{X}]^{\mathrm{H}}$. Also introducing $\mathbf{y} = \bar{\mathbf{w}} \otimes \mathbf{w}$ and $\mathbf{1} = [1, \ldots, 1]^{\mathrm{T}}$, we can write (6.21) as

$$\begin{aligned} \frac{1}{N}\sum_k (|\hat{s}_k|^2 - 1)^2 &= \frac{1}{N}\sum_k [(\bar{\mathbf{x}}_k \otimes \mathbf{x}_k)^{\mathrm{H}}(\bar{\mathbf{w}} \otimes \mathbf{w}) - 1]^2 \\ &= \frac{1}{N}\|\mathbf{P}\mathbf{y} - \mathbf{1}\|^2. \end{aligned} \tag{6.22}$$

[4]The material in Sections 6.5.1–6.5.3 was presented in similar form in [28, 29].

Thus, the CMA(2,2) optimization problem asks for the least squares solution of a linear system of equations, subject to a quadratic constraint, namely $\mathbf{y} = \bar{\mathbf{w}} \otimes \mathbf{w}$.

The linear system can be conveniently rewritten as follows. Let $\mathbf{Q}$ be any unitary matrix such that $\mathbf{Q}\mathbf{1} = \sqrt{N}\begin{bmatrix} 1 \\ 0 \end{bmatrix}$, for example, found by computing a QR factorization of $[\mathbf{1} \quad \mathbf{P}]$. Apply $\mathbf{Q}$ to $[\mathbf{1} \quad \mathbf{P}]$ and partition the result as

$$\mathbf{Q}[\mathbf{1} \quad \mathbf{P}] =: \sqrt{N}\begin{bmatrix} \mathbf{1} & \hat{\mathbf{p}}^{\mathrm{H}} \\ \mathbf{0} & \mathbf{G} \end{bmatrix}. \tag{6.23}$$

Then

$$\mathbf{P}\mathbf{y} = \mathbf{1} \iff \mathbf{Q}[\mathbf{1} \quad \mathbf{P}]\begin{bmatrix} -1 \\ \mathbf{y} \end{bmatrix} = \mathbf{0} \iff \begin{cases} \hat{\mathbf{p}}^{\mathrm{H}}\mathbf{y} = 1 \\ \mathbf{G}\mathbf{y} = \mathbf{0} \end{cases} \tag{6.24}$$

and therefore (6.22) can be written as

$$\frac{1}{N}\sum_k (|\hat{s}_k|^2 - 1)^2 = \left|\hat{\mathbf{p}}^{\mathrm{H}}\mathbf{y} - 1\right|^2 + \|\mathbf{G}\mathbf{y}\|^2. \tag{6.25}$$

The second term of this expression requires $\mathbf{y} = \bar{\mathbf{w}} \otimes \mathbf{w}$ to be in the nullspace of matrix $\mathbf{G}$, as much as possible. The first term puts a constraint which avoids the trivial solution $\mathbf{y} = \mathbf{0}$. By squaring (6.23) to eliminate $\mathbf{Q}$, we obtain explicit expressions for $\hat{\mathbf{p}}$ and a matrix $\hat{\mathbf{C}} := \mathbf{G}^{\mathrm{H}}\mathbf{G}$:

$$\hat{\mathbf{p}} = \frac{1}{N}\mathbf{P}^{\mathrm{H}}\mathbf{1} = \frac{1}{N}\sum\nolimits_k \bar{\mathbf{x}}_k \otimes \mathbf{x}_k \tag{6.26}$$

$$\begin{aligned} \hat{\mathbf{C}} := \mathbf{G}^{\mathrm{H}}\mathbf{G} &= \frac{1}{N}\mathbf{P}^{\mathrm{H}}\mathbf{P} - \frac{1}{N}\mathbf{P}^{\mathrm{H}}\mathbf{1}\cdot\frac{1}{N}\mathbf{1}^{\mathrm{H}}\mathbf{P} \\ &= \frac{1}{N}\sum\nolimits_k (\bar{\mathbf{x}}_k \otimes \mathbf{x}_k)(\bar{\mathbf{x}}_k \otimes \mathbf{x}_k)^{\mathrm{H}} - \left[\frac{1}{N}\sum\nolimits_k \bar{\mathbf{x}}_k \otimes \mathbf{x}_k\right]\left[\frac{1}{N}\sum\nolimits_k \bar{\mathbf{x}}_k \otimes \mathbf{x}_k\right]^{\mathrm{H}}. \end{aligned} \tag{6.27}$$

The former expression shows that $\hat{\mathbf{p}} = \frac{1}{N}\sum \bar{\mathbf{x}}_k \otimes \mathbf{x}_k = \mathrm{vec}(\frac{1}{N}\sum \mathbf{x}_k\mathbf{x}_k^{\mathrm{H}}) = \mathrm{vec}(\hat{\mathbf{R}}_{\mathbf{x}})$, where $\hat{\mathbf{R}}_{\mathbf{x}}$ is the sample covariance matrix of the data. Thus, (for $\mathbf{y} = \bar{\mathbf{w}} \otimes \mathbf{w}$)

$$\hat{\mathbf{p}}^{\mathrm{H}}\mathbf{y} = \mathbf{w}^{\mathrm{H}}\hat{\mathbf{R}}_{\mathbf{x}}\mathbf{w} \tag{6.28}$$

and we see that the condition $\hat{\mathbf{p}}^{\mathrm{H}}\mathbf{y} = 1$ in (6.24) is equal to requiring that the average output power of the beamformer is 1.

Returning to (6.25), let $\hat{\mathbf{y}}$ be the (structured) minimizer of this expression, and define $\beta = \hat{\mathbf{p}}^{\mathrm{H}}\hat{\mathbf{y}}$. Equation (6.28) shows that it is the output power of the beamformer corresponding to $\hat{\mathbf{y}}$ and hence $\beta > 0$. Regarding β as some known fixed constant, we can add a condition that $\hat{\mathbf{p}}^{\mathrm{H}}\mathbf{y} = \beta$ to the optimization problem without changing the outcome:

$$\hat{\mathbf{y}} = \arg\min_{\substack{\mathbf{y}=\bar{\mathbf{w}}\otimes\mathbf{w} \\ \hat{\mathbf{p}}^{\mathrm{H}}\mathbf{y}=\beta}} |\hat{\mathbf{p}}^{\mathrm{H}}\mathbf{y} - 1|^2 + \|\mathbf{G}\mathbf{y}\|^2 = \arg\min_{\substack{\mathbf{y}=\bar{\mathbf{w}}\otimes\mathbf{w} \\ \hat{\mathbf{p}}^{\mathrm{H}}\mathbf{y}=\beta}} \quad |\beta - 1|^2 + \|\mathbf{G}\mathbf{y}\|^2$$

$$= \arg\min_{\substack{\mathbf{y}=\bar{\mathbf{w}}\otimes\mathbf{w} \\ \hat{\mathbf{p}}^{\mathrm{H}}\mathbf{y}=\beta}} \|\mathbf{G}\mathbf{y}\|^2$$

Since β is real and positive, replacing β by 1 will only scale the solution $\hat{\mathbf{y}}$ to $\frac{1}{\beta}\hat{\mathbf{y}}$, and does not affect the fact that it has a Kronecker structure. Therefore, it is possible to drop the first term in the optimization problem (6.25) and replace it by a constraint $\hat{\mathbf{p}}^{\mathrm{H}}\mathbf{y} = 1$.

This constraint in turn motivates in a natural way the choice of a prewhitening filter $\mathbf{F} = \hat{\mathbf{U}}_s\hat{\boldsymbol{\Sigma}}_s^{-1}$ as given in (6.15). Indeed, we derived in (6.28) that $\hat{\mathbf{p}}^{\mathrm{H}}\mathbf{y} = \mathbf{w}^{\mathrm{H}}\hat{\mathbf{R}}_x\mathbf{w}$. If we change variables to $\underline{\mathbf{x}} = \mathbf{F}^{\mathrm{H}}\mathbf{x}$ and $\mathbf{w} = \mathbf{F}\mathbf{t}$, then $\hat{\underline{\mathbf{R}}}_x = \mathbf{I}$ and

$$\hat{\mathbf{p}}^{\mathrm{H}}\mathbf{y} = \mathbf{w}^{\mathrm{H}}\hat{\mathbf{R}}_x\mathbf{w} = \mathbf{t}^{\mathrm{H}}\mathbf{t} = \|\mathbf{t}\|^2.$$

Moreover, $(\bar{\mathbf{t}}\otimes\mathbf{t})^{\mathrm{H}}(\bar{\mathbf{t}}\otimes\mathbf{t}) = \bar{\mathbf{t}}^{\mathrm{H}}\bar{\mathbf{t}}\otimes\mathbf{t}^{\mathrm{H}}\mathbf{t} = \|\mathbf{t}\|^4$. It thus follows that $\hat{\mathbf{p}}^{\mathrm{H}}\mathbf{y} = 1 \Leftrightarrow \|\bar{\mathbf{t}}\otimes\mathbf{t}\| = 1$. Hence, up to a scaling which is not important, the CMA(2,2) optimization problem is equivalent to solving

$$\mathbf{t} = \arg\min_{\substack{\mathbf{y}=\bar{\mathbf{t}}\otimes\mathbf{t} \\ \|\mathbf{y}\|=1}} \|\underline{\mathbf{G}}\,\mathbf{y}\|^2 = \arg\min_{\substack{\mathbf{y}=\bar{\mathbf{t}}\otimes\mathbf{t} \\ \|\mathbf{y}\|=1}} \mathbf{y}^{\mathrm{H}}\hat{\underline{\mathbf{C}}}\,\mathbf{y} \tag{6.29}$$

and setting $\mathbf{w} = \mathbf{F}\mathbf{t}$.

As discussed in Section 6.3, it is important to also incorporate a dimension reduction in the prewhitening filter. Without dimension reduction, the nullspace of $\mathbf{G}$ and $\mathbf{C}$ contain additional vectors which do not correspond to valid solutions. The prefilter $\mathbf{F}$ reduces the dimension of $\mathbf{x}_k$ from M to d, and the corresponding matrix $\underline{\mathbf{P}} := [\bar{\underline{\mathbf{X}}}\circ\underline{\mathbf{X}}]^{\mathrm{H}}$ has size $N\times d^2$. We assume that $N > d^2$ so that $\underline{\mathbf{P}}$ is "tall," and $\underline{\mathbf{G}}$ and $\hat{\underline{\mathbf{C}}}$ have a well-defined (i.e., overdetermined) nullspace. It can be shown [30] that for sufficiently large N (at least $N > d^2$) the nullspace has dimension d and does not contain other solutions.

6.5.2 Solving the CMA(2,2) Optimization Problem

To solve the CMA(2,2) optimization problem in (6.29), it is required to numerically optimize the minimization problem and find d independent solutions. The solutions will be unit-norm vectors $\mathbf{y}$ that have the required Kronecker structure and minimize $\|\underline{\mathbf{G}}\,\mathbf{y}\|^2$. With noise, the solutions will not exactly be in the approximate nullspace of $\underline{\mathbf{G}}$ since in general this space will not contain vectors with the Kronecker structure.

Instead of solving this problem, an alternative approach is to first find an orthonormal basis $\mathbf{Y} = [\mathbf{y}_1, \ldots, \mathbf{y}_d]$ for the d-dimensional approximate nullspace of $\underline{\mathbf{G}}$ (or $\hat{\underline{\mathbf{C}}}$),

$$\mathbf{Y} = \arg\min_{\mathbf{Y}^{\mathrm{H}}\mathbf{Y}=\mathbf{I}} \|\underline{\mathbf{G}}\mathbf{Y}\|_{\mathrm{F}}^2 = \arg\min_{\mathbf{Y}^{\mathrm{H}}\mathbf{Y}=\mathbf{I}} \sum \mathbf{y}_i^{\mathrm{H}} \hat{\underline{\mathbf{C}}}\, \mathbf{y}_i, \tag{6.30}$$

whose solution is the set of d least dominant eigenvectors of $\hat{\underline{\mathbf{C}}}$. We can subsequently look for a set of d unit-norm vectors $\bar{\mathbf{t}}_i \otimes \mathbf{t}_i$ that best spans the same subspace,

$$\mathbf{T} = \arg\min_{\mathbf{T},\mathbf{m}\Lambda} \|\mathbf{Y} - (\bar{\mathbf{T}} \circ \mathbf{T})\mathbf{\Lambda}\|_{\mathrm{F}}^2.$$

where $\mathbf{T} = [\mathbf{t}_1, \ldots, \mathbf{t}_d]$ is the set of beamformers in the whitened domain, $\bar{\mathbf{T}} \circ \mathbf{T} := [\bar{\mathbf{t}}_1 \otimes \mathbf{t}_1, \ldots, \bar{\mathbf{t}}_d \otimes \mathbf{t}_d]$ denotes the Khatri–Rao product, and $\mathbf{\Lambda}$ is a full rank $d \times d$ matrix that relates the two bases of the subspace.

Alternatively, we can formulate this as a joint diagonalization problem, since [using (6.3)–(6.6)]

$$\|\mathbf{Y} - (\bar{\mathbf{T}} \circ \mathbf{T})\mathbf{\Lambda}\|_{\mathbf{F}}^2 = \sum_i \|\mathbf{y}_i - (\bar{\mathbf{T}} \circ \mathbf{T})\boldsymbol{\lambda}_i\|_{\mathbf{F}}^2 = \sum_i \|\mathbf{Y}_i - \mathbf{T}\mathbf{\Lambda}_i\mathbf{T}^{\mathrm{H}}\|_{\mathbf{F}}^2$$

where $\boldsymbol{\lambda}_i$ is the ith column of $\mathbf{\Lambda}$, $\mathbf{\Lambda}_i = \mathrm{diag}(\boldsymbol{\lambda}_i)$ is a diagonal matrix constructed from this vector, and $\mathbf{Y}_i$ is a matrix obtained by unstacking $\mathbf{y}_i$ into a square $d \times d$ matrix such that $\mathrm{vec}(\mathbf{Y}_i) = \mathbf{y}_i$; we have also used (6.6). The latter equation shows that all $\mathbf{Y}_i$ can be diagonalized by the same matrix $\mathbf{T}$. The resulting joint diagonalization problem is a generalization of the standard eigenvalue decomposition problem, and can be solved iteratively (see Section 6.5.4 ahead).

The preceding two-step approach gives an approximate but closed-form solution to (6.29); the corresponding algorithm is the Analytical CMA (ACMA) [4]. An important advantage of this algorithm is that it provides the complete set of beamformers as the solution of the joint diagonalization problem. A second advantage is that in the noise-free case and with $N \geq d^2$, the algorithm produces the exact solution $\mathbf{W} = \mathbf{A}^{\dagger\mathrm{H}}$. The algorithm is summarized in Figure 6.7. The main complexity is in the construction of $\hat{\underline{\mathbf{C}}}$.

Given data $\mathbf{X}$, compute beamformers $\mathbf{W}$ and output $\hat{\mathbf{S}} = \mathbf{W}^H\mathbf{X}$:

1. SVD: $\mathbf{X} =: \hat{\mathbf{U}}\hat{\Sigma}\hat{\mathbf{V}}$ — (M^2N)
   Prefiltering: $\underline{\mathbf{X}} = \hat{\Sigma}_s^{-1}\hat{\mathbf{U}}_s^{\mathrm{H}}\mathbf{X} = \hat{\mathbf{V}}_s$
   $\hat{\underline{\mathbf{C}}} = \frac{1}{N}\sum(\bar{\underline{\mathbf{x}}}_k \otimes \underline{\mathbf{x}}_k)(\hat{\underline{\mathbf{x}}}_k \otimes \underline{\mathbf{x}}_k)^{\mathrm{H}} - [\frac{1}{N}\sum \bar{\underline{\mathbf{x}}}_k \otimes \underline{\mathbf{x}}_k][\frac{1}{N}\sum \bar{\underline{\mathbf{x}}}_k \otimes \underline{\mathbf{x}}_k]^{\mathrm{H}}$ — (d^4N)
   EVD of $\hat{\underline{\mathbf{C}}}$, let $\{\mathbf{y}_i\}$ be the d least dominant eigenvectors. — $(\leq d^6)$
2. $\mathbf{Y}_i = \mathrm{vec}^{-1}\mathbf{y}_i (i = 1, \ldots, d)$
   Find $\mathbf{T}$ to jointly diagonalize $\{\mathbf{Y}_i\}$ as $\mathbf{Y}_i = \mathbf{T}\mathbf{\Lambda}_i\mathbf{T}^{\mathrm{H}} (i = 1, \ldots, d)$ — (d^4)
3. Scale each column of $\mathbf{T}$ to norm 1.
   Set $\mathbf{W} = \hat{\mathbf{U}}_s\hat{\Sigma}_s^{-1}\mathbf{T}$ and $\hat{\mathbf{S}} = \mathbf{T}^{\mathrm{H}}\underline{\mathbf{X}}$ — (d^2N)

Figure 6.7 Summary of ACMA.

6.5.3 Asymptotic Analysis of ACMA

A detailed performance analysis of ACMA has appeared in [29, 31]. The most important finding is that asymptotically, the solution converges to the Wiener beamformer, for a large number of samples or for a high SNR. The derivation is summarized in this subsection.

Recall the definition of $\hat{\underline{\mathbf{C}}}$ in equation (6.27). To analyze the nullspace of $\hat{\underline{\mathbf{C}}}$, we consider a large number of samples, so that $\hat{\underline{\mathbf{C}}}$ converges to $\underline{\mathbf{C}}$,

$$\underline{\mathbf{C}} = \mathrm{E}(\underline{\bar{\mathbf{x}}} \otimes \underline{\mathbf{x}})(\underline{\bar{\mathbf{x}}} \otimes \underline{\mathbf{x}})^{\mathrm{H}} - \mathrm{E}[\underline{\bar{\mathbf{x}}} \otimes \bar{\mathbf{x}}]\mathrm{E}[\underline{\bar{\mathbf{x}}} \otimes \underline{\mathbf{x}}]^{\mathrm{H}}. \tag{6.31}$$

where $\underline{\mathbf{x}}_k = \underline{\mathbf{A}}\mathbf{s}_k + \underline{\mathbf{n}}_k$. Thus, $\underline{\mathbf{C}}$ involves both fourth order and second order statistics of the data.

In general, for a zero mean random vector $\mathbf{x} = [x_i]$, the fourth order cumulant matrix is defined as

$$\mathbf{K}_{\mathbf{x}} = \sum_{abcd} (\mathbf{i}_b \otimes \mathbf{i}_a)(\mathbf{i}_c \otimes \mathbf{i}_d)^{\mathrm{H}} \mathrm{cum}(x_a, \bar{x}_b, x_c, \bar{x}_d), \tag{6.32}$$

where $\mathbf{i}_i$ is the ith column of the identity matrix $\mathbf{I}$, and, for circularly symmetric variables, the cumulant function is defined by

$$\mathrm{cum}(x_a, \bar{x}_b, x_c, \bar{x}_d) = \mathrm{E}(x_a\bar{x}_b x_c\bar{x}_d) - \mathrm{E}(x_a\bar{x}_b)\mathrm{E}(x_c\bar{x}_d) - \mathrm{E}(x_a\bar{x}_d)\mathrm{E}(\bar{x}_b x_c).$$

Therefore, (6.32) can be written compactly as

$$\mathbf{K}_{\mathbf{x}} = \mathrm{E}(\bar{\mathbf{x}} \otimes \mathbf{x})(\bar{\mathbf{x}} \otimes \mathbf{x})^{\mathrm{H}} - \mathrm{E}(\bar{\mathbf{x}} \otimes \mathbf{x})\mathrm{E}(\bar{\mathbf{x}} \otimes \mathbf{x})^{\mathrm{H}} - \mathrm{E}(\bar{\mathbf{x}}\bar{\mathbf{x}}^{\mathrm{H}}) \otimes \mathrm{E}(\mathbf{x}\mathbf{x}^{\mathrm{H}}) \tag{6.33}$$

Cumulants have several important properties: for sums of independent sources they are additive, the fourth order cumulant of Gaussian sources is $\mathbf{K}_{\mathbf{n}} = 0$, the fourth order cumulant of independent CM sources is $\mathbf{K}_{\mathbf{s}} = -\mathbf{I}$. For the data model $\underline{\mathbf{x}}_k = \underline{\mathbf{A}}\mathbf{s}_k + \underline{\mathbf{n}}_k$, it thus follows that

$$\mathbf{K}_{\underline{\mathbf{x}}} = [\underline{\bar{\mathbf{A}}} \otimes \underline{\mathbf{A}}](-\mathbf{I})[\underline{\bar{\mathbf{A}}} \otimes \underline{\mathbf{A}}]^{\mathrm{H}}$$

and hence by combining (6.31) with (6.33), $\underline{\mathbf{C}}$ is given by[5]

$$\begin{aligned} \underline{\mathbf{C}} &= \mathbf{K}_{\underline{\mathbf{x}}} + \mathrm{E}(\underline{\bar{\mathbf{x}}}\underline{\bar{\mathbf{x}}}^{\mathrm{H}}) \otimes \mathrm{E}(\underline{\mathbf{x}}\underline{\mathbf{x}}^{\mathrm{H}}) \\ &= [\underline{\bar{\mathbf{A}}} \circ \underline{\mathbf{A}}](-\mathbf{I})[\underline{\bar{\mathbf{A}}} \circ \underline{\mathbf{A}}]^{\mathrm{H}} + \mathbf{I}. \end{aligned} \tag{6.34}$$

[5]This analysis is not valid for BPSK sources, because they are not circularly symmetric. For such sources, the ACMA has to be modified [32].

where we also used that after prewhitening $\mathrm{E}(\underline{\mathbf{x}\mathbf{x}}^{\mathrm{H}}) = \mathbf{I}$. Consequently, the CMA(2,2) cost function (6.29) becomes asymptotically

$$\begin{aligned}\arg\min_{\mathbf{y}=\bar{\mathbf{t}}\otimes\mathbf{t},\ \|\mathbf{y}\|=1} \mathbf{y}^{\mathrm{H}}\underline{\mathbf{C}}\mathbf{y} &= \arg\min_{\mathbf{y}=\bar{\mathbf{t}}\otimes\mathbf{t},\ \|\mathbf{y}\|=1} \mathbf{y}^{\mathrm{H}}\{[\underline{\bar{\mathbf{A}}}\circ\underline{\mathbf{A}}](-\mathbf{I})[\underline{\bar{\mathbf{A}}}\circ\underline{\mathbf{A}}]^{\mathrm{H}} + \mathbf{I}\}\mathbf{y} \\ &= \arg\max_{\mathbf{y}=\bar{\mathbf{t}}\otimes\mathbf{t},\ \|\mathbf{y}\|=1} \mathbf{y}^{\mathrm{H}}\{[\underline{\bar{\mathbf{A}}}\circ\underline{\mathbf{A}}][\underline{\bar{\mathbf{A}}}\circ\underline{\mathbf{A}}]^{\mathrm{H}}\}\mathbf{y}. \end{aligned} \tag{6.35}$$

ACMA first constructs a basis $\{\mathbf{y}_1, \ldots, \mathbf{y}_d\}$ for the d nullspace vectors of $\underline{\mathbf{C}}$ without the constraint $\mathbf{y} = \bar{\mathbf{t}} \otimes \mathbf{t}$, which clearly are given by

$$\mathrm{span}\{\mathbf{y}_1, \ldots, \mathbf{y}_d\} = \mathrm{span}[\underline{\bar{\mathbf{A}}} \circ \underline{\mathbf{A}}] = \mathrm{span}\{\underline{\bar{\mathbf{a}}}_1 \otimes \underline{\mathbf{a}}_1, \ldots, \underline{\bar{\mathbf{a}}}_d \otimes \underline{\mathbf{a}}_d\}.$$

As a second step, the joint diagonalization procedure is used to replace the unstructured basis by one that has the required Kronecker product structure, that is, d independent vectors of the form $\bar{\mathbf{t}} \otimes \mathbf{t}$ within this column span. From the above equation, we see that the unique solution is $\bar{\mathbf{t}}_i \otimes \mathbf{t}_i = \underline{\bar{\mathbf{a}}}_i \otimes \underline{\mathbf{a}}_i$ (up to a scaling to make $\mathbf{t}_i$ have unit norm), and thus

$$\mathbf{t}_i = \underline{\mathbf{a}}_i, \quad i = 1, \ldots, d.$$

Similar to Section 6.3.3, the solution in the original (unwhitened) domain is $\mathbf{w}_i = \mathbf{R}_{\mathbf{x}}^{-1}\mathbf{a}_i$. We have just shown that as $N \to \infty$, the beamformers provided by ACMA converge to the Wiener receivers (6.16). This is partly due to the choice in prewhitening scheme, and partly due to the two-step solution of the CMA(2,2) problem (6.29). In general, the exact solution of CMA(2,2) does not lead to the Wiener solution, although it has already been observed by Godard that that the solutions are often close. Quantitative evidence for this is given in [33–35].

6.5.4 Joint Diagonalization via Iterative Projection

Since the problem of joint diagonalization shows up in several other signal processing applications, it has actually been well studied over the past decade [4, 5, 36–46]. The algorithms are iterative, for example, using extensions of Jacobi iterations or QZ iterations. Generally the performance of these algorithms is satisfactory although the speed of convergence is often linear, and without an exact solution (i.e., in the noisy case) the convergence to local minima cannot be avoided. The reason for good performance is that a good starting point is available from the solution for only two matrices, which reduces to a standard eigenvalue problem.

The joint diagonalization step in the ACMA can be written as

$$\mathbf{T} = \arg\min_{\|\mathbf{t}_i\|=1} \|\mathbf{V}_n - (\bar{\mathbf{T}} \circ \mathbf{T})\boldsymbol{\Lambda}\|_{\mathrm{F}}^2 \tag{6.36}$$

where $\mathbf{V}_n$ is an orthogonal basis for the nullspace of $\mathbf{G}$. The unit-norm constraint on the columns of $\mathbf{T}$ avoids a scaling indeterminacy between $\mathbf{T}$ and $\mathbf{\Lambda}$. Since $\mathbf{\Lambda}$ is invertible, the problem can also be written as

$$\mathbf{T} = \underset{\substack{\|\mathbf{t}_i\|=1 \\ \mathbf{M}\text{ invertible}}}{\arg\min} \; \|\mathbf{V}_n\mathbf{M} - (\bar{\mathbf{T}} \circ \mathbf{T})\|_F^2. \tag{6.37}$$

This is not entirely equivalent, but almost so since $\mathbf{\Lambda}$ is close to unitary. However, a condition that $\mathbf{M}$ is invertible is needed to avoid finding repeated solutions $\mathbf{t}_i = \mathbf{t}_j$.

As inspired by [44], the problem (6.37) can be solved iteratively using alternating least squares: (1) given an initial value for $\mathbf{T}$, solve for $\mathbf{M}$; (2) for the new value of $\mathbf{M}$, solve for $\mathbf{T}$. Both steps are simple to implement: for fixed $\mathbf{T}$, solving for $\mathbf{M}$ gives

$$\mathbf{M} = \mathbf{V}_n^{\mathrm{H}}(\bar{\mathbf{T}} \circ \mathbf{T}),$$

whereas for fixed $\mathbf{M}$, solving for $\mathbf{T}$ gives

$$\mathbf{T} = \underset{\|\mathbf{t}_i\|=1}{\arg\min}\, \|\mathbf{V}_n\mathbf{M} - (\bar{\mathbf{T}} \circ \mathbf{T})\|^2 = \underset{\|\mathbf{t}_i\|=1}{\arg\min}\, \|\mathbf{Y} - (\bar{\mathbf{T}} \circ \mathbf{T})\|^2,$$

where $\mathbf{Y} = \mathbf{V}_n\mathbf{M} = \mathbf{V}_n\mathbf{V}_n^{\mathrm{H}}(\bar{\mathbf{T}} \circ \mathbf{T})$ is the projection of the previous solution onto the estimated subspace. The problem decouples into solving for the individual columns of $\mathbf{T}$:

$$\mathbf{t}_i = \underset{\|\mathbf{t}_i\|=1}{\arg\min}\, \|\mathbf{y}_i - (\bar{\mathbf{t}}_i \otimes \mathbf{t}_i)\|^2 = \underset{\|\mathbf{t}_i\|=1}{\arg\min}\, \|\mathbf{Y}_i - \mathbf{t}_i\mathbf{t}_i^{\mathrm{H}}\|^2,$$

where $\mathbf{Y}_i = \mathrm{vec}^{-1}(\mathbf{y}_i)$. The solution is given by an SVD of $\mathbf{Y}_i$ and retaining the dominant component, that is

$$\mathbf{Y}_i =: \sum_j \sigma_j \mathbf{u}_j \mathbf{u}_j^{\mathrm{H}}, \quad \mathbf{t}_i = \mathbf{u}_1 =: \pi_1(\mathbf{y}_i). \tag{6.38}$$

We will denote this projection onto rank-1 operation by $\mathbf{t}_i = \pi_1(\mathbf{y}_i)$.

The algorithm thus projects a prior solution $\bar{\mathbf{t}}_i \otimes \mathbf{t}_i$ onto the estimated subspace $\mathbf{V}_n$, and then projects the result back onto the Kronecker structure (the latter projection is nonlinear). After each projection, the error is decreasing, therefore the algorithm converges monotonically to a local minimum. As with most alternating projection algorithms, the speed of convergence is only linear. Nonetheless, experience shows that in practice only a few iterations are needed (two or three): the minimum error depends on the noise level and is quickly reached.

The columns of $\mathbf{T}$ are processed independently. Therefore, there is a risk that they converge to the same solution. To avoid this situation, the independence of the

Given data $\mathbf{X} = [\mathbf{x}_1, \ldots, \mathbf{x}_N]$, compute beamformer $\mathbf{W}$ and output $\mathbf{s}_k = \mathbf{W}_k^{\mathrm{H}}\mathbf{x}_k$:	
1. Compute $\hat{\mathbf{R}}_{\mathbf{x}} = \frac{1}{N}\sum_k \mathbf{x}_k\mathbf{x}_k^{\mathrm{H}}$	(M^2N)
Compute the EVD $\hat{\mathbf{R}}_{\mathbf{x}} = \mathbf{U}\Sigma^2\mathbf{U}^{\mathrm{H}}$	(M^3)
Set the prewhitening filter $\mathbf{F} = \hat{\mathbf{U}}_s\hat{\Sigma}_s^{-1/2}$	(dM)
2. Prefilter the data, $\underline{\mathbf{x}}_k = \mathbf{F}^{\mathrm{H}}\mathbf{x}_k$	(dMN)
$\hat{\underline{\mathbf{C}}} = \frac{1}{N}\sum_k (\bar{\underline{\mathbf{x}}}_k \otimes \underline{\mathbf{x}}_k)(\bar{\underline{\mathbf{x}}}_k \otimes \underline{\mathbf{x}}_k)^{\mathrm{H}} - \mathrm{vec}(\hat{\underline{\mathbf{R}}}_{\mathbf{x}})\mathrm{vec}(\hat{\underline{\mathbf{R}}}_{\mathbf{x}})^{\mathrm{H}}$	(d^4N)
3. Compute $\mathbf{V}_n$, the nullspace of $\hat{\underline{\mathbf{C}}}$	$(\leq d^6)$
4. Initialize $\mathbf{T} = \mathbf{I}$	
Until convergence:	
$\mathbf{M} = \mathbf{V}_n^{\mathrm{H}}(\bar{\mathbf{T}} \circ \mathbf{T})$	(d^4)
$\mathbf{M}' = \mathrm{recond}(\mathbf{M})$, equation (6.19)	(d^3)
$\mathbf{Y} = \mathbf{V}_n\mathbf{M}'$	(d^4)
$\mathbf{t}_i = \pi_1(\mathbf{y}_i), i = 1, \ldots, d$, equation (6.38)	(d^4)
5. Set $\mathbf{W} = \mathbf{FT}$ and $\hat{\mathbf{s}}_k = \mathbf{W}^{\mathrm{H}}\mathbf{x}_k$	(dMN)

Figure 6.8 ACMA using iterative implementation of the joint diagonalization.

solutions has to be monitored and corrected, if needed. The easiest point to do this is to compute an SVD of $\mathbf{M}$ and recondition it if needed [see equation (6.19)]. The motivation for this is that, after prewhitening, $\mathbf{T}$ is nearly unitary, hence the columns of $\bar{\mathbf{T}} \circ \mathbf{T}$ are nearly orthogonal. Therefore, $\mathbf{M} = \mathbf{V}_n^{\mathrm{H}}(\bar{\mathbf{T}} \circ \mathbf{T})$ is expected to be nearly unitary: its singular values are close to 1. On the other hand, if two columns of $\mathbf{T}$ converge to the same solution, then a singular value of $\mathbf{M}$ gets close to 0.

The resulting iterative ACMA is summarized in Figure 6.8. Also the computational complexity of each step is indicated. Most of the work is done in two steps: the SVD of $\mathbf{X}$ which has a complexity of M^2N, and the construction of $\hat{\underline{\mathbf{C}}}$, which has a complexity of d^4N. The joint diagonalization step has a complexity of order d^4 (independent of its implementation), which can be neglected. Therefore, the total complexity is about $(M^2 + d^4)N$.

The computational complexity of the LS-CMA (Figure 6.6) is dominated by the two large inner products, $\mathbf{T}^{\mathrm{H}}\mathbf{X}$ and $\mathbf{S}\underline{\mathbf{X}}^{\dagger}$, each of complexity d^2N. The whitening step has a complexity of M^2N. With 25 iterations, the total complexity is about $(M^2 + 50d^2)N$. In comparison, the ACMA is more efficient than the LS-CMA algorithm if $d \leq 7$.

6.5.5 Comparison of ACMA to LS-CMA

Some performance results for the block-iterative methods are shown in Figure 6.9. In the simulations, we took a uniform linear array with $M = 4$ antennas spaced at half wavelengths, and $d = 3$ constant-modulus sources with directions $[-10°, 20°, 30°]$ and amplitudes [1, 0.8, 0.9], respectively. (Recall that the algorithms do not use knowledge of the array structure—the above information just serves to specify the simulation data.) The noise power is determined by the signal to noise ratio (SNR), which is the SNR per antenna for the strongest user.

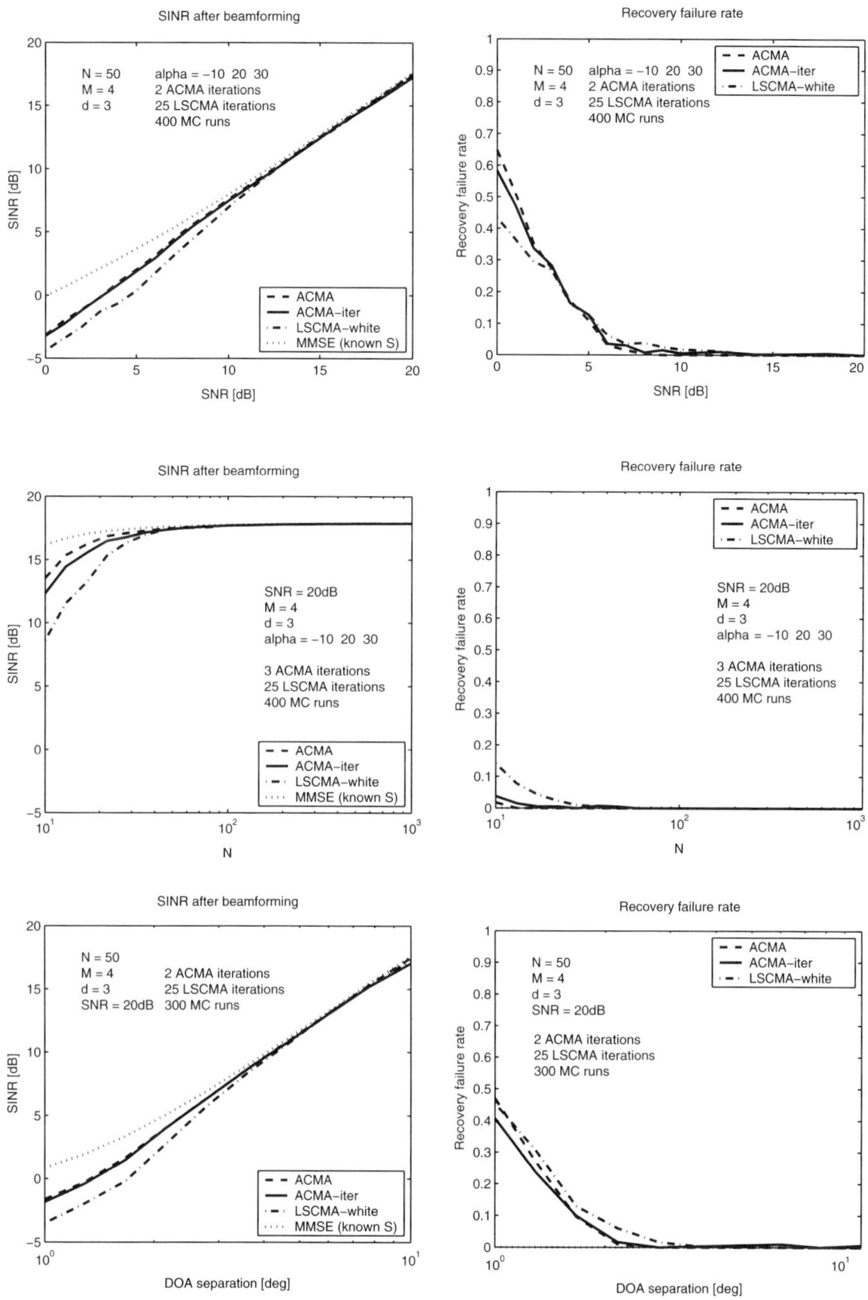

Figure 6.9 SINR performance and failure rate of ACMA, iterative ACMA, and whitened LS-CMA, as function of SNR, *N* and DOA separation.

We compare the performance of the prewhitened LS-CMA as in Figure 6.6 (25 iterations), ACMA as in Figure 6.7, and iterative ACMA as in Figure 6.8 (two or three iterations). For reference, we also show the performance of the sample data LMMSE receiver, $\hat{\mathbf{W}} = (\mathbf{S}\mathbf{X}^{\dagger})^{H}$ with known $\mathbf{S}$. The performance measure is the residual signal to interference plus noise ratio (SINR) at the output of the beamformers. We only consider the SINR of the worst output channel. The graphs show the performance as a function of signal to noise ratio (SNR), number of samples N, and angular separation between the second and third source, respectively.

It is seen in Figure 6.9 that all algorithms converge to the LMMSE receiver for sufficiently large N or SNR and source separation. The panel at the right shows the fraction of times that not all independent beamformers were recovered (this was established by first verifying for each beamformer which signal it recovers most strongly; these failures were omitted from the SINR statistics). The algorithms have similar failure rate performance, which is very small for sufficiently large N, SNR and source separation.

Figure 6.10 shows the convergence of the CMA(2,2) cost function (averaged over all sources and over 800 Monte Carlo runs), for various SNR levels. It is seen that the LS-CMA converges much slower, and for large SNR may never reach the minimum cost. Also note that LS-CMA does not aim to minimize the CMA(2,2) cost, rather it was derived to minimize the CMA(1,2) cost [2]. This explains that the convergence graph in Figure 6.10 can be nonmonotonic.

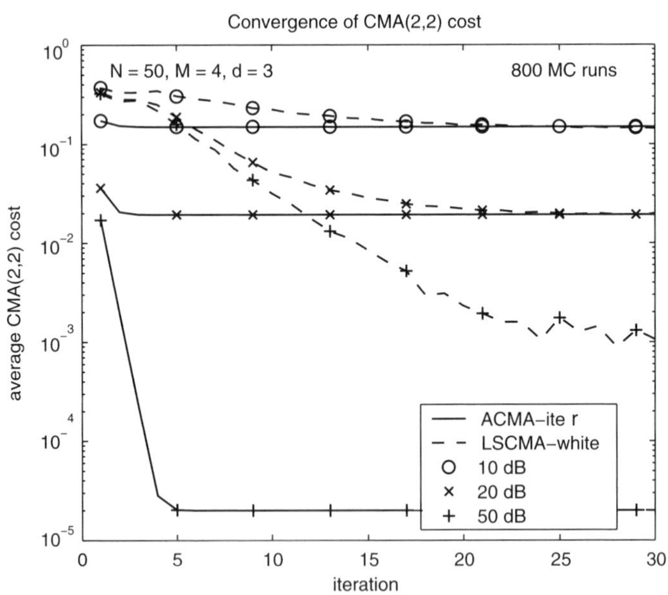

Figure 6.10 Average convergence of iterative ACMA and whitened LS-CMA iterations, for various SNR levels.

6.6 ADAPTIVE PREWHITENING

In the previous sections, the importance of the prewhitening step was highlighted. This has been recognized in the blind source separation community, and algorithms have been devised to produce the filtering matrix $\mathbf{F}$ adaptively. In particular, the prewhitening step requires tracking the inverse of a Cholesky factor of the data covariance matrix $\mathbf{R}_\mathbf{x}$. Most of the proposed prewhitening algorithms, however, consider the case $d = M$ and therefore do not implement the dimension reduction step. With dimension reduction, also the dominant subspace has to be tracked: the d-dimensional column span of $\mathbf{F}$ should span the d-dimensional principal column span of the data matrix $\mathbf{X}$. Again, there are many adaptive algorithms for subspace tracking, the most popular being the PAST algorithm [47] and derivatives of it (a 1990 comprehensive overview of subspace tracking algorithms is [48]). Apparently the only paper where these two steps are explicitly combined is a conference paper by Douglas [49]. The derivation is as follows (throughout, it is assumed that the subspace dimension d is known and fixed).

As explained in [47], PAST is based on the idea that the minimizer $\mathbf{U}$ of

$$J(\mathbf{U}) = \mathrm{E}\|\mathbf{x} - \mathbf{U}\mathbf{U}^\mathrm{H}\mathbf{x}\|^2, \quad \mathbf{U}: M \times d$$

is equal to an orthonormal matrix whose columns span the principal subspace of $\mathrm{E}(\mathbf{x}\mathbf{x}^\mathrm{H})$. To map this into a practical algorithm, a number of modifications are made. First, the problem is converted into a least squares problem, $\min \sum_i \|\mathbf{x}_i - \mathbf{U}\mathbf{U}^\mathrm{H}\mathbf{x}_i\|^2$. Next, to obtain an adaptive algorithm, the contributions of the received samples to the cost function are exponentially scaled down by a scalar λ, where $0 < \lambda \leq 1$, which gives after k samples

$$J(\mathbf{U}_k) = \sum_{i=1}^{k} \lambda^{k-i} \left\|\mathbf{x}_i - \mathbf{U}_k\mathbf{U}_k^\mathrm{H}\mathbf{x}_i\right\|^2.$$

Finally, to enable a simple solution to this problem, the vector $\mathbf{U}_k^\mathrm{H}\mathbf{x}_i$ which depends on the unknown $\mathbf{U}_k$ is replaced by $\mathbf{v}_i = \mathbf{U}_{i-1}^\mathrm{H}\mathbf{x}_i$, which depends on the previous estimate of $\mathbf{U}$ and does not adapt with k. Thus, the cost function used in PAST for subspace tracking is

$$J(\mathbf{U}_k) = \sum_{i=1}^{k} \lambda^{k-i} \|\mathbf{x}_i - \mathbf{U}_k\mathbf{v}_i\|^2.$$

This has the form of a standard adaptive least squares problem. The solution $\mathbf{U}_k$ minimizing this cost is

$$\mathbf{U}_k = \hat{\mathbf{R}}_{\mathbf{xv}}(k)\hat{\mathbf{R}}_\mathbf{v}^{-1}(k)$$

where

$$\hat{\mathbf{R}}_{\mathbf{xv}}(k) = \sum_{i=1}^{k} \lambda^{k-i}\mathbf{x}_i\mathbf{v}_i^{\mathrm{H}}, \quad \hat{\mathbf{R}}_{\mathbf{v}}(k) = \sum_{i=1}^{k} \lambda^{k-i}\mathbf{v}_i\mathbf{v}_i^{\mathrm{H}} \tag{6.39}$$

and can be found by the RLS-like algorithm [47]

$$\begin{aligned}
\mathbf{v}_k &= \mathbf{U}_{k-1}^{\mathrm{H}}\mathbf{x}_k \\
\mathbf{e}_k &= \mathbf{x}_k - \mathbf{U}_{k-1}\mathbf{v}_k & &\text{(the subspace error)} \\
\mathbf{k}_k &= \frac{\hat{\mathbf{R}}_{\mathbf{v}}^{-1}(k-1)\mathbf{v}_k}{\lambda + \mathbf{v}_k^{\mathrm{H}}\hat{\mathbf{R}}_{\mathbf{v}}^{-1}(k-1)\mathbf{v}_k} & &\text{(Kalman gain)} \\
\hat{\mathbf{R}}_{\mathbf{v}}^{-1}(k) &= \frac{1}{\lambda}\left[\hat{\mathbf{R}}_{\mathbf{v}}^{-1}(k-1) - \mathbf{k}_k\mathbf{v}_k^{\mathrm{H}}\hat{\mathbf{R}}_{\mathbf{v}}^{-1}(k-1)\right] \\
\mathbf{U}_k &= \mathbf{U}_{k-1} + \mathbf{e}_k\mathbf{k}_k^{\mathrm{H}}.
\end{aligned} \tag{6.40}$$

This is the PAST algorithm for principal subspace tracking; its complexity is of order dM. Similar to other subspace tracking algorithms of this complexity, the update of $\mathbf{U}_{k-1}$ consists of a rank-1 update in the direction of the subspace error vector $\mathbf{e}_k$, that is, the component of $\mathbf{x}_k$ outside the estimated subspace.

Although $\mathbf{U}_k$ will converge to an orthonormal matrix for $k \to \infty$ and $\lambda = 1$, this does not ensure the orthonormality of $\mathbf{U}_k$ at every step. As shown in [24], the update step for $\mathbf{U}_k$ can be slightly modified with a second-order term such that the update takes the form of a Householder transformation, which ensures that $\mathbf{U}_k^{\mathrm{H}}\mathbf{U}_k = \mathbf{U}_{k-1}^{\mathrm{H}}\mathbf{U}_{k-1}$. Indeed, let

$$\mathbf{z}_k = \mathbf{e}_k - \frac{\|\mathbf{e}_k\|^2}{2}\mathbf{U}_{k-1}\mathbf{k}_k$$

and define the Householder transformation

$$\mathbf{U}_k = \left(\mathbf{I} - 2\frac{\mathbf{z}_k\mathbf{z}_k^{\mathrm{H}}}{\|\mathbf{z}_k\|^2}\right)\mathbf{U}_{k-1}$$

then (since $\mathbf{e}_k^{\mathrm{H}}\mathbf{U}_{k-1} = \mathbf{0}^{\mathrm{H}}$)

$$\mathbf{U}_k = \mathbf{U}_{k-1} + \frac{\mathbf{z}_k\mathbf{k}_k^{\mathrm{H}}}{1 + 0.25\|\mathbf{e}_k\|^2\|\mathbf{k}_k\|^2}$$

which is close to the update in (6.40), but ensures the orthonormality of $\mathbf{U}_k$ provided the iteration is started with an orthonormal matrix $\mathbf{U}_0$.

The output of the subspace tracking step is a sequence of d-dimensional vectors $\mathbf{v}_k = \mathbf{U}_{k-1}^{\mathrm{H}}\mathbf{x}_k$, where $\mathbf{U}_k$ is an $M \times d$-dimensional matrix whose columns span the estimated principal subspace of $\mathbf{X}$. At convergence, the updates on $\mathbf{U}_k$ are only small, of order $\|\mathbf{e}_k\|^2$. Therefore, $\mathbf{v}_k$ is approximately stationary and $\hat{\mathbf{R}}_{\mathbf{v}}(k)$ is a meaningful estimate of the covariance matrix of $\mathbf{v}_k$.

Subsequently, as described in [49], $\mathbf{v}_k$ can be whitened by an adaptive filter $\mathbf{W}_k$ of size $d \times d$ such that $\mathbf{y}_k := \mathbf{W}_{k-1}^{\mathrm{H}}\mathbf{v}_k$ has covariance matrix $\mathrm{E}(\mathbf{y}_k\mathbf{y}_k^{\mathrm{H}}) = \mathbf{I}$, that is, $\mathbf{W}_k^{\mathrm{H}}\mathbf{R}_{\mathbf{v}}(k)\mathbf{W}_k = \mathbf{I}$ or, with estimated quantities,

$$\mathbf{W}_k\mathbf{W}_k^{\mathrm{H}} = \hat{\mathbf{R}}_{\mathbf{v}}^{-1}(k). \tag{6.41}$$

Therefore, $\mathbf{W}_k$ is a square-root factor of $\hat{\mathbf{R}}_{\mathbf{v}}^{-1}(k)$. Since $\hat{\mathbf{R}}_{\mathbf{v}}^{-1}(k)$ is obtained via a rank-1 update (6.40), a rank-1 update for $\mathbf{W}_k$ can be found in closed form as well [50]. Indeed, substituting (6.41) in the update for $\hat{\mathbf{R}}_{\mathbf{v}}^{-1}(k)$ gives

$$\mathbf{W}_k\mathbf{W}_k^{\mathrm{H}} = \frac{1}{\lambda}\mathbf{W}_{k-1}\left(\mathbf{I} - \frac{\mathbf{y}_k\mathbf{y}_k^{\mathrm{H}}}{\lambda + \|\mathbf{y}_k\|^2}\right)\mathbf{W}_{k-1}^{\mathrm{H}} \tag{6.42}$$

where $\mathbf{y}_k = \mathbf{W}_{k-1}^{\mathrm{H}}\mathbf{v}_k$. Let $\mathbf{B}_k$ be a symmetric square-root factor of the term in braces, then (6.42) implies

$$\mathbf{W}_k = \frac{1}{\sqrt{\lambda}}\mathbf{W}_{k-1}\mathbf{B}_k.$$

The square-root factor $\mathbf{B}_k$ is found as

$$\mathbf{B}_k = \left(\mathbf{I} - \frac{\mathbf{y}_k\mathbf{y}_k^{\mathrm{H}}}{\|\mathbf{y}_k\|^2}\right) + \sqrt{\frac{\lambda}{\lambda + \|\mathbf{y}_k\|^2}} \cdot \frac{\mathbf{y}_k\mathbf{y}_k^{\mathrm{H}}}{\|\mathbf{y}_k\|^2},$$

and together this gives the following update [50]:

$$\begin{aligned}
\mathbf{u}_k &= \mathbf{W}_{k-1}\mathbf{y}_k \\
\zeta_k &= \frac{1}{\|\mathbf{y}_k\|^2}\left(1 - \sqrt{\frac{\lambda}{\lambda + \|\mathbf{y}_k\|^2}}\right) = \frac{1}{\lambda + \|\mathbf{y}_k\|^2 + \sqrt{\lambda}\sqrt{\lambda + \|\mathbf{y}_k\|^2}} \\
\mathbf{W}_k &= \frac{1}{\sqrt{\lambda}}(\mathbf{W}_{k-1} - \zeta_k\mathbf{u}_k\mathbf{y}_k^{\mathrm{H}}).
\end{aligned}$$

A final point is that, due to the definition of $\hat{\mathbf{R}}_{\mathbf{v}}(k)$ in (6.39), it converges to $1/(1-\lambda)\mathbf{R}_{\mathbf{v}}$ instead of $\mathbf{R}_{\mathbf{v}}$. Rather than modifying all the above equations to take this into account, we can simply scale the resulting whitened output vector $\mathbf{y}_k$ by

Given $\mathbf{X} = [\mathbf{x}_1, \mathbf{x}_2, \ldots]$ and an exponential forgetting factor $0 < \lambda < 1$, adaptively compute a prewhitening filter $\mathbf{F}_k = \alpha_k \mathbf{U}_k \mathbf{W}_k$, and a prewhitened output $\underline{\mathbf{X}} = [\underline{\mathbf{x}}_1, \underline{\mathbf{x}}_2, \ldots]$, where $\underline{\mathbf{x}}_k = \mathbf{F}_k^{\mathrm{H}} \mathbf{x}_k$, $\mathbf{U}_k$ is an orthonormal basis tracking the d-dimensional column span of $\mathbf{X}$, and $\alpha_k^2 \mathbf{W}_k \mathbf{W}_k^{\mathrm{H}} \to (\mathbf{U}^{\mathrm{H}} \mathbf{R}_{\mathbf{x}} \mathbf{U})^{-1}$

Initialize:

$\mathbf{U} = \mathbf{I}_{m \times d}$, $\mathbf{W} = \delta^{-1} \mathbf{I}_{d \times d}$, where δ is very small

$\alpha^2 = 0$

Update:

for $k = 1, 2, \cdots$ do

1. $\mathbf{v} = \mathbf{U}^{\mathrm{H}} \mathbf{x}$ $\quad (dM)$

   $\mathbf{y} = \mathbf{W}^{\mathrm{H}} \mathbf{v}$ $\quad (d^2)$

   $\mathbf{u} = \mathbf{W} \mathbf{y}$ $\quad (d^2)$

2. $\mathbf{k} = \dfrac{1}{\lambda + \|\mathbf{y}\|^2} \mathbf{u}$ $\quad (d)$

   $\mathbf{e} = \mathbf{x} - \mathbf{U} \mathbf{v}$ $\quad (dM)$

   $\mathbf{z} = \mathbf{e} - \dfrac{\|\mathbf{e}\|^2}{2} \mathbf{U} \mathbf{k}$ $\quad (dM)$

   $\mathbf{U} = \mathbf{U} + \dfrac{1}{1 + .25\|\mathbf{e}\|^2 \|\mathbf{k}\|^2} \mathbf{z} \mathbf{k}^{\mathrm{H}}$ $\quad (dM)$

3. $\zeta = \dfrac{1}{\lambda + \|\mathbf{y}\|^2 + \sqrt{\lambda}\sqrt{\lambda + \|\mathbf{y}\|^2}}$ $\quad (1)$

   $\mathbf{W} = \dfrac{1}{\sqrt{\lambda}} (\mathbf{W} - \zeta \mathbf{u} \mathbf{y}^{\mathrm{H}})$ $\quad (d^2)$

4. $\alpha^2 = \lambda \alpha^2 + 1$

   $\underline{\mathbf{x}}_k = \alpha \mathbf{y}$ $\quad \overline{4dM + 3d^2}$

Figure 6.11 Adaptive prewhitening filter [49].

a factor α_k, where α_k^2 is computed recursively in a similar way as $\mathbf{R}_{\mathbf{v}}$, namely

$$\alpha_k^2 = \lambda \alpha_{k-1}^2 + 1, \quad \alpha_0 = 0.$$

The resulting algorithm is summarized in Figure 6.11. Its complexity is of order dM. The output of the filter is $\underline{\mathbf{x}}_k = \alpha_k \mathbf{W}_k^{\mathrm{H}} \mathbf{U}_k^{\mathrm{H}} \mathbf{x}_k$.

6.7 ADAPTIVE ACMA

Recall Figure 6.8, which summarizes the iterative version of ACMA. To make this block algorithm adaptive, the following ingredients are needed:

1. Adaptive implementation of the prewhitening filter $\mathbf{F}$,
2. Adaptive tracking of the nullspace of $\hat{\underline{\mathbf{C}}}$,

3. Adaptive update of the joint diagonalization, or alternatively, of the rank-1 mapping of each subspace vector.

The adaptive prewhitening was discussed in Section 6.6.

For the second item, ideally we would track the unwhitened $\hat{\mathbf{C}}$ and apply a prewhitening operation $\mathbf{F}$ to $\hat{\mathbf{C}}$ after each update to find a consistent estimate of $\underline{\hat{\mathbf{C}}}$ and its nullspace. An update of $\hat{\mathbf{C}}$ is straightforward to derive from its definition in equation (6.27), but it would cost order M^4 per update, which is too much. Applying $(\bar{\mathbf{F}}_k \otimes \mathbf{F}_k)$ to the left and right of this matrix after each update would be even more costly. Therefore, we have to assume that the prewhitening filter $\mathbf{F}_k$ changes only slowly, and track the whitened $\underline{\hat{\mathbf{C}}}$ using whitened update vectors $\bar{\underline{\mathbf{x}}}_k \otimes \underline{\mathbf{x}}_k$. This leads to a complexity of order d^4, which is still too much in comparison to the existing adaptive CMAs which have complexity d^2. Thus, we will avoid to construct and store $\underline{\hat{\mathbf{C}}}$, but will directly update its nullspace using the update vectors of $\underline{\hat{\mathbf{C}}}$.

6.7.1 Adaptive Tracking of $\underline{\hat{\mathbf{C}}}$

First we derive an update equation for $\underline{\hat{\mathbf{C}}}$. So far, $\underline{\hat{\mathbf{C}}}$ was defined only in terms of a given batch of N samples, but we would like to convert this into an exponential window (λ-scaling). As before, let $\mathbf{P}$ be an $N \times d^2$ dimensional matrix with rows $\mathbf{y}_k^{\mathrm{H}} := (\bar{\underline{\mathbf{x}}}_k \otimes \underline{\mathbf{x}}_k)^{\mathrm{H}}$, where $\underline{\mathbf{x}}_k$ is the received (whitened) sample at time k. For simplicity of notation, we will drop the underscores in this subsection from now on, since all data will be in the whitened domain. According to equation (6.27)

$$\hat{\mathbf{C}} := \frac{1}{N}\mathbf{P}^{\mathrm{H}}\mathbf{P} - \frac{1}{N^2}\mathbf{P}^{\mathrm{H}}\mathbf{1}\,\mathbf{1}^{\mathrm{H}}\mathbf{P}.$$

Comparing this matrix to

$$\mathbf{M} := \sum_{k=1}^{N} \begin{bmatrix} 1 \\ \mathbf{y}_k \end{bmatrix} [1 \quad \mathbf{y}_k^{\mathrm{H}}] = [\mathbf{1} \quad \mathbf{P}]^{\mathrm{H}} [\mathbf{1} \quad \mathbf{P}] = \begin{bmatrix} \mathbf{1}^{\mathrm{H}}\mathbf{1} & \mathbf{1}^{\mathrm{H}}\mathbf{P} \\ \mathbf{P}^{\mathrm{H}}\mathbf{1} & \mathbf{P}^{\mathrm{H}}\mathbf{P} \end{bmatrix} \tag{6.43}$$

we see that $N\,\hat{\mathbf{C}}$ is equal to the Schur complement $\mathbf{M}_{2,2} - \mathbf{M}_{2,1}\mathbf{M}_{1,1}^{-1}\mathbf{M}_{1,2}$.

An adaptive update rule for $\hat{\mathbf{C}}$ can be derived from an adaptive update rule for $\mathbf{M}$, where we scale previous estimates with λ, with $0 < \lambda < 1$. Thus let $\mathbf{M}_k$ and $\mathbf{C}_k$ be defined as

$$\mathbf{M}_k := (1-\lambda)\sum_{i=1}^{k} \lambda^{k-i} \begin{bmatrix} 1 \\ \mathbf{y}_i \end{bmatrix} [1 \quad \mathbf{y}_i^{\mathrm{H}}] =: \begin{bmatrix} \alpha_k & \mathbf{p}_k^{\mathrm{H}} \\ \mathbf{p}_k & \mathbf{N}_k \end{bmatrix}, \quad \mathbf{C}_k := \mathbf{N}_k - \frac{\mathbf{p}_k\mathbf{p}_k^{\mathrm{H}}}{\alpha_k}$$

then $\mathbf{M}_k$ is an unbiased estimate of $\mathbf{M}$, due to multiplication with the factor $(1-\lambda)$, and $\mathbf{C}_k$ is an unbiased estimate of $\hat{\mathbf{C}}$. The update rule for $\mathbf{M}_k$ which follows from this equation is

$$\mathbf{M}_k = \lambda \mathbf{M}_{k-1} + (1-\lambda)\begin{bmatrix} 1 \\ \mathbf{y}_k \end{bmatrix}[1 \quad \mathbf{y}_k^{\mathrm{H}}]$$

so that

$$\begin{bmatrix} \alpha_k & \mathbf{p}_k^{\mathrm{H}} \\ \mathbf{p}_k & \mathbf{N}_k \end{bmatrix} = \begin{bmatrix} \lambda\alpha_{k-1} + (1-\lambda) & \lambda\mathbf{p}_{k-1}^{\mathrm{H}} + (1-\lambda)\mathbf{y}_k^{\mathrm{H}} \\ \lambda\mathbf{p}_{k-1} + (1-\lambda)\mathbf{y}_k & \lambda\mathbf{N}_{k-1} + (1-\lambda)\mathbf{y}_k\mathbf{y}_k^{\mathrm{H}} \end{bmatrix}$$

and

$$\begin{aligned} \mathbf{C}_k &= \mathbf{N}_k - \frac{\mathbf{p}_k\mathbf{p}_k^{\mathrm{H}}}{\alpha_k} \\ &= \lambda\mathbf{C}_{k-1} + \frac{\lambda}{\alpha_{k-1}}\mathbf{p}_{k-1}\mathbf{p}_{k-1}^{\mathrm{H}} + (1-\lambda)\mathbf{y}_k\mathbf{y}_k^{\mathrm{H}} \\ &\quad - \frac{1}{\alpha_k}[\lambda\mathbf{p}_{k-1} + (1-\lambda)\mathbf{y}_k][\lambda\mathbf{p}_{k-1} + (1-\lambda)\mathbf{y}_k]^{\mathrm{H}} \\ &= \lambda\mathbf{C}_{k-1} + [\mathbf{y}_k \quad \mathbf{p}_{k-1}]\begin{bmatrix} (1-\lambda) - \dfrac{(1-\lambda)^2}{\alpha_k} & -\dfrac{\lambda(1-\lambda)}{\alpha_k} \\ -\dfrac{\lambda(1-\lambda)}{\alpha_k} & \dfrac{\lambda}{\alpha_{k-1}} - \dfrac{\lambda^2}{\alpha_k} \end{bmatrix}\begin{bmatrix} \mathbf{y}_k^{\mathrm{H}} \\ \mathbf{p}_{k-1}^{\mathrm{H}} \end{bmatrix}. \end{aligned}$$

Using $\alpha_k = \lambda\alpha_{k-1} + (1-\lambda)$, it follows that the 2×2 matrix in the update is actually of rank 1, namely equal to

$$\frac{1-\lambda}{\alpha_k}\begin{bmatrix} \alpha_k - (1-\lambda) & -\lambda \\ -\lambda & \dfrac{\lambda}{\alpha_{k-1}} \end{bmatrix} = \frac{\alpha_{k-1}}{\alpha_k}\lambda(1-\lambda)\begin{bmatrix} 1 & -\dfrac{1}{\alpha_{k-1}} \\ -\dfrac{1}{\alpha_{k-1}} & \dfrac{1}{\alpha_{k-1}^2} \end{bmatrix}$$

so that we remain with a one-dimensional update

$$\begin{aligned} \mathbf{C}_k &= \lambda\mathbf{C}_{k-1} + \frac{\alpha_{k-1}}{\alpha_k}\lambda(1-\lambda)\cdot(\mathbf{y}_k - \mathbf{p}_{k-1}/\alpha_{k-1})(\mathbf{y}_k - \mathbf{p}_{k-1}/\alpha_{k-1})^{\mathrm{H}} \\ &=: \lambda\mathbf{C}_{k-1} + \beta_k\mathbf{c}_k\mathbf{c}_k^{\mathrm{H}}. \end{aligned} \tag{6.44}$$

Therefore, the vector by which to update $\mathbf{C}_{k-1}$ is equal to a scaling of the modified data vector

$$\mathbf{c}_k := \bar{\underline{\mathbf{x}}}_k \otimes \underline{\mathbf{x}}_k - \mathbf{p}_{k-1}/\alpha_{k-1} \tag{6.45}$$

where $\mathbf{p}_{k-1}$ and α_{k-1} are updated as

$$\mathbf{p}_k = \lambda \mathbf{p}_{k-1} + (1-\lambda)\bar{\underline{\mathbf{x}}}_k \otimes \underline{\mathbf{x}}_k, \qquad \alpha_k = \lambda\alpha_{k-1} + (1-\lambda). \tag{6.46}$$

Note that $\mathbf{p}_k$ is an exponentially-weighted unbiased estimate of the mean of $\bar{\underline{\mathbf{x}}}_k \otimes \underline{\mathbf{x}}_k$, which matches its use in equation (6.26), whereas α_k converges to 1 and is only relevant during the initialization phase of the algorithm.

6.7.2 Adaptive Tracking of the Nullspace of $\hat{\underline{\mathbf{C}}}$

Subspace tracking is well studied in signal processing, and a variety of algorithms has been derived. Algorithms can be classified in several ways:

1. *Subspace:* principal components versus minor subspace (or nullspace) tracking, or both.
2. *Rank determination:* a specified subspace dimension or a specified error threshold.
3. *Complexity:* for d-dimensional subspaces in an M-dimensional space, algorithms of order M^2d, d^2M, down to dM have been reported.
4. *Strategy:* exact (deterministic), gradient descent (stochastic), and so on.

The paper [48] gives an overview; the most reliable algorithm seems to be Karasalo's, with complexity d^2M.

In the case of an adaptive version for ACMA, we have to track the nullspace of $\hat{\underline{\mathbf{C}}}$, which is a d-dimensional nullspace in a d^2-dimensional space. The lowest complexity that can be achieved for this case is of order d^3. To remain competitive with CMA and MUK, a higher complexity cannot be tolerated. Therefore only nullspace tracking algorithms of order dM are applicable. The most popular algorithms in this context are based on the PAST algorithm [47], which as described in Section 6.6 is derived from an iterative optimization of the cost function

$$\begin{aligned} J(\mathbf{V}) &= \mathrm{E}\|\mathbf{c} - \mathbf{V}\mathbf{V}^{\mathrm{H}}\mathbf{c}\|^2 \\ &= \mathrm{Tr}(\mathbf{C}) - 2\mathrm{Tr}(\mathbf{V}^{\mathrm{H}}\mathbf{C}\mathbf{V}) + \mathrm{Tr}(\mathbf{V}^{\mathrm{H}}\mathbf{C}\mathbf{V}\mathbf{V}^{\mathrm{H}}\mathbf{V}) \end{aligned} \tag{6.47}$$

where $\mathbf{V}$ is the estimated subspace, $\mathbf{c}$ is the data vector, in our context given by equation (6.45), and $\mathbf{C} = \mathrm{E}(\mathbf{c}\mathbf{c}^{\mathrm{H}})$ is the matrix from which the subspace has to be determined. Minimization of the cost function will produce an orthogonal basis for the principal subspace, whereas maximization leads to a basis of the nullspace (in this case normalization constraints on $\mathbf{V}$ are needed to avoid trivial solutions).

As discussed in Section 6.6, the PAST algorithm follows from an alternating least squares implementation of the optimization problem: compute the vector $\mathbf{y}_k = \mathbf{V}_{k-1}^{\mathrm{H}}\mathbf{c}_k$ using the previous estimate of $\mathbf{V}$, then optimize $\sum_{i=1}^{k} \lambda^{k-i}\|\mathbf{c}_i - \mathbf{V}\mathbf{y}_i\|^2$ over $\mathbf{V}$, where λ is the forgetting factor. This resulted in an RLS-type algorithm.

In an alternative approach in [51], a gradient descent algorithm is derived from (6.47):

$$\mathbf{V}_k := \mathbf{V}_{k-1} - \beta_k \nabla_J(\mathbf{V}_{k-1}) \tag{6.48}$$

where β_k is a step size and $\nabla_J(\mathbf{V}_{k-1})$ the gradient of $J(\mathbf{V})$ evaluated at $\mathbf{V}_{k-1}$:

$$\nabla_J(\mathbf{V}) = (-2\mathbf{C} + \mathbf{C}\mathbf{V}\mathbf{V}^{\mathrm{H}} + \mathbf{V}\mathbf{V}^{\mathrm{H}}\mathbf{C})\mathbf{V}.$$

By taking $\beta_k < 0$, the cost function is maximized and an estimate of the nullspace is obtained. In [51], $\mathbf{V}$ is constrained to be orthonormal ($\mathbf{V}^{\mathrm{H}}\mathbf{V} = \mathbf{I}$), and a variable step size β_k is selected such that it optimizes the cost in every step (since the cost function (6.47) with $\mathbf{V}$ replaced by $\mathbf{V}_k$ from (6.48) is quadratic in β_k, it can be computed in closed form). Subsequently, the gradient is approximated by replacing $\mathbf{C}$ by an estimate based on the current sample, $\mathbf{c}_k\mathbf{c}_k^{\mathrm{H}}$, which leads to

$$\mathbf{V}_k := \mathbf{V}_{k-1} - \beta_{\mathrm{OPT},k}(-\mathbf{c}_k + \mathbf{V}_{k-1}\mathbf{y}_k)\mathbf{y}_k^{\mathrm{H}}. \tag{6.49}$$

For a sequence of M-dimensional vectors $\mathbf{c}_1, \mathbf{c}_2, \ldots$, compute adaptive estimates of an orthonormal basis $\mathbf{V}_k$ of the d-dimensional nullspace:

Initialize $\mathbf{V}_1 = [\mathbf{I}_{d\times d} \quad \mathbf{0}_{d\times(m-d)}]^{\mathrm{H}}$

for $k = 1, 2, \ldots$ do

- $\mathbf{y} = \mathbf{V}_k^{\mathrm{H}}\mathbf{c}_k$ $\qquad (dM)$
- $\mathbf{z} = \mathbf{V}_k\mathbf{y}$ $\qquad (dM)$
- $\mathbf{p} = \mathbf{c}_k - \mathbf{z}$
- $\beta = \dfrac{1}{\|\mathbf{c}_k\|^2 - \|\mathbf{y}\|^2 + 0.2}$
- $\phi = \dfrac{1}{\sqrt{1 + \beta^2\|\mathbf{p}\|^2\|\mathbf{y}\|^2}}$
- $\tau = \dfrac{\phi - 1}{\|\mathbf{y}\|^2}$
- $\mathbf{u} = -\tau/\beta\mathbf{z} + \phi\mathbf{p}$
- $\mathbf{V}_{k+1} = \mathbf{V}_k - 2\dfrac{\mathbf{u}\mathbf{u}^{\mathrm{H}}}{\|\mathbf{u}\|^2}\mathbf{V}_k$ $\qquad (2dM)$

$\qquad 4dM$

end

Figure 6.12 Adaptive nullspace tracking using NOOJA [51].

After this update step, the new basis $\mathbf{V}_k$ needs to be orthonormalized again, which is done by setting

$$\mathbf{V}_k := \mathbf{V}_k(\mathbf{V}_k^{\mathrm{H}}\mathbf{V}_k)^{-1/2}.$$

Since (6.49) represents a rank-1 update, the normalization can be computed efficiently in closed form using a Householder reflection, similar as in Section 6.6. The resulting algorithm is called the normalized orthogonal OJA (NOOJA) [51], and is summarized in Figure 6.12. It is interesting to note that it is of complexity $4dM$, and scaling-independent: if the input data $\mathbf{c}_k$ is multiplied by a scalar, then the resulting $\mathbf{V}_k$ is unchanged. Although there are a few alternative algorithms for minor subspace tracking [52–55], this algorithm currently seems to be one of the fastest and more reliable (e.g., some of the other algorithms perform poor for high SNR). A disadvantage is that the subspace estimate remains jittery in steady state, because the large stepsize tends to emphasize the instantaneous noise. Therefore, instead of taking the maximal stepsize $\beta_{\mathrm{OPT},k}$, typically a fraction of this step size is used.

6.7.3 Adaptive Update of the Joint Diagonalization

Using the preceding nullspace tracking algorithm in our application, we can update the basis $\mathbf{V}_n$ of the nullspace of $\underline{\hat{\mathbf{C}}}$ given the update vector $\mathbf{c}_k$ in (6.45). The final step is an efficient implementation of the joint diagonalization. After the subspace update, the iterative ACMA (Figure 6.8) uses the previous estimate of the beamformers, $\mathbf{T}$, and continues with the following three steps:

$$\begin{aligned}
\mathbf{M} &= \mathbf{V}_n^{\mathrm{H}}(\bar{\mathbf{T}} \circ \mathbf{T}) \\
\mathbf{Y} &= \mathbf{V}_n\mathbf{M} = \mathbf{V}_n\mathbf{V}_n^{\mathrm{H}}(\bar{\mathbf{T}} \circ \mathbf{T}) \\
\mathbf{t}_i &= \pi_1(\mathbf{y}_i), \quad i = 1, \ldots, d.
\end{aligned}$$

This projects $\bar{\mathbf{T}} \circ \mathbf{T}$ onto the estimated subspace, resulting in $\mathbf{Y}$, and subsequently maps the columns of $\mathbf{Y}$ back to the Kronecker-product structure, resulting in new estimates of the columns $\mathbf{t}_i$ of $\mathbf{T}$. However, the complexity of the projection is too high (order d^4 instead of d^3).

Therefore, the following modification is introduced: instead of updating the basis $\mathbf{V}_n$, we compute $\bar{\mathbf{T}} \circ \mathbf{T}$ and regard it as the current estimate of the subspace basis (i.e., set $\mathbf{V}_n = \bar{\mathbf{T}} \circ \mathbf{T}$). Using this basis, the subspace update is performed, giving $\mathbf{Y}$, and then the result is mapped back to the Kronecker-product structure. In this context, the update performed by the NOOJA algorithm is interpreted as a Householder reflection which tries to make $\bar{\mathbf{T}} \circ \mathbf{T}$ orthogonal to the current update vector $\mathbf{c}_k$.

The last step of the algorithm is the mapping of the columns $\mathbf{y}_i$ of $\mathbf{Y}$ to a Kronecker-product structure, $\mathbf{y}_i =: \bar{\mathbf{t}}_i \otimes \mathbf{t}_i$, or equivalently,

$$\mathbf{Y}_i = \mathbf{t}_i \mathbf{t}_i^{\mathrm{H}}$$

where $\mathrm{vec}(\mathbf{Y}_i) = \mathbf{y}_i$. An SVD can be used to estimate $\mathbf{t}_i$ as the dominant singular vector, as was done in the nonadaptive version in Figure 6.8, but it would cost order d^3 per subspace vector, or d^4 in total. Since we need only the dominant singular vector, we can instead apply a power iteration [23]. The general form of an iterative step in the power iteration is

$$\mathbf{v}_{k+1} = \mathbf{Y}\mathbf{v}_k, \quad k = 1, 2, \ldots$$

where the iteration is initialized by a randomly selected vector $\mathbf{v}_0$. However, the best choice of an initial point is the previous estimate for $\mathbf{t}_i$, and in this case, a single step of the iteration is sufficient to give a good improvement of the estimate. The complexity of one update step is d^2 per subspace vector, or d^3 in total.

Given data $\mathbf{X} = [\mathbf{x}_1, \mathbf{x}_2, \ldots]$, compute beamformers $\mathbf{W}_k = \mathbf{F}_k \mathbf{T}_k$ and output $\hat{\mathbf{s}}_k = \mathbf{W}_k^{\mathrm{H}} \mathbf{x}_k$:

Initialize prewhitening filter $\mathbf{F}$ using about $5M$ prior input samples;
$\mathbf{T} = \mathbf{I}_{d \times d}$, $\mathbf{p} = \mathbf{0}$, $\alpha = 0$

for $k = 1, 2, \ldots$ do

1. Update $\mathbf{F}$, the prewhitening filter, using $\mathbf{x}_k$ (Fig. 6.11) — $(4dM + 3d^2)$
   $\underline{\mathbf{x}} = \mathbf{F}^{\mathrm{H}} \mathbf{x}_k$, the prewhitened input vector
2. Compute the update vector $\mathbf{c}$ for $\hat{\underline{\mathbf{C}}}$: — (d^2)
   - $\mathbf{c} = \bar{\underline{\mathbf{x}}} \otimes \underline{\mathbf{x}} - \mathbf{p}/\alpha$
   - $\mathbf{p} = \lambda \mathbf{p} + (1 - \lambda) \bar{\underline{\mathbf{x}}} \otimes \underline{\mathbf{x}}$
   - $\alpha = \lambda \alpha + (1 - \lambda)$

   Compute $\mathbf{Y} = \bar{\mathbf{T}} \circ \mathbf{T}$ — (d^3)
   Regard $\mathbf{Y}$ as a basis of the nullspace of $\hat{\underline{\mathbf{C}}}$, and update it using $\mathbf{c}$ (Fig. 6.12) — $(4d^3)$
3. for $i = 1, \ldots, d$ do — (d^3)
   - $\mathbf{Y}_i = \mathrm{vec}^{-1}(\mathbf{y}_i)$
   - $\mathbf{t}_i = \mathbf{Y}_i \mathbf{t}_i$ (one step of a power iteration)
   - $\mathbf{t}_i = \mathbf{t}_i / \|\mathbf{t}_i\|$

   end
4. $\mathbf{T} = \mathrm{recond}(\mathbf{T})$, equation (6.19) — (d^3)
5. $\hat{\mathbf{s}}_k = \mathbf{T}^{\mathrm{H}} \underline{\mathbf{x}}$ — (d^2)

Total: $4dM + 6d^3$

Figure 6.13 Adaptive implementation of ACMA.

6.7.4 Summary of the Resulting Algorithm

The resulting algorithm is summarized in Figure 6.13. As indicated, the complexity of the algorithm is of order $4dM + 6d^3$. This is comparable to the complexity of OCMA and MUK (Figure 6.5), which was computed as order $4dM + d^3 + 5d^2$, where the term d^3 is contributed by the reorthogonalization step, which perhaps can be implemented cheaper. Therefore, the complexity is at most a factor d worse, where d is small (typically $d \leq 5$).

6.7.5 Comparison of MUK with Adaptive-ACMA

To compare the performance of the MUK algorithm with the adaptive-ACMA derived in this section, we show two sets of simulations. In the first, a stationary scenario is considered, whereas in the second, the source powers and array response vectors are time varying.

6.7.5.1 Stationary Channel. Entirely similar to Section 6.5.5, we take a uniform linear array with $M = 4$ antennas, $d = 3$ constant-modulus sources with directions $[-10^\circ, 20^\circ, 30^\circ]$ and amplitudes [1, 0.8, 0.9]. The SNR is 10 dB per antenna for the strongest user.

Figure 6.14(a) shows the worst SINR among the users, averaged over 600 monte-carlo runs, as a function of the sample index. Since not always all independent users are recovered, Figure 6.14(b) shows the percentage of failed cases. These cases are not used in the SINR statistics. The dotted reference curve is formed by ACMA acting on a growing block of samples (no forgetting factor λ).

For adaptive-ACMA, the first 20 samples are used to initialize the adaptive prewhitening filter. The convergence speed depends on λ, the plot shows the results for $\lambda = 0.995$. A smaller λ gives faster initial convergence but a lower steady-state performance. The final SINR level is also determined by the nullspace tracking

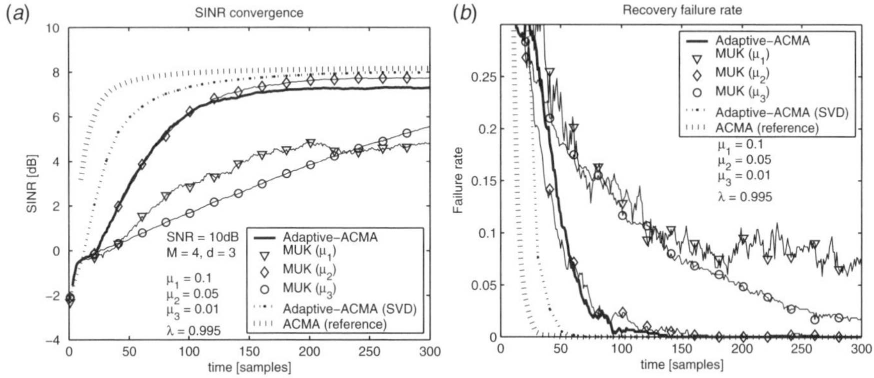

Figure 6.14 Average convergence and failure rate of MUK and adaptive-ACMA—stationary channel.

algorithm. NOOJA with optimal step size is a "greedy" algorithm which quickly tracks the subspace, but this also results in a more noisy output. Instead of the optimal step size $\beta_{\text{OPT},k}$, we have used $0.3\beta_{\text{OPT},k}$ which gives a smoother performance at the expense of initial tracking speed. To verify that the nullspace tracking error is a limiting factor in the steady state, we also implemented a version where $\hat{\mathbf{C}}_k$ is formed—as in (6.44) and its nullspace computed using SVD [dotted reference curve adaptive-ACMA(SVD) in the figure; for this algorithm also the rank-1 truncation is performed using SVD].

For MUK, the performance behavior depends on λ due to the adaptive prewhitening, and also strongly depends on the value of the step size μ, therefore three values are shown; the value $\mu = 0.05$ gives a remarkably similar performance to adaptive-ACMA in this scenario both in convergence speed and in asymptotic SNR. The experience is that for higher SNRs, adaptive-ACMA will outperform MUK by a few dB.

6.7.5.2 Time-Varying Channel. To test the tracking behavior of the adaptive algorithms, the preceding scenario is made time-varying. Specifically, the source amplitudes β_i are varied in sinusoidal patterns with randomly selected periods, with a maximum of three periods over the simulated time interval ($N = 1500$ samples). An example is shown in Figure 6.15(a). The direction vectors $\{\mathbf{a}_i\}$ of each source are not selected on an array manifold, instead each entry of each $\mathbf{a}_i$ is a unimodular complex number with a linear phase progression, causing at most

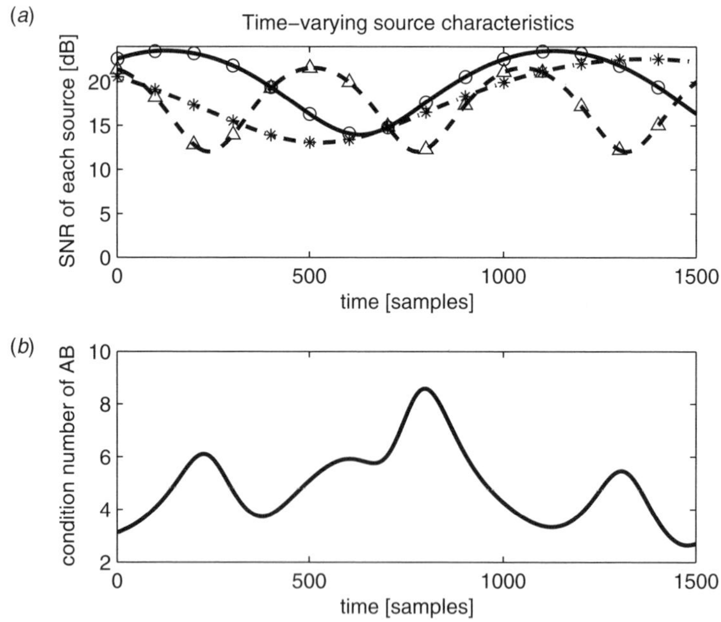

Figure 6.15 (a) SNR and (b) conditioning of a time-varying channel (example).

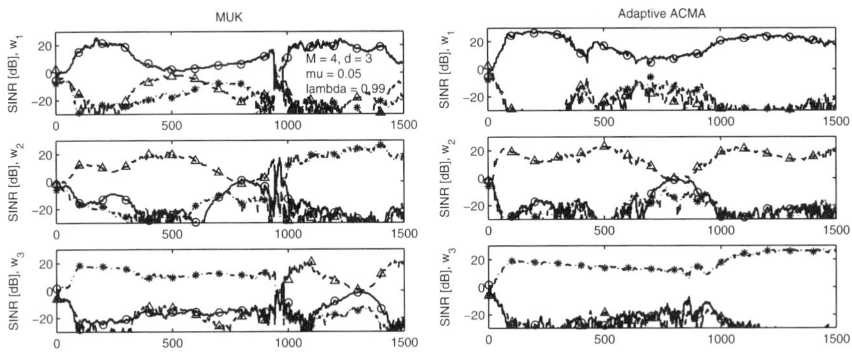

Figure 6.16 Example tracking behavior of MUK and adaptive-ACMA—time-varying channel. For each beamformer the output SINR corresponding to each source is shown.

one cycle per interval. The condition number of the resulting channel matrix $\mathbf{AB} = [\mathbf{a}_1\beta_1, \ldots, \mathbf{a}_d\beta_d]$ is plotted in Figure 6.15(*b*).

The output SINR for each beamformer is shown in Figure 6.16 for MUK and adaptive-ACMA, respectively. Each panel corresponds to a beamformer $\mathbf{w}_i, i = 1, \ldots, d$, and the *j*th curve in a panel corresponds to the SINR behavior of the *j*th source, that is, related to $\mathbf{a}_j\beta_j$. For this example, MUK experienced a case of *port swapping* around sample number 1000, that is, a beamformer suddenly starts to track a different source. This can occur if two sources come too close or if the scenario changes faster than the algorithm can track. The fluctuations in the output SINR of the tracked source obviously also follow the fluctuations in the input SNR.

Figure 6.17(*a*) shows the average output SINR of each source for MUK and adaptive-ACMA as a function of SNR, where the average is over time, over the

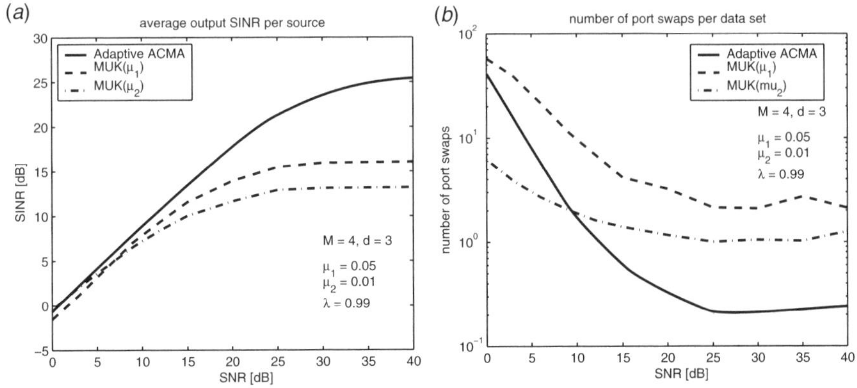

Figure 6.17 Performance of MUK and adaptive-ACMA in a time-varying scenario: (*a*) average output SINR, (*b*) average number of port swaps per interval (1500 samples).

three sources, and over 300 Monte Carlo runs, each with different randomly varying channels (there has been no attempt to detect and remove failed cases). Finally, Figure 6.17(*b*) shows the average number of times that a port swap occurred in a data run of N samples. The performance of MUK is sensitive to the choice of stepsize μ: a larger μ gives faster tracking and resulted in better output SINR, but also gave more port swaps: usually at least once per data set. Both algorithms are performance limited at high SNR due to the time-variation in the observation window, but overall, adaptive-ACMA is better in this particular scenario. Since this is a new algorithm, it is unclear whether this is observation holds in general.

6.8 DOA ASSISTED BEAMFORMING OF CONSTANT MODULUS SIGNALS

In some applications the sensor array is assumed to be calibrated, that is, it is known parametrically as a function of a vector parameter $\boldsymbol{\theta}$. One typical example is when $\boldsymbol{\theta}$ represents the direction of arrivals (DOAs) of the received signals. DOA estimation has been thoroughly studied for arbitrary signals. For a good overview of DOA estimation for arbitrary signals we refer the reader to [56] and [57], or indeed some of the other chapters in this book. When the signals have a constant modulus the number of nuisance parameters (i.e., the signal parameters) is reduced by a factor of two, since the amplitudes are constant. As we will see this enables a much more accurate estimation of the direction of arrival of the signals and consequently leads to a better separation.

The literature on DOA estimation of CM signals is relatively sparse. The most common approach is to separate the sources based on the CM property, followed by estimating the direction of each recovered signal. Initially this has been done using the CM Array [19, 58, 59], but as mentioned in preceding sections, the CM Array is recursive in nature and requires many hundreds of samples to obtain convergence. Very good results on small numbers of samples have been obtained by first estimating the channel matrix using ACMA, and subsequently obtaining a decoupled DOA estimation problem where the estimated steering vectors are projected onto the model based steering vectors [60].

Maximum likelihood estimation for the joint CM signals estimation and DOA estimation is the optimal way. For the trivial case of a single source it had been shown to be equivalent to L_1 beamforming, which is more robust than the classical L_2 beamformer [61]. For more than a single source a computationally attractive approach has been suggested in [62, 63].

The Cramer–Rao bound (CRB) gives a lower bound on the variance of any unbiased estimator, and is an important measure for the efficiency of estimators. It has been widely used for estimating the performance bound of DOA estimation. The additional information brought by the CM assumption has been computed in [60, 64]. In this section we present some of these results: the Cramer–Rao lower bound, a description of the maximum likelihood algorithm, and simulation results presenting the various methods and the robustness to mismodeling.

6.8.1 Data Model

For the purpose of this section, we have to extend the previously used data model $\mathbf{x}(t) = \mathbf{A}\mathbf{s}(t) + \mathbf{n}(t)$ to include a parametrically known channel matrix $\mathbf{A}$. Since now we cannot place the unknown source amplitudes into $\mathbf{A}$, we also have to introduce a gain factor $\mathbf{B}$, which leads to

$$\mathbf{x}(t) = \mathbf{A}(\boldsymbol{\theta})\mathbf{B}\mathbf{s}(t) + \mathbf{n}(t) \tag{6.50}$$

where

$\mathbf{A}(\boldsymbol{\theta}) = [\mathbf{a}(\theta_1), \ldots, \mathbf{a}(\theta_d)]$, where $\mathbf{a}(\theta)$ is the array response vector for a signal from direction θ, and $\boldsymbol{\theta} = [\theta_1, \ldots, \theta_d]$ is the DOA vector of the sources (for simplicity, we assume 1 parameter per source),

$\mathbf{B} = \mathrm{diag}(\boldsymbol{\beta})$ is the gain matrix, with parameters $\boldsymbol{\beta} = [\beta_1, \ldots, \beta_d]^{\mathrm{T}}$, where $\beta_i \in \mathbb{R}^+$ is the amplitude of the ith signal as received by the array.

As usual in DOA estimation, we require that the array manifold satisfies the uniqueness condition, that is, every collection of M vectors on the manifold is linearly independent.

As before, we assume that all sources have constant modulus. Unequal source powers are absorbed in the gain matrix $\mathbf{B}$. Phase offsets of the sources after demodulation are part of the $s_i(t)$. Thus we can write $s_i(t) = e^{j\phi_i(t)}$, where $\phi_i(t)$ is the unknown phase modulation for source i, and we define $\boldsymbol{\phi}(t) = [\phi_1(t), \ldots, \phi_d(t)]^{\mathrm{T}}$ as the phase vector for all sources at time t. We further assume that the noise is Gaussian and spatially white with covariance matrix $\mathbf{R}_{nn} = \sigma^2\mathbf{I} = \nu\mathbf{I}$, where ν is the known noise variance as received on a single antenna.

6.8.2 Likelihood Function

Based on the model and assuming N received sample vectors collected in a matrix $\mathbf{X}$, we can derive the likelihood function. For deterministic CM signals in white Gaussian noise the likelihood function is given by

$$L(\mathbf{X}|\boldsymbol{\Phi}, \boldsymbol{\theta}, \boldsymbol{\beta}) = \frac{1}{(2\pi)^N \left(\frac{\nu}{2}\right)^{pN}} \exp\left\{\frac{1}{\nu}\sum_{k=1}^{N} \mathbf{e}^{\mathrm{H}}(k)\mathbf{e}(k)\right\}, \tag{6.51}$$

where

$$\mathbf{e}(k) = \mathbf{x}(k) - \mathbf{A}\mathbf{B}\mathbf{s}(k), \tag{6.52}$$

and

$$\boldsymbol{\Phi} = [\boldsymbol{\phi}(1), \ldots, \boldsymbol{\phi}(N)]. \tag{6.53}$$

Let $\mathcal{L}(\mathbf{X}|\mathbf{\Phi}, \boldsymbol{\theta}) = \log L(\mathbf{X}|\mathbf{\Phi}, \boldsymbol{\theta}, \boldsymbol{\beta})$. After omitting constants we obtain the log-likelihood function

$$\mathcal{L}(\mathbf{X}|\mathbf{\Phi}, \boldsymbol{\theta}, \boldsymbol{\beta}) = -\frac{1}{\nu}\sum_{k=1}^{N} \|\mathbf{e}(k)\|^2. \tag{6.54}$$

6.8.3 Cramer–Rao Bound

The Cramer–Rao bound (CRB) is a lower bound on the estimation variance of any unbiased estimator. Its derivation from the log-likelihood function follows along standard lines [65]. Indeed, the CRB is given by the main diagonal of the inverse of the Fisher information matrix (FIM). In turn, the FIM specifies the sensitivity of the log-likelihood function (regarded as cost function) to changes in the parameters,

$$\mathbf{F}_N = \mathrm{E}\left\{\frac{\partial\mathcal{L}}{\partial\boldsymbol{\rho}} \cdot \left(\frac{\partial\mathcal{L}}{\partial\boldsymbol{\rho}}\right)^{\mathrm{T}}\right\}$$

where $\boldsymbol{\rho}$ is a vector which collects all parameters,

$$\boldsymbol{\rho} = \left[\boldsymbol{\phi}(1)^{\mathrm{T}}, \ldots, \boldsymbol{\phi}(N)^{\mathrm{T}},\ \boldsymbol{\theta}^{\mathrm{T}},\ \boldsymbol{\beta}^{\mathrm{T}}\right]^{\mathrm{T}}.$$

For the case at hand, this can be worked out in closed form as follows [60]. Partition the FIM as

$$\mathbf{F}_N = \begin{bmatrix} \mathbf{F}_{11} & \mathbf{F}_{12} \\ \mathbf{F}_{21} & \mathbf{F}_{22} \end{bmatrix} \tag{6.55}$$

where the partitioning follows the partitioning of $\boldsymbol{\rho}$ into $\mathrm{vec}(\mathbf{\Phi})$ followed by $[\boldsymbol{\theta};\boldsymbol{\beta}]$. Then

$$\mathbf{F}_{11} = \begin{bmatrix} \mathbf{H}_1 & & \mathbf{0} \\ & \ddots & \\ \mathbf{0} & & \mathbf{H}_N \end{bmatrix}, \quad \mathbf{F}_{12} = \begin{bmatrix} \mathbf{\Delta}_1^{\mathrm{T}} & \mathbf{E}_1^{\mathrm{T}} \\ \vdots & \vdots \\ \mathbf{\Delta}_N^{\mathrm{T}} & \mathbf{E}_N^{\mathrm{T}} \end{bmatrix} \tag{6.56}$$

$$\mathbf{F}_{21} = \begin{bmatrix} \mathbf{\Delta}_1, \ldots, & \mathbf{\Delta}_N \\ \mathbf{E}_1, \ldots, & \mathbf{E}_N \end{bmatrix}, \quad \mathbf{F}_{22} = \begin{bmatrix} \mathbf{\Gamma} & \mathbf{\Lambda}^{\mathrm{T}} \\ \mathbf{\Lambda} & \mathbf{\Upsilon} \end{bmatrix}, \tag{6.57}$$

where

$$\begin{aligned}
\mathbf{H}_k &:= \mathrm{E}\frac{\partial\mathcal{L}}{\partial\boldsymbol{\phi}(k)}\left(\frac{\partial\mathcal{L}}{\partial\boldsymbol{\phi}(k)}\right)^{\mathrm{T}} = \frac{2}{\nu}\mathrm{Re}(\mathbf{S}_k^{\mathrm{H}}\mathbf{B}^{\mathrm{H}}\mathbf{A}^{\mathrm{H}}\mathbf{A}\mathbf{B}\mathbf{S}_k) \\
\boldsymbol{\Delta}_k &:= \mathrm{E}\frac{\partial\mathcal{L}}{\partial\boldsymbol{\theta}}\left(\frac{\partial\mathcal{L}}{\partial\boldsymbol{\phi}(k)}\right)^{\mathrm{T}} = -\frac{2}{\nu}\mathrm{Im}(\mathbf{S}_k^{\mathrm{H}}\mathbf{B}^{\mathrm{H}}\mathbf{D}^{\mathrm{H}}\mathbf{A}\mathbf{B}\mathbf{S}_k) \\
\mathbf{E}_k &:= \mathrm{E}\frac{\partial\mathcal{L}}{\partial\boldsymbol{\beta}}\left(\frac{\partial\mathcal{L}}{\partial\boldsymbol{\phi}(k)}\right)^{\mathrm{T}} = -\frac{2}{\nu}\mathrm{Im}(\mathbf{S}_k^{\mathrm{H}}\mathbf{A}^{\mathrm{H}}\mathbf{A}\mathbf{B}\mathbf{S}_k) \\
\boldsymbol{\Gamma} &:= \mathrm{E}\frac{\partial\mathcal{L}}{\partial\boldsymbol{\theta}}\left(\frac{\partial\mathcal{L}}{\partial\boldsymbol{\theta}}\right)^{\mathrm{T}} = \frac{2}{\nu}\sum_{k=1}^{N}\mathrm{Re}(\mathbf{S}_k^{\mathrm{H}}\mathbf{B}^{\mathrm{H}}\mathbf{D}^{\mathrm{H}}\mathbf{D}\mathbf{B}\mathbf{S}_k) \\
\boldsymbol{\Lambda} &:= \mathrm{E}\frac{\partial\mathcal{L}}{\partial\boldsymbol{\beta}}\left(\frac{\partial\mathcal{L}}{\partial\boldsymbol{\theta}}\right)^{\mathrm{T}} = \frac{2}{\nu}\sum_{k=1}^{N}\mathrm{Re}(\mathbf{S}_k^{\mathrm{H}}\mathbf{A}^{\mathrm{H}}\mathbf{D}\mathbf{B}\mathbf{S}_k) \\
\boldsymbol{\gamma} &:= \mathrm{E}\frac{\partial\mathcal{L}}{\partial\boldsymbol{\beta}}\left(\frac{\partial\mathcal{L}}{\partial\boldsymbol{\beta}}\right)^{\mathrm{T}} = \frac{2}{\nu}\sum_{\mathrm{k}=1}^{\mathrm{N}}\mathrm{Re}(\mathbf{S}_{\mathrm{k}}^{\mathrm{H}}\mathbf{A}^{\mathrm{H}}\mathbf{A}\mathbf{S}_{\mathrm{k}})
\end{aligned}$$

and

$$\mathbf{S}_k = \mathrm{diag}(\mathbf{s}(k)), \quad \mathbf{D} = \left[\frac{d\mathbf{a}}{d\theta}(\theta_1), \ldots, \frac{d\mathbf{a}}{d\theta}(\theta_d)\right].$$

To give closed-form expressions for the inverse of the FIM, we use the following general result for block-partitioned matrices (which can be easily derived by inverting an LDU factorization):

$$\begin{bmatrix} \mathbf{F}_{11} & \mathbf{F}_{12} \\ \mathbf{F}_{21} & \mathbf{F}_{22} \end{bmatrix}^{-1} = \begin{bmatrix} \mathbf{I} & -\mathbf{F}_{11}^{-1}\mathbf{F}_{12} \\ \mathbf{0} & \mathbf{I} \end{bmatrix} \begin{bmatrix} \mathbf{F}_{11}^{-1} & \\ & (\mathbf{F}_{22} - \mathbf{F}_{21}\mathbf{F}_{11}^{-1}\mathbf{F}_{12})^{-1} \end{bmatrix} \times \begin{bmatrix} \mathbf{I} & \mathbf{0} \\ -\mathbf{F}_{21}\mathbf{F}_{11}^{-1} & \mathbf{I} \end{bmatrix}$$

Thus let

$$\begin{bmatrix} \boldsymbol{\Xi}_{11} & \boldsymbol{\Xi}_{12} \\ \boldsymbol{\Xi}_{21} & \boldsymbol{\Xi}_{22} \end{bmatrix} := \mathbf{F}_{21}\mathbf{F}_{11}^{-1}\mathbf{F}_{12} = \begin{bmatrix} \sum_{k=1}^{N}\boldsymbol{\Delta}_k\mathbf{H}_k^{-1}\boldsymbol{\Delta}_k^{\mathrm{T}} & \sum_{k=1}^{N}\boldsymbol{\Delta}_k\mathbf{H}_k^{-1}\mathbf{E}_k^{\mathrm{T}} \\ \sum_{k=1}^{N}\mathbf{E}_k\mathbf{H}_k^{-1}\boldsymbol{\Delta}_k^{\mathrm{T}} & \sum_{k=1}^{N}\mathbf{E}_k\mathbf{H}_k^{-1}\mathbf{E}_k^{\mathrm{T}} \end{bmatrix}$$

and

$$\boldsymbol{\Psi} := \mathbf{F}_{22} - \mathbf{F}_{21}\mathbf{F}_{11}^{-1}\mathbf{F}_{12} = \begin{bmatrix} \boldsymbol{\Gamma} & \boldsymbol{\Lambda}^{\mathrm{T}} \\ \boldsymbol{\Lambda} & \boldsymbol{\gamma} \end{bmatrix} - \begin{bmatrix} \boldsymbol{\Xi}_{11} & \boldsymbol{\Xi}_{12} \\ \boldsymbol{\Xi}_{21} & \boldsymbol{\Xi}_{22} \end{bmatrix} \tag{6.58}$$

so that

$$\left(\mathbf{F}_N^{-1}\right)_{11} = \begin{bmatrix} \mathbf{H}_1^{-1} & & \mathbf{0} \\ & \ddots & \\ \mathbf{0} & & \mathbf{H}_N^{-1} \end{bmatrix} + \begin{bmatrix} \mathbf{H}_1^{-1}\boldsymbol{\Delta}_1^{\mathrm{T}} & \mathbf{H}_1^{-1}\mathbf{E}_1^{\mathrm{T}} \\ \vdots & \vdots \\ \mathbf{H}_N^{-1}\boldsymbol{\Delta}_N^{\mathrm{T}} & \mathbf{H}_N^{-1}\mathbf{E}_N^{\mathrm{T}} \end{bmatrix} \times \boldsymbol{\Psi}^{-1} \begin{bmatrix} \boldsymbol{\Delta}_1\mathbf{H}_1^{-1}, \ldots, \boldsymbol{\Delta}_N\mathbf{H}_N^{-1} \\ \mathbf{E}_1\mathbf{H}_1^{-1}, \ldots, \mathbf{E}_N\mathbf{H}_N^{-1} \end{bmatrix} \tag{6.59}$$

$$\left(\mathbf{F}_N^{-1}\right)_{12} = -\begin{bmatrix} \mathbf{H}_1^{-1}\boldsymbol{\Delta}_1^{\mathrm{T}} & \mathbf{H}_1^{-1}\mathbf{E}_1^{\mathrm{T}} \\ \vdots & \vdots \\ \mathbf{H}_N^{-1}\boldsymbol{\Delta}_N^{\mathrm{T}} & \mathbf{H}_N^{-1}\mathbf{E}_N^{\mathrm{T}} \end{bmatrix} \boldsymbol{\Psi}^{-1} \tag{6.60}$$

$$\left(\mathbf{F}_N^{-1}\right)_{21} = -\boldsymbol{\Psi}^{-1} \begin{bmatrix} \boldsymbol{\Delta}_1\mathbf{H}_1^{-1}, \ldots, \boldsymbol{\Delta}_N\mathbf{H}_N^{-1} \\ \mathbf{E}_1\mathbf{H}_1^{-1}, \ldots, \mathbf{E}_N\mathbf{H}_N^{-1} \end{bmatrix} \tag{6.61}$$

$$\left(\mathbf{F}_N^{-1}\right)_{22} = \boldsymbol{\Psi}^{-1} = \left(\begin{bmatrix} \boldsymbol{\Gamma} & \boldsymbol{\Lambda}^{\mathrm{T}} \\ \boldsymbol{\Lambda} & \boldsymbol{\gamma} \end{bmatrix} - \begin{bmatrix} \boldsymbol{\Xi}_{11} & \boldsymbol{\Xi}_{12} \\ \boldsymbol{\Xi}_{21} & \boldsymbol{\Xi}_{22} \end{bmatrix} \right)^{-1} \tag{6.62}$$

We assumed that the $\mathbf{H}_k$ are invertible, which would follow from the independence condition on the array manifold and the independence of the sources.

The CRB on the parameters is given by the diagonal elements of $\mathbf{F}_N^{-1}$. Using the partitioned matrix inversion formula again on (6.62), the CRB for DOAs and amplitudes follows as

$$\begin{aligned} \mathrm{CRB}_N(\boldsymbol{\theta}) &= \mathrm{diag}(\boldsymbol{\Psi}^{-1})_{11} \\ &= \mathrm{diag}[(\boldsymbol{\Gamma} - \boldsymbol{\Xi}_{11}) - (\boldsymbol{\Lambda}^{\mathrm{T}} - \boldsymbol{\Xi}_{12})(\boldsymbol{\Upsilon} - \boldsymbol{\Xi}_{22})^{-1}(\boldsymbol{\Lambda} - \boldsymbol{\Xi}_{21})]^{-1} \end{aligned} \tag{6.63}$$

and

$$\begin{aligned} \mathrm{CRB}_N(\boldsymbol{\beta}) &= \mathrm{diag}(\boldsymbol{\Psi}^{-1})_{22} \\ &= \mathrm{diag}[(\boldsymbol{\Upsilon} - \boldsymbol{\Xi}_{22}) - (\boldsymbol{\Lambda} - \boldsymbol{\Xi}_{21})(\boldsymbol{\Gamma} - \boldsymbol{\Xi}_{11})^{-1}(\boldsymbol{\Lambda}^{\mathrm{T}} - \boldsymbol{\Xi}_{12})]^{-1}. \end{aligned} \tag{6.64}$$

Similarly, the bound on the estimation variance of the signal phases follows as

$$\mathrm{CRB}_N(\boldsymbol{\phi}(k)) = \mathrm{diag}\left\{\mathbf{H}_k^{-1}\left[\mathbf{I} + [\boldsymbol{\Delta}_k^{\mathrm{T}} \ \ \mathbf{E}_k^{\mathrm{T}}]\boldsymbol{\Psi}^{-1}\begin{bmatrix}\boldsymbol{\Delta}_k \\ \mathbf{E}_k\end{bmatrix}\mathbf{H}_k^{-1}\right]\right\}. \tag{6.65}$$

Note that the number of samples and the quality of DOA estimation affects the bound only through the matrix $\boldsymbol{\Psi}^{-1}$.

6.8.4 CM-DOA Algorithm Based on ACMA Initialization

To estimate the parameters of the model, we have to minimize the negative log-likelihood function (6.54), which is a least squares problem:

$$\min \|\mathbf{X} - \mathbf{A}(\boldsymbol{\theta})\mathbf{B}(\boldsymbol{\beta})\mathbf{S}(\boldsymbol{\Phi})\|_F^2.$$

In spite of the simple appearance, it cannot be solved in closed form. A simple technique in such cases is to revert to alternating least squares types of algorithms (as in some of the preceding sections): estimate $\mathbf{S}(\boldsymbol{\Phi})$, then estimate $\mathbf{A}(\boldsymbol{\theta})$, and so on. Based on this idea, the following ad-hoc technique gives surprisingly good results:

1. Blindly estimate a matrix $\hat{\mathbf{A}}$, using the CM assumption on $\mathbf{S}$. This step can be done by ACMA;
2. Estimate the directions which best fit the matrix $\hat{\mathbf{A}}$.

The second step can be carried out for each column of $\hat{\mathbf{A}}$ separately: let $\hat{\mathbf{A}} = [\hat{\mathbf{a}}_1, \ldots, \hat{\mathbf{a}}_d]$ be the estimate of the mixing matrix, then solve for each source i

$$\hat{\theta}_i = \arg\min_{\theta,\beta} \|\hat{\mathbf{a}}_i - \mathbf{a}(\theta)\beta\|^2$$

This problem can be decoupled. The optimal value for β is $\hat{\beta} = \mathbf{a}(\theta)^{\dagger}\hat{\mathbf{a}}_i$, and after eliminating β, the estimate for θ_i is given by

$$\hat{\theta}_i = \arg\min_{\theta} \|(\mathbf{I} - \mathbf{a}(\theta)\mathbf{a}(\theta)^{\dagger})\hat{\mathbf{a}}_i\|$$

which can be converted into

$$\hat{\theta}_i = \arg\max_{\theta} \frac{|\hat{\mathbf{a}}_i^{\mathrm{H}}\mathbf{a}(\theta)|}{\|\mathbf{a}(\theta)\|}. \tag{6.66}$$

Equation (6.66) simply describes a maximization of the projection of the estimated vector onto the array manifold. The computational complexity of this method is not very large compared to other DOA estimation methods, and can be ignored in view of the complexity of the first step.

Various other suboptimal methods for combining existing DOA and CM estimators have been proposed. For example, in [66] the ESPRIT algorithm for DOA estimation is suboptimally combined with ACMA. The problem in such an approach is the choice of weighting of the two properties: without a good weighting, the solution may be worse than the best single method by itself, and finding the proper weighting is a hard problem. Often, the simple two-step approach gives equally good results. Another advantage of the simple CM-DOA method is that it is applicable to arbitrary array configurations: it does not use the special array structure required by ESPRIT.

6.8.5 Deterministic Maximum Likelihood Techniques

In general, maximum likelihood techniques are more complex but are expected to give better results. A deterministic ML approach for DOA estimation of CM signals was derived in [63]. It is based on the least squares formulation of the log likelihood in equation (6.54):

$$\hat{\boldsymbol{\rho}} = \arg\min_{\boldsymbol{\rho}} \sum_{k=1}^{N} \|\mathbf{e}(k)\|^2 \tag{6.67}$$

The Newton scoring method (where we replace the Hessian by its expected value for better numerical performance) to find the minimum is the iteration

$$\boldsymbol{\rho}^{(n+1)} = \boldsymbol{\rho}^{(n)} - \lambda \mathbf{F}_N^{-1}\big(\boldsymbol{\rho}^{(n)}\big)\nabla\mathcal{L}\big(\boldsymbol{\rho}^{(n)}\big)$$

where $\lambda \leq 1$ is a suitable step size, $\mathbf{F}_N$ is the FIM, which is the expected value of the Hessian, and $\nabla\mathcal{L}(\boldsymbol{\rho}^{(n)})$ is the gradient, with components $\nabla_{\boldsymbol{\phi}(k)}\mathcal{L}$, $\nabla_{\boldsymbol{\theta}}\mathcal{L}$, and $\nabla_{\boldsymbol{\beta}}\mathcal{L}$. These are given by

$$\nabla_{\boldsymbol{\phi}(k)}\mathcal{L} = \frac{\partial\mathcal{L}}{\partial\boldsymbol{\phi}(k)} = \frac{2}{\nu}\,\mathrm{Im}\big(\mathbf{S}(k)^{\mathrm{H}}\mathbf{B}^{\mathrm{H}}\mathbf{A}^{\mathrm{H}}\mathbf{e}(k)\big). \tag{6.68}$$

$$\nabla_{\boldsymbol{\theta}}\mathcal{L} = \frac{\partial\mathcal{L}}{\partial\boldsymbol{\theta}} = \frac{2}{\nu}\sum_{k=1}^{N}\mathrm{Re}\big(\mathbf{S}^{\mathrm{H}}(k)\mathbf{B}^{\mathrm{H}}\mathbf{D}^{\mathrm{H}}\mathbf{e}(k)\big) \tag{6.69}$$

and

$$\nabla_{\boldsymbol{\beta}}\mathcal{L} = \frac{\partial\mathcal{L}}{\partial\boldsymbol{\beta}} = \frac{2}{\nu}\sum_{k=1}^{N}\mathrm{Re}\big(\mathbf{S}^{\mathrm{H}}(k)\mathbf{A}^{\mathrm{H}}\mathbf{e}(k)\big). \tag{6.70}$$

The Newton update direction $\mathbf{v}$ is given by $\mathbf{v} = \mathbf{F}_N^{-1}\nabla\mathcal{L}$. This is a $d(N+2) \times 1$ vector function of the parameters. Explicit expressions for the components of $\mathbf{v}$

are given by:

$$
\begin{aligned}
(\mathbf{v})_{\boldsymbol{\phi}(k)} &= \mathbf{H}_k^{-1}\nabla_{\boldsymbol{\phi}(k)}\mathcal{L} + \left[\mathbf{H}_k^{-1}\boldsymbol{\Delta}_k^{\mathrm{T}}\mathbf{H}_k^{-1}\mathbf{E}_k^{\mathrm{T}}\right]\boldsymbol{\Psi}^{-1} \\
&\quad \times \begin{bmatrix} \sum_{i=1}^{N} \boldsymbol{\Delta}_i \mathbf{H}_i^{-1}\nabla_{\boldsymbol{\phi}(i)}\mathcal{L} - \nabla_{\boldsymbol{\theta}}\mathcal{L} \\ \sum_{i=1}^{N} \mathbf{E}_i \mathbf{H}_i^{-1}\nabla_{\boldsymbol{\phi}(i)}\mathcal{L} - \nabla_{\boldsymbol{\beta}}\mathcal{L} \end{bmatrix} \\
(\mathbf{v})_{\boldsymbol{\theta},\boldsymbol{\beta}} &= \Psi^{-1}\begin{bmatrix} \nabla_{\boldsymbol{\theta}}\mathcal{L} - \sum_{i=1}^{N} \boldsymbol{\Delta}_i \mathbf{H}_i^{-1}\nabla_{\boldsymbol{\phi}(i)}\mathcal{L} \\ \nabla_{\boldsymbol{\beta}}\mathcal{L} - \sum_{i=1}^{N} \mathbf{E}_i \mathbf{H}_i^{-1}\nabla_{\boldsymbol{\phi}(i)}\mathcal{L} \end{bmatrix}
\end{aligned}
\tag{6.71}
$$

where $(\mathbf{v})_{\boldsymbol{\phi}(k)}$ are the components related to the phase parameters, and $(\mathbf{v})_{\boldsymbol{\theta},\boldsymbol{\beta}}$ are the components related to the DOAs and signal power parameters. Note that the matrices $\mathbf{H}_k$ are of size $d \times d$ and therefore their inversion is simple, and that there is no dependence of $\mathbf{v}$ on the noise variance ν.

We thus arrive at the algorithm in Figure 6.18 [63]. The step size parameter λ in the algorithm is selected such that it minimizes the likelihood function, and can be obtained by a standard one-dimensional optimization method (cf. [67]), using the good initialization $\lambda_0 = 1$ which is optimal for the quadratic approximation of the likelihood.

The computational complexity of an update step of the scoring algorithm can be estimated as $O((d^3 + Md)N)$ [63], and can be reduced by optimizing the order of operations in the computation. This puts the algorithm in the class of moderate complexity algorithms: harder than eigenspace methods or polynomial rooting algorithms, but lower than multidimensional search methods. Since it can be used with an arbitrary array geometry, it is appealing in many cases where specific assumptions on the array geometry are not valid.

6.8.6 Simulation Results

We finish the section by some simulations demonstrating the efficiency of the CM-DOA estimation methods. We have used an $M = 7$-element uniform linear array (ULA), and $d = 3$ sources with varying angle separations.

The first experiment tests the performance as function of the input signal to noise ratio (SNR). The three sources had equal power and were located at $-5°$, $0°$, $5°$, the number of samples was $N = 30$, and the SNR was varied from 5 to 50 dB. Figure 6.19 shows the results for various techniques: ACMA followed by a one-dimensional DOA

1. Find an initial estimate $\boldsymbol{\rho}_0$ for the parameters using the suboptimal algorithm in Section 6.8.4.
2. For $n = 0, 1, \ldots$ until convergence do
   a. Estimate the Newton direction $\mathbf{v}$ using (6.71)
   b. Compute λ by $\lambda = \arg\min_{\mu} -\mathcal{L}(\boldsymbol{\rho}^{(n)} - \mu\mathbf{v})$
   c. Update the parameters using $\boldsymbol{\rho}_{n+1} = \boldsymbol{\rho}^{(n)} - \lambda\mathbf{v}$

   end

Figure 6.18 Deterministic ML DOA estimation and source separation for constant modulus sources.

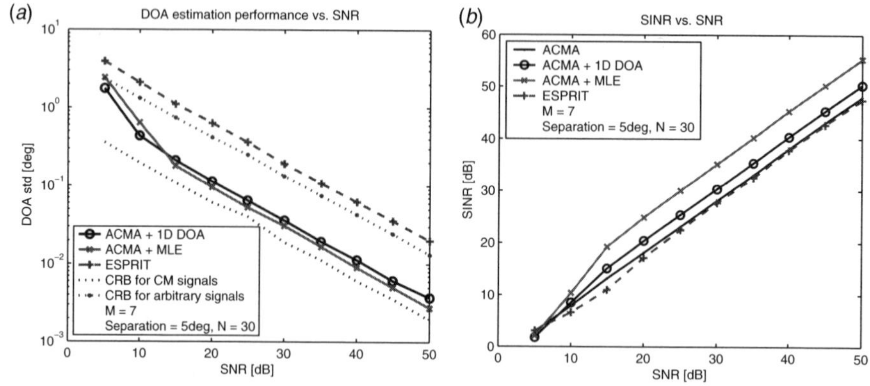

Figure 6.19 First experiment: (*a*) DOA estimation accuracy versus SNR, (*b*) SINR versus SNR.

search as in Section 6.8.4, the same technique followed by the ML estimation as described in Figure 6.18, and (in dashed lines) the ESPRIT algorithm [68] which uses the ULA structure. As seen in Figure 6.19, the use of the CM property gives an order of magnitude improvement in the DOA estimate. In terms of output SINR, ACMA and ESPRIT are about equal, but the MLE is 3–5 dB better. The number of iterations required for the MLE was about five for large SNRs, and 15–20 for smaller SNRs (10 dB and below).

A second experiment tests the performance as a function of the angle separation between the sources, in a case with near–far problems. We have used $N = 40$ samples. The central source is fixed at 0° while the two other sources are located at $-\Delta^\circ$, Δ°, where Δ is changed from 4° to 30°. The SIR of the central source is at -20 dB below the strong sources, and the SNR for the weak source is 20 dB. This tests the near–far robustness of the method. Figure 6.20 presents the standard deviation of the DOA estimate and the output SINR for the weak source.

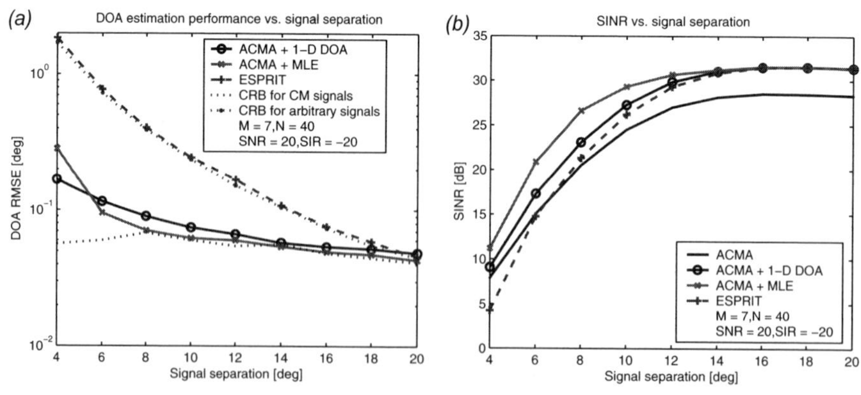

Figure 6.20 Second experiment: (*a*) DOA estimation accuracy versus separation, (*b*) SINR versus separation.

For small separations, the CM-initialized techniques lead to better DOA estimates than ESPRIT. However, for large separations the ESPRIT algorithm tends to have better output SINR performance than ACMA. The MLE outperforms the other techniques, with a significant 3–5 dB advantage at small separations, which demonstrates the importance of exploiting both the array structure and the CM property.

6.9 CONCLUDING REMARKS

In this chapter, we studied algorithms for the blind separation of multiple constant modulus signals. The challenge was to find the complete set of *all* beamformers (one for each impinging signal). For a small batch of samples, ACMA is currently the only algorithm that can reliably do this. For a moving window of samples (sample-adaptive techniques), only algorithms which adaptively prewhiten the data and recondition the orthogonality of the beamformers are reliable. One example is the MUK algorithm. We have derived an adaptive implementation of ACMA which was shown to be more reliable than MUK in a rapidly time-varying scenario, at a similar computational complexity. The robustness and performance of this algorithm in more general cases (e.g., varying number of sources) still needs to be established.

If the aim is to estimate the directions of arrival of the sources, algorithms which also exploit the constant modulus property have been shown to give a significant performance improvement at reasonable computational costs over algorithms which only consider the array manifold (e.g., the ESPRIT algorithm). This is in particular true at small angular separations of the sources.

ACKNOWLEDGMENT

This work was supported in part by the Dutch Min. Econ. Affairs under TSIT 1025 'Beyond-3G,' and by NWO-STW under the VICI program (DTC.5893).

REFERENCES

1. J. Treichler and B. Agee, "A new approach to multipath correction of constant modulus signals," *IEEE Trans. Acoust., Speech, Signal Processing*, 31, 459–471 (1983).
2. R. Gooch and J. Lundell, "The CM array: An adaptive beamformer for constant modulus signals," in *IEEE Intl. Conf. Acoustics, Speech, Signal Processing (ICASSP)* (Tokyo), pp. 2523–2526, 1986.
3. J. Treichler and M. Larimore, "New processing techniques based on constant modulus adaptive algorithm," *IEEE Trans. Acoust., Speech, Signal Processing*, 33, 420–431 (1985).
4. A. van der Veen and A. Paulraj, "An analytical constant modulus algorithm," *IEEE Trans. Signal Processing*, 44, 1136–1155 (1996).

5. J. Cardoso and A. Souloumiac, "Blind beamforming for non-Gaussian signals," *IEE Proc. F (Radar and Signal Processing)*, 140, 362–370 (1993).
6. C. Papadias, "Globally convergent blind source separation based on a multiuser kurtosis maximization criterion," *IEEE Trans. Signal Processing*, 48, 3508–3519 (2000).
7. B. Widrow and S. Stearns, *Adaptive Signal Processing*. Prentice-Hall, Englewood Cliffs, NJ, 1985.
8. M. Larimore and J. Treichler, "Convergence behavior of the constant modulus algorithm," in *IEEE Intl. Conf. Acoustics, Speech, Signal Processing (ICASSP)*, pp. 13–16, Vol. 1, 1983.
9. Y. Sato, "A method of self-recovering equalization for multilevel amplitude-modulation systems," *IEEE Trans. Communications*, 23, 679–682 (1975).
10. D. Godard, "Self-recovering equalization and carrier tracking in two-dimensional data communication systems," *IEEE Trans. Communications*, 28, 1867–1875 (1980).
11. S. Haykin, Ed., *Blind Deconvolution*. Prentice Hall, Englewood Cliffs, NJ, 1994.
12. R. Johnson, P. Schniter, T. Endres, J. Behm, D. Brown, and R. Casas, "Blind equalization using the constant modulus criterion: a review," *Proceedings of the IEEE*, 86, 1927–1950 (1998).
13. J. Treichler, M. Larimore, and J. Harp, "Practical blind demodulators for high-order QAM signals," *Proceedings of the IEEE*, 86, 1907–1926 (1998).
14. K. Hilal and P. Duhamel, "A convergence study of the constant modulus algorithm leading to a normalized-CMA and a block-normalized-CMA," in J. Vandewalle, R. Boite, M. Moonen, and A. Oosterlinck, Eds., *Signal Processing VI–Theories and Applications. Proceedings of EUSIPCO-92, Sixth European Signal Processing Conference* (Brussels), pp. 135–138, Vol. 1, Elsevier, 1992.
15. S. Haykin, *Adaptive Filter Theory*. Prentice-Hall, Englewood Cliffs, NJ, 1991.
16. R. Pickholtz and K. Elbarbary, "The recursive constant modulus algorithm: A new approach for real-time array processing," in *27th Asilomar Conf. Signals, Syst. Comp.*, pp. 627–632, IEEE, 1993.
17. A. Keerthi, A. Mathur, and J. Shynk, "Misadjustment and tracking analysis of the constant modulus array," *IEEE Trans. Signal Processing*, 46, 51–58 (1998).
18. A. Mathur, A. Keerthi, J. Shynk, and R. Gooch, "Convergence properties of the multistage constant modulus array for correlated sources," *IEEE Trans. Signal Processing*, 45, 280–286 (1997).
19. J. Shynk and R. Gooch, "The constant modulus array for cochannel signal copy and direction finding," *IEEE Trans. Signal Processing*, 44, 652–660 (1996).
20. T. Nguyen and Z. Ding, "Blind CMA beamforming for narrowband signals with multipath arrivals," *Int. J. Adaptive Control and Signal Processing*, 12, 157–172 (1998).
21. C. Papadias and A. Paulraj, "A constant modulus algorithm for multiuser signal separation in presence of delay spread using antenna arrays," *IEEE Signal Processing Letters*, 4, 178–181 (1997).
22. O. Shalvi and E. Weinstein, "New criteria for blind deconvolution of non-minimum phase systems (channels)," *IEEE Trans. Information Theory*, 36, 312–321 (1990).
23. G. Golub and C. Van Loan, *Matrix Computations*. The Johns Hopkins University Press, Baltimore, MD, 1989.

24. S. Douglas, "Numerically-robust adaptive subspace tracking using householder transformations," in *First IEEE Sensor Array and Multichannel Signal Processing Workshop* (Boston, MA), pp. 499–503, March 2000.
25. B. Agee, "The least-squares CMA: A new technique for rapid correction of constant modulus signals," in *IEEE Intl. Conf. Acoustics, Speech, Signal Processing (ICASSP)* (Tokyo), pp. 953–956, 1986.
26. R. Gerchberg and W. Saxton, "A practical algorithm for the determination of phase from image and diffraction plane pictures," *Optik*, 35, 237–246 (1972).
27. Y. Wang, Y. Pati, Y. Cho, A. Paulraj, and T. Kailath, "A matrix factorization approach to signal copy of constant modulus signals arriving at an antenna array," in *Proc. 28th Conf. on Informat. Sciences and Systems* (Princeton, NJ), March 1994.
28. A. van der Veen, "Algebraic constant modulus algorithms," in G. Giannakis, Ed., *Signal Processing Advances in Wireless and Mobile Communications*, Chapter 3, Prentice Hall, Englewood Cliffs, NJ, 2000.
29. A. van der Veen, "Asymptotic properties of the algebraic constant modulus algorithm," *IEEE Trans. Signal Processing*, 49, 1796–1807 (2001).
30. A. Leshem, N. Petrochilos, and A. van der Veen, "Finite sample identifiability of multiple constant modulus sources," *IEEE Trans. Information Theory*, 49, 2314–2319 (2003).
31. A. van der Veen, "Statistical performance analysis of the algebraic constant modulus algorithm," *IEEE Trans. Signal Processing*, 50, 3083–3097 (2002).
32. A. van der Veen, "Analytical method for blind binary signal separation," *IEEE Trans. Signal Processing*, 45, 1078–1082 (1997).
33. M. Gu and L. Tong, "Geometrical characterizations of constant modulus receivers," *IEEE Trans. Signal Processing*, 47, 2745–2756 (1999).
34. H. Zeng, L. Tong, and C. Johnson, "Relationships between the constant modulus and Wiener receivers," *IEEE Trans. Inform. Theory*, 44, 1523–1538 (1998).
35. H. Zeng, L. Tong, and C. Johnson, "An analysis of constant modulus receivers," *IEEE Trans. Signal Processing*, 47, 2990–2999 (1999).
36. A. Belouchrani, K. Abed-Meraim, J.-F. Cardoso, and E. Moulines, "A blind source separation technique using second-order statistics," *IEEE Trans. Signal Processing*, 45, 434–444 (1997).
37. P. Binding, "Simultaneous diagonalization of several Hermitian matrices," *SIAM J. Matrix Anal. Appl.*, 4(11), 531–536 (1990).
38. A. Bunse-Gerstner, R. Byers, and V. Mehrmann, "Numerical methods for simultaneous diagonalization," *SIAM J. Matrix Anal. Appl.*, 4, 927–949 (1993).
39. J.-F. Cardoso and A. Souloumiac, "Jacobi angles for simultaneous diagonalization," *SIAM J. Matrix Anal. Appl.*, 17(1), 161–164 (1996).
40. M. Chu, "A continuous Jacobi-like approach to the simultaneous reduction of real matrices," *Lin. Alg. Appl.*, 147, 75–96 (1991).
41. B. Flury and B. Neuenschwander, "Simultaneous diagonalization algorithms with applications in multivariate statistics," in R. Zahar, Ed., *Approximation and Computation*, pp. 179–205, Birkhäuser, Basel, 1995.
42. M. Haardt and J. Nossek, "Simultaneous Schur decomposition of several nonsymmetric matrices to achieve automatic pairing in multidimensional harmonic retrieveal problems," *IEEE Trans. Signal Processing*, 46, 161–169 (1998).

43. L. D. Lathauwer, B. D. Moor, and J. Vandewalle, "Independent component analysis based on higher-order statistics only," in *Proc. IEEE SP Workshop on Stat. Signal Array Processing* (Corfu, Greece), pp. 356–359, 1996.
44. N. Sidiropoulos, G. Giannakis, and R. Bro, "Parallel factor analysis in sensor array processing," *IEEE Trans. Signal Processing*, 48, 2377–2388 (2000).
45. M. Wax and J. Sheinvald, "A least-squares approach to joint diagonalization," *IEEE Signal Processing Letters*, 4, 52–53 (1997).
46. A. Yeredor, "Non-orthogonal joint diagonalization in the least-squares sense with application in blind source separation," *IEEE Trans. Signal Processing*, 50, (2002).
47. B. Yang, "Projection approximation subspace tracking," *IEEE Trans. Signal Processing*, 43, 95–107 (1995).
48. P. Comon and G. Golub, "Tracking a few extreme singular values and vectors in signal processing," *Proc. IEEE*, 78, 1327–1343 (1990).
49. S. Douglas, "Combined subspace tracking, prewhitening, and contrast optimization for noisy blind signal separation," in *Proc. 2nd Int. Workshop Indept. Component Anal. Source Sep.* (Helsinki Finland), pp. 579–584, June 2000.
50. S. Douglas, "Numerically-robust $O(N^2)$ RLS algorithms using least-squares prewhitening," in *IEEE Intl. Conf. Acoustics, Speech, Signal Processing (ICASSP)* (Istanbul, Turkey), pp. 412–415, June 2000.
51. S. Attallah and K. Abed-Meraim, "Fast algorithms for subspace tracking," *IEEE Signal Processing Letters*, 8, 203–206 (2001).
52. K. Abed-Meraim, S. Attallah, A. Chkeif, and Y. Hua, "Orthogonal Oja algorithm," *IEEE Signal Processing Letters*, 7, 116–119 (2000).
53. S. Douglas, S.-Y. Kung, and S. Amari, "A self-stabilized minor subspace rule," *IEEE Signal Processing Letters*, 5, 328–330 (1998).
54. S. Douglas and X. Sun, "A projection approximation minor subspace tracking algorithm," in *9th IEEE DSP Workshop* (Hunt, TX), October 2000.
55. E. Oja, "Principal components, minor components, and linear neural networks," *Neural Networks*, 5, 927–935 (1992).
56. H. Krim and M. Viberg, "Two decades of array signal processing research: The parametric approach," *IEEE Signal Processing Magazine*, 13, 67–94 (1996).
57. H. van Trees, *Optimum Array Processing (Part IV): Detection, Estimation, and Modulation Theory*. Wiley Interscience, New York, 2002.
58. A. Keerthi, A. Mathur, and J. Shynk, "Direction-finding performance of the multistage CMA array," in *28th Asilomar Conf. Signals, Syst. Comp.*, Vol. 2, pp. 847–852, 1994.
59. J. Shynk, A. Keerthi, and A. Mathur, "Steady state analysis of the multistage constant modulus array," *IEEE Trans. Signal Processing*, 44, 948–962 (1996).
60. A. Leshem and A. van der Veen, "Direction-of-arival estimation for constant modulus signals," *IEEE Trans. Signal Processing*, 47 (1999).
61. P. Stoica and O. Besson, "Maximum likelihood DOA estimation for constant-modulus signal," *Electronics Letters*, 36, 849–851 (2000).
62. A. Leshem, "Maximum likelihood separation of phase modulated signals," in *IEEE Intl. Conf. Acoustics, Speech, Signal Processing (ICASSP)*, March 1999.
63. A. Leshem, "Maximum likelihood estimation of constant modulus signals," *IEEE Trans. Signal Processing*, 48, 2948–2952 (2000).

64. B. Sadler, R. Kozick, and T. Moore, "Bounds on bearing and symbol estimation with side information," *IEEE Trans. Signal Processing*, 49, 822–834 (2001).

65. P. Stoica and A. Nehorai, "MUSIC, maximum likelihood, and Cramer–Rao bound," *IEEE Trans. Acoust., Speech, Signal Processing*, 37, 720–743 (1989).

66. A. van der Veen, "Blind source separation based on combined direction finding and constant modulus properties," in *IEEE SP Workshop on Stat. Signal Array Processing* (Portland, OR), September 1998.

67. P. Gill, W. Murray, and M. Wright, *Practical Optimization*. Academic Press, NY, 1981.

68. R. Roy, A. Paulraj, and T. Kailath, "ESPRIT—A subspace rotation approach to estimation of parameters of cisoids in noise," *IEEE Trans. Acoust., Speech, Signal Processing*, 34, 1340–1342 (1986).

7

ROBUST WIDEBAND BEAMFORMING

Elio D. Di Claudio and Raffaele Parisi
INFOCOM Department, University of Roma "La Sapienza," Roma, Italy

"Numquam ponenda est pluralitas sine necessitate (Never assume more without necessity)."

—William of Ockham [1]

7.1 INTRODUCTION

The standard narrowband array model assumes that the array response and the signal and noise spectra are almost constant within the sensor bandwidth [2].

However, in many practical applications involving sensor arrays, such as speech analysis [3], acoustic surveillance [4], seismic prospecting, sonar [5] and ultrasound [6], the *signal of interest* (SOI) is intrinsically wideband, spanning up to several octaves in frequency, and is characterized by a high *spectral variability* [7]. Moreover, if the SOI bandwidth after pass-band pre-filtering of sensor outputs exceeds a few percent of its center frequency, the array signal model in the time domain is not anymore expressed by an instantaneous mixture, but becomes a multichannel convolution [8].

Wideband adaptive beamforming is typically used either to recover an undistorted copy of the SOI radiated by a wavefield source, optimally cleaned from noise and interference [9], or to perform imaging duties [10]. These tasks generally require a rather accurate knowledge of the propagation model and a high computational power to deal with the impressive data rates expected in typical real time operations.

Robust Adaptive Beamforming, Edited by Jian Li and Petre Stoica

In fact the wideband array model is much more complicated than the narrowband one and demands for precise solutions of the wave equations and an extensive database for calibration data [5].

Moreover, the signal propagation is often affected by largely unpredictable perturbations, like sensor calibration errors, mutual coupling [11], multipath, reverberation and scattering [12], that reduce the reliability of prior knowledge of the array response. This happens, for example, in teleconferencing and acoustic surveillance applications, when people enters or exits from a closed room [3].

Classical wideband adaptive processors [7, 13], based on the constrained minimum variance (MV) criterion, suffer from severe cancellation [14, 15] of the SOI at the beamformer output in the presence of even small model mismatches.

This phenomenon can be easily understood with reference to the classical *generalized sidelobe canceller* (GSC) formulation of MV adaptive beamforming [16]. The SOI leaks into the *blocking subspace*, which should be orthogonal to the array response, as a consequence of model mismatches. If this spurious signal becomes comparable to noise and independent interference components at the output of the sidelobe canceller, it predicts (i.e., cancels out) the SOI component passing through the *quiescent beamformer* path [15].

A similar cancellation mechanism arises as a consequence of the finite size of the training data set [17], because estimation errors of the required spatial covariance matrices introduce spurious correlations between the quiescent beamformer path and the sidelobe canceller output. In this regard, the key parameter to minimize is the ratio between the numbers of the adjustable parameters of the beamformer weight vector and of the training snapshots [17].

In wideband environments, the cancellation problem is complicated by the fact that the SOI often exhibits a strong temporal correlation as a consequence of its spectral variability. In the time domain, this means that unmodeled multipath entering into the blocking subspace can cancel out the desired signal if its delay of arrival does not largely exceed the SOI correlation time [14, 18].

Moreover, array calibration errors are strongly amplified near the spectral peaks of the SOI, leading to a further degradation of beamforming performance [19].

Example. The correlation time of unvoiced speech is around 25 ms [20]. The typical reverberation time of a living room is $T_{60} \approx 0.5$ s.[1] Therefore echoes arriving with a delay less than about 0.5 s from the direct (e.g., the shortest) path can be considered as temporally correlated and may induce SOI cancellation.

Since the complex wideband propagation model of this room may require the use of a very long *delay and sum* (DS) beamformer [7] to suppress noise and interference [3], finite sample errors and reverberation can easily produce a strong (and often frequency selective) cancellation of the recorded voice.

[1]The reverberation time T_{60} is defined as the time required before the ensemble averaged instantaneous power of room impulse responses is reduced by 60 dB [12].

The described issue is an example of a likely catastrophic failure mechanism of adaptive beamforming, quite unacceptable in engineering applications. A *robust wideband beamformer* should prevent the occurrence of such a catastrophe as long as possible, while retaining a nearly optimal performance in an ideal environment [21].

Because of the complexity of involved phenomena, an universal solution to robust wideband beamforming may not be easily foreseen. In particular, concepts originally developed in the narrowband case may become inadequate in wideband environments, without proper extensions and modifications. For these reasons and the intrinsic difficulty in handling the huge amounts of data required for modeling and signal processing purposes, robust wideband beamforming received comparatively less attention from researchers than its narrowband counterpart.

Existing robust wideband approaches can be grouped into three main categories:

- Formulation of a proper set of *linear or quadratic constraints* on the weight vector, on the basis on a previous wavefield analysis [13, 22]. Constraints are designed to avoid cancellation or heavy SOI distortion in the presence of specific model mismatches. This approach can be viewed as a direct extension of existing narrowband techniques [15, 23] and its efficacy is strictly tied to the quality of starting assumptions. In particular, these methods do not take into account robustness issues deriving from spectral variability, nonstationarity and non-Gaussianity of signals and noise.
- Formulation of a *random model* for the array response around a baseline wavefield solution predicted by analytical methods or numerical simulation. This approach has been traditionally followed in the field of underwater acoustics [24], where propagation is intrinsically multimodal and depends on uncertain *environmental parameters* [5]. In these scenarios, gross estimation errors and SOI cancellation may result in the presence of unmodeled multipath and reverberation. In particular, the robust *matched-field* (MF) beamforming techniques [5, 24] take into account all of the propagation modes and define suitable constraints and optimal *quiescent weight vectors* [16], in order to statistically minimize performance losses [5, 24].
- Synthesis of wideband beamformers, characterized by a low number of adaptive parameters, such as the *steered adaptive beamformer* (STBF), introduced in [10]. Nonadaptive STBFs were already used in bioengineering [6] and radar and are characterized by the preliminary focusing [25, 26] of the array response with respect to the SOI onto a single steering vector over the entire analysis bandwidth [10, 26]. Adaptive STBFs are known for their capability of efficiently separating multiple incoming wavefronts in the spatial domain, by using a single weight vector for all frequencies of interests. Moreover, they are capable of suppressing multipath and reverberation through a *frequency smoothing* process [10, 14].

In the foregoing a very promising STBF (formerly introduced in [14]) is described. It is based on a new, constrained *stochastic maximum likelihood* (ML) error functional for adapting the weight vector within a GSC formulation [16]. This ML-STBF was derived in the frequency domain within a Gaussian framework, assuming statistically independent SOI, interference and background noise [14]. Even if Gaussian ML estimators are not generally considered as *statistically robust* [21], the resulting ML-STBF exhibits a set of interesting properties, very useful in real world application, that are summarized below:

- The ML-STBF is derived from a manageable and well founded Gaussian model for the frequency transformed array signals [27], largely independent of the SOI probability distribution function (PDF). Most deviations from this assumption can be afforded by classical *robust statistics* [21] tools.
- The ML-STBF delivers a very high rate of independent snapshots in the frequency domain and its weight vector features a minimal number of free parameters, the same as in the narrowband case. Therefore, finite sample errors are minimized, even for very short data records [28].
- The preliminary focusing stage of the wideband SOI, typical of all STBFs [10, 28], is very flexible and can compensate for many kinds of model errors. In particular, focusing techniques inspired to the MF approach [5] will be presented in the foregoing to deal with strongly reverberant environments.
- The ML-STBF weight vector is nearly independent of any prefiltering, simultaneously applied to all sensor outputs, provided that the filter transfer function has not zeroes at any frequency within the analysis bandwidth [14].
- The implicit prewhitening [29] of the beamformer output [14], induced by the ML functional, greatly reduces the negative effects of a temporally correlated SOI. Thanks to the smoothing [26] allowed by the frequency domain formulation of the ML-STBF, reflections can be effectively *decorrelated* from the direct path and nulled out, if their relative delays of arrival are roughly greater than the reciprocal of the analysis bandwidth [18].
- Linear constraints can be freely imposed on the ML-STBF weight vector to increase robustness.
- A norm constraint on the weight vector can be easily incorporated to prevent cancellation in uncertain environments [28].
- Optimization of the ML-STBF can be effectively accomplished by a fast *modified Newton algorithm* [30], originally developed for neural network training [31], which requires the iterative solution of a quadratic *ridge regression* [32] problem. The typical computational cost of the ML-STBF in practical applications is roughly twice the cost of the classical MV-STBF [10], excluding the common data preprocessing and focusing. Since these stages are by far the most demanding ones, the ML-STBF solution can be preferred in many cases for the increased performance and robustness.

This chapter is organized as follows. The wideband array model and the classical beamforming architectures are briefly reviewed in Sections 7.3 and 7.4, respectively. In Section 7.5 the robustness issues of wideband beamforming are stated and analyzed.

The STBF concept is described in Section 7.6. The ML-STBF is developed in detail in the next Section 7.7. Efficient techniques for training the ML-STBF error functional are described in Section 7.8.

Special topics for improving the ML-STBF performance are treated in Section 7.9. They include techniques for reducing the focusing error [25] and for the sound selection of the quiescent weight vector [16] in reverberant environments.

The results of experiments with simulated and real-world data using the ML-STBF are described in Section 7.10. In some cases, the classical MV-STBF [10] was used as a benchmark to point out the performance and the robustness of the ML-STBF in difficult environments. Finally, conclusions are drawn in Section 7.11.

7.2 NOTATION

Throughout the present work, matrices will be indicated by capital boldface letters, and vectors by lower-case boldface letters. Other symbols are as follows:

$(.)^T$	Matrix transpose.
$(.)^H$	Hermitian matrix transpose.
$(.)^*$	Complex conjugate of the argument.
$\|.\|$	Absolute value of a real number, or the modulus of a complex number.
$\|\|.\|\|_2$	L_2 norm of the vector or matrix argument [32].
$\|\|.\|\|_F$	Frobenius norm of the matrix argument [32].
span(.)	The subspace spanned by the columns of the matrix argument.
det(.)	Determinant of the matrix argument.
$\mathbf{I}_m$	$(m \times m)$ identity matrix.
$\mathbf{0}$	Matrix or vector of zeroes.
$\mathbf{1}$	Column vector of 1s.
$E[.]$	Statistical expectation operator.
$u_0(n)$	Discrete time impulse function.
$u_{-1}(n)$	Discrete time step function: $u_{-1}(n) = 1$ for $n \geq 0$ and zero elsewhere.
ω	Angular frequency in continuous time.
i	Imaginary unit.
$\odot$	Schur–Hadamard (element-wise) product of matrices.
$\max\{a_1, a_2, \ldots\}$	The maximum of the argument list.

In the foregoing, sample quantities will be marked by a tilde superscript to avoid any ambiguity.

7.3 WIDEBAND ARRAY SIGNAL MODEL

A sensor array having M sensors receives the SOI $s(t)$ radiated by a *point source*. The source location is described by the generic coordinate vector $\mathbf{p}$ (e.g., azimuth, elevation, range, depth, cartesian coordinates, etc.). The propagating medium and the sensors are assumed linear and generally dispersive.

Under these assumptions, the $(M \times 1)$ *snapshot* vector $\mathbf{x}(t)$, which stacks the sensor outputs at time t, obeys the convolutive model

$$\mathbf{x}(t) = \int_{t_0}^{+\infty} \mathbf{h}(\tau, \mathbf{p}) s(t - \tau) d\tau + \mathbf{v}(t) \tag{7.1}$$

where $\mathbf{h}(\tau, \mathbf{p})$ is the global M-channel impulse response of the array with respect to the SOI, including multipath, t_0 is the wave propagation time from the source to the array and the $(M \times 1)$ vector $\mathbf{v}(t)$ conveniently collects all the interference and background noise statistically independent of $s(t)$. This setting does not impair generality for the beamforming problem [23].

It is initially assumed that $\mathbf{h}(\tau, \mathbf{p})$ is known for all locations of interest and does not change during the observation time. These hypotheses will be relaxed in the foregoing to include different kinds of model mismatches relevant to robustness.

The general model (7.1) can accurately represent most environments encountered in typical array processing applications in the fields of underwater acoustics [24], telecommunications [33], ultrasound [6], multimedia [3] and remote sensing [34].

7.3.1 Discrete Time Model

Array outputs are converted to the baseband and sampled with period T. The *discrete time snapshot* $\mathbf{x}(n)$, for $n = 1, 2, \ldots, N$, under very general assumptions, obeys the M-channel FIR model

$$\mathbf{x}(n) = \sum_{k=N_0}^{N_f} \mathbf{h}(k, \mathbf{p}) s(n - k) + \mathbf{v}(n) \tag{7.2}$$

where the symbol meanings can be directly derived from (7.1) and the impulse response has been considered negligible for $t > N_f T$. However, due to aliasing and sampling issues, the derivation of $\mathbf{h}(n, \mathbf{p})$ from its continuous time counterpart $\mathbf{h}(\tau, \mathbf{p})$ is not straightforward and depends upon the receiver architecture.

Differently from the standard narrowband model [35], the length N_h of the impulse response with respect to the SOI[2] exceeds the sampling period and is at least equal to the transit time of the wavefront through the array. The SOI $s(n)$ and the noise plus interference term $\mathbf{v}(n)$ are assumed, without much loss of

[2] $N_h = N_f - N_0 + 1$ from (7.2).

generality, as realizations of independent, circular, MA processes [29] of maximum order N_s, characterized by zero mean and finite second and fourth order moments.

Under the above hypotheses, the M-channel (spatial) correlation function of $\mathbf{x}(n)$, defined as $\mathbf{R}_{xx}(m) = E[\mathbf{x}(n)\mathbf{x}^H(n+m)]$, is finite and nonzero for $m < N_c = N_h + N_s$. It will be shown in the foregoing that the *snapshot correlation time* N_c plays an important role for the adaptive beamforming performance in multipath fields.

Signal Prewhitening. In some applications, such as speech, the SOI spectrum is characterized by sharp peaks. Moreover, standing waves within closed rooms give origin to highly correlated common modes around a countable set of eigenfrequencies [36]. In these cases, the correlation time N_c can be quite high and might be shortened by a proper multichannel prewhitening of $\mathbf{x}(n)$, based on autoregressive (AR) or ARMA models [3, 29].

However, it is worth to point out that the impulse response $\mathbf{h}(n, \mathbf{p})$ is generated by multiple arrivals, whose relative delays generally are not integer multiples of T, and is therefore hardly invertible by prewhitening, even in the mean square sense.

7.3.2 Frequency Domain Model

The sequence $\mathbf{x}(n)$ is partitioned into L consecutive blocks[3] of length J, that are separately processed by a M-channel, J-point windowed DFT [10].

Under the hypothesis that $J \gg N_c$ it holds that:

- Each *subband snapshot* $\mathbf{x}(\omega_j, l)$ $(l = 1, 2, \ldots, L;\ j = 1, \ldots, J)$, obtained by DFT processing and referred to the angular frequency ω_j, satisfies the *narrowband* model

$$\mathbf{x}(\omega_j, l) = \mathbf{h}(\omega_j, \mathbf{p})s(\omega_j, l) + \mathbf{v}(\omega_j, l) \tag{7.3}$$

  where the array response $\mathbf{h}(\omega_j, \mathbf{p})$ with respect to the SOI component $s(\omega_j, l)$ spans a *rank-one subspace* and is commonly referred to as the *steering vector* [35];
- DFT filtering can isolate single SOI spectral peaks and reverberation eigenfrequencies [36]. Around these critical frequencies the model (7.3) abruptly changes;
- Each $\mathbf{x}(\omega_j, l)$ can be considered as a multivariate, zero mean, circular random variable, independent with respect to both time and frequency indexes. Moreover, under the hypothesis of ergodicity, it is approximately Gaussian distributed [27].

The cross-sensor covariance matrix (CSCM) $\mathbf{R}_{xx}(\omega_j) = E[\mathbf{x}(\omega_j, l)\mathbf{x}^H(\omega_j, l)]$ obeys the single source narrowband model [35]

$$\mathbf{R}_{xx}(\omega_j) = P_s(\omega_j)\mathbf{h}(\omega_j, \mathbf{p})\mathbf{h}^H(\omega_j, \mathbf{p}) + \mathbf{R}_{vv}(\omega_j) \tag{7.4}$$

[3]Blocks can be assumed nonoverlapping to make the statistical analysis easier. In practice a reasonable overlap may be beneficial to reduce the observation time and the statistical impact of single snapshots [21].

where $P_s(\omega_j)$ is the SOI power spectrum at ω_j and $\mathbf{R}_{vv}(\omega_j) = E[\mathbf{v}(\omega_j, l)\mathbf{v}^H(\omega_j, l)]$ is the noise plus interference spatial covariance matrix.

In the following the analysis bandwidth will be supposed to span the DFT bins $\{j_1, j_1 + 1, \ldots, j_2\}$, with $\omega_{j_1} < \omega_{j_1+1} < \cdots < \omega_{j_2}$.

7.3.3 Limitations of Wideband Arrays

In wideband arrays, the model significantly changes with frequency. This fact introduces an upper limitation on the relative bandwidth that can be effectively covered by the array. At the highest frequencies, the average inter-sensor distance may become too large to ensure a nonambiguous sampling of the incident wavefronts. As a consequence, *grating lobes* appear, characterized by the relationship

$$\frac{|\mathbf{h}^H(\omega_j, \mathbf{p}_1)\mathbf{h}(\omega_j, \mathbf{p}_2)|}{\|\mathbf{h}(\omega_j, \mathbf{p}_1)\|_2 \|\mathbf{h}(\omega_j, \mathbf{p}_2)\|_2} \approx 1 \tag{7.5}$$

for distinct source positions $\mathbf{p}_1$ and $\mathbf{p}_2$. Under these conditions, no beamformer can separate the two source signals [37] at ω_j. In general this happens for most array geometries when the minimum intersensor spacing approaches half wavelength [37]. At the lower end of the bandwidth problems arise due to the strong spatial correlation of background noise, which makes the source detection more difficult. In addition, steering vectors for close source positions approach the same ambiguity condition (7.5), because of the reduction of angular resolution [38]. As a further consequence, all steering vectors are numerically confined within a subspace of dimension much smaller than M [37, 39].

These problems limit the effective relative bandwidth of the array to about one octave [28] and can be overcome by the use of compound arrays made up of multiple subarrays. Each subarray has the same geometry but different intersensor spacing and covers a specific subband. An example of this arrangement is the multiring circular array [4].

7.3.4 Multimodal Propagation

In many environments, the propagating wave can be decomposed as the sum of several modes or paths, each carrying a delayed and linearly distorted replica of $s(n)$.

Several models have been developed in the past for describing multipath propagation in specific scenarios, starting from the general properties of the wave equation, subject to the proper boundary conditions.

For example, in radar [34] and telecommunications [33, 40] it is customary to model multipath propagation as the sum of plane (or spherical) waves, originated by specular reflections on large, smooth surfaces and scattering from small objects and rough surfaces.

Propagation within bounded media, such as waveguides, subsoil, shallow water [24] and closed rooms [3] is naturally characterized as the linear combination of spatial eigenfunctions, that form an orthogonal basis for wavefield solutions. Each eigenfunction satisfies the prescribed boundary conditions, is excited by the SOI source and is spatially sampled by the array sensors [5].

In all these cases, the discrete time array impulse response can be written in the form

$$\mathbf{h}(n, \mathbf{p}) = \sum_{q=1}^{Q} A_q(\mathbf{p})\mathbf{h}_q(n, \mathbf{p})u_{-1}(n - N_q) \tag{7.6}$$

for $n = 0, 1, \ldots, N_f$, where $\mathbf{h}_q(n, \mathbf{p})$ is the impulse response associated to the generic qth mode ($q = 1, 2, \ldots, Q$) and starting at time $N_q \geq N_0$, and $A_q(\mathbf{p})$ is the corresponding mode amplitude coefficient.

The frequency domain counterpart of (7.6) is given by[4]

$$\mathbf{h}(\omega_j, \mathbf{p}) = \mathbf{H}(\omega_j, \mathbf{p})\mathbf{a}(\mathbf{p}) \tag{7.7}$$

where the columns of the $(M \times Q)$ matrix $\mathbf{H}(\omega_j, \mathbf{p})$ are the steering vectors $\mathbf{h}_q(\omega_j, \mathbf{p})$ ($q = 1, \ldots, Q$) of each mode [5] and $\mathbf{a}(\mathbf{p}) = [A_1(\mathbf{p}), \ldots, A_Q(\mathbf{p})]^T$ is the $(Q \times 1)$ vector collecting mode amplitudes.

In the present framework, array responses are obtained by means of analytical calculus or empirical measurements, collected either in a separate calibration phase or on line, with the aid of a proper training sequence [33].

Because of medium inhomogeneity and internal losses, sensor mismatches, imperfect specification of boundary conditions, environment nonstationarity and numerical or estimation errors, (7.6) and (7.7) are inherently characterized by an intrinsic limited accuracy. Therefore, it is natural to develop appropriate error models for each scenario within a probabilistic framework and analyze the corresponding impact on beamforming performance. In the following, some of most useful settings are briefly reviewed.

1. *Perturbed steering vector.* In this approach, the nominal steering vector in (7.3) is perturbed as

$$\tilde{\mathbf{h}}(\omega_j, \mathbf{p}) = \mathbf{h}(\omega_j, \mathbf{p}) + \mathbf{h}_p(\omega_j) \tag{7.8}$$

being $\mathbf{h}_p$ a random $(M \times 1)$ vector, with $\|\mathbf{h}_p(\omega_j)\|_2 \ll \|\mathbf{h}(\omega_j, \mathbf{p})\|_2$ [14, 41].

This model is convenient for developing universal robust beamforming strategies, based on the optimization of the worst case [14] performance, and can cope with some kinds of nonstationarities [41].

However, it is worth noting that the direction of $\tilde{\mathbf{h}}(\omega_j, \mathbf{p})$ is mainly relevant for beamforming performance, while norm changes and *phase rolling* of the

[4]This will be the preferred expansion in the foregoing, since wave equations are commonly and conveniently solved in frequency by both analytical and numerical techniques [24].

steering vector (e.g., global phase rotations of the type $\tilde{\mathbf{h}}(\omega_j, \mathbf{p}) = \mathbf{h}(\omega_j, \mathbf{p})e^{i\theta}$) are interpreted within (7.3) as linear distortions (or prefiltering) of the received SOI. These distortions are hardly identifiable after beamforming.

2. *Fading channel.* This modeling is very common in telecommunications to describe moving sources and channel nonstationarities (e.g., *shadowing*) [42] and assumes that the elements of $\mathbf{h}(\omega_j, \mathbf{p})$ are complex random variables, statistically independent of the SOI and $\mathbf{v}(\omega_j, l)$, and characterized by a prescribed joint PDF $p_{\mathbf{h}}(\mathbf{h})$ (e.g., Rayleigh). Effects of random fluctuations on (7.7) are different if they take place within each observation time (*fast fading*) or among different acquisitions (*slow fading*).

   In the first case, under the hypothesis of ergodicity, the CSCM (7.4) becomes

$$\begin{aligned}\mathbf{R}_{xx}(\omega_j) &= P_s(\omega_j)E[\mathbf{h}(\omega_j, \mathbf{p})\mathbf{h}^H(\omega_j, \mathbf{p})] + \mathbf{R}_{vv}(\omega_j) \\ &= P_s(\omega_j)\mathbf{R}_{hh}(\omega_j, \mathbf{p}) + \mathbf{R}_{vv}(\omega_j)\end{aligned} \tag{7.9}$$

   where $\mathbf{R}_{hh}(\omega_j, \mathbf{p})$ is the ensemble $(M \times M)$ Hermitian *spatial correlation matrix* of $\mathbf{h}(\omega_j, \mathbf{p})$[5] at ω_j. It is evident that $\mathbf{R}_{hh}(\omega_j, \mathbf{p})$ is characterized by a *numerical rank*[6] greater than one.

   In the slow fading case, a different $\mathbf{h}(\omega_j, \mathbf{p})$ holds at each acquisition and the model essentially coincides with the perturbed steering vector one.

3. *Random environment.* In this setting, common in the field of underwater acoustics [24, 43], $\mathbf{h}(\omega_j, \mathbf{p})$ is dependent upon a random vector $\mathbf{b}$ of uncertain environmental parameters (e.g., sound speed, temperature, bottom depth and reflectivity, etc.), characterized by a prescribed joint PDF $p_{\mathbf{b}}(\mathbf{b})$. Therefore, for each value of $\mathbf{b}$, a different $\mathbf{h}(\omega_j, \mathbf{p}, \mathbf{b})$ is obtained. This mapping is generally non-linear and generates a ball of responses when $\mathbf{b}$ fluctuates around its nominal value $\mathbf{b}_0$ [24].

   The resulting CSCM model is similar to (7.9), with

$$\mathbf{R}_{hh}(\omega_j, \mathbf{p}) = \int_{\mathbf{b}\in B(\mathbf{b}_0)} \mathbf{h}(\omega_j, \mathbf{p}, \mathbf{b})\mathbf{h}^H(\omega_j, \mathbf{p}, \mathbf{b})p_{\mathbf{b}}(\mathbf{b})d\mathbf{b} \tag{7.10}$$

   where $B(\mathbf{b}_0)$ represents the environmental parameter space. An estimate of $\mathbf{R}_{hh}(\omega_j, \mathbf{p})$ can be obtained by running a numerical wavefield simulator for many random values of $\mathbf{b}$ in a Monte Carlo fashion and numerically approximating the integral (7.10) [5, 24].

   A representative steering vector of the propagation can be chosen as a scaled replica of the dominant eigenvector of $\mathbf{R}_{hh}(\omega_j, \mathbf{p})$ [5]. Other reasonable solutions can be envisaged by analyzing the a posteriori localization performance of the array, in terms of the ambiguity function (7.5), the sidelobe

[5]It is in general $E[\mathbf{h}(\omega_j, \mathbf{p})] \neq \mathbf{0}$

[6]The numerical rank of a matrix is given by the number of its nonnegligible singular values [32].

level [24], the Cramer–Rao lower bound (CRB) [44] and the Barankin bound [45].

A distinct property of the model (7.10) is the capability of locating the SOI source in 3-D using a single array, thanks to the contributes of several modes [5]. However, the resulting beampatterns (7.5) are often characterized by high sidelobes that make Bartlett beamforming critical [24]. For this reason and because of uncertainties, robust adaptive beamformers are highly desirable in ocean acoustics [5]. In addition, simplified random array models are interesting for closed-room multimedia applications, as it will be shown in Section 7.10.

4. *Partially known response.* This model assumes that the direct path propagation is reasonably well modeled by analytical techniques or can be empirically calibrated in controlled environments (e.g., free space, anechoic chambers). In particular, the time separation among the direct path impulse response, referred to as $\mathbf{h}_d(n, \mathbf{p})$ ($N_0 \leq n \leq N_d$), and the early reflections is assumed much larger than T [14].

   Reflections and reverberation components are instead hard to model in time-varying environments [3]. The corresponding impulse response $\mathbf{h}_r(n, \mathbf{p})$ for $N_r \leq n \leq N_f$ (with $N_r > N_d + 1$) is considered as deterministic, but unknown, and is characterized only by generic, global parameters, such as the total energy and the overall length [14].

   The time and frequency domain models corresponding to the above assumptions are respectively

$$\mathbf{h}(n, \mathbf{p}) = \mathbf{h}_d(n, \mathbf{p})u_{-1}(n - N_0) + \mathbf{h}_r(n, \mathbf{p})u_{-1}(n - N_r) \tag{7.11}$$

$$\mathbf{h}(\omega_j, \mathbf{p}) = \mathbf{h}_d(\omega_j, \mathbf{p}) + \mathbf{h}_r(\omega_j, \mathbf{p}) \tag{7.12}$$

   A possible robust beamforming strategy assumes the direct path as the *reference model* and attempts to bound in a statistical sense the adverse effects of unpredicted multipath [14, 46].

   The partitioning between $\mathbf{h}_d(n, \mathbf{p})$ and $\mathbf{h}_r(n, \mathbf{p})$ is somewhat arbitrary in some applications [46]. Depending upon the degree of uncertainty in estimating the various paths, it may be convenient to include some of them in $\mathbf{h}_d(n, \mathbf{p})$, thus approaching the random response model.

   The partially known response approach was derived from robust regression techniques [21] and requires a minimal computational effort for wavefield analysis or simulation. Its usefulness is tied to the capability of adaptive beamforming of separating incoming wavefronts in time and space, which likely requires large arrays and analysis bandwidths [14].

7.4 WIDEBAND BEAMFORMING

Adaptive wideband beamforming attempts to recover an undistorted copy of the SOI through a linear, convolutive operation applied to array outputs, either in time [7] or in frequency [13]. The *distortionless response* (DR) constraint is very important in several wideband applications, such as speech processing [3], high fidelity audio and radar [34]. In fact, SOI waveform distortion may impair matched filtering, detection and pattern recognition operations after beamforming.

It is worth to remark that distortion does not affect narrowband beamformers, since they are characterized by unique steering and weight vectors for the entire bandwidth [16], that merely produce a phase shift on the recovered SOI.

Another subtle performance issue in wideband applications is related to the spectrum of noise and interference residuals, measured at the beamformer output. In fact, classical performance indexes, like the *mean square error* (MSE) and the signal-to-noise ratio (SNR) may be inadequate for applications, if applied to the entire analysis bandwidth, as shown by the following example.

Example. A wideband SOI, embedded in colored noise, is characterized by a uniform, normalized power spectral density (PSD) of 30 dB within the 20 percent of the analysis bandwidth, referred to as Region 1, and by a PSD of 0 dB elsewhere (Region 2). The global source power is $0.2 \times 1000.0 + 0.8 \times 1.0 = 200.8$. Array outputs are applied to two DR beamformers (indicated as A and B), characterized by the same output power of 2.008 due to noise and interference residuals and, therefore, by the identical average SNR of 20 dB.

Beamformer A exhibits a residual noise PSD of 10 dB within Region 1 and -20 dB in Region 2 ($0.2 \times 10.0 + 0.8 \times 0.01 = 2.008$). Beamformer B has a noise PSD of 7.81 dB in Region 1 and 0 dB in Region 2 ($0.2 \times 6.04 + 0.8 \times 1.0 = 2.008$).

In particular, beamformer B achieves a local SNR of 22.19 dB within Region 1 and only 0 dB in Region 2. As a result, the SOI is actually lost over the 80 percent of the analysis bandwidth. In contrast, beamformer A exhibits an uniform local SNR of 20 dB over all the bandwidth, which appears adequate for masking noise in most applications.

The local SNR behavior of adaptive beamformers strongly depends upon the architecture and the error functional and will be analyzed in the following sections.

7.4.1 Time Domain Wideband Beamforming

The classical time domain wideband beamformer linearly combines delayed versions of $\mathbf{x}(n)$ [7] to get the SOI estimate $y(n, \{\mathbf{w}(k)\})(k = 0, 1, \ldots, D)$, according to the equation

$$y(n, \{\mathbf{w}(k)\}) = \sum_{k=0}^{D} \mathbf{w}^H(k)\mathbf{x}(n - k) \tag{7.13}$$

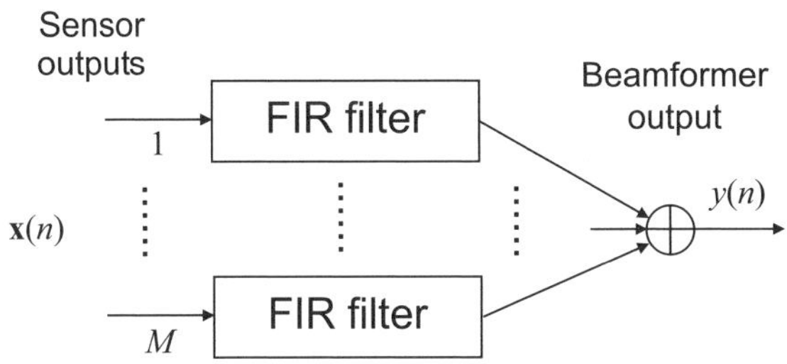

Figure 7.1 Architecture of the DS beamformer.

for $n = 1, 2, \ldots, N$, where $\mathbf{w}(k)$s are $(M \times 1)$ weight vectors. After introducing the $(M(D+1) \times 1)$ *wideband snapshot*

$$\mathbf{x}_{ST}(n) = [\mathbf{x}^T(n), \mathbf{x}^T(n-1), \ldots, \mathbf{x}^T(n-D)]^T \tag{7.14}$$

and the $(M(D+1) \times 1)$ *wideband weight vector*

$$\mathbf{w} = [\mathbf{w}^T(0), \mathbf{w}^T(1), \ldots, \mathbf{w}^T(D)]^T \tag{7.15}$$

equation (7.13) can be compactly rewritten as [7]

$$y(n, \mathbf{w}) = \mathbf{w}^H \mathbf{x}_{ST}(n) \tag{7.16}$$

for $n = 1, 2, \ldots, N$. The architecture of this delay and sum (DS) beamformer is depicted in Figure 7.1.

The MV-DS beamformer solves the constrained least squares (LS) [32] optimization problem

$$\tilde{\mathbf{w}} = \arg\min_{\mathbf{w}} \frac{1}{N-D} \sum_{n=1}^{N-D} |y(n, \mathbf{w})|^2 \tag{7.17}$$

subject to the DR constraint

$$\sum_{k=0}^{D} \mathbf{w}^H(k)\mathbf{h}(n-k, \mathbf{p}) = u_0(n - N_t) \tag{7.18}$$

where $D \gg N_h$ and $N_t \geq N_0$ is a prespecified target delay. It is worth to remark that the numerical algorithms for (7.13), (7.17), and (7.18) work on very large block-Hankel (or block-Toeplitz) matrices and can be accelerated by the use of the FFT [29].

In particular, the solution of (7.18) is rarely exact, since the array impulse response may not be invertible. Moreover, the corresponding system matrix usually is *rank deficient* and requires a reduced rank, SVD-based LS solution [32].[7]

[7] As a rule of thumb, the rank of the system matrix of (7.18) can be assumed equal to N_h and hence depends upon $\mathbf{p}$.

In addition, after imposing this constraint, not all solutions are admissible for $\mathbf{w}$ [32]. In particular, indicating with $\mathbf{w}_0$ a particular solution of (7.18), usually the minimum norm one [32], all admissible weight vectors are obtained as

$$\mathbf{w} = \mathbf{w}_0 + \mathbf{w}_1 \tag{7.19}$$

where $\mathbf{w}_1 = [\mathbf{w}_1^T(0), \ldots, \mathbf{w}_1^T(D)]^T$ satisfies

$$\sum_{k=0}^{D} \mathbf{w}_1^H(k)\mathbf{h}(n-k, \mathbf{p}) \approx 0 \tag{7.20}$$

for all n. This is a particular and numerically approximated case of the GSC formulation [16]. The huge size and the intrinsically ill-posed nature of DS beamforming require dedicated iterative algorithms, such as the Lanczos's ones [32].

For the same reasons, it is difficult to incorporate in the DS beamformer linear and quadratic constraints [15, 22] and statistically optimal error functionals [14]. In particular, the PDF of $y(n, \mathbf{w})$ is essentially determined by the SOI distribution when the beamformer is pointed to the wavefield source. Since wideband SOIs are generally non-Gaussian and temporally correlated [3], the unweighted MV functional (7.17) can rarely be considered as optimal in the ML sense and it is well known that approaches based on the online estimation of the actual SOI PDF are characterized by very slow statistical convergence and a general lack of robustness with respect to nonstationarities [21].

In conclusion, the DS beamformer does not appear as a good basis for developing robust wideband beamforming strategies. Better solutions may be envisaged in the frequency domain.

7.4.2 Frequency Domain Wideband Beamforming

In classical frequency domain wideband beamforming (FDBF) [38], a different $(M \times 1)$ *subband weight vector* $\mathbf{w}_j$ $(j = j_1, \ldots, j_2)$ is used for each DFT bin to get the estimate $y(\omega_j, l, \mathbf{w}_j)$ $(l = 1, \ldots, L)$ of the SOI component $s(\omega_j, l)$ as

$$y(\omega_j, l, \mathbf{w}_j) = \mathbf{w}_j^H \mathbf{x}(\omega_j, l) \tag{7.21}$$

The overall architecture of a frequency domain beamformer is depicted in Figure 7.2.

Since DFT outputs $\mathbf{x}(\omega_j, l)$ are assumed Gaussian and statistically independent across frequency bins and blocks within the model (7.3), the ML-FDBF coincides with the *linearly constrained MV beamformer* (LCMV) [23], which solves independently for each ω_j the LS problem

$$\tilde{\mathbf{w}}_j = \arg\min_{\mathbf{w}_j} \frac{1}{L} \sum_{l=1}^{L} \left| y(\omega_j, l, \mathbf{w}_j) \right|^2 \tag{7.22}$$

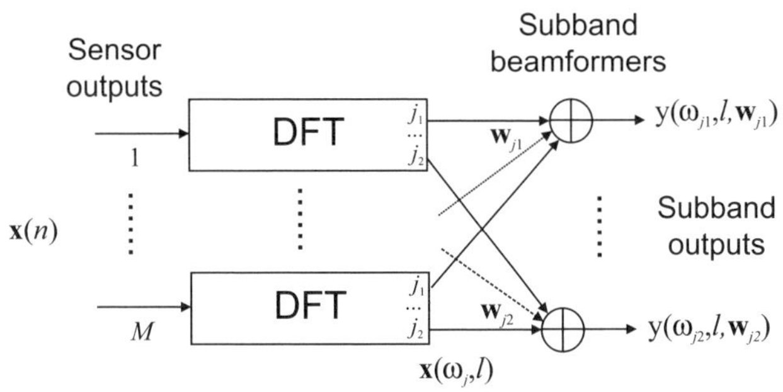

Figure 7.2 Architecture of the classical frequency domain beamformer.

subject to $1 \leq M_c < M$ linear constraints, compactly expressed as

$$\mathbf{C}^H(\omega_j, \mathbf{p})\mathbf{w}_j = \mathbf{f}(\omega_j, \mathbf{p}) \tag{7.23}$$

being $\mathbf{f}(\omega_j, \mathbf{p})$ a prespecified $(M_c \times 1)$ target vector [13, 38]. In the foregoing, this wideband LCMV beamformer will be referred to as the MV-FDBF.

In particular, the DR condition (7.18) with respect to the SOI is rewritten in the frequency domain as

$$\mathbf{w}_j^H \mathbf{h}(\omega_j, \mathbf{p}) = e^{-i\omega_j N_t} \tag{7.24}$$

for $j = j_1, \ldots, j_2$.

In this work, for the sake of clarity, the GSC formulation of (7.22) is adopted, which decomposes $\mathbf{w}_j$ as

$$\mathbf{w}_j = \mathbf{w}_{0,j} + \mathbf{C}_\perp(\omega_j, \mathbf{p})\mathbf{w}_{1,j} \tag{7.25}$$

where the $(M \times 1)$ *quiescent weight vector* $\mathbf{w}_{0,j}$ satisfies

$$\mathbf{C}^H(\omega_j, \mathbf{p})\mathbf{w}_{0,j} = \mathbf{f}(\omega_j, \mathbf{p}) \tag{7.26}$$

and $\mathbf{w}_{1,j}$ is the *adaptive weight vector*, of length $M_b = M - M_c$ [16]. The columns of the $(M \times M_b)$ orthogonal *blocking matrix* $\mathbf{C}_\perp(\omega_j, \mathbf{p})$ span the orthogonal complement of the span$[\mathbf{C}(\omega_j, \mathbf{p})]$.[8]

Substitution of (7.25) and (7.21) into (7.22) leads to the unconstrained LS problem

$$\tilde{\mathbf{w}}_{1,j} = \arg\min_{\mathbf{w}_{1,j}} \frac{1}{L}\sum_{l=1}^{L} \left|y(\omega_j, l, \mathbf{w}_{1,j})\right|^2 \tag{7.27}$$

[8]It is preferable in actual computations to choose the minimum norm solution of (7.26) with the use of the SVD [32].

for $j = j_1, \ldots, j_2$, where

$$\begin{aligned} y(\omega_j, l, \mathbf{w}_{1,j}) &= y_0(\omega_j, l) + \mathbf{w}_{1,j}^H \mathbf{y}_1(\omega_j, l) \\ y_0(\omega_j, l) &= \mathbf{w}_{0,j}^H \mathbf{x}(\omega_j, l) \\ \mathbf{y}_1(\omega_j, l) &= \mathbf{C}_\perp^H(\omega_j, \mathbf{p})\mathbf{x}(\omega_j, l) \end{aligned} \tag{7.28}$$

In particular, the use of the DR constraint (7.24) makes $\mathbf{C}_\perp(\omega_j, \mathbf{p})\mathbf{w}_{1,j}$ orthogonal to $\mathbf{h}(\omega_j, \mathbf{p})$, so that the *sidelobe canceller* output $\mathbf{y}_1(\omega_j, l)$ does not contain any SOI component [16]. So, under the hypotheses of perfect estimation of $\mathbf{w}_{1,j}$ and exact knowledge of $\mathbf{h}(\omega_j, \mathbf{p})$, the MV-FDBF processor (7.27) suppresses the independent noise and interference component $\mathbf{w}_{0,j}^H \mathbf{v}(\omega_j, l)$ collected by the quiescent beamformer output $y_0(\omega_j, l)$, without affecting the SOI [23].

REMARK. The GSC decomposition (7.25) of the weight vector is independent of the particular choice of the MV functional (7.27) and is indeed typical of linearly constrained optimization problems [32, 47]. Therefore, any error functional expressed in terms of $y(\omega_j, l, \mathbf{w}_j)$ can be adopted. This property allows to optimize beamforming in the case of deviations from the Gaussian assumption on $\mathbf{x}(\omega_j, l)$ [27].

7.4.3 Matched Field Beamforming

Adaptive wideband MF beamforming can be regarded as a special case of the FDBF, casted on the random multimodal model (7.10) described in Section 7.3.4. The MF approach assumes the detailed knowledge of the global $\mathbf{h}(\omega_j, \mathbf{p}, \mathbf{b})$ to locate the SOI in the three-dimensional space [5, 43].

However, when propagation characteristics randomly change with time, $\mathbf{h}(\omega_j, \mathbf{p}, \mathbf{b})$ deviates from its nominal value [5]. Therefore, $\mathbf{y}_1(\omega_j, l)$ in (7.28) contains a component correlated with $s(\omega_j, l)$, which is interpreted by the MV processor (7.27) as interference to be suppressed [15]. To avoid this problem, it should be $\mathbf{C}_\perp^H(\omega_j, \mathbf{p})\mathbf{h}(\omega_j, \mathbf{p}, \mathbf{b}) = 0$ for any likely $\mathbf{h}(\omega_j, \mathbf{p}, \mathbf{b})$ [5].

The mathematical formulation of the adaptive MF beamformer is based on a proper combination of (7.10), (7.27), and (7.28). The quiescent vector $\mathbf{w}_{0,j}$ can be conveniently chosen as proportional to the dominant eigenvector of $\mathbf{R}_{hh}(\omega_j, \mathbf{p})$ [5]. Its phase and amplitude can be adjusted on the basis of (7.24), if SOI reconstruction is required.

The columns of $\mathbf{C}_\perp(\omega_j, \mathbf{p})$ are formed by the orthonormal eigenvectors of the same matrix, corresponding to the M_b numerically negligible eigenvalues [32], that indicate the directions having the minimal SOI content. Alternative formulations of MF beamforming exploit the subspace spanned by the partial derivatives of the steering vector with respect to environmental parameters [5].

Drawbacks of the original MF approach are rather evident:

- The formulation is essentially narrowband and cannot effectively exploit the time delay of arrival to discriminate reflections from the direct path [14].

- Since the mode configuration may significantly change with respect to $\mathbf{p}$ and ω, $\mathbf{w}_{0,j}$ and $\mathbf{C}_{\perp}(\omega_j, \mathbf{p})$ have to be recomputed on a very fine grid of frequencies and locations of interest, thus increasing the computational burden.
- A reliable wave-field simulator should be used to bootstrap the MF beamformer. The resulting overhead and sometimes the modeling accuracy may be unacceptable in critical applications.
- Most existing simulators do not properly account for errors in the sensor transfer functions, due to gain, phase and position mismatches, mutual coupling and scattering.
- The number M_b of complex free parameters of each $\mathbf{w}_{1,j}$ may be severely reduced in small-sized arrays, thus impairing the interference and noise suppression capabilities of the beamformer.
- Typical applications of MF beamforming (e.g., sonar) involve short observation times [28], since the environment is nonstationary and it is not possible or convenient to repeat observations. In these cases, a large number of parameters has to be estimated on the basis of few subband snapshots (7.3), leading to high misadjustment error and poor SOI reconstruction capability [48].

In the past, robust MF techniques were derived on the basis of narrowband array processing and wavefield modeling concepts [5, 24]. However, the continuity of the array manifold [35] with respect to frequency and novel interpretations of the multimodal propagation itself[9] might allow for a simplified modeling of $\mathbf{h}(\omega_j, \mathbf{p}, \mathbf{b})$ and the subsequent use of structured beamformers, characterized by a low number of free parameters. In Sections 7.10.3 and 7.10.4 some preliminary experiments are reported, that exploit similar concepts in reverberant room applications [3, 36].

7.5 ROBUSTNESS

Robustness of beamforming can be generally defined as the capability of coping with reasonable and unpredicted deviations from initial assumptions [21] about the propagation model and the signal and noise statistics, while maintaining a stable reconstruction of the SOI and of the related spectral features for imaging and recognition purposes.

In particular, following the ideas established in [21], a robust adaptive beamformer should

- Be nearly optimal in the case of known environment;
- Exhibit a bounded and progressive performance degradation in the presence of relatively small deviations from the theoretical assumptions;

[9]Modes can be sometimes viewed as distorted planar, cylindrical or spherical wavefronts in the proximity of the array [14, 43].

- Avoid catastrophic performance breakdowns (e.g., SOI cancellation, break of target tracking, gross noise amplification) in the case of strong model and statistical mismatches.

The following examples describe practical aspects of beamforming robustness.

Examples. The main motivation for the introduction of adaptive beamforming in speech processing, teleconferencing and audio recording [3, 49] was the idea of replacing near-field microphones with a centralized device (the array) to save manpower and time for reaching the talkers. However, the key for the success of this idea for end users lies in achieving a perceived audio quality similar to the one of classical near-field devices. In particular, reverberation and fluctuations of level and tone of speech and music are terribly annoying for listeners.

In digital telecommunications [33, 50], the main objective of beamforming is to reliably increase the link range and/or the useful data rate. However, intermittent cancellation of the SOI or abrupt reductions of the SNR at the beamformer output may likely lead to the total loss of the link and its associated capacity.

In towed passive array sonar, the contact with the target should be maintained across transient environment changes, induced, for instance, by underwater explosions and jamming, or in the presence of sensor misalignment, due to ship manoeuvres and ocean streams.

In general, array processing performance is determined by the combination of several parameters, related both to the array (and the underlying wavefield) and to the signal environment [44]. In the same way, catastrophic failure of beamforming may happen because of multiple error mechanisms simultaneously acting in a specific environment.

In the foregoing, most likely error sources will be reviewed and their impact on beamforming performance pointed out. In addition, some classical techniques for enhancing the robustness of beamforming will be described. Historically, they were designed to cope with very specific issues (e.g., pointing errors, short data records, etc.) and do not completely fulfill the concept of robustness developed in this chapter. In addition, combined effects of multiple error sources may not be easily predicted, except within a small perturbations assumption [19].

Nevertheless, the following analysis will be useful to develop intrinsically robust beamforming architectures, capable of working in real world environment with minimal design and tuning requirements.

Considerations will be mainly referred to the MV-FDBF defined by (7.27) and (7.28), but they are valid even for the MV-DS beamformer (7.17), through the straightforward redefinition of some quantities. In particular, $\mathbf{x}(\omega_j, l)$ $(j = j_1, \ldots, j_2,\ l = 1, \ldots, L)$ in the MV-DS is replaced by $\mathbf{x}_{ST}(n)$ $(n = 1, \ldots, N)$ and the narrowband CSCM $\mathbf{R}_{xx}(\omega_j)$ by the $(M(D+1) \times M(D+1))$ *space-time covariance matrix* $\mathbf{R}_{ST} = E[\mathbf{x}_{ST}(n)\mathbf{x}_{ST}^H(n)]$ [7, 13].

In addition, basic error mechanisms can be heuristically extended to other functionals [14], because of the generality of the GSC formulation of the optimization process [47].

7.5.1 Optimal MV-FDBF Weight Vectors

The optimal weight vectors of the MV-FDBF are given by [32, 48]

$$\begin{aligned}\tilde{\mathbf{w}}_j = \mathbf{w}_{j,0} - \mathbf{C}_\perp(\omega_j, \mathbf{p})\left[\mathbf{C}_\perp^H(\omega_j, \mathbf{p})\tilde{\mathbf{R}}_{xx}(\omega_j)\mathbf{C}_\perp(\omega_j, \mathbf{p})\right]^{-1} \\ \times \mathbf{C}_\perp^H(\omega_j, \mathbf{p})\tilde{\mathbf{R}}_{xx}(\omega_j)\mathbf{w}_{j,0}\end{aligned} \tag{7.29}$$

for $j = j_1, \ldots, j_2$, being

$$\tilde{\mathbf{R}}_{xx}(\omega_j) = \frac{1}{L}\sum_{l=1}^{L}\mathbf{x}(\omega_j, l)\mathbf{x}^H(\omega_j, l) \tag{7.30}$$

an estimate of $\mathbf{R}_{xx}(\omega_j)$ using L subband snapshots.[10]

In (7.29), $\mathbf{C}_\perp^H(\omega_j, \mathbf{p})\tilde{\mathbf{R}}_{xx}(\omega_j)\mathbf{w}_{j,0}$ is recognized as the sample cross-covariance between the outputs of the sidelobe canceller $\mathbf{y}_1(\omega_j, l)$ and of the quiescent beamformer $y_0(\omega_j, l)$, and $\mathbf{C}_\perp^H(\omega_j, \mathbf{p})\tilde{\mathbf{R}}_{xx}(\omega_j)\mathbf{C}_\perp(\omega_j, \mathbf{p})$ is the sample covariance of $\mathbf{y}_1(\omega_j, l)$.

These terms play a major role for beamforming robustness. In fact, using (7.28) it is immediate to see that $\mathbf{C}_\perp^H(\omega_j, \mathbf{p})\mathbf{h}(\omega_j, \mathbf{p}) = \mathbf{0}$ and $\mathbf{y}_1(\omega_j, l)$ does not contain any SOI component. Therefore $\tilde{\mathbf{w}}_j$ depends only upon the noise field [48]. However, because of model mismatches and estimation errors in $\tilde{\mathbf{R}}_{xx}(\omega_j)$, the sample weight vector may be heavily influenced by the SOI presence.

7.5.2 Model Related Errors

A first class of errors heavily affecting the reliability of beamforming depends upon the imperfect knowledge of $\mathbf{h}(\omega_j, \mathbf{p})$. In fact, the quality of the solution of the wave equations is tied to the exact specification of the boundary conditions, of the properties of the medium and of the sensor transfer functions [5]. In addition, several approximations have to be made to get handy analytical or numerical solutions [43]. In an alternative approach, the array response can be empirically measured on a grid of locations of interest, using test data sets of finite size.

All these procedures are affected by systematic and random errors. In turn, the wrong estimate of $\mathbf{h}(\omega_j, \mathbf{p})$ leads to the misspecification of beamformer constraints (7.26) and of the corresponding blocking subspace $\mathbf{C}_\perp(\omega_j, \mathbf{p})$ [16].

Effects of modeling mismatches can be isolated by considering that, under the hypothesis of ergodicity, $\tilde{\mathbf{R}}_{xx}(\omega_j) \rightarrow \mathbf{R}_{xx}(\omega_j)$ with probability one (w.p.1) in the limit of $L \rightarrow \infty$ [19].

The general perturbed response model (7.8) is adopted, assuming $\|\mathbf{h}(\omega_j, \mathbf{p})\|_2 = 1$ through a proper scaling of $s(\omega_j, l)$ [14].

The single DR constraint (7.24) is imposed, leading to $\mathbf{w}_{j,0} = \mathbf{h}(\omega_j, \mathbf{p})$ by (7.26). Therefore, the columns of $\mathbf{C}_\perp(\omega_j, \mathbf{p})$ span the $(M-1)$th dimensional orthogonal

[10] $\tilde{\mathbf{R}}_{xx}(\omega_j)$ is actually the Gaussian ML estimate of $\mathbf{R}_{xx}(\omega_j)$ and, in general, a nonparametric estimate of the CSCM for all distributions having finite second- and fourth-order moments [19, 21].

complement of $\mathbf{h}(\omega_j, \mathbf{p})$. In addition, the perturbation $\mathbf{h}_p(\omega_j, \mathbf{p})$ in (7.8) is conveniently decomposed onto the orthogonal basis $[\mathbf{h}(\omega_j, \mathbf{p})\ \mathbf{C}_\perp(\omega_j, \mathbf{p})]$ as

$$\mathbf{h}_p(\omega_j, \mathbf{p}) = \mathbf{h}(\omega_j, \mathbf{p})\gamma_j + \mathbf{C}_\perp(\omega_j, \mathbf{p})\mathbf{a}_{p,j} \tag{7.31}$$

being γ_j a complex-valued scalar and $\mathbf{a}_{p,j}$ a $(M-1)$th dimensional vector. Inserting (7.31) into (7.8) and (7.28) leads to

$$\begin{aligned}
\mathbf{x}(\omega_j, l) &= \tilde{\mathbf{h}}(\omega_j, \mathbf{p})s(\omega_j, l) + \mathbf{v}(\omega_j, l)\\
&= [\mathbf{h}(\omega_j, \mathbf{p})(1+\gamma_j) + \mathbf{C}_\perp(\omega_j, \mathbf{p})\mathbf{a}_{p,j}]s(\omega_j, l) + \mathbf{v}(\omega_j, l)\\
y_0(\omega_j, l) &= (1+\gamma_j)s(\omega_j, l) + \mathbf{w}_{j,0}^H\mathbf{v}(\omega_j, l)\\
\mathbf{y}_1(\omega_j, l) &= \mathbf{a}_{p,j}s(\omega_j, l) + \mathbf{C}_\perp^H(\omega_j, \mathbf{p})\mathbf{v}(\omega_j, l)
\end{aligned} \tag{7.32}$$

Under the above assumptions and after substituting (7.32) into (7.29), it is found that $\tilde{\mathbf{w}}_j$ converges w.p.1 for $L \to \infty$ to the vector

$$\begin{aligned}
\mathbf{w}_j = \mathbf{w}_{j,0} - \mathbf{C}_\perp(\omega_j, \mathbf{p})\big[P_s(\omega_j)\mathbf{a}_{p,j}\mathbf{a}_{p,j}^H + \mathbf{C}_\perp^H(\omega_j, \mathbf{p})\mathbf{R}_{vv}(\omega_j)\mathbf{C}_\perp(\omega_j, \mathbf{p})\big]^{-1}\\
\times \big[P_s(\omega_j)(1+\gamma_j*)\mathbf{a}_{p,j} + \mathbf{C}_\perp^H(\omega_j, \mathbf{p})^H\mathbf{R}_{vv}(\omega_j)\mathbf{w}_{j,0}\big]
\end{aligned} \tag{7.33}$$

To get an idea of the effects of the perturbation on the MV-FDBF, let $\mathbf{R}_{vv}(\omega_j) = \sigma_v^2(\omega_j)\mathbf{I}_M$ (e.g., spatially white noise background), $\|\mathbf{a}_{p,j}\|_2 = \varepsilon_j$ and define the *nominal array SNR* as $\rho_j^2 = P_s(\omega_j)/\sigma_v^2(\omega_j)$. The variances $P_{sy}(\omega_j)$ and $P_{vy}(\omega_j)$ of the SOI and the noise components at the subband output $y(\omega_j, l, \mathbf{w}_j)$ are easily computed as [48]

$$P_{sy}(\omega_j) = P_s(\omega_j)\frac{|1+\gamma_j|^2}{(1+\varepsilon_j^2\rho_j^2)^2} \tag{7.34}$$

$$\begin{aligned}
P_{vy}(\omega_j) &= \mathbf{w}_j^H\mathbf{R}_{vv}(\omega_j)\mathbf{w}_j = \sigma_v^2(\omega_j)\|\mathbf{w}_j\|_2^2\\
&= \sigma_v^2(\omega_j)\left[1 + \frac{\rho_j^4\varepsilon_j^2|1+\gamma_j|^2}{(1+\varepsilon_j^2\rho_j^2)^2}\right]
\end{aligned} \tag{7.35}$$

A close look at these results shows that

- When the SNR $\varepsilon_j^2\rho_j^2$ of the SOI component in the blocking subspace, given by $\mathbf{C}_\perp(\omega_j, \mathbf{p})\mathbf{a}_{p,j}s(\omega_j, l)$, exceeds about 0.5 (e.g., −3 dB), the SOI is abruptly cancelled out from $y(\omega_j, l, \mathbf{w}_j)$ [22];
- The corresponding $\|\mathbf{w}_j\|_2$ increases well beyond unity, leading to a strong noise amplification, and the output SNR of the beamformer may became lower than the SNR of a single sensor;

- γ_j induces linear distortion on the quiescent path response, evident after wideband SOI reconstruction in the time domain. This kind of distortion is mainly induced by multipath entering through the quiescent path or mismatches in the sensor transfer functions. Its reduction calls for better sensor calibration and the use of larger arrays, to enhance the spatial resolution of multipath under constraint (7.24).

In a wideband setting and in the presence of colored SOI and noise spectra, a frequency selective cancellation or distortion may result. This effect is usually most evident around SOI spectral peaks, so that the recovered wideband signal at the beamformer output appears as temporally whitened [51]. For example, in speech recognition, sonar and high fidelity audio the SOI may not be recognized or may lose intelligibility [14, 20]. In digital communications strong *inter symbol interference* may arise [50].

It is worth noting that cancellation arising from model mismatches is a systematic error and cannot be contrasted by increasing the number L of subband snapshots. Classical cures are represented by the formulation of proper linear or quadratic constraints on each $\mathbf{w}_j$ [5, 15, 22].

7.5.2.1 Linear Constraints.

Linear constraints are generally linked to specific error models. Some remarkable solutions are listed below.

Pointing Errors. In this scenario, it is assumed that the SOI source is located at position $\mathbf{p}+\mathbf{p}\boldsymbol{\delta}$, slightly different than expected. Around $\mathbf{p}$, the array manifold [35] can be expanded in Taylor series as [52]

$$\mathbf{h}(\omega_j, \mathbf{p}+\mathbf{p}\boldsymbol{\delta}) = \mathbf{h}(\omega_j, \mathbf{p}) + \mathbf{D}(\omega_j, \mathbf{p})\mathbf{p}\boldsymbol{\delta} + \cdots \tag{7.36}$$

where the columns of the matrix $\mathbf{D}(\omega_j, \mathbf{p})$ are the vector derivatives [44] of $\mathbf{h}(\omega_j, \mathbf{p})$ with respect to each location parameter, evaluated at ω_j and $\mathbf{p}$. Therefore a derivative constrained FDBF can be constructed by imposing [52]

$$[\mathbf{h}(\omega_j, \mathbf{p})\ \mathbf{D}(\omega_j, \mathbf{p})]^H \mathbf{w}_j = \begin{bmatrix} e^{i\omega_j N_t} \\ \mathbf{0} \end{bmatrix} \tag{7.37}$$

for $j = j_1, \ldots, j_2$, so that $\mathbf{w}_j^H \mathbf{h}(\omega_j, \mathbf{p}+\mathbf{p}\boldsymbol{\delta}) \approx e^{-i\omega_j N_t}$ by (7.36). Alternatively, system (7.37) may be extended with the use of higher-order derivatives of $\mathbf{h}(\omega_j, \mathbf{p})$ or directly rewritten in terms of the array response as [5]

$$[\mathbf{h}(\omega_j, \mathbf{p}+\mathbf{p}\boldsymbol{\delta}_1)\ \cdots\ \mathbf{h}(\omega_j, \mathbf{p}+\mathbf{p}\boldsymbol{\delta}_{M_c})]^H \mathbf{w}_j = \mathbf{1} e^{i\omega_j N_t} \tag{7.38}$$

subject to the DR constraint (7.24). The trial locations $\mathbf{p}+\boldsymbol{\delta}\mathbf{p}_k$ $(k = 1, 2, \ldots, M_c)$ should be chosen according to the array CRB for direction finding [44]. This approach can be directly extended to spatially distributed sources [3, 6] and near-source scattering [40].

Sensor Position Errors. Errors in sensor positions can be treated similarly to pointing errors by using the proper derivatives with respect to sensor coordinates.

Gain and Phase Errors. Gain errors can be modeled by the introduction of a proper random vector $\mathbf{e}_g$, so that the perturbed response $\tilde{\mathbf{h}}(\omega_j, \mathbf{p})$ obeys the multiplicative model [40]

$$\tilde{\mathbf{h}}(\omega_j, \mathbf{p}) = \mathbf{h}(\omega_j, \mathbf{p}) \odot (\mathbf{1} + \mathbf{e}_g) \tag{7.39}$$

In this case, the Hermitian correlation matrix $\mathbf{R}_{hh}(\omega_j, \mathbf{p})$ is constructed according to (7.10). The condition $\mathbf{C}_{\perp}^{H}(\omega_j, \mathbf{p})\mathbf{R}_{hh}(\omega_j, \mathbf{p}) \approx \mathbf{0}$ is ensured by choosing the columns of $\mathbf{C}_{\perp}^{H}(\omega_j, \mathbf{p})$ as the orthonormal eigenvectors of $\mathbf{R}_{hh}(\omega_j, \mathbf{p})$ corresponding to its negligible eigenvalues.

Mutual Coupling. Sensor mutual coupling can be taken into account by a proper symmetric *transformation matrix* $\mathbf{M}(\omega_j, \mathbf{p})$ [11], acting on the steering vector as

$$\hat{\mathbf{h}}(\omega_j, \mathbf{p}) = \mathbf{M}(\omega_j, \mathbf{p})\mathbf{h}(\omega_j, \mathbf{p}) \tag{7.40}$$

This error requires preliminary array calibration. Matrix $\mathbf{M}(\omega_j, \mathbf{p})$ weakly depends on $\mathbf{p}$ in the case of wire and aperture antennas. In a wideband setting, $\mathbf{M}(\omega_j, \mathbf{p})$ is a structured matrix, since it is derived by multiple reflections within the array that are closely spaced in time.

Known Multipath and Interference. It is possible to steer one or more nulls toward the positions $\mathbf{p}_1, \ldots, \mathbf{p}_Q$ of known sources of multipath or even statistically independent interference [22, 53]. The constraint system looks like

$$[\mathbf{h}(\omega_j, \mathbf{p}), \mathbf{h}(\omega_j, \mathbf{p}_1), \ldots, \mathbf{h}(\omega_j, \mathbf{p}_Q)]^H \mathbf{w}_j = \begin{bmatrix} e^{i\omega_j N_t} \\ \mathbf{0} \end{bmatrix} \tag{7.41}$$

for each ω_j. In particular, forcing nulls at multiple positions around the detected interference prevents cancellation or SNR losses because of estimation errors and relative motion between the SOI source and the array [22].

In some cases, the system matrix of (7.41) is characterized by a numerical rank lower than M_c. This occurrence means that some constraints are linearly dependent (e.g., redundant) and requires the use of a subset selection procedure [32].

7.5.2.2 Quadratic Constraints. The use of linear constraints restricts the length M_b of the adaptive vector $\mathbf{w}_{1,j}$, thus limiting the suppression of independent interferences. Moreover, the extent of the perturbation $\mathbf{h}_p(\omega_j, \mathbf{p})$ within the model

(7.8) may be too large and unpredictable to develop an effective set of linear constraints [14]. For example, $\|\mathbf{h}_p(\omega_j, \mathbf{p})\|_2 \gg \|\mathbf{h}(\omega_j, \mathbf{p})\|_2$ frequently occurs near the eigenfrequencies of a closed room [36].

When model uncertainties are relevant, *quadratic constraints* based on (7.33) [15] may offer a better protection and preserve more degrees of freedom for beamforming.

Diagonal Loading. Diagonal loading (DL) is probably the most widespread quadratic approach [23, 54]. It is based on the idea of limiting a priori the SNR of the SOI component leaking in the blocking subspace, as observed by the MV-FDBF processor (7.33), by adding a *synthetic noise* contribution to $\tilde{\mathbf{R}}_{xx}(\omega, j)$ in (7.29) as

$$\tilde{\mathbf{R}}_{xx}(\omega_j, \mu_j) = \tilde{\mathbf{R}}_{xx}(\omega_j) + \mu_j^2 \mathbf{I}_M \tag{7.42}$$

being μ_j a real-valued, positive scalar. DL implies that $\mathbf{w}_{1,j}$ solves the regularized optimization problem [47]

$$\tilde{\mathbf{w}}_{1,j} = \arg\min_{\mathbf{w}_{1,j}} \left[\frac{1}{L} \sum_{l=1}^{L} |y(\omega_j, l, \mathbf{w}_{1,j})|^2 + \mu_j^2 \|\mathbf{w}_{1,j}\|_2^2 \right] \tag{7.43}$$

for $j = j_1, \ldots, j_2$. This functional penalizes the increase of $\|\mathbf{w}_{1,j}\|_2^2$, which represents the noise amplification factor in (7.35) and a by-product of SOI cancellation [48]. The regularization parameter μ_j can be independently adjusted for each subband to avoid cancellation. In particular, (7.33) suggests choosing $\mu_j^2 \propto P_s(\omega_j)$. However, (7.34) shows that very large μ_js may be required in the case of significant model mismatches, thus reducing the sensitivity of the beamformer versus statistically independent interference.

DL is also useful to widen the nulls generated by the MV-FDBF processor in order to cope with moving interferers [53] and the multiple rank SOI model (7.9) [54].

Norm Bounds. Norm bounds imposed on the MV-FDBF functional (7.27) are effective in avoiding SOI cancellation and the associated phenomena [14, 28]. In fact, assuming for simplicity $\gamma_j = 0$ in (7.32), the SOI component at $y(\omega_j, l, \mathbf{w}_{1,j})$, given by the *conditional mean* $y_s(\omega_j, l, \mathbf{w}_{1,j}) = E[y(\omega_j, l, \mathbf{w}_{1,j}) \mid s(\omega_j, l)]$ [48], becomes

$$\begin{aligned} y_s(\omega_j, l, \mathbf{w}_{1,j}) &= [\mathbf{w}_{0,j} + \mathbf{C}_\perp(\omega_j, \mathbf{p})\mathbf{w}_{1,j}]^H [\mathbf{h}(\omega_j, \mathbf{p}) + \mathbf{C}_\perp(\omega_j, \mathbf{p})\mathbf{a}_{p,j}] s(\omega_j, l) \\ &= \left[1 + \mathbf{w}_{1,j}^H \mathbf{a}_{p,j}\right] s(\omega_j, l) \end{aligned} \tag{7.44}$$

for $j = j_1, \ldots, j_2$. A robust bound on SOI cancellation can be imposed by requiring that

$$\left|1 + \mathbf{w}_{1,j}^H \mathbf{a}_{p,j}\right| \geq \delta \tag{7.45}$$

for a prespecified threshold $0 < \delta < 1$ and *any* $\|\mathbf{a}_{p,j}\|_2 \leq \varepsilon_{\max}$ [14]. Coherently with the principles of robustness [21], the positive scalar $\varepsilon_{\max}$ represents the maximum expected size of the perturbation and can be established by previous wavefield analysis. By the Cauchy–Schwartz inequality for vectors, the worst case perturbation is given by $\mathbf{a}_{p,j} = -\varepsilon_{\max}\mathbf{w}_{1,j}/\|\mathbf{w}_{1,j}\|_2$ and leads to

$$\|\mathbf{w}_{1,j}\|_2^2 \leq \left(\frac{1-\delta}{\varepsilon_{\max}}\right)^2 \tag{7.46}$$

This bound appears very tight, since the sidelobes of the beam generated by the MV-FDBF solution (7.29) in the case of SOI multipath are steered toward multipath sources, with phase roughly opposite with respect to the main beam [15, 22].

Adding the constraint (7.46) to (7.29) yields a classical *ridge regression* (RR) problem, well known in statistics [32]. Its effective solution requires the *reduced size* SVD (R-SVD) of the $(L \times M_b)$ sample subband matrix $[\mathbf{y}_1(\omega_j, 1), \ldots, \mathbf{y}_1(\omega_j, L)]^H$ and the iterative solution of a secular equation for each ω_j [32].

Since the RR can be regarded as an enhanced, data-adaptive DL [32], it can be sometimes approximated by a line search of optimal regularization parameters μ_j^2 using (7.43) and checking (7.46) [28].

When a preliminary wavefield analysis is available, a more general error functional can be derived as [15]

$$\tilde{\mathbf{w}}_{1,j} = \arg\min_{\mathbf{w}_{1,j}} \frac{1}{L}\sum_{l=1}^{L} \left|y(\omega_j, l, \mathbf{w}_{1,j})\right|^2 \tag{7.47}$$

for $j = j_1, \ldots, j_2$, subject to $\|\mathbf{C}_q\mathbf{w}_{1,j}\|_2^2 \leq \eta_j^2$, being $\mathbf{C}_q$ the $(M_b \times M_b)$ quadratic constraint matrix and $\{\eta_j; j = j_1, \ldots, j_2\}$ a proper set of positive real numbers.

Imposing linear or quadratic constraints on the MV-FDBF within each subband highly increases the overall computational burden. Moreover, the formulation remains essentially narrowband and does not allow for a sufficient degree of control of the SOI distortion. In fact, misspecification of constraints within only few subbands can still cause strong, selective cancellations that impair the reconstruction of the SOI waveform, very important for speech recognition [3] and digital data transmission [50] purposes.

7.5.3 Signal Related Errors

The intrinsic weakness of the MV-FDBF [13] is confirmed by the observation that only L independent snapshots $\mathbf{x}(\omega_j, l)$ of the overall $(j_2 - j_1 + 1)L$ are available for optimizing each $\mathbf{w}_{1,j}$, thus leading to a poor CSCM estimate $\tilde{\mathbf{R}}_{xx}(\omega_j)$ in (7.29). In many applications, L may be uncomfortably low, as demonstrated by the the following example.

Example. A microphone array with $M = 20$ sensors operates in a reverberant room, covering the bandwidth between 400 and 1200 Hz. The baseband sampling period is $T = 1.25$ ms. Since $TN_c = 500$ ms, due to reverberation, a DFT with at least $J = 400 = 500/1.25$ points should be selected [14]. Within a (perfectly reasonable) observation time of two seconds, only four independent snapshots are collected from each DFT bin to adjust 19 complex free parameters. A reasonable 50 percent overlap [29] across DFT blocks leads only to $L = 7$. The situation does not change with the MV-DS beamformer [7], which still requires $D \approx 400$ and delivers $N = 1600$ samples, in front of about $MD = 400 \times 20 = 8000$ free complex parameters.

Therefore, statistical estimation errors of $\tilde{\mathbf{w}}_{1,j}$ must be considered as a major threat to the applicability of wideband adaptive beamforming [28].

7.5.3.1 Finite Sample Errors.

Estimation errors of $\tilde{\mathbf{R}}_{xx}(\omega_j)$ generated by finite sample size induce random misadjustment on each $\tilde{\mathbf{w}}_{1,j}$ in (7.29) [28]. This effect can be separately analyzed by assuming perfect knowledge of $\mathbf{h}(\omega_j, \mathbf{p})$ and of the related DR constraint (7.24). By the *orthogonality principle* [55], the output sequence $y(\omega_j, l, \tilde{\mathbf{w}}_{1,j})$ $(l = 1, \ldots, L)$, is given by the projection of $y_0(\omega_j, l)$ $(l = 1, \ldots, L)$ onto the random $(L - M_b)$th dimensional subspace, orthogonal to the row space of the sample matrix $[\mathbf{y}_1(\omega_j, 1), \ldots, \mathbf{y}_1(\omega_j, L)]$ [48].

Qualitative effects are similar to those generated by model mismatches. In particular, a systematic SOI cancellation occurs, characterized by [48]

$$
\begin{aligned}
y_s(\omega_j, l, \tilde{\mathbf{w}}_{1,j}) &= E[\,y(\omega_j, l, \tilde{\mathbf{w}}_{1,j}) \mid s(\omega_j, l)] \\
&= \left(1 - \frac{M_b}{L}\right) s(\omega_j, l)
\end{aligned} \tag{7.48}
$$

for $L > M_b$. Even other quality measures of the output, such as the MSE, get worse as a function of the ratio M_b/L [48].

If $L < M_b$, each subband LS system (7.27) exhibits infinite solutions characterized by zero output. This occurrence can be avoided by quadratically constraining $\tilde{\mathbf{w}}_{1,j}$ using DL or similar techniques [14]. In fact, $\tilde{\mathbf{R}}_{xx}(\omega_j, \mu_j)$ is positive definite for $\mu_j^2 > 0$.

Alternatively, in the *eigenvalue thresholding* approach [23], the eigenvector decomposition (EVD) [32]

$$
\tilde{\mathbf{R}}_{xx}(\omega_j) = \mathbf{V}_j \boldsymbol{\Lambda}_j \mathbf{V}_j^H \tag{7.49}
$$

is first computed for $j = j_1, \ldots, j_2$, where $\mathbf{V}_j$ is the unitary $(M \times M)$ eigenvector matrix and $\boldsymbol{\Lambda}_j$ is the diagonal $(M \times M)$ eigenvalue matrix with real and nonnegative entries $\lambda_{j,m}(j = j_1, \ldots, j_2$ and $m = 1, \ldots, M)$ [32]. Then any eigenvalue is compared against a threshold $\lambda_0 > 0$. A regularized, positive definite CSCM estimate is obtained by replacing each eigenvalue in (7.49) with

$$
\hat{\lambda}_{j,m} = \max\{\lambda_{j,m}, \lambda_0\} \tag{7.50}
$$

Eigenvalue thresholding has been interpreted as the Gaussian ML estimate of $\mathbf{R}_{xx}(\omega_j)$ over the set of all positive definite Hermitian matrices, whose eigenvalues are constrained to be greater than or equal to λ_0 [23].

A decisive improvement to the finite sample behavior of the MV-FDBF would be obtained by developing a structured beamformer having a number of free parameters much smaller than the total number of available snapshots [28]. This goal will be pursued in the foregoing.

7.5.3.2 Nonstationarity.

Nonstationarity may arise as an effect of fast fading, moving sources and SOI properties and invalidate the assumption of identically distributed subband snapshots. Historically this problem was faced in two different ways. A first approach considers nonstationarities as model uncertainties and applies constraints to the weight vectors [41] or *tapering*[11] to $\tilde{\mathbf{R}}_{xx}(\omega_j)$ [53]. This approach is somewhat limited since equations within each subband system (7.27) do not have the same statistical impact [21, 32]. In particular, time slots where the SOI and/or the noise are stronger heavily influence the synthesis of MV-FDBF weight vectors (7.27). Depending on the particular environment, some ad hoc techniques were developed:

- The beamformer is adapted only in the absence of the SOI [53] or using a previous data batch [48]. This approach is justified by the fact that the MV-FDBF solution (7.29) depends only upon the noise field under ideal conditions and allows to avoid the cancellation issue [48]. This solution is hampered by the requirement of accurate SOI detection, but it may be interesting in radar applications [34].
- An alternative theoretical solution derives from noting that the MV-FDBF output is nonstationary and can be modeled as a non-Gaussian process [27]. This leads to a ML solution which departs from the MV-FDBF one (7.27) [21]. Beside the problem of finding an alternative statistical characterization of the output, beamforming is a regression [21] which fits random data onto random data. Robust regression is best afforded in the framework of robust CSCM estimation [21, 27].

7.5.3.3 Statistical Robustness.

Since the MV-FDBF is casted as a Wiener filter [55] in the presence of an elliptical signal PDF, characterized by a rotationally invariant[12] ML CSCM estimate, a viable solution to nonstationarity can be given by robust *pseudo-covariance* approaches [21, 27].

[11]Matrix tapers weight each element of $\tilde{\mathbf{R}}_{xx}(\omega_j)$ and can be viewed as the result of a rank-one approximation to $\mathbf{R}_{hh}(\omega_j, \mathbf{p})$ within the fast fading model (7.9). They are intrinsic in some wideband beamformers [10].

[12]Rotational invariance allows to interchange snapshot transformation by unitary matrices and CSCM estimate [19, 27].

In these techniques, $\mathbf{x}(\omega_j, l)$ is stochastically characterized by the zero-mean elliptical PDF

$$f_{\mathbf{x}}(\mathbf{x}(\omega_j, l)) = \frac{1}{\det[\mathbf{R}_p(\omega_j)]} f(\mathbf{x}(\omega_j, l)^H \mathbf{R}_p^{-1}(\omega_j)\mathbf{x}(\omega_j, l)) \tag{7.51}$$

being $\mathbf{R}_p(\omega_j) = (\mathbf{Z}_j^H \mathbf{Z}_j)^{-1}$ the $(M \times M)$ Hermitian pseudo-covariance or *scatter matrix* of $\mathbf{x}(\omega_j, l)$, and $f(x) = f(|x|)$ a properly chosen *spherically symmetric* PDF [21].

A M-estimate (e.g., ML-type [21]) $\tilde{\mathbf{Z}}_j$ of the $(M \times M)$ matrix $\mathbf{Z}_j$ is obtained by solving the following set of implicit equations

$$\begin{cases} \tilde{\mathbf{Z}}_j \tilde{\mathbf{z}}_{l,j} = \mathbf{x}(\omega_j, l) \quad \text{for } l = 1, \ldots, L \\ \sum_{l=1}^{L} g(\|\tilde{\mathbf{z}}_{l,j}\|_2)^2 \tilde{\mathbf{z}}_{l,j} \tilde{\mathbf{z}}_{l,j}^H = \sum_{m=1}^{L} g(\|\tilde{\mathbf{z}}_{m,j}\|_2)^2 \mathbf{I}_M \end{cases} \tag{7.52}$$

for $L > M$ and $j = j_1, \ldots, j_2$, being $g(x)$ a prespecified scalar, real-valued and positive weighting function [19]. The $(M \times 1)$ vectors $\tilde{\mathbf{z}}_{l,j} (l = 1, \ldots, L)$ are recognized as the subband snapshots collected at ω_j and *spatially whitened* by $\tilde{\mathbf{Z}}_j$[13] [19]. By (7.52), the sample pseudo-covariance $\tilde{\mathbf{R}}_p(\omega_j)$ is rewritten as

$$\tilde{\mathbf{R}}_p(\omega_j) = \left(\tilde{\mathbf{Z}}_j^H \tilde{\mathbf{Z}}_j\right)^{-1} = \sum_{l=1}^{L} \left[\frac{g(\|\tilde{\mathbf{z}}_{l,j}\|_2)^2}{\sum_{m=1}^{L} g(\|\tilde{\mathbf{z}}_{m,j}\|_2)^2} \right] \mathbf{x}(\omega_j, l)\mathbf{x}(\omega_j, l)^H \tag{7.53}$$

and interpreted as a *weighted CSCM estimate*, which shares the same structure of $\tilde{\mathbf{R}}_{xx}(\omega_j)$ [19]. Optimal weights depend upon the stochastic impact $\|\tilde{\mathbf{z}}_{l,j}\|_2$ of each $\mathbf{x}(\omega_j, l)$ and are obtained through the iterative solution of (7.52).

A statistically robust solution of (7.52) exists if $g(x)$ is finite for $x \geq 0$ and decays asymptotically as $O(x^{-1})$ for $x \to +\infty$ [21]. The choice

$$g(x) = \left(x + a\sqrt{\frac{M}{L}}\right)^{-1} \tag{7.54}$$

being a a real, positive constant in the range (0.05, 0.5) was recommended for statistical performance and numerical speed of convergence [19]. Using (7.54), at the equilibrium all the weighted snapshots are characterized by nearly the same stochastic impact and therefore have the same importance in the formation of $\tilde{\mathbf{R}}_p(\omega_j)$ by (7.53).

[13]The matrix $\mathbf{Z}_j$ is identifiable up to an unitary left factor [21].

The pseudo-covariance can also be regarded as a data driven cross validation of received snapshots [32]. In fact, those observations that depart from the spatial model described by $\tilde{\mathbf{R}}_p(\omega_j)$ are generally under-weighted in (7.53).

A stable and reasonably fast algorithm for solving (7.52) uses an iteratively reweighted R-SVD of the sample matrices $[\mathbf{x}(\omega_j, l), \ldots, \mathbf{x}(\omega_j, L)]^H (j = j_1, \ldots, j_2)$ and is given in [19].

Many interesting uses of the pseudo-covariance concept can be envisaged for robust wideband beamforming in nonstationary environments:

- *Automatic rejection or balancing of snapshots characterized by slightly different underlying signal models.* For example, in CFAR radar applications [34] typically only few resolution cells contain a target at a given time. Returns from the surrounding cells are used to adapt the beamformer on the background noise and clutter field, as described in Section 7.5.3.2. The pseudo-covariance of *all* snapshots collected from a block of neighboring cells automatically under-weights returns from targets and generates a more reliable statistic of the background field from the remaining cells.
- *A generally better statistical estimate of the CSCM in the presence of non-Gaussian SOI, noise and interference* [27]. In fact, the function $g(x)$ suggested in (7.54) is derived from the ML estimator of the scatter matrix of an elliptical distribution, characterized by a rather heavy-tailed PDF $f(x)$ [21]. The structure of (7.51) does not require modifications to array processing algorithms developed under the Gaussian assumption [27] and the statistical performance of the pseudo-covariance is nearly optimal even in the Gaussian case [19].
- *Better averaging* of *universal spatial covariances, built from snapshots collected at different frequencies and/or times* [25, 56, 57]. These snapshots are generally characterized by a different signal content, due to the temporal and/or spectral variability of sources, but by a rather invariant propagation model (e.g., steering vectors remain nearly the same). In these cases, all snapshots are stacked into a single data matrix and the overall pseudo-covariance is computed.
- *Optimal weighting of MV-FDBF systems* (7.27). In this case, the pseudo-covariance of $\mathbf{y}_1(\omega_j, l)(l = 1, \ldots, L)$ is estimated for each ω_j. The weights

$$g_{l,j} = \frac{g(\|\tilde{\mathbf{z}}_{l,j}\|_2)}{\sqrt{\sum_{l=1}^{L} g(\|\tilde{\mathbf{z}}_{l,j}\|_2)^2}} \tag{7.55}$$

are computed through (7.52), after replacing $\mathbf{x}(\omega_j, l)$ with $\mathbf{y}_1(\omega_j, l)$, and then applied to the MV-FDBF functional (7.27), which becomes

$$\tilde{\mathbf{w}}_{1,j} = \arg\min_{\mathbf{w}_{1,j}} \frac{1}{L} \sum_{l=1}^{L} |g_{l,j}|^2 |y(\omega_j, l, \mathbf{w}_{1,j})|^2 \tag{7.56}$$

for $j = j_1, \ldots, j_2$. This procedure reduces the effect of modeling errors, affecting the blocking subspace in the presence of a strong SOI [14].

- *Outlier search.* The pseudo-covariance can localize the presence of spurious data (or outliers [21]) in the snapshots, frequently due to hardware and control faults (e.g., A/D converter saturation, electromagnetic pulses, etc.), by examining $\tilde{\mathbf{z}}_{l,j}$ [21] and the associated weights. Outliers can be removed from the sample, and the pseudo-covariance recomputed from the remaining snapshots.[14] It is worth noting that the pseudo-covariance alone can cope only with a fraction of outlier-contaminated snapshots smaller than about M^{-1} [21].

The pseudo-covariance performed well in our experiments with real world data and appears as a valuable tool in wideband beamforming, which is characterized by the simultaneous presence of colored sources and a complex and uncertain propagation model. Its widespread use seems mainly hampered by the high computational cost (about 10 times the cost of a SVD of the same matrix [19]) and by the intrinsic difficulty of evaluating and optimizing its performance in specific environments [21].

7.5.3.4 Discussion. Robust wideband beamforming is an active topic of research. However, new approaches should take into account the different roles played by modeling errors, that induce systematic bias and gross errors [21] on $\tilde{\mathbf{w}}_{1,j}$s, and signal related errors, due to the imperfect probabilistic description of array signals, that cause a random misadjustment, highly dependent on the sample size.

Rather than attempting to marginally improve the MV-FDBF against specific error source, a truly robust wideband beamforming strategy should be based on a different architecture, characterized by a drastic reduction of the number of free parameters, in order to improve the performance with short data records [28], and by an intrinsically robust error functional, though derived within a ML framework to optimize statistical performance.

7.6 STEERED ADAPTIVE BEAMFORMING

The wideband steered adaptive beamformer (STBF) was first introduced in [10] and is characterized by a low number of free parameters and the use of a single weight vector for all frequencies. Thanks to this strong constraint, the STBF is capable of decorrelating in time delayed multipath and canceling it in the element space [14].

The basic idea of the STBF stems from the consideration that M identical sensors, characterized by a flat magnitude response and a linear-phase response over the analysis bandwidth, collect M delayed copies of the SOI in a nondispersive wavefield [10]. Therefore, it is possible to align in time these waveforms, by adding a proper delay to each sensor output. After this correction, the baseband impulse

[14] A robust CSCM estimator should be used, since $f_{\mathbf{x}}(\mathbf{x})$ has been modified by the outlier excision.

response of the array becomes

$$\mathbf{h}(n, \mathbf{p}) \propto \mathbf{1}\, u_0(n - N_t) \tag{7.57}$$

and *any* Mth dimensional weight vector $\mathbf{w}$ applied to the time domain snapshots $\mathbf{x}(n)$ $(n = 1, \ldots, N)$ satisfies the DR constraint (7.18) except for a complex gain factor, since $\mathbf{w}^H \mathbf{h}(n, \mathbf{p}) \propto \mathbf{w}^H \mathbf{1}\, u_0(n - N_t)$. The remaining architecture of the STBF corresponds to that of a narrowband MV beamformer, with the exception that the wideband SOI and the background noise and interference generally are temporally correlated.

In a more general and flexible approach, required to take into account multipath and the associated robustness issues, the STBF is recasted in the frequency domain, under the partially known response array model (7.12) introduced in Section 7.3.4 [14]. In particular, the compensation of relative delays of arrival by a DS scheme is replaced by a more sophisticated focusing stage [26].

7.6.1 Focusing

The array is first presteered or focused to the location $\mathbf{p}$ of interest by applying a $(M \times M)$ *focusing matrix* $\mathbf{T}_j$ to each subband snapshot $\mathbf{x}(\omega_j, l)$ to get the corresponding *focused snapshot* [26]

$$\mathbf{x}_f(\omega_j, l) = \mathbf{T}_j \mathbf{x}(\omega_j, l) \tag{7.58}$$

for $l = 1, \ldots, L$ and $j = j_1, \ldots, j_2$. In particular, $\mathbf{T}_j$ aligns the direct path response $\mathbf{h}_d(\omega_j, \mathbf{p})$ at each ω_j onto the unit norm steering vector $\mathbf{a}_d(\mathbf{p})$ of a virtual narrowband array, according to the equation

$$\mathbf{T}_j \mathbf{h}_d(\omega_j, \mathbf{p}) = \mathbf{a}_d(\mathbf{p}) g_d(\omega_j, \mathbf{p}) + \mathbf{e}_f(\omega_j, \mathbf{p}) \tag{7.59}$$

where $\mathbf{e}_f(\omega_j, \mathbf{p})$ is the *focusing error* [19] and

$$g_d(\omega_j, \mathbf{p}) = \|\mathbf{h}_d(\omega_j, \mathbf{p})\|_2 e^{-i\omega_j N_t} \tag{7.60}$$

is a preselected target transfer function[15] [14].

Focusing can be interpreted in the time domain as a *partial inversion* of the direct path impulse response $\mathbf{h}_d(n, \mathbf{p})$, which often becomes nearly proportional to $\mathbf{a}_d(\mathbf{p}) u_0(n - N_t)$ [6].

Substituting (7.3) and (7.12) into (7.58) leads to the signal model

$$\mathbf{x}_f(\omega_j, l) = \mathbf{T}_j[\mathbf{h}_d(\omega_j, \mathbf{p}) + \mathbf{h}_r(\omega_j, \mathbf{p})] s(\omega_j, l) + \mathbf{T}_j \mathbf{v}(\omega_j, l) \tag{7.61}$$

[15]The described focusing procedure is not influenced by a circular time shift applied to $\mathbf{h}(n, \mathbf{p})$. This property may be useful for array calibration without synchronization with the SOI source.

After inserting (7.59) into (7.61) and defining the following auxiliary $(M \times 1)$ vectors:

- The response at ω_j with respect to the SOI, due to transformed multipath and focusing errors, given by

$$\mathbf{h}_{fr}(\omega_j, \mathbf{p}) = \mathbf{T}_j \mathbf{h}_r(\omega_j, \mathbf{p}) + \mathbf{e}_f(\omega_j) \tag{7.62}$$

- The transformed subband component, due to noise and independent interference

$$\mathbf{v}_f(\omega_j, l) = \mathbf{T}_j \mathbf{v}(\omega_j, l) \tag{7.63}$$

the model (7.61) is rewritten as

$$\mathbf{x}_f(\omega_j, l) = [\mathbf{a}_d g_d(\omega_j, \mathbf{p}) + \mathbf{h}_{fr}(\omega_j, \mathbf{p})] s(\omega_j, l) + \mathbf{v}_f(\omega_j, l) \tag{7.64}$$

A single $(M \times 1)$ weight vector $\mathbf{w}$ is used for the entire analysis bandwidth [10] and adapted, subject to the $1 \leq M_c < M$ linear constraints [16]

$$\mathbf{C}^H(\mathbf{p})\mathbf{w} = \mathbf{f}(\mathbf{p}) \tag{7.65}$$

Each subband output of the STBF is defined as

$$y(\omega_j, l, \mathbf{w}) = \mathbf{w}^H \mathbf{x}_f(\omega_j, l) \tag{7.66}$$

and furnishes an estimate of $s(\omega_j, l)$ for $j = j_1, \ldots, j_2$ and $l = 1, 2, \ldots, L$. The overall STBF architecture is depicted in Figure 7.3.

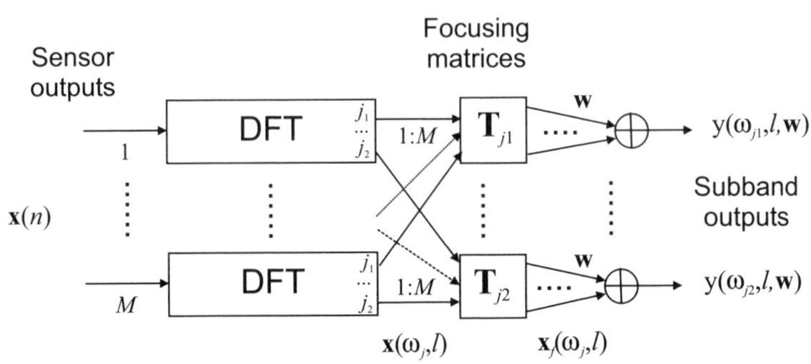

Figure 7.3 Frequency domain STBF.

In the GSC formulation [16] of the STBF it is possible to drop the frequency index in (7.25), (7.26) and (7.28), obtaining, respectively

$$\mathbf{w} = \mathbf{w}_0 + \mathbf{C}_{\perp}(\mathbf{p})\mathbf{w}_1 \tag{7.67}$$

$$\mathbf{C}^H(\mathbf{p})\mathbf{w}_0 = \mathbf{f}(\mathbf{p}) \tag{7.68}$$

$$\begin{aligned} y(\omega_j, l, \mathbf{w}_1) &= y_0(\omega_j, l) + \mathbf{w}_1^H \mathbf{y}_1(\omega_j, l) \\ y_0(\omega_j, l) &= \mathbf{w}_0^H \mathbf{x}_f(\omega_j, l) \\ \mathbf{y}_1(\omega_j, l) &= \mathbf{C}_{\perp}^H(\mathbf{p})\mathbf{x}_f(\omega_j, l) \end{aligned} \tag{7.69}$$

7.6.1.1 Unitary Focusing Matrices. The choice of the focusing matrices $\mathbf{T}_j$s $(j = j_1, \ldots, j_2)$ is not unique [25]. Several possible solutions were proposed in the past. In particular, unitary $\mathbf{T}_j$s [32] were often recommended, because they preserve:

- The covariance of spatially white noise characterized by $\mathbf{R}_{vv}(\omega_j) = \sigma_v^2(\omega_j)\mathbf{I}_M$ (being $\mathbf{T}_j\mathbf{R}_{vv}(\omega_j)\mathbf{T}_j^H = \sigma_v^2(\omega_j)\mathbf{I}_M$) and the overall statistical information [25];
- The orthogonality between constraint and blocking subspaces in (7.67) at any frequency [14].

In general, it is not possible to focus the array with small errors over the entire field of view, due to geometric constraints [25, 57]. Especially unitary focusing is limited to angular sectors smaller than one beamwidth [25].

Anyway, sufficient focusing should be ensured for a ball of positions around $\mathbf{p}$ [25] to keep the SOI distortion low even in the case of small pointing errors of the beamformer [14]. Useful choices of unitary focusing matrices are reviewed below.

Diagonal, Unitary Matrices. Diagonal and unitary focusing matrices are the simplest choice and generalize the idea of separately delaying sensor outputs to temporally align the SOI waveforms [10, 25]. According to the conventions of (7.59), the diagonal entries of each $\mathbf{T}_j$ are given by

$$\mathbf{T}_j(m, m) = \frac{\mathbf{a}_d(\mathbf{p})(m)\mathbf{h}_d^*(\omega_j, \mathbf{p})(m)}{|\mathbf{a}_d(\mathbf{p})(m)||\mathbf{h}_d(\omega_j, \mathbf{p})(m)|} \tag{7.70}$$

for $m = 1, \ldots, M$. Diagonal matrices are computationally efficient (they require only M complex multiplications for each subband snapshots), but exhibit strong focusing errors in the cases of non-identical sensor gains and of interference sources widely spaced from $\mathbf{p}$.

Rotational, Unitary Matrices. Unitary focusing matrices can be obtained by the solution of the following orthogonal *Procrustes problem* [32]

$$\mathbf{T}_j = \arg\min_{\mathbf{T}} \left\| \mathbf{A}_d - \mathbf{T}\mathbf{A}_j \right\|_F \tag{7.71}$$

for $j = j_1, \ldots, j_2$, where $\mathbf{A}_d = [\mathbf{a}_d(\mathbf{p}_1) \cdots \mathbf{a}_d(\mathbf{p}_Q)]$, with all auxiliary locations $\mathbf{p}_q (q = 1, 2, \ldots, Q)$ lying within about half beamwidth from $\mathbf{p}$ [25], and $\mathbf{A}_j = [\mathbf{a}_{j,1} \cdots \mathbf{a}_{j,Q}]$ with

$$\mathbf{a}_{j,q} = \frac{\mathbf{h}_d(\omega_j, \mathbf{p}_q)}{\|\mathbf{h}_d(\omega_j, \mathbf{p}_q)\|_2} e^{i\varphi_q} \tag{7.72}$$

The normalization and the phase rotation $e^{i\varphi_q}$ applied to each steering vector in $\mathbf{A}_j$ have been introduced in (7.71) in order to avoid *irreducible fitting errors*. In fact, an unitary transformation cannot modify the L_2 norm of a vector, while a proper choice of the phase factor can compensate for synchronization errors between the source and the array during calibration and for systematic deviations from the nominal sensor transfer functions [11].

The solution of (7.71) is given by

$$\mathbf{T}_j = \mathbf{U}_j \mathbf{V}_j^H \tag{7.73}$$

where $\mathbf{U}_j$ and $\mathbf{V}_j$ are the left and right singular vector matrices of $\mathbf{A}_d \mathbf{A}_j^H$ [25, 32]. From a close analysis of (7.73) and of the final focusing error $\mathbf{A}_d - \mathbf{T}_j \mathbf{A}_j$ [57], it follows that:

- Steering vectors drawn from a narrow angular sector tend to be linearly dependent, so that the numerical ranks [32] of $\mathbf{A}_d$ and $\mathbf{A}_j$ are generally much smaller than M [57]. In this case, by (7.73), $\mathbf{T}_j$ maps the numerical span($\mathbf{A}_j$) onto the numerical span($\mathbf{A}_d$) [57];
- A good focusing is attainable if the singular value spectra [32] of $\mathbf{A}_d$ and $\mathbf{A}_j$ are similar. Therefore each $\mathbf{a}_d(\mathbf{p}_q)$ is often chosen as the normalized array response at the center frequency of the analysis bandwidth [25]. This choice is not mandatory and there is some freedom for selecting the reference virtual array geometry [58];
- The spatial resolution capabilities of the actual and the virtual array, as measured by the CRB [44], should be close each other. Since the array CRB is function of the frequency, an effective focusing is possible only if the analysis bandwidth spans less than about one octave [28];

- Compound arrays covering several octaves [4] can be focused by mapping the response of the single subarray dedicated to each subband onto the same $\mathbf{A}_d$ [28];
- Minimization of the fitting errors of (7.73) requires the maximization of the sum of singular values of $\mathbf{A}_d\mathbf{A}_j^H$ [32]. In particular, the phase factors $e^{i\varphi_q}$ may be varied to this purpose. This is a very nonlinear problem. Heuristically, error minimization can be made by maximizing $\|\mathbf{A}_d\mathbf{A}_j^H\|_F$ with respect to $e^{i\varphi_q}$, which is more tractable. Choosing different phase factors for each subband leads to a very low error, but it increases SOI linear distortion, acceptable only for imaging and energy detection purposes. Otherwise, the structured solution $e^{i\varphi_q} = e^{i(\varphi_{0,q}-\omega_j\tau_q)}$ should be chosen in each subband. The estimator of the starting phase $\varphi_{0,q}$ and of the group delay τ_q is simplified by the use of the FFT.

Further error weighting and corrections of the SOI distortion can be embedded into (7.71), as demonstrated by the experiments reported in Section 7.10.

7.6.1.2 Focusing FIR Filters. When the wideband SOI reconstruction in the time domain is required, frequency domain focusing may not be proper, because of end effects and circular aliasing of DFT processing [59].

This problem can be circumvented in the STBF by considering the (m, n)th entry of $\mathbf{T}_j$ as the transfer function at ω_j of a FIR filter of length $j_2 - j_1 + 1$. The baseband impulse response of this filter can be recovered by classical frequency sampling approaches using the inverse DFT [59] or LS fitting [60]. The resulting multichannel filter-bank properly focuses the SOI in the time domain.

Assuming that (7.59) is capable of inverting the direct path response with small error, so that for any $\mathbf{p}$ of interest $g_d(\omega_j, \mathbf{p}) \approx g_d e^{-i\omega_j N_t}$, where g_d is a complex-valued constant, the $(M \times 1)$ *focused wideband snapshot* $\mathbf{x}_f(n)$ obeys the model [14]

$$\mathbf{x}_f(n) = \mathbf{a}_d(\mathbf{p}) g_d s(n - N_t) + \sum_{k=N_{rf}}^{N_f} \mathbf{h}_{rf}(k) s(n-k) + \mathbf{v}_f(n) \tag{7.74}$$

for $n = 1, \ldots, N$. The symbol meaning can be directly deduced from (7.64) and N_{rf} is the starting time of the multipath related response after focusing.[16]

The weight vector decomposition (7.69) and the constraint equation (7.68) are still valid in the time domain. In particular, if the DR condition

$$\mathbf{a}_d(\mathbf{p})^H \mathbf{w} = 1 \tag{7.75}$$

[16]Causal focusing filters must introduce a non-negative group delay in order to invert the direct path response, so that $N_{rf} \geq N_r$, as defined in (7.11).

is imposed, starting from (7.69) the STBF output $y(n)$ in the time domain is expressed as

$$\begin{aligned} y(n, \mathbf{w}_1) &= y_0(n) + \mathbf{w}_1^H \mathbf{y}_1(n) \\ y_0(n) &= \mathbf{w}_0^H \mathbf{x}_f(n) \\ &= g_d s(n - N_t) + \sum_{k=N_{rf}}^{N_f} \mathbf{w}_0^H \mathbf{h}_{rf}(k) s(n-k) + \mathbf{w}_0^H \mathbf{v}_f(n) \\ \mathbf{y}_1(n) &= \mathbf{C}_\perp^H(\mathbf{p}) \mathbf{x}_f(n) \\ &= \sum_{k=N_{rf}}^{N_f} \mathbf{C}_\perp^H(\mathbf{p}) \mathbf{h}_{rf}(k) s(n-k) + \mathbf{C}_\perp^H(\mathbf{p}) \mathbf{v}_f(n) \end{aligned} \tag{7.76}$$

for $n = 1, \ldots, N$. It is worth noting that $\mathbf{y}_1(n)$, which is the output of the sidelobe canceller [16], does not contain any direct path SOI component, under the hypothesis of exact specification of (7.75). Moreover, the multipath SOI component of $\mathbf{y}_1(n)$ is delayed by at least $N_{rf} - N_t$ samples from the direct path itself [14]. As shown in the foregoing, this property makes the STBF capable of robust multipath suppression and reduces the SOI distortion at the beamformer output.

7.6.2 Minimum Variance STBF

The MV criterion (7.27) is reformulated for the STBF as [10]

$$\tilde{\mathbf{w}}_{1MV} = \arg\min_{\mathbf{w}_1} \sum_{j=j_1}^{j_2} \tilde{\sigma}_j^2(\mathbf{w}_1) \tag{7.77}$$

being

$$\tilde{\sigma}_j^2(\mathbf{w}_1) = \frac{1}{L} \sum_{l=1}^{L} \left| y(\omega_j, l, \mathbf{w}_1) \right|^2 \tag{7.78}$$

The equivalent time domain formulation of this *minimum variance STBF* (MV-STBF) is [13]

$$\tilde{\mathbf{w}}_{1MV} = \arg\min_{\mathbf{w}_1} \frac{1}{N} \sum_{n=1}^{N} \left| y(n, \mathbf{w}_1) \right|^2 \tag{7.79}$$

Therefore the MV-STBF solves a single LS system containing $N \approx (j_2 - j_1 + 1)L$ equations in only $M_b = M - M_c$ complex unknowns. In contrast, the equivalent MV-FDBF (7.27) has to optimize $(j_2 - j_1 + 1)M_b$ complex free parameters using the same number of equations. This impressive parameter economy makes the finite sample misadjustment of the MV-STBF negligible even for very short data records [28]. In fact, according to (7.48) [17], the finite sample SOI cancellation is $M_b/[(j_2 - j_1 + 1)L]$ in the MV-STBF, under the simplifying hypothesis of statistically independent snapshots [14].

From the robust beamforming viewpoint, this property indicates that the MV-STBF performance is mainly threatened by the presence of systematic model errors, in particular those induced by unpredicted multipath and reverberation [14].

To get insight about this issue, $\tilde{\mathbf{w}}_{1MV}$ is conveniently expressed from (7.79) as the solution of the following linear system of M_b normal equations [32]

$$\left[\frac{1}{N}\sum_{n=1}^{N}\mathbf{y}_1(n)\mathbf{y}_1^H(n)\right]\tilde{\mathbf{w}}_{1MV} = \left[\frac{1}{N}\sum_{n=1}^{N}\mathbf{y}_1(n)y_0(n)^H\right] \tag{7.80}$$

Taking the expected value of both sides of (7.80) under the hypothesis of ergodicity leads to the Wiener solution [51]

$$E\big[\mathbf{y}_1(n)\mathbf{y}_1^H(n)\big]\mathbf{w}_{1MV} = E\big[\mathbf{y}_1(n)y_0(n)^H\big] \tag{7.81}$$

Recalling (7.76), $E[\mathbf{y}_1(n)y_0(n)^H]$ does not contain any contribute from the SOI direct path if

$$N_{rf} - N_t > N_s \tag{7.82}$$

that is, if the smaller multipath delay exceeds the SOI correlation time N_s. Therefore, if (7.82) holds, $\mathbf{w}_{1MV}$ cannot cancel the direct path. Moreover, multipath components appear in both sides of (7.81) just like independent interference and can be cancelled out by $\mathbf{w}_{1MV}$ from $y(n, \mathbf{w}_{1MV})$.

REMARKS. An effective multipath suppression imposes other requirements on the array and the environment, beside the satisfaction of (7.82). In particular:

- The array must spatially resolve the direct path from multipath arrivals under the DR constraint (7.75). Otherwise, the multipath collected by the quiescent beamformer $\mathbf{w}_0$ cannot be cancelled and gives origin to linear, frequency selective distortion of the reconstructed SOI.
- The size M of the array must largely exceed the number of significant multipath arrivals. In particular, focusing errors of multipath and independent interference generally induce broad nulls in the MV-STBF beampattern. Therefore multiple, closely spaced nulls are required to adequately suppress single strong multipath sources, thus requiring a larger $\mathbf{w}_1$.
- Other sources of misadjustment, especially pointing errors, must be adequately controlled by a suitable set of linear and quadratic constraints. In particular,

focusing errors may not satisfy (7.82), so requiring the use of norm bounds on $\mathbf{w}_1$ [14, 28].

- Array calibration errors and other model mismatches are amplified near the spectral peaks of the SOI [19]. Therefore, the MV-STBF performance may be worse than expected and a pseudo-covariance based cross-validation of equations may be useful. Regretfully, the computational cost of the pseudo-covariance applied to the large systems (7.77) and (7.79) makes this approach impractical in many circumstances.

7.7 MAXIMUM LIKELIHOOD STBF

The original MV-STBF [10] was an important step ahead in terms of robustness versus finite sample errors and multipath. However, later experiments showed significant amplification of background noise in the presence of colored sources [28].

This behavior can be explained by observing that SOI, interference and noise in (7.76) are *temporally correlated* and often non-Gaussian. Therefore, the MV-STBF defined by (7.77) and (7.79)[17] cannot be claimed as statistically optimal.

In particular, restricting the attention on the frequency domain MV-STBF, it can be observed that when the beamformer is steered off-source and properly suppresses both the SOI and the statistically independent interference, the $y(\omega_j, l, \mathbf{w}_{1MV})$s are approximately Gaussian distributed [27] and essentially contain background noise residuals, that can be often considered as temporally white. In this case, the MV-STBF approaches the ML estimator [14].

On the contrary, when the STBF is steered toward a colored SOI source, the $y(\omega_j, l, \mathbf{w}_{1MV})$s can still be considered Gaussian for a sufficiently large DFT size J, but are characterized by widely different variances at each frequency. Therefore, the MV-STBF is no longer optimal. In this case, nonoptimality leads to two nasty side effects:

- The statistical impact of the equations in (7.77) strongly changes with frequency, according to the varying spectral levels of signals and noise. In particular, the MV-STBF solution is mainly determined by those bins near the spectral peaks of the SOI. As a consequence, high misadjustment of the beamformer generally occurs at the spectral "valleys" of the SOI, leading in turn to a low local output SNR. This issue is typical of LS fitting [51] and well known in speech coding [20], where it may heavily affect the intelligibility of the decoded signal [14]. Similar problems are expected in spectral matching recognition techniques, spread-spectrum telecommunications and high-fidelity audio recording [3].
- The amplification of model mismatches in the high-SNR regions often produces a frequency selective SOI cancellation around the spectral peaks, which impairs signal reconstruction and recognition [14].

[17] After time domain focusing, both MV solutions are essentially identical [13].

SOI prewhitening can heuristically cope with all these phenomena [3], by equalizing the impact of MV-STBF equations and reducing the risk of accidental SOI cancellations due to model mismatches. In particular, more importance will be given to interference suppression near the SOI spectral valleys, where it is more necessary for ensuring SOI intelligibility. A possible performance loss near SOI spectral peaks is less important, since the resulting SNR can still be adequate for masking noise and interference residuals [14].

An attractive side effect of prewhitening would be the reduction of the effective SOI correlation time N_s in (7.82), thus allowing an effective temporal decorrelation of multipath, down to one sampling period [18].

However, in the multichannel array environment is not clear how to optimally synthesize the prewhitening filter, since the SOI spectrum cannot be consistently estimated from the STBF output [23].

7.7.1 Stochastic Model

A theoretically sound solution to the SOI reconstruction problem is given by the ML estimator of the weight vector (ML-STBF), based on a proper *stochastic modeling* of $y(\omega_j, l, \mathbf{w}_1)$ for $j = j_1, \ldots, j_2$. In particular, under the hypotheses given in Section 3.2, each sequence $y(\omega_j, l, \mathbf{w}_1)$ $(l = 1, \ldots, L)$ can be considered by the central limit theorem as the realization of a zero-mean, ergodic, white, complex, circular Gaussian process, characterized by the subband variance $\sigma_j^2(\mathbf{w}_1)$ [27]. Moreover, subband processes can be mutually considered as statistically independent for $J \gg N_c$.

Therefore, the scaled, negative log-likelihood of the STBF output can be written as

$$\mathcal{L}(\mathbf{w}_1) = \sum_{j=j_1}^{j_2} \left[\ln[\sigma_j^2(\mathbf{w}_1)] + \frac{\sum_{l=1}^{L} |y(\omega_j, l, \mathbf{w}_1)|^2}{L\sigma_j^2(\mathbf{w}_1)} \right] \tag{7.83}$$

after neglecting inessential additive constants [14, 23]. The subband variances can be considered as nuisance parameters and eliminated from (7.83) by imposing the necessary conditions

$$\frac{\partial \mathcal{L}(\mathbf{w}_1)}{\partial \sigma_j^2(\mathbf{w}_1)} = 0 \tag{7.84}$$

for $j = j_1, \ldots, j_2$. The solution of these equations is given by (7.78). In particular, the quantities $\tilde{\sigma}_j^2(\mathbf{w}_1)$s are the *spectrum estimates* of $y(n, \mathbf{w}_1)$, obtained by a Welch periodogram approach [29].

If $\tilde{\sigma}_j^2(\mathbf{w}_1) > 0$ for $j = j_1, \ldots, j_2$, substituting (7.78) into (7.83) and neglecting constants leads to the optimal ML-STBF weight vector

$$\tilde{\mathbf{w}}_{1ML} = \arg\min_{\mathbf{w}_1} \mathcal{L}_c(\mathbf{w}_1)$$

$$\mathcal{L}_c(\mathbf{w}_1) = \sum_{j=j_1}^{j_2} \ln\left[\tilde{\sigma}_j^2(\mathbf{w}_1)\right] \tag{7.85}$$

[14]. Differently from the MV-FDBF (7.27) and the MV-STBF (7.77), the ML-STBF requires the solution of a highly nonlinear optimization problem.

7.7.2 Properties

It is interesting to observe that, for known $\sigma_j^2(\mathbf{w}_1)$s, (7.83) would be minimized by solving the weighted linear LS optimization problem

$$\tilde{\mathbf{w}}_{1ML} = \arg\min_{\mathbf{w}_1} \sum_{j=j_1}^{j_2} \frac{\tilde{\sigma}_j^2(\mathbf{w}_1)}{\sigma_j^2(\mathbf{w}_1)} \tag{7.86}$$

Therefore the ML-STBF (7.85) can be interpreted as a particular MV-STBF, whose weight vector is computed after the output has been previously whitened. This processing is implicit in the optimization of (7.85) and does not affect the actual beamformer output [14].

As a consequence, the ML-STBF behavior changes, depending on the SOI content at the output, measured after adaptation. One of the following cases may happen in practice.

- *Wideband SOI* spectrum $P_s(\omega_j)$ is significant in all analyzed DFT bins and the ML-STBF converges to a good solution. If the final output SNR is much larger than unity at each frequency, $y(\omega_j, l, \mathbf{w}_{1ML})$ is essentially formed by the recovered SOI. Therefore, the ML-STBF has been computed after near-perfect SOI whitening, which results in an effective $N_s \approx 1$ in (7.82). By (7.81), the ML-STBF suppresses as independent interference any perturbation which translates into a SOI replica, whose relative delay with respect to the direct path roughly exceeds the reciprocal of the analysis bandwidth, given by

$$T_r = \frac{JT}{j_2 - j_1 + 1} \tag{7.87}$$

- *Narrowband SOI* embedded in wideband noise and interference is generally significant only in a few bins, while the remaining ones contain essentially noise and independent interference. The ML-STBF implicitly attenuates the SOI and adjusts to the background. As explained in Section 7.5.1, this occurrence avoids risks of SOI cancellation. However, since SOI bins have been essentially ruled out from the estimator, unexpected focusing errors may slightly affect interference nulling.

Probably the most significant property of the ML-STBF is its insensitivity to a common pre-filtering applied to sensor outputs. This property can be assessed by considering that any scaling of $\mathbf{x}_f(\omega_j, l, \mathbf{w}_{1ML})$ by a nonzero complex constant results into an additive term in $\mathcal{L}_c(\mathbf{w}_1)$, which does not depend on $\mathbf{w}_1$ and does not modify the solution of (7.85) [14].

Since DFT scaling can be viewed as a circular convolution, this property translates into the insensitivity to any common sensor filtering by a transfer function having no zeros on the unit circle [14]. This insensitivity implies that the ML-STBF works by steering deep and sharp nulls toward spatially correlated interference and minimizing the norm of the weight vector which determines the amplification of the background noise.

From a statistical point of view, the ML-STBF cost functional (7.85) is intrinsically designed to optimally cope with a very spreaded and heavy-tailed PDF of output DFT coefficients, since it grows very slowly with the subband variances [21].

7.7.3 Regularization

The ML-STBF cost functional (7.85) is not convex far from the global minimum, due to the presence of the logarithm operator. Anyway the *trust region* [47] appears quite wide and is generally characterized by a smooth and well-directed gradient. However the functional is not lower bounded when some $\tilde{\sigma}_j^2(\mathbf{w}_1) = 0$ and may have deep local minima when $\tilde{\sigma}_j^2(\mathbf{w}_1) \approx 0$. This drawback is due to the fact that the output can be exactly cancelled out by some admissible weight vector and is surely avoided if each data matrix $\left[\mathbf{x}_f(\omega_j, 1) \cdots \mathbf{x}_f(\omega_j, L)\right]$ is full-rank. In particular DL or eigenvalue thresholding [23] can be effectively applied when $L \leq M_b$ [14].

In addition to subband regularization, it is advisable to impose the global norm constraint

$$\|\mathbf{w}_1\|_2^2 \leq \eta^2 = \left(\frac{1-\delta}{\varepsilon_{\max}}\right)^2 \tag{7.88}$$

to cope with effects of focusing errors and unexpected model mismatches [14, 28], with η selected as in (7.46). However, the typical value of $\varepsilon_{\max}$ for the use with the ML-STBF is much smaller than in the MV-FDBF, since

- Only perturbations affecting the direct path response $\mathbf{h}_d(\omega_j, \mathbf{p})$ (e.g., near scattering or sensor mismatches) are relevant for the ML-STBF, because the others are adaptively suppressed by the combined effects of the SOI prewhitening (7.86) and the temporal decorrelation of delayed arrivals (7.80);
- Amplification of model errors is negligible in the ML-STBF because of SOI prewhitening;

- The ML-STBF weight vector generally has a much smaller norm than the one of the MV-FDBF or the MV-STBF.

When (7.88) is applied to the ML-STBF functional (7.85), it gives origin to a nonquadratic RR optimization problem, which requires nonlinear programming techniques [47].

In addition, when (7.88) becomes active during optimization, it generally slows down the convergence rate. This occurrence is a clear symptom of gross model mismatches, that should be corrected through more accurate theoretical analysis or empirical calibration. Even in such extreme cases, the ML-STBF coupled with (7.88) often furnished acceptable results, while the MV-FDBF and the MV-STBF generally failed [14].

7.7.4 Link to Homomorphic Processing

Functional (7.85) can be interpreted as the sum of the log-spectrum of the outputs. In contrast the MV-STBF (7.77) sums the spectrum samples. This recalls the homomorphic processing concept [59]. In particular the insensitivity to the sensor output pre-filtering is analogous to the deconvolution property of *cepstral analysis* [61].

The capability of homomorphic processing of separating multipath contributions is well-known [59]. In addition, the ML-STBF employs a cepstrum-related measure, but synthesizes a linear processor. This is a clear advantage in those applications, like high fidelity recording [3] and SOI recognition problems, where nonlinear SOI distortion must be avoided as much as possible.

7.7.5 Link to Image Processing

Interestingly, (7.85) exhibits a striking resemblance with the ML detector and estimator of the slope and the offset of statistically independent straight patterns in an image [43], embedded in Gaussian noise. This problem arises, for example, in time-frequency analysis of chirp signals [62] and wavefield imaging. The optimal image pre-processor uses a preliminary Radon transform [63] to convert each straight pattern into a 1-D signal. Array focusing (7.59) acts in a similar way on the spatial covariance matrices, measured at different frequencies [10]. Moreover, in both the array and the image cases, the output of a 1-D whitening filter, applied to the aligned signal, furnishes the sufficient statistic for ML parameter estimation [14, 62].

7.8 ML-STBF OPTIMIZATION

7.8.1 Least Squares Coordinate Descent

The ML functional $\mathcal{L}_c(\mathbf{w}_1)$ depends upon two subsets of unknowns, namely the adaptive weight vector $\mathbf{w}_1$ and the subband variances $\tilde{\sigma}_j^2$. In this case minimization can be performed by cyclically optimizing $\mathbf{w}_1^{[k+1]}$ after recomputing each $\tilde{\sigma}_j^2(\mathbf{w}_1^{[k]})$ at the kth iteration. This practice is a simplified version of the *coordinate descent* method [47] and it turns out to be an iteratively reweighted LS (IRLS) method, frequently adopted in robust regressions [21]. Weights are different at each bin frequency and equal to $\tilde{\sigma}_j^{-2}(\mathbf{w}_1^{[k]})$.

As described by (7.80) and (7.86), this weighting at the equilibrium whitens the ML-STBF output, within the DFT spectral approximation. As a consequence, weighted output samples (i.e., the fitting residuals) can be considered as identically distributed and mutually uncorrelated.

The advantage of IRLS is the fast convergence near a local minimum [21], due to the excellent approximation of the Hessian. In contrast, convergence can be hampered by the sample approximations of weights and the introduction of a norm constraint on $\mathbf{w}_1$ requires a SVD recomputation of the weighted system matrix at each iteration [32]. This may be awkward with the large arrays and number of samples typically encountered in STBF applications [28].

7.8.2 Modified Newton Method

The Newton method is a well-known technique for solving unconstrained optimization problems [47]. It is based on a local quadratic approximation of the functional $\mathcal{L}_c(\mathbf{w}_1)$ defined in (7.83), which must be minimized. Its pure form is

$$\mathbf{w}_1^{[k+1]} = \mathbf{w}_1^{[k]} - \mathbf{H}^{-1}\left(\mathbf{w}_1^{[k]}\right)\nabla_{\mathbf{w}_1^{[k]}}\mathcal{L}_c(\mathbf{w}_1)$$

where $\mathbf{H}(\mathbf{w}_1^{[k]})$ and $\nabla_{\mathbf{w}_1^{[k]}}\mathcal{L}_c(\mathbf{w}_1)$ are respectively the Hessian matrix and the gradient of the functional $\mathcal{L}_c(\mathbf{w}_1)$, evaluated at the solution $\mathbf{w}_1^{[k]}$ obtained at the kth iteration. Newton method has order of convergence two,[18] when started sufficiently close to a solution point [47].

A *modified Newton's method* is any method expressed by

$$\mathbf{w}_1^{[k+1]} = \mathbf{w}_1^{[k]} - \alpha^{[k]}\mathbf{S}^{[k]}\nabla_{\mathbf{w}_1^{[k]}}\mathcal{L}_c(\mathbf{w}_1)$$

where $\mathbf{S}^{[k]}$ is a symmetric matrix and the *step size* $\alpha^{[k]}$ is chosen to minimize $\mathcal{L}_c(\mathbf{w}_1^{[k+1]})$. The steepest descent method and the pure Newton method correspond to $\mathbf{S}^{[k]} = \mathbf{I}$ and $\mathbf{S}^{[k]} = \mathbf{H}^{-1}(\mathbf{w}_1^{[k]})$ respectively. In order to yield a descent direction, $\mathbf{S}^{[k]}$ must be a positive definite matrix [32]. Ideally $\mathbf{S}^{[k]}$ should be as close as possible to $\mathbf{H}^{-1}(\mathbf{w}_1^{[k]})$ in the proximity of a local minimum. Unfortunately in the presence of highly nonlinear functionals the Hessian stays nonnegative definite or nearly singular. In ML estimation this means that some parameters cannot be uniquely identified. In these cases a proper choice of $\mathbf{S}^{[k]}$ can help in regularizing the functional and leading to an acceptable solution.

[18]Given a sequence of numbers $\{x_k\}$ converging to $\bar{x}$, its order of convergence is defined as the supremum of the non negative numbers p satisfying $0 \leq \lim_{k\to\infty} (|x_{k+1} - \bar{x}|/|x_k - \bar{x}|^p) < \infty$.

7.8.2.1 Descent in the Neuron Space.

The STBF can be viewed as a linear perceptron [31] with two layers (focusing and beamforming) and constrained weights. Since the output layer of the STBF computes a linear combination of focused snapshots, it turns out that the gradient of the ML functional can be expressed as the product between a matrix, whose columns are all of the snapshots $\mathbf{y}_1(\omega_j, l)$s ($j = j_1, \ldots, j_2$ and $l = 1, \ldots, L$) collected at the sidelobe canceller output, and the gradient vector of $\mathcal{L}_c(\mathbf{w}_1)$ with respect to each output sample $-y(\omega_j, l, \mathbf{w}_1^{[k]})$ [14]. This particular stochastic gradient was referred to as *gradient in the neuron space* and was initially introduced as a learning paradigm in a neural network framework [30].

From a signal processing point of view, the gradient in the neuron space represents the stochastic impact of each observation on the error functional. At a local minimum the gradient vector must be statistically orthogonal to the combiner input vectors. Thanks to the orthogonality principle [55], this property can be restated by affirming that the L_2 norm of the gradient in the neuron space must be minimum at the equilibrium [14]. As shown in the foregoing, this concept yields an iterative LS algorithm which can be interpreted as a modified Newton descent.

7.8.2.2 Data Preconditioning.

A square root training algorithm [32] is desirable for the ML-STBF, because of the numerical ill-conditioning, typical of real world beamforming problems.

In addition, proper preconditioning of data is required in order to guarantee that system matrices are numerically well balanced and that a good local approximation of the Hessian is actually employed. This task can be accomplished by the use of a reduced-size QR decomposition (R-QRD) [32], applied to the STBF output (7.69). More specifically, the following compact representation is introduced for the focused snapshots

$$\mathbf{Y}_{1,j}^H = [\mathbf{y}_1(\omega_j, 1) \quad \cdots \quad \mathbf{y}_1(\omega_j, L)] \tag{7.89}$$

$$\mathbf{y}_{0,j}^H = [y_0(\omega_j, 1) \quad \cdots \quad y_0(\omega_j, L)] \tag{7.90}$$

for $j = j_1, \ldots, j_2$. Since matrices $\mathbf{Y}_{1,j}$ are involved in an iterative optimization algorithm, it is essential to perform a proper size reduction. To this purpose, the R-QRD is applied to the $L \times (M_b + 1)$ matrix $\mathbf{Y}_j = [\mathbf{Y}_{1,j} \; \mathbf{y}_{0,j}]$ giving

$$\mathbf{Q}_{j,0}\mathbf{R}_{j,0} = \mathbf{Y}_j \tag{7.91}$$

where $\mathbf{R}_{j,0}$ is a square upper triangular matrix [32] of size $M_b + 1$. The R-QRD retains all the information about the CSCM required by the ML-STBF, in particular the subband MSE.

It is easy to recognize possible near-singularities of (7.85) from the presence of negligible entries on the main diagonal of any $\mathbf{R}_{j,0}$ [32]. Following a square root diagonal loading approach [23], near-singularity can be effectively dealt with by the further R-QRD

$$\mathbf{Q}_{j,\mu}\mathbf{R}_{j,\mu} = \begin{bmatrix} \mathbf{R}_{j,0} \\ \mu_j \mathbf{I}_{M_b+1} \end{bmatrix} \tag{7.92}$$

The real-valued parameter μ_j, which represents the standard deviation of the spatially white regularization noise in (7.42), can be independently set for each bin. The choice of μ_js is generally made on the basis of previous experiments, depending on the specific environment. It is important to remark that this choice is not critical in the ML-STBF approach, because of the very large number of equations and the simultaneous presence of linear and quadratic constraints [15, 28] on the weight vector. Alternative regularization schemes are also possible, such as a square root version of the eigenvalue thresholding algorithm (7.50) [23], through the R-SVD of $\mathbf{Y}_j$ [32].

7.8.2.3 Problem Setup.

The system matrix $\mathbf{F}$ and the target vector $\mathbf{g}$ are obtained by stacking weighted submatrices of $\mathbf{R}_{j,\mu_j}$ as follows

$$\mathbf{F} = \begin{bmatrix} \rho_{j_1}\mathbf{R}_{j_1,\mu_{j_1}}(1 : M_b + 1, 1{:}M_b) \\ \vdots \\ \rho_{j_2}\mathbf{R}_{j_2,\mu_{j_2}}(1 : M_b + 1, 1{:}M_b) \end{bmatrix} = \begin{bmatrix} \mathbf{F}_{j_1} \\ \vdots \\ \mathbf{F}_{j_2} \end{bmatrix} \tag{7.93}$$

$$\mathbf{g} = -\begin{bmatrix} \rho_{j_1}\mathbf{R}_{j_1,\mu_{j_1}}(1 : M_b + 1, M_b + 1) \\ \vdots \\ \rho_{j_2}\mathbf{R}_{j_2,\mu_{j_2}}(1 : M_b + 1, M_b + 1) \end{bmatrix} = \begin{bmatrix} \mathbf{g}_{j_1} \\ \vdots \\ \mathbf{g}_{j_2} \end{bmatrix}. \tag{7.94}$$

Weights ρ_j are chosen to approximate the Hessian of (7.85) as close as possible. In particular in the case of very good focusing and in the absence of unmodeled reflections, the ML-STBF should approach the same MSE $\sigma_j^2(\mathbf{w}_{1,j})$ as the subband MV-FDBF (7.27) at any frequency bin.[19] Since the quantity $\tilde{\sigma}_j^2(\mathbf{w}_{1,j}) = |\mathbf{R}_{j,\mu}(M_{b+1}, M_{b+1})|^2/L$ is an estimate of $\sigma_j^2(\mathbf{w}_{1,j})$ [32], a convenient weighting turns out to be $\rho_j \propto |\mathbf{R}_{j,\mu}(M_{b+1}, M_{b+1})'|^{-1}$.

The presence of temporally coherent, unexpected multipath [15] leads instead to $\tilde{\sigma}_j^2(\mathbf{w}_{1,j}) \ll \tilde{\sigma}_j^2(\mathbf{w}_1)$ and requires a slightly different weighting strategy [14].

[19] Deviations between the two methods are mainly due to the different number of independent snapshots employed [48], that generate a greater SOI cancellation in the MV-FDBF.

7.8.2.4 Optimization Loop. The following vector is computed at the generic kth iteration

$$\mathbf{z}\left(\mathbf{w}_1^{[k-1]}\right) = \mathbf{F}\mathbf{w}_1^{[k-1]} \tag{7.95}$$

The gradient in the neuron space $\nabla_z \mathcal{L}_c(\mathbf{w}_1^{[k-1]})$ is obtained as follows, by using complex derivatives satisfying the chain rule [55]. The fitting error is first computed as

$$\mathbf{e}\left(\mathbf{w}_1^{[k-1]}\right) = \mathbf{g} - \mathbf{z}\left(\mathbf{w}_1^{[k-1]}\right) = \begin{bmatrix} \mathbf{e}_{j_1}^{[k-1]} \\ \vdots \\ \mathbf{e}_{j_2}^{[k-1]} \end{bmatrix} \tag{7.96}$$

Each subvector $\mathbf{e}_j^{[k-1]}$ $(j = j_1, \ldots, j_2)$, of length M, contains the residuals at ω_j. In particular, the subband sample variances at the STBF output are given by $\tilde{\sigma}_j^2\left(\mathbf{w}_1^{[k-1]}\right) = \left\|\mathbf{e}_j^{[k-1]}\right\|_2^2$ for $j = j_1, \ldots, j_2$ [32]. Then

$$\nabla_z \mathcal{L}_c\left(\mathbf{w}_1^{[k-1]}\right) = \frac{\partial \mathcal{L}_c\left(\mathbf{w}_1^{[k-1]}\right)}{\partial \mathbf{z}(\mathbf{w}_1^{[k-1]})} = -\frac{2}{L}\begin{bmatrix} \dfrac{\mathbf{e}_{j_1}^{[k-1]}}{\tilde{\sigma}_{j_1}^2\left(\mathbf{w}_1^{[k-1]}\right)} \\ \vdots \\ \dfrac{\mathbf{e}_{j_2}^{[k-1]}}{\tilde{\sigma}_{j_2}^2\left(\mathbf{w}_1^{[k-1]}\right)} \end{bmatrix} \tag{7.97}$$

Finally, the weight vector $\mathbf{w}_1^{[k]}$ is found by solving the quadratic RR problem [32]

$$\mathbf{F}\mathbf{w}_1^{[k]} \approx \mathbf{z}\left(\mathbf{w}_1^{[k-1]}\right) - \alpha^{[k]} \nabla_z \mathcal{L}_c\left(\mathbf{w}_1^{[k-1]}\right) \tag{7.98}$$

subject to $\left\|\mathbf{w}_1^{[k]}\right\|_2^2 \leq \eta^2$, according to (7.88).

The algorithm is initialized by an arbitrary weight vector $\mathbf{w}_1^{[0]}$ and for a sufficiently small step size $\alpha^{[k]} > 0$ is proven to locally converge [30]. Moreover, since the desired solution is characterized by a small norm, $\mathbf{w}_1^{[0]} = \mathbf{0}$ is conveniently chosen and never failed in our experiments.

The factors of the R-SVD of $\mathbf{F}$, required by the RR approach [32], are computed and stored before starting the cycle. This trick greatly speeds up the execution of the loop (7.98), which requires only matrix-vector multiplications and the solution of a nonlinear secular equation in a single variable [32].

The algorithm is stopped when $\|\mathbf{w}_1^{[k]} - \mathbf{w}_1^{[k-1]}\|_2 \leq tol \times \|\mathbf{w}_1^{[k-1]}\|_2$, being tol a small positive constant (e.g., 10^{-4}). In practical applications a variable step size should be employed to ensure the optimal descent of (7.85) [47]. However, during simulations of the ML-STBF, the choice

$$\alpha^{[k]} = \frac{L}{2}\left(\frac{1}{j_2 - j_1 + 1}\sum_{j=j_1}^{j_2}\frac{1}{\tilde{\sigma}_j^4(\mathbf{w}_1^{[k-1]})}\right)^{-1/2} \tag{7.99}$$

proved to be very effective, since the convergence was mostly achieved within 20 iterations.

The variable step size may be effective especially when the norm constraint is operating. In fact, in these cases the convergence of the ML-STBF was found to become only linear [47], because the RR regularization made the Hessian replacement matrix of (7.98) closer to $\mathbf{I}_{M_b}$, as typical of steepest descent [30].

The modified Newton algorithm is preferable to the IRLS approach because of the higher speed of the inner loop. In the vast majority of cases the presented algorithm required only a few more iterations than the coordinate descent counterpart, in spite of possible Hessian mismatches near convergence [47].

In addition, the proposed algorithm is very flexible since it allows to incorporate other kinds of constraints and adaptive updates of sample matrices [30].

7.8.2.5 Algorithm Summary.

The full ML-STBF algorithm is summarized below.

Step 1. Collect $N = LJ$ wideband snapshots $\mathbf{x}(n)$, for $n = 1, \ldots, N$.

Step 2. Compute subband snapshots $\mathbf{x}(\omega_j, l)$ for $l = 1, \ldots, L$ and $j = 1, \ldots, J$, using a J-point windowed FFT, applied to L consecutive blocks of $\mathbf{x}(n)$.

Step 3. For $j = j_1, \ldots, j_2$ and $l = 1, \ldots, L$, apply focusing matrices $\mathbf{T}_j$s, obtaining the focused subband snapshots $\mathbf{x}_f(\omega_j, l) = \mathbf{T}_j\mathbf{x}(\omega_j, l)$.

Step 4. For $j = j_1, \ldots, j_2$ and $l = 1, \ldots, L$, build the targets $y_0(\omega_j, l) = \mathbf{w}_0^H\mathbf{x}_f(\omega_j, l)$ and the sidelobe canceller output vectors $\mathbf{y}_1(\omega_j, l) = \mathbf{C}_\perp^H(\mathbf{p})\mathbf{x}_f(\omega_j, l)$.

Step 5. For $j = j_1, \ldots, j_2$, build regularized matrices $\mathbf{R}_{j,\mu}$ through (7.89), (7.90), (7.91) and (7.92).

Step 6. Compute the system matrix $\mathbf{F}$ and the target vector $\mathbf{g}$, according to (7.93) and (7.94).

Step 7. Initialize $\mathbf{w}_1^{[0]}$ with all zeros or small complex random values [30].

Step 8. For $k = 1, 2, \ldots$, iterate until convergence to $\tilde{\mathbf{w}}_{1ML}$, using sequentially (7.95), (7.99), (7.99) and solving the RR problem (7.98), subject to $\|\mathbf{w}_1^{[k]}\|_2^2 \le \eta^2$.

Step 9. Compute the optimal weight vector $\tilde{\mathbf{w}}_{ML}$ using (7.67) and/or the output sequence $y(\omega_j, l, \tilde{\mathbf{w}}_{1:ML})$ by (7.69).

7.8.3 Computational Cost

The ML-STBF, like all steered beamformers [10], has a high computational cost, since focusing must be repeated for each direction of interest. This is the price paid for a robust estimator, acting on a particular slice of received data. Robustness is intrinsic to the slice operating mode, which allows to design simple estimators working in a single dimensional space [21]. This robustness can be extended to elliptical multivariate distributions only with severe difficulty [19, 27].

In the slice approach the observed space is reconstructed through the analysis of several directions [21]. In the ML-STBF case the directions are not arbitrary but are set by the candidate source positions.

Working on several directions suggests to apply focusing matrices after the QRD compression (7.91) of raw snapshots $\mathbf{x}(\omega_j, l)$. The use of a general focusing matrix with P directions implies a cost of about $(2 + P)M^2L(j_2 - j_1 + 1)$ complex flops (CFLOPs) [32]. This cost must be added to the one of initial FFTs, which is $\frac{LMJ}{2}\log_2\left(\frac{J}{2}\right)$ CFLOPs [59]. Moreover, for each direction of interest a tiny R-SVD of size $(j_2 - j_1 + 1)M \times M_b$ must be computed, globally leading to $3P(j_2 - j_1 + 1)MM_b^2$ CFLOPs.

These costs are common to both the MV-STBF and the ML-STBF using RR. In addition the ML-STBF requires two matrix-vector multiplications for look direction and iteration, equivalent to $2P(j_2 - j_1 + 1)MM_b$ CFLOPS, times the average number of iterations. In practice, the cost of the common pre-processing largely dominates [14].

Given the power of available computing resources, the intrinsic pipelineability of many operations and the cost associated with imaging or SOI reconstruction and analysis, the ML-STBF overhead with respect to the MV-STBF is totally tolerable.

7.9 SPECIAL TOPICS

The performance of adaptive STBFs in a specific application can be improved by solving side problems according to robust estimation paradigms [21].

7.9.1 Unitary Matched Field Focusing

MF concepts can be incorporated into the STBF by properly redefining the direct path response $\mathbf{h}_d(\omega_j, \mathbf{p})$ in (7.59), after including multiple modes, obtained by theoretical analysis or numerical simulation [5, 24].

Recalling the model (7.7) introduced in Section 7.3.4, a subset of Q_0 *dominant modes* is selected for focusing. The relative propagation delays of these modes should fall within a few T_r, while the subset should be large enough to ensure 3-D localization capabilities [24].

The array response now obeys the hybrid MF/partial response model

$$\tilde{\mathbf{h}}(\omega_j, \mathbf{p}) = \mathbf{H}(\omega_j, \mathbf{p})\mathbf{a}(\mathbf{p}) + \mathbf{h}_r(\omega_j, \mathbf{p}) \tag{7.100}$$

where the residual multipath response $\mathbf{h}_r(\omega_j, \mathbf{p})$ is considered as a random quantity.

The focusing operation (7.59) is properly redefined as

$$\mathbf{T}_j\mathbf{H}(\omega_j, \mathbf{p})\mathbf{a}(\mathbf{p}) = \mathbf{a}_d(\mathbf{p})g_d(\omega_j, \mathbf{p}) + \mathbf{e}_f(\omega_j, \mathbf{p}) \tag{7.101}$$

This focusing adds up the contributions of multiple modes, potentially enhancing the SNR and equalizing the beamformer response to the SOI. The target response $\mathbf{a}_d(\mathbf{p})$ should be drawn from an array capable of 3-D localization in the specific environment, to avoid identifiability losses. Finally, unitary focusing matrices are computed according to (7.71).

An intriguing alternative to (7.101) is to focus the entire array response $\tilde{\mathbf{h}}(\omega_j, \mathbf{p})$ onto $\mathbf{a}_d(\mathbf{p})$ over a grid of nearby locations by (7.71). This idea is based on a set of loose assumptions:

- The steering vector $\mathbf{H}(\omega_j, \mathbf{p})\mathbf{a}(\mathbf{p})$ numerically spans a subspace of dimension $Q_1 \ll M$ when $\mathbf{p}$ varies within a ball $B(\mathbf{p}_0)$ centered on the reference position $\mathbf{p}_0$;
- The numerical rank Q_1 can be estimated from the dimension of the subspace spanned by $\mathbf{a}_d(\mathbf{p})$, for $\mathbf{p} \in B(\mathbf{p}_0)$. This hypothesis implies a low expected focusing error;
- $\mathbf{h}_r(\omega_j, \mathbf{p})$ quickly changes with $\mathbf{p}$. In particular it should be $E[\mathbf{h}_r(\omega_j, \mathbf{p})] = 0$ over $B(\mathbf{p}_0)$. The elements of $\mathbf{h}_r(\omega_j, \mathbf{p})$ can be assumed as identically distributed (i.i.d.) circular variables, characterized by $E[\mathbf{h}_r(\omega_j, \mathbf{p})\mathbf{h}_r^H(\omega_j, \mathbf{p})] = \sigma_r^2(\omega_j, \mathbf{p}_0)\mathbf{I}_M$ [46]. This hypothesis is reasonable in the case of significant diffraction [12];
- $\|\mathbf{h}(\omega_j, \mathbf{p})\|_2$ is almost constant within $B(\mathbf{p}_0)$.

The SVD factors $\mathbf{U}_j$ and $\mathbf{V}_j$ in (7.73) are respectively given by the eigenvectors of the Hermitian matrices $\mathbf{A}_0\tilde{\mathbf{A}}_j^H\tilde{\mathbf{A}}_j\mathbf{A}_0^H$ and $\tilde{\mathbf{A}}_j\mathbf{A}_0^H\mathbf{A}_0\tilde{\mathbf{A}}_j^H$ [57]. Under the given assumptions, $E[\tilde{\mathbf{A}}_j] = \mathbf{A}_j$ and $E[(\tilde{\mathbf{A}}_j - \mathbf{A}_j)(\tilde{\mathbf{A}}_j - \mathbf{A}_j)^H] = \beta_j^2\mathbf{I}_M \propto \sigma_r^2(\omega_j, \mathbf{p}_0)\mathbf{I}_M$ are obtained. As a consequence

$$\begin{aligned} E\left[\mathbf{A}_0\tilde{\mathbf{A}}_j^H\tilde{\mathbf{A}}_j\mathbf{A}_0^H\right] &= \mathbf{A}_0\mathbf{A}_j^H\mathbf{A}_j\mathbf{A}_0^H + \beta_j^2\mathbf{A}_0\mathbf{A}_0^H \\ E\left[\tilde{\mathbf{A}}_j\mathbf{A}_0^H\mathbf{A}_0\tilde{\mathbf{A}}_j^H\right] &= \mathbf{A}_j\mathbf{A}_0^H\mathbf{A}_0\mathbf{A}_j^H + \beta_j^2\mathbf{I}_M \end{aligned} \tag{7.102}$$

It turns out that the estimate of $\mathbf{V}_j$ is unbiased while the bias of $\mathbf{U}_j$ can be removed by computing the EVD of $\tilde{\mathbf{A}}_j^H\tilde{\mathbf{A}}_j$ followed by noise subtraction [35]. Since the rank of $\mathbf{A}_j^H\mathbf{A}_j$ is known, the variance β_j^2 can be estimated by averaging the eigenvalues of the resulting noise subspace of size $M - Q_1$ [64]. The validity of this MF-based focusing will be demonstrated in the experiments described in Sections 7.10.3 and 7.10.4.

7.9.2 Quiescent Vector Selection

Focusing errors can be viewed as random perturbations of the steering vector $\mathbf{a}_d(\mathbf{p})$ with respect to frequency. According to the MF paradigm, this problem should be afforded within the random model (7.10), by finding the appropriate response correlation matrix $\mathbf{R}_{hh}(\mathbf{p})$ [5]. The eigenvalues and the eigenvectors of this matrix can be found by the SVD of the $(M \times (j_2 - j_1 + 1))$ matrix

$$\mathbf{Z}_T(\mathbf{p}) = \left[\mathbf{T}_{j_1}\mathbf{h}(\omega_{j_1}, \mathbf{p}) \ \cdots \ \mathbf{T}_{j_2}\mathbf{h}(\omega_{j_2}, \mathbf{p})\right] \tag{7.103}$$

In particular, the quiescent weight vector can be selected as the dominant left singular vector of $\mathbf{Z}_T(\mathbf{p})$.

The blocking matrix $\mathbf{C}_\perp(\mathbf{p})$ is defined by the orthonormal set of left singular vectors of $\mathbf{Z}_T(\mathbf{p})$, corresponding to negligible singular values [5].

This approach can be refined by weighting each term $\mathbf{T}_j\mathbf{h}(\omega_j, \mathbf{p})$ $(j = j_1, \ldots, j_2)$ within a robust pseudo-covariance solution (WAVES [19]).

7.9.3 Nonstationary Processing

Processing of nonstationary (even narrowband) signals is naturally accomplished by a straightforward modification of the ML-STBF approach. In particular, for each ω_j, the subband snapshot sequence $\mathbf{x}_f(\omega_j, l)$ $(l = 1, 2, \ldots, L \cdot L_2)$ containing nonstationary SOI or interference is partitioned into L_2 consecutive blocks of length L. Each block is considered as derived from a different subband, statistically characterized by the variance $\sigma^2_{j,r}(\mathbf{w}_1)$ for $j = j_1, \ldots, j_2$ and $r = 1, \ldots, L_2$.

The ML-STBF cost functional (7.85) becomes

$$\mathcal{L}_c(\mathbf{w}_1) = \sum_{j=j_1}^{j_2}\sum_{r=1}^{L_2} \ln\left[\tilde{\sigma}^2_{j,r}(\mathbf{w}_1)\right]$$
$$\tilde{\sigma}^2_{j,r}(\mathbf{w}_1) = \frac{1}{L}\sum_{l=(r-1)L+1}^{rL} \left|y(\omega_j, l, \mathbf{w}_1)\right|^2 \tag{7.104}$$

and a compromise $\tilde{\mathbf{w}}_{1ML}$ is computed for the entire batch.

7.10 EXPERIMENTS

The importance of robustness and the validity of the ML-STBF will be demonstrated by the following experiments, using both simulated and live data.

7.10.1 Colored Sources

In this experiment, the ML-STBF and the MV-STBF were compared in a computer simulated scenario, involving far-field and colored sources.

A uniform linear array (ULA) of 10 omnidirectional sensors received the signals radiated by two far-field AR sources [29], driven by independent, equipowered, Gaussian white processes and located at 7° and 15°, referred to broadside.

The AR transfer functions were respectively $H_7(z) = \left[(1 - 0.9e^{-i\pi/3}z^{-1})(1 - 0.8e^{i\pi/12}z^{-1})\right]^{-1}$ and $H_{15}(z) = \left(1 - 0.9e^{i\pi/4}z^{-1}\right)^{-1}$. The additive background noise was Gaussian, temporally and spatially white. The SNR was 20 dB, referred to the driving noise power of each source at a single sensor. The sensor pass-band was 60–140 Hz and the array was steered to the direction of interest using diagonal, unitary focusing matrices (7.70), with $\mathbf{a}_d(\mathbf{p}) = (\sqrt{10})^{-1}\mathbf{1}$ [10]. Sensor spacing was half wavelength at the focusing frequency of 100 Hz [26].

Baseband array outputs, sampled at 80 Hz, were processed by a 64-point FFT, using a rectangular window, with no overlap between consecutive blocks. $L = 100$ subband snapshots from 33 bins, corresponding to the analog band 80–120 Hz, were used for adaptation. The quiescent vector was chosen according to the WAVES approach described in Section 7.9.2 and $\eta^2 = 1$ was imposed to both STBFs [28].

Figure 7.4 compares the output sample spectra of the two STBFs, averaged over 100 independent realizations and plotted versus the look angle. The steering angle was sampled from −5° to 25° in steps of 1°.

It is evident that source spectra were better preserved by the ML-STBF, which *captured* the nearest signal and effectively suppressed the other one, considered as interference. The hand-over between the two sources abruptly occurred near 12°. Spectral levels of each SOI remained almost constant within about half beamwidth.

In contrast the MV-STBF exhibited significant interference residuals, except when the array was steered exactly toward one source, and canceled out the SOI even for small pointing errors.

Residuals of the signal impinging from 7° showed up at steering angles between 20° and 25° with both beamformers. This is due to the relatively high focusing errors of diagonal matrices [28].

As a further result, the different performance between the two STBFs held even for $L < M$ subband snapshots, thanks to the use of DL regularization [14].

7.10.2 Sonar Data

This experiment used the vertical array data, recorded off the North coast of the island of Elba, in front of the West coast of Italy, on October 26, 1993, by the NATO SACLANT Centre of La Spezia, Italy [5]. Original data and documentation files were obtained from the IEEE Signal Processing Information Base (SPIB) library.

A moored vertical ULA of 48 omnidirectional hydrophones, interspaced by 2 m, was immersed in shallow water (127 m depth). The top hydrophone was at a depth of 18.7 m. Array outputs were originally sampled at 1 KHz and each record contained 65,536 consecutive snapshots.

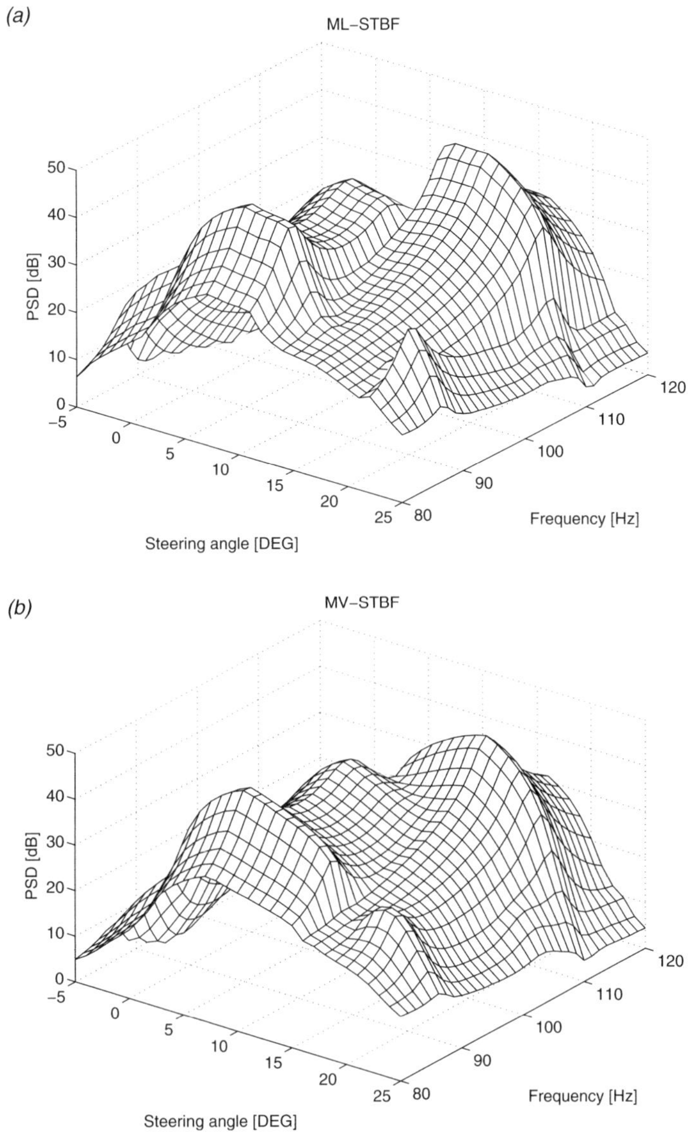

Figure 7.4 Plot of the average output spectra versus the steering angle for two colored, far-field sources. (*a*) ML-STBF; (*b*) MV-STBF.

This environment was characterized by a multimodal waveguide type propagation, with the presence of multipath, reverberation and diffraction [5]. In particular, the STBF was applied to separate single propagation modes, roughly modeled as partially coherent and spatially perturbed plane waves, embedded in colored noise.

The data file No. 9 of the first SACLANT experiment was analyzed. A stationary narrow-band source radiating at 170 Hz was located at 5.8 km from the array, at a depth of 79 m. Results from file No. 1 were previously reported in [14]. The entire file was processed by a 512-point FFT, using a Hann window and 25 percent overlap between blocks. An average sound speed of 1515 m/s was assumed and 35 frequency bins were used for adaptation in the band 136.72–203.13 Hz.

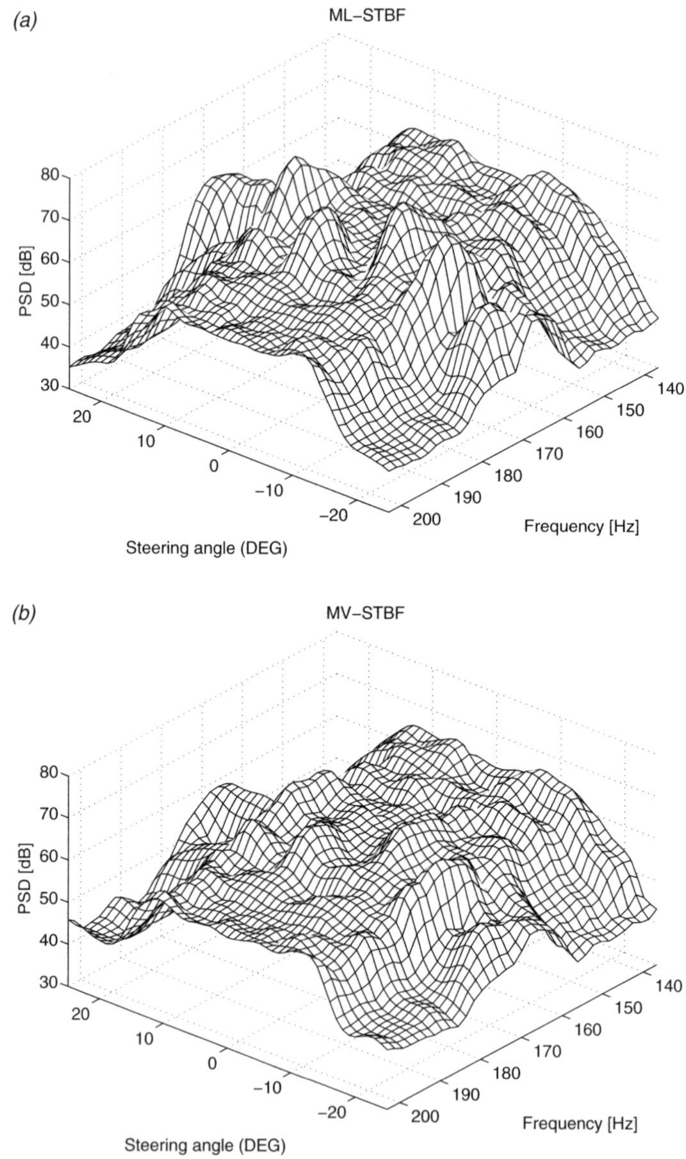

Figure 7.5 Plot of the sample output spectra versus the steering angle for the Mediterranean Vertical Array Data. File No. 9 of the first session. (*a*) ML-STBF; (*b*) MV-STBF.

Figure 7.6 The eight microphone array at the INFOCOM Department of the University of Rome "La Sapienza."

The array was divided into two subarrays formed by even- and odd-numbered hydrophones to mitigate ill-conditioning due to spatial oversampling [39]. The two subarray snapshots were stacked into each matrix $\mathbf{Y}_{1,j}$ and into the corresponding vector $\mathbf{y}_{0,j}$ to ensure some forward spatial smoothing [11] and to compensate for systematic calibration errors through the reduction of the statistical impact of each observation [21].

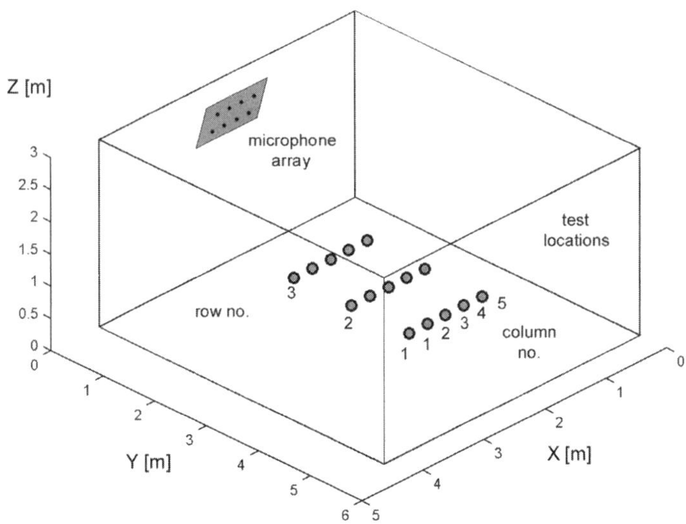

Figure 7.7 Test sources and array positions within the INFOCOM laboratory.

Analytical far-field ULA steering vectors were assumed for focusing and the quiescent vector was computed through the WAVES algorithm [19]. Though the array was perturbed by sea stream, $\eta^2 = 1$ and the single DR constraint (7.75) were used.

Figure 7.5 shows the sample output spectra of the two STBFs versus the look angle, varying between -25° and 25° referred to broadside, in steps of 1°. Positive angles look toward the sea bottom. The wavenumber scan of the ML-STBF

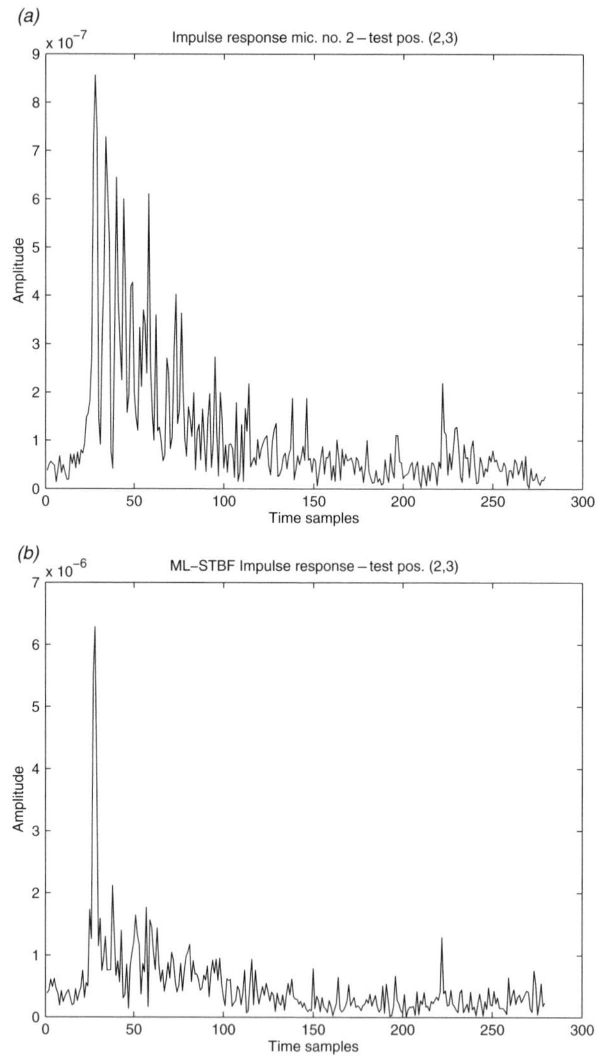

Figure 7.8 Plot of the amplitudes of measured impulse responses. The SOI is located at (2.10, 3.11, 0.82 m) within the room. (*a*) Single microphone output. (*b*) ML-STBF output.

exhibited sharp and well resolved peaks at the source frequency, likely corresponding to propagating modes. In contrast, the same modes were barely distinguishable from the background noise using the MV-STBF. In addition the SOI spectrum was not recognizable, due to near-complete cancellation near the spectral peak.

The lower norm of the ML-STBF weight vector resulted in a noise floor about 10 dB lower at the extreme look angles.

7.10.3 Dereverberation

The next two tests were performed in the Circuit Theory Laboratory of the INFOCOM Dept. of the University of Rome "La Sapienza," using an eight-element rectangular (two rows by four columns) microphone array, in a room of size $4.23 \times 5.5 \times 2.93$ m. Intersensor spacing was 0.2 m. AKG *C5625M* condenser microphones having hemispherical directivity pattern were mounted on a damped wooden panel of size 0.6×1 m, at position (1.78, 0.195, 2.40) m. The array is shown in Figure 7.6. The reverberation time of the room was relatively long ($T_{60} = 0.45$ s [12]).

The array was calibrated on 15 positions, depicted in Figure 7.7 and located within one beamwidth. Sensor signals were sampled at 44100 Hz, 16 bit. They were converted to the baseband, low-pass filtered and decimated at 1764 Hz. Two sets of 512 unitary focusing matrices were synthesized, one set using the direct path analytical array steering vectors and the other one using the MF focusing described in Section 7.9.1 and calibrated impulse responses. Focusing filterbanks

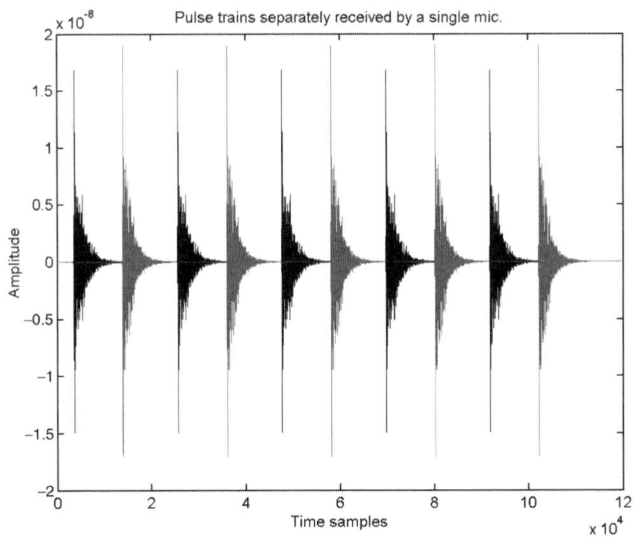

Figure 7.9 Trasmitted pulse trains. Light gray: source at position A (1.80, 4.22, 0.82) m. Dark gray: source at position B (2.70, 3.11, 0.82) m.

were obtained from the two sets of focusing matrices in the baseband and applied to the signal. A 512-point FFT with Hann window and 25 percent overlap was applied and the bins corresponding to the band 375–1125 Hz were processed. So the array had no ambiguities in the analyzed sector. In all experiments the SOI level was adjusted to get an average SNR of 40 dB, referred to each sensor.

In the first experiment a loudspeaker was located at position (2.10, 3.11, 0.82) m, at a distance of about 3.3 m from the array geometrical center, and played a musical piece. The microphone panel was vertically tilted toward the source (24.5° with respect to the vertical axis).

The ML-STBF was trained for about five seconds and its output impulse response computed using the MF focusing filters, a weighted SVD quiescent vector and $\eta^2 = 1$.

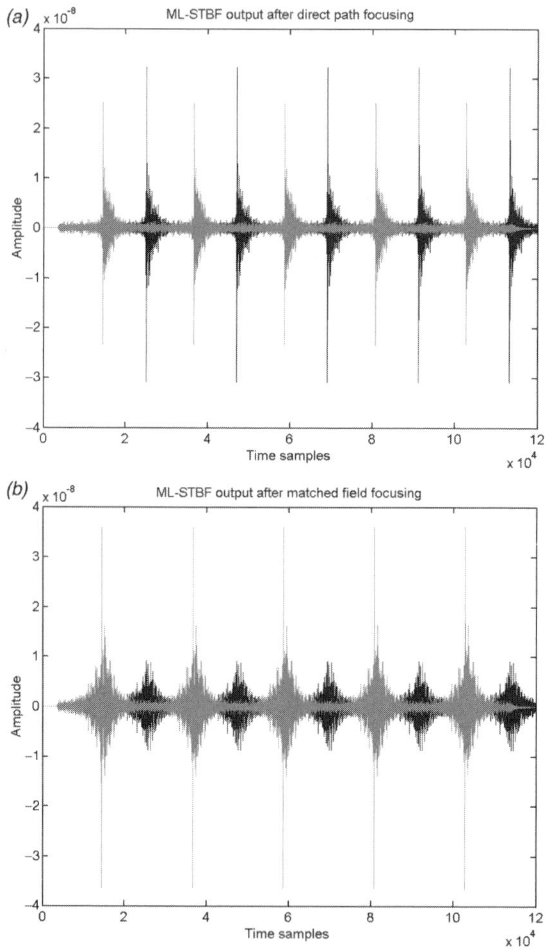

Figure 7.10 ML-STBF discrete time outputs of the two test sequences after adaptation. (*a*) Direct path focusing. (*b*) MF focusing. Light gray: source at position A (1.80, 4.22, 0.82) m. Dark gray: source at position B (2.70, 3.11, 0.82) m.

Figure 7.8 shows the amplitudes of the ML-STBF impulse response and of the impulse response of a single microphone, measured at the baseband and sampled at 750 Hz. The array gain and the multipath suppression are clearly visible.

7.10.4 Acoustic Source Separation

The second experiment at the INFOCOM Laboratory used the same arrangement and settings of the first experiment. In this case the array was trained to separate two acoustic sources located at positions A (1.80, 4.22, 0.82) m and B (2.70, 3.11, 0.82) m and playing a musical piece. To make visualization of results easier, pulse trains spaced by one second were passed through the adapted ML-STBF and then interpolated up to 22,050 Hz. The ML-STBF was first steered on source A. Waveforms received by a single microphone are superposed in Figure 7.9 (source A in light gray). Figure 7.10 shows the separated signals using the direct path and the MF focusing. It is clear that the direct path focusing does not work within one beamwidth. The MF focusing obtained a separation of about 10 dB between source signals. This is a remarkable result in spite of the reverberated sound reentering through the quiescent path. In particular, playback from the cancelled source was not intelligible, being essentially composed by diffraction residuals. In addition, it can be seen that MF focusing with phase correction approximately led to a zero phase impulse response with respect to the SOI [59].

To complete the experiment, Figure 7.11 shows the ML-STBF beampattern obtained when the array is steered to source B, under the assumption of perfect focusing [19]. The depth of the nulls and the height of the main lobe are remarkable, despite the small size of the array.

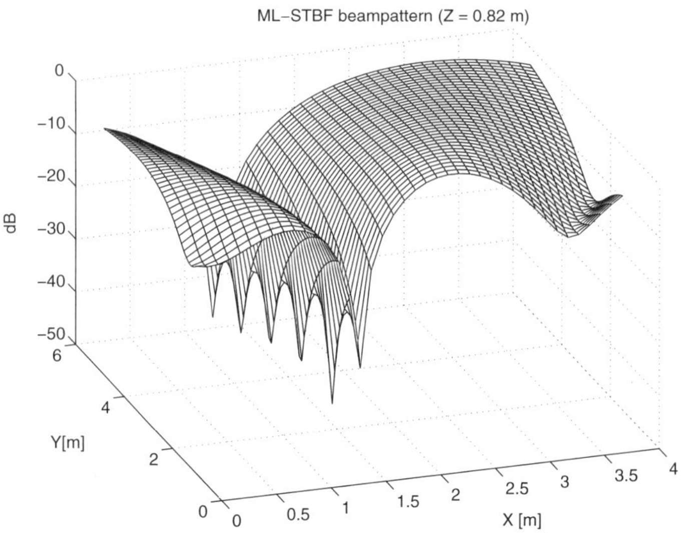

Figure 7.11 Estimated beampattern for the ML-STBF steered to position B (2.70, 3.11, 0.82) m.

7.11 SUMMARY

Wideband beamforming can exploit one more dimension than its narrowband counterpart, that is, the frequency. While this generally implies a higher computational complexity, it can be usefully employed to reduce finite data effects and discriminate among model mismatches induced by multipath and reverberation. In particular, parsimonious wideband architectures, such as the ML-STBF, can combine statistical optimality and intrinsic robustness versus the vast majority of catastrophic errors.4

GLOSSARY

Coherence. Two wavefronts impinging onto an array are said to be coherent when they carry temporally correlated replicas of the same signal [18]. This implies that their relative delay of arrival is smaller than the impulse response length plus the correlation time [14]. Coherent arrivals cannot be separated in space without detailed knowledge of the propagation model and the use of special algorithms [65]. In wideband array processing coherence may be influenced by signal pre-filtering [61].

Condition number. The condition number of a matrix is the ratio between its maximum and minimum singular values [32]. It furnishes a bound on the accuracy of many numerical algorithms running on a computer with finite precision arithmetic.

Coordinate descent. Iterative minimization technique of a multivariate functional. The functional $F(\mathbf{x}, \mathbf{y})$ is alternatively minimized with respect to the two subsets of parameters $\mathbf{x}$ and $\mathbf{y}$. Under proper conditions [47], this iterative process converges to a local minimum of F.

Frobenius norm. Frobenius norm of a matrix. It is defined as the Euclidean norm of a vector stacking all the columns of a matrix [32].

Hessian. The Hessian of a functional $F(\mathbf{x})$, where $\mathbf{x}$ is the vector stacking the real-valued parameters $\{x_k;\ k = 1, 2, \ldots, K\}$, is the square, symmetric matrix $\mathbf{H}$ with entries $\mathbf{H}_{k,l} = (\partial^2 F(\mathbf{x})/\partial x_k \partial x_l)$.

Modified Newton method. An optimization algorithm which replaces the Hessian matrix $\mathbf{H}$ in the Newton descent formula [47] with a properly chosen positive definite matrix.

Multipath. Multipath indicates propagation characterized by the presence of a few reflections originated by surfaces whose size is much larger than the wavelength of the incident signal. Each reflection carries an undistorted replica of the original signal. The reflected wavefronts are well approximated by plane waves, even at short distance.

Mutual coupling. Mutual coupling represents the interference between pairs of nearby sensors within an array [11]. At a macroscopic level, it can be viewed as the effect of the waves back-scattered by sensors. Mutual coupling linearly transforms the ideal steering vector through a symmetric (not Hermitian) matrix.

Neuron. A neuron is a processing element formed by a linear combiner, followed by a nonlinear activation function [31]. The neuron space [30] is formed by the outputs of linear combiners within a neuron layer [31].

Procrustes rotation. Orthogonal Procrustes rotation. This optimization problem finds the unitary $(n \times n)$ matrix $\mathbf{Q}$ which minimizes $\|\mathbf{A} - \mathbf{QB}\|_F$, being $\mathbf{A}$ and $\mathbf{B}$ two complex-valued $(n \times m)$ matrices [32]. The solution is given by $\mathbf{Q} = \mathbf{UV}^H$, where $\mathbf{U}$ and $\mathbf{V}^H$ are the unitary matrices containing, respectively, the left and right singular vectors singular vectors of $\mathbf{AB}^H$.

QRD. QR decomposition. The full size QRD decomposes the $(m \times n)$ matrix $\mathbf{A}$, with $m \geq n$, as $\mathbf{A} = \mathbf{QR}$. The $(m \times m)$ matrix $\mathbf{Q}$ is unitary and $\mathbf{R}$ is an upper triangular $(m \times n)$ matrix [32]. The reduced size QRD is defined as $\mathbf{A} = \mathbf{Q}_1\mathbf{R}_1$, being $\mathbf{Q}_1$ the orthogonal $(m \times n)$ matrix [32], made up by the first n columns of $\mathbf{Q}$, and $\mathbf{R}_1$ is the upper $(n \times n)$ submatrix of $\mathbf{R}$.

Reverberation. Reverberation is a complex phenomenon, due to the superposition of a large number of wavefronts reflected by rough surfaces (diffuse field) and edges (diffracted waves) [12].

Ridge regression. The (standard) ridge regression problem finds the vector $\mathbf{x}$ which minimizes $\|\mathbf{Ax} - \mathbf{b}\|_2$, subject to the constraint $\|\mathbf{x}\|_2^2 < \alpha^2$, being α a real-valued constant [32]. Using the SVD, the ridge regression requires the solution of a nonlinear secular equation in a single unknown, inducing a marginal computational overhead with respect to the unconstrained LS solution [32]. In this chapter, an extension of the ridge regression to a more general error functional is actually used.

Scattering. Scattering indicates reflections from obstacles whose size is smaller than the wavelength. In this case the reflected wavefield is well approximated by the superposition of cylindrical or spherical waves [40].

SVD. Singular value decomposition. The full size SVD decomposes the $(m \times n)$ matrix $\mathbf{A}$ as $\mathbf{A} = \mathbf{U\Sigma V}^H$ [32]. $\mathbf{U}$ is the unitary $(m \times m)$ matrix having the left singular vectors as columns. $\mathbf{V}$ is the unitary $(n \times n)$ matrix of right singular vectors. $\mathbf{\Sigma}$ is a diagonal $(m \times n)$ matrix, having nonnegative diagonal entries, called the *singular values* [32], ordered in a nonincreasing manner from the left-top corner. The reduced size SVD is defined for $m > n$ as $\mathbf{A} = \mathbf{U}_1\mathbf{\Sigma}_1\mathbf{V}^H$, where $\mathbf{U}_1$ is the orthogonal $(m \times n)$ matrix containing the first n columns of $\mathbf{U}$ and $\mathbf{\Sigma}_1$ is the upper $(n \times n)$ diagonal submatrix of $\mathbf{\Sigma}$.

ACKNOWLEDGMENTS

This work was supported in part by the Italian Ministry for Education, University and Research (M.I.U.R.).

The authors wish to thank Dr. M. Moreschini for performing some of the presented experiments with real world acoustic data at the INFOCOM Department.

Authors thank the NATO SACLANT Centre of La Spezia, Italy, for publishing the vertical array data employed for the experiment described in Section 7.10.2.

REFERENCES

1. William of Ockham. *Super Quattuor Libros Sententiarum*. Lugd., 1495. i, dist. 27, qu. 2, K.
2. P. M. Schultheiss and H. Messer. Optimal and suboptimal broad-band source location estimation. *IEEE Trans. on Signal Processing*, 41(9), Sept. 1993.
3. M. Brandstein and D. B. Ward, Eds. *Microphone Arrays: Techniques and Applications*. Springer-Verlag, May 2001.
4. Y. Li, K. C. Ho, and C. Kwan. Design of broad-band circular ring microphone array for speech acquisition in 3-D. In *Proc. of the 2003 IEEE Int. Conf. Acoustics, Speech, and Signal Processing*, Volume V, pp. 221–224, Apr. 6, 10, 2003.
5. J. L. Krolik. The performance of matched-field beamformers with Mediterranean vertical array data. *IEEE Trans. on Signal Processing*, 44(10):2605–2611, Oct. 1996.
6. G. Jacovitti, A. Neri, and G. Scarano. Pulsed response of focused arrays in echographic B-mode systems. *IEEE Trans. on Sonics and Ultrasonics*, SU-32(6):850–860, Nov. 1985.
7. O. L. Frost. An algorithm for linearly constrained adaptive antenna array processing. *Proc. IEEE*, 60:926–935, Aug. 1972.
8. G. Su and M. Morf. Signal subspace approach for multiple wide-band emitter location. *IEEE Trans. on Acoustics Speech and Signal Processing*, 31:1502–1522, Dec. 1983.
9. S. F. Boll. Suppression of acoustic noise in speech using spectral subtraction. *IEEE Trans. on Acoustics, Speech and Signal Processing*, ASSP-27:113–120, Apr. 1979.
10. J. L. Krolik and D. N. Swingler. Multiple wide-band source location using steered covariance matrices. *IEEE Trans. on Acoustics Speech and Signal Processing*, 37(10):1481–1494, Oct. 1989.
11. M. Wax and J. Sheinvald. Direction finding of coherent signals via spatial smoothing for uniform circular arrays. *IEEE Trans. on Antennas and Propagation*, 42(5):613–620, May 1994.
12. H. Kuttruff. *Room Acoustics*. Elsevier Applied Science, 3rd edition, 1991.
13. L. C. Godara and M. R. S. Jahromi. Limitations and capabilities of frequency domain broadband constrained beamforming schemes. *IEEE Trans. on Signal Processing*, 47(9):2386–2395, Sept. 1999.
14. E. D. Di Claudio and R. Parisi. Robust ML wide-band beamforming in reverberant fields. *IEEE Trans. on Signal Processing*, 51(2):338–349, Feb. 2003.
15. F. Quian and B. D. Van Veen. Quadratically constrained adaptive beamforming for coherent signals and interference. *IEEE Trans. on Signal Processing*, 43(8):1890–1900, Aug. 1995.
16. L. J. Griffiths and C. W. Jim. An altermative approach to linearly constrained adaptive beamforming. *IEEE Trans. Antennas and Propagation*, 30(1):27–34, Jan. 1982.
17. J. L. Krolik and D. N. Swingler. On the mean-square error performance of adaptive minimum variance beamformers based on the sample covariance matrix. *IEEE Trans. on Signal Processing*, 42(2):445–448, Feb. 1994.
18. G. Clifford Carter, Ed. *Coherence and Time Delay Estimation*. IEEE Press, 1993.
19. E. D. Di Claudio and R. Parisi. WAVES: Weighted AVErage of signal Subspaces for robust wideband direction finding. *IEEE Trans. on Signal Processing*, 49(10):2179–2191, Oct. 2001.

20. B. S. Atal, V. Cuperman, and A. Gersho, Eds. *Advances in Speech Coding*, volume 114. Kluwer Academic Publishers, Kluwer International Series in Engineering and Computer Science, Jan. 1991.
21. P. J. Huber. *Robust Statistics*. John Wiley, New York, 1981.
22. M. Agrawal and S. Prasad. Robust adaptive beamforming for wide-band, moving and coherent jammers via uniform linear arrays. *IEEE Trans. on Signal Processing*, 47(8):1267–1275, Aug. 1999.
23. K. Harmanci, J. Tabrikian, and J. L. Krolik. Relationships between adaptive minimum variance beamforming and optimal source localization. *IEEE Trans. on Signal Processing*, 48(1):1–12, Jan. 2000.
24. D. F. Gingras. Robust broadband matched-field processing performance in shallow water. *IEEE Journal of Oceanic Engineering*, 18(3):253–264, July 1993.
25. H. Hung and M. Kaveh. Focussing matrices for coherent signal-subspace processing. *IEEE Trans. on Acoustics Speech and Signal Processing*, 36(8):1272–1281, Aug. 1988.
26. H. Wang and M. Kaveh. Coherent signal-subspace processing for the detection and estimation of angles of arrival of multiple wide-band sources. *IEEE Trans. on Acoustics, Speech and Signal Processing*, ASSP-33(4):823–831, Aug. 1985.
27. D. B. Williams and D. H. Johnson. Robust estimation of structured covariance matrices. *IEEE Trans. on Signal Processing*, 41(9):2891–2906, Sept. 1993.
28. D. N. Swingler. A low complexity MVDR beamformer for use with short observation times. *IEEE Trans. on Signal Processing*, 47(4):1154–1160, Apr. 1999.
29. S. L. Marple. *Digital Spectral Analysis*. Prentice Hall, Upper Saddle River, NJ, 1987.
30. R. Parisi, E. D. Di Claudio, G. Orlandi, and B.D. Rao. A generalized learning paradigm exploiting the structure of feedforward neural networks. *IEEE Trans. on Neural Networks*, 7(6):1450–1460, Nov. 1996.
31. S. Haykin. *Neural Networks–A Comprehensive Foundation*, 2nd edition, Prentice Hall, Upper Saddle River, NJ, 1999.
32. G. H. Golub and C. F. Van Loan. *Matrix Computations*, 2nd edition, John Hopkins University Press, Baltimore, MD, 1989.
33. G. Xu, H. P. Lin, S. S. Jeng, and W. J. Vogel. Experimental studies of spatial signature variation at 900 MHz for smart antenna systems. *IEEE Trans. Antennas and Propagation*, 46(7):953–962, July 1998.
34. Haykin, Litva, and Shepherd, editors. *Radar Array Processing*. Springer-Verlag, Berlin, 1993.
35. R. O. Schmidt. Multiple emitter location and signal parameter estimation. *IEEE Trans. on Antennas and Propagation*, 34(3):276–280, Mar. 1986.
36. Y. Haneda, S. Makino, and Y. Kaneda. Common acoustical pole and zero modeling of room transfer functions. *IEEE Trans. on Speech and Audio Processing*, 2(2):320–328, Apr. 1994.
37. M. A. Doron and E. Doron. Wavefield modeling and array processing, Part I–Spatial sampling. *IEEE Trans. on Signal Processing*, 42(10):2549–2559, Oct. 1994.
38. H. L. Van Trees. *Detection, Estimation, and Modulation Theory, Part IV: Optimum Array Processing*. John Wiley, New York, 2002.
39. M. A. Doron and E. Doron. Reduced rank processing for oversampled arrays. *IEEE Trans. on Signal Processing*, 44(4):900–911, Apr. 1996.

40. M. Ghogho, O. Besson, and A. Swami. Estimation of directions of arrival of multiple scattered sources. *IEEE Trans. on Signal Processing*, 49(11):2467–2480, Nov. 2001.
41. S. A. Vorobyov, A. B. Gershman, Zhi-Quan Luo, and Ning Ma. Adaptive beamforming with joint robustness against mismatched signal steering vector and interference non-stationarity. *IEEE Signal Processing Letters*, 11(2):108–111, Feb. 2004.
42. G. B. Giannakis, Y. Hua, P. Stoica, and L. Tong, Eds. *Signal Processing Advances in Wireless and Mobile Communications*, Volume 2. Chapter I, Prentice Hall, NJ, 2000.
43. D. Lee and M. H. Schultz. *Numerical Ocean Acoustic Propagation in Three Dimensions*, 2nd edition. World Scientific Co., Singapore, 1998.
44. P. Stoica, E. G. Larsson, and A. B. Gershman. The stochastic CRB for array processing: a textbook derivation. *IEEE Sig. Proc. Letters*, 8(5):148–150, May 2001.
45. J. Tabrikian and J. L. Krolik. Barankin bounds for source localization in an uncertain ocean environment. *IEEE Trans. on Signal Processing*, 47(11):2917–2927, Nov. 1999.
46. T. Gustafsson, B. D. Rao, and M. Trivedi. Analysis of time-delay estimation in reverberant environments. In *Proc. of the 2002 Int. Conf. Acoustics, Speech and Signal Processing*, Volume II, pp. 2097–2100, Orlando, FL, May 13–17 2002.
47. D. G. Luenberger. *Linear and Nonlinear Programming*, 2nd edition. Addison Wesley, 1989.
48. J. L. Krolik. Matched-field minimum variance beamforming in a random ocean channel. *J. Acoust. Soc. Amer.*, 92:1408–1419, Sept. 1992.
49. O. Hoshuyama, A. Sugiyama, and A. Hirano. A robust adaptive beamformer for microphone arrays with a blocking matrix using constrained adaptive filters. *IEEE Trans. on Signal Processing*, 47(10):2677–2684, Oct. 1999.
50. D. P. Palomar, J. M. Cioffi, and M. A. Lagunas. Joint TX-RX beamforming design for multicarrier MIMO channels: A unified framework for convex optimization. *IEEE Trans. on Signal Processing*, 51(9):2381–2401, Sept. 2003.
51. S. M. Kay and S. L. Marple (Jr.). Spectrum analysis–A modern perspective. *Proc. IEEE*, 64:1380–1419, Nov. 1981.
52. I. Thug, A. Cantoni, and Y. H. Leung. Derivative constrained optimum broad band antenna arrays. *IEEE Trans. Signal Processing*, 41(7):2376–2388, July 1993.
53. A. B. Gershman, E. Nemeth, and J. F. Bohme. Experimental performance of adaptive beamforming in a sonar environment with a towed array and moving interfering sources. *IEEE Trans. on Signal Processing*, 48(1):246–250, Jan. 2000.
54. S. Shahbazpanahi, A. B. Gershman, Zhi-Quan Luo, and Kon Max Wong. Robust adaptive beamforming for general-rank signal models. *IEEE Trans. on Signal Processing*, 51(9):2257–2269, Sept. 2003.
55. S. Haykin. *Adaptive Filter Theory*, 3rd edition. Prentice Hall, 1996.
56. M. Agrawal and S. Prasad. Broadband DOA estimation using spatial-only modeling of array data. *IEEE Trans. on Signal Processing*, 48(3):663–670, Mar. 2000.
57. M. A. Doron and A. J. Weiss. On focusing matrices for wide-band array processing. *IEEE Trans. on Signal Processing*, 40(6):1295–1302, June 1992.
58. B. Friedlander and A. J. Weiss. Direction finding for wide-band signals using an interpolated array. *IEEE Trans. Signal Processing*, 41(4):1618–1635, Apr. 1993.
59. A. V. Oppenheim and R. W. Schafer. *Discrete-Time Signal Processing*. Prentice Hall, Upper Saddle River, NJ, 1989.

60. D. B. Ward, Zhi Ding, and R. A. Kennedy. Broadband DOA estimation using frequency invariant beamforming. *IEEE Trans. on Signal Processing*, 46(5):1463–1469, May 1998.

61. R. Parisi, R. Gazzetta, and E. D. Di Claudio. Prefiltering approaches for time delay estimation in reverberant environments. In *Proceedings of the 2002 IEEE Int. Conf. on Acoustics, Speech and Signal Processing*, Volume III, pp. 2997–3000, Orlando FL, May 13–17 2002.

62. A. Neri. Optimal detection and estimation of straight patterns. *IEEE Trans. on Image Processing*, 5(5):787–792, May 1996.

63. S. R. Deans. Hough transform from the Radon transform. *IEEE Trans. on Patt. Anal. and Machine Intell.*, 3(2):185–188, March 1981.

64. M. Wax and T. Kailath. Detection of signals by information theoretic criteria. *IEEE Trans. on Acoustics Speech and Signal Processing*, 33(2):387–392, Apr. 1985.

65. J. A. Cadzow. Multiple source location–the signal subspace approach. *IEEE Trans. on Acoustics Speech and Signal Processing*, 38(7):1110–1125, July 1990.

INDEX

Robust Adaptive Beamforming, Edited by Jian Li and Petre Stoica

イリス フェルミーン

J. Fermin

[illegible] April 2008